1983

Essential College Algebra

HOUGHTON MIFFLIN COMPANY   BOSTON

Dallas   Geneva, Illinois   Hopewell, New Jersey   Palo Alto   London

Essential College Algebra

Doris S. Stockton • *University of Massachusetts, Amherst*

Acknowledgments

The author is grateful to the following people for their invaluable help: Natalie F. Noble, for typing so much and so well; Teresa Farnum, Veronica Gold, Fred R. Stockton, and Fred D. Stockton, for working exercises and tests; Thomas Stockton, for providing youthful perspective at meetings; and Kenneth Klopfenstein, Colorado State University; Henry Harmeling, North Shore Community College; and William D. Arendt, University of Missouri, for reviewing the manuscript.

COVER AND PART OPENING ILLUSTRATIONS BY JOHN C. HILL

Library of Congress Catalog Card Number: 78-69526

ISBN: 0-395-26544-4

Contents

Part 3 Higher Algebra

Appendixes

Answers to Quizzes, Tests, and Review Exercises

Odd-Numbered Exercise Answers, Selected Graphs and Proofs

Index

Preface

The purpose of this book is to provide the essentials of college algebra. A review of basic algebra and also some optional topics concerning elementary functions and analytic geometry are included.

The topics are arranged in three major parts of ten or eleven sections each. The three parts are

Part 1 Basic Algebra A detailed review of basic algebraic manipulations and definitions.

Part 2 Functions and Graphs An introduction to the function concept and analytic geometry.

Part 3 Higher Algebra Topics from college algebra that are especially useful in mathematics and its applications.

All sections of Part 2 except for the first two sections are starred as optional. To increase flexibility, the material in Part 3 has been written so that students can go directly from the second section of Part 2 to Part 3. None of Part 3 depends on the optional sections in Part 2.

Numerous exercises are provided for each section. A set of review exercises with answers is provided for each major part. Answers to most odd-numbered exercises as well as a generous selection of graphs and proofs are included. Illustrative examples are given in the exercise sets prior to most new types of exercises as well as being given in the body of the text.

A self-scoring quiz appears at the end of each section, and a final test appears at the end of each part. Additional final tests, a placement test, an algebra diagnostic test, and all exercise and test answers that do not appear in the text are available in an instructor's manual.

Other pedagogical features include an informal statement of objectives at the beginning of each section and a check list of objectives for the student at the end of each section. Study-guide features such as the section quizzes, student check lists, final tests, review exercises, and answers to quizzes, tests, and review exercises appear on pages tinted with a second color. The second color is used throughout the text as a pedagogical tool. The format is usable in either self-paced courses or standard lecture courses.

Tables of logarithms are included. These tables are used in optional sections of the text along with appropriate references to the comparative efficiency and accuracy of hand calculators. Computations in these optional sections are made both with tables and with a calculator.

Students also needing to study trigonometry could use the companion volume *Essential Trigonometry* or one of the comprehensive volumes

Essential Algebra and Trigonometry or *Essential Precalculus,* all by this author and published by Houghton Mifflin Company. Algebraic topics reviewed in those texts and in Part 1 of this text are treated in detail for the beginning algebra student in Doris S. Stockton, *Essential Algebra* (Glenview, Illinois: Scott, Foresman and Company, 1973). The author is grateful to Scott, Foresman and Company for permitting her to use bits of that material.

It is indeed a pleasure to thank my sons, Fred and Tom, and my husband, Fred D. Stockton, for their continued encouragement and understanding, and for their cheerful companionship. It is likewise a privilege to salute my parents, Mary and Roland Skillman, for their fifty-eight plus years of marriage and to thank them for all they have done for me.

D.S.S.

Note to the Student

This book is designed to provide you with the essentials of college algebra. The material is arranged in three major parts subdivided into sections. At the end of each section there is a set of exercises and a check list of objectives for you. There are illustrative examples throughout each section and within each exercise set. References in the margin refer to pertinent exercises.

There is a self-scoring quiz at the end of each section and a set of review exercises together with a final test at the end of each major part. Answers to most odd-numbered exercises, including many graphs, are given at the end of the book. For your convenience, the pages containing the answers to the section quizzes as well as the pages containing the quizzes, your check lists, the review exercises, and the final tests are tinted in a second color. Answers to the review exercises and the final tests also appear on tinted pages.

One way to use this book is as follows.

1. With pencil or pen in hand, study the text and work through the illustrative examples until you encounter a reference to exercises in the margin.

2. Try some of the recommended exercises; check your answers in the back of the book.

3. When you reach the end of a section, do more exercises, selecting them at random from the various types.

4. Read the check list; go back and review any areas in which you feel insecure.

5. Take and score the section quiz. If necessary, review, and then correct your answers.

6. When you reach the end of a major part, do the review exercises, check your answers, and then take the final test.

I want this book to be as helpful and as useful as possible. Please feel free to address comments or criticism to me in care of the publisher.

My best wishes to you.

D.S.S.

Essential College Algebra

Part 1 Basic Algebra

College algebra students need to be able to perform basic algebraic manipulations with ease. Part 1 of this text consists of a detailed review of fundamental manipulations and definitions that are frequently used.

1.1 Common Sets of Numbers

The purpose of this section is to review the most commonly used sets of numbers.

The numbers usually used in counting are called the **counting numbers,** the **natural numbers,** or the **positive integers.** The set of all the positive integers may be written

$$\{1, 2, 3, 4, \ldots\}. \qquad \text{The positive integers}$$

The three dots within the braces indicate that all the rest of the positive integers are also included in the set. The numeral for a positive integer might be prefixed with a plus sign for emphasis. The set $\{1, 2, 3, 4, \ldots\}$ might be written

$$\{+1, +2, +3, +4, \ldots\}.$$

The set consisting of the positive integers together with zero, 0, is called the set of **whole numbers** and may be written

$$\{0, 1, 2, 3, 4, \ldots\}. \qquad \text{The whole numbers}$$

Assiciated with each positive integer there is a number called its *negative.* The **negative** of a positive integer is designated by writing the numeral for the positive integer with a negative or minus sign in front of it. For example, the negative of 3 is written -3 and is read ''negative three.'' Although this is also sometimes read ''minus three,'' it is advisable to reserve the use of the word *minus* for cases in which the symbol $-$ denotes the operation subtraction rather than the negative of a number. The negatives of all the positive integers constitute the set of **negative integers.**

$$\{-1, -2, -3, -4, \ldots\} \qquad \text{The negative integers}$$

The set consisting of all the positive integers, zero, and all the negative integers is called the set of **integers.**

$$\{\ldots, -4, -3, -2, -1, 0, 1, 2, 3, 4, \ldots\} \qquad \text{The integers}$$

The integers may be conveniently represented by points on a line called a *number line.* Some integers are shown on a number line in Figure 1.1. It is understood that the line continues indefinitely both to the

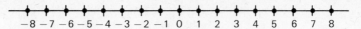

FIGURE 1.1   Integers on a number line

left and to the right so that for each integer there is exactly one point on the number line.

The set of integers evenly divisible by the integer 2 is called the set of *even integers.* The set of integers other than the even integers is called the set of *odd integers.* Note that 0 is an even integer (0 divided by 2 is 0).

$$\{\ldots, -6, -4, -2, 0, 2, 4, 6, \ldots\} \qquad \text{The even integers}$$

$$\{\ldots, -5, -3, -1, 1, 3, 5, \ldots\} \qquad \text{The odd integers}$$

Any even integer can be considered to be of the form $2k$ where k is an integer. As indicated in the next display, the entire set of even integers can be generated by computing $2k$ as k successively takes on the values of all the integers:

$$k: \quad \ldots, -4, -3, -2, -1, 0, 1, 2, 3, 4, \ldots$$

$$2k: \quad \ldots, -8, -6, -4, -2, 0, 2, 4, 6, 8, \ldots$$

Any odd integer can be considered to be of the form $2k + 1$ or $2k - 1$ where k is an integer.

The entire set of odd integers can be generated by computing $2k + 1$ or $2k - 1$ as k successively takes on the values of all the integers.

$$k: \quad \ldots, -4, -3, -2, -1, \quad 0, 1, 2, 3, 4, \ldots$$

$$2k + 1: \quad \ldots, -7, -5, -3, -1, \quad 1, 3, 5, 7, 9, \ldots$$

$$2k - 1: \quad \ldots, -9, -7, -5, -3, -1, 1, 3, 5, 7, \ldots$$

A **rational number** is any number that can be expressed in the form m/n where m and n are integers and n is not zero. The symbols

$$\tfrac{2}{3}, \tfrac{17}{5}, \tfrac{-3}{7}, \tfrac{0}{6}, \tfrac{0}{1}, \tfrac{5}{1}, \text{ and } \tfrac{-8}{1}$$

represent rational numbers.

The word **fraction** is commonly used for any symbol of the form P/D where P and D are mathematical expressions. The P is called the **numerator** and the D is called the **denominator**. In this sense of the word *fraction,* any rational number can be written as a fraction whose numerator and denominator are integers and whose denominator is not zero.

Any rational number or any fraction can be written in more than one form. Two fractions P/D and R/S with nonzero denominators are said to be **equal** if the product PS equals the product DR. This definition leads to the so-called **fundamental principle of fractions:**

$$\frac{PN}{DN} = \frac{P}{D}, \qquad \text{for nonzero } N \text{ and } D.$$

$$\frac{5xy}{13y} = \frac{5x}{13}, \qquad \text{if } y \neq 0.$$

We can consider the integers to be rational numbers by identifying each integer with a rational number whose denominator is 1. For example, identify the integer 3 with the rational number 3/1; identify -7 with $-7/1$; identify 0 with 0/1.

For any integer n, let $n = n/1$.

We can represent any rational number as either a terminating or a repeating decimal by formally dividing its denominator into its numerator. For example,

$$\tfrac{3}{4} = 0.75 \qquad\qquad \tfrac{1}{3} = 0.333\ldots$$

$$\tfrac{-9}{2} = -4.5 \qquad\qquad \tfrac{-14}{11} = -1.272727\ldots$$

(In a repeating decimal, it is customary to designate the shortest complete repeating pattern by drawing a horizontal bar over it so that $0.333\ldots$ could also be written $0.\overline{3}$, and $-1.272727\ldots$ could also be written $-1.\overline{27}$.) Likewise, each terminating or repeating decimal represents a rational number. For example, $1.5 = \tfrac{3}{2}$, $2.0 = 2$, and $-0.666\ldots = \tfrac{-2}{3}$. Thus we can say that the set of all rational numbers is the same as the set of all terminating or repeating decimals.

Each rational number is represented by exactly one point on a number line. Some rational numbers are shown on a number line in Figure 1.2.

FIGURE 1.2   Rational numbers on a number line

There do exist points on the number line that do not represent rational numbers. These points represent numbers called *irrational numbers.* An **irrational number** is not a rational number; it cannot be expressed in the form m/n where m and n are integers and n is not zero. An irrational number cannot be represented by a terminating or repeating decimal. Instead, any irrational number can be represented by a nonterminating, nonrepeating decimal. Likewise, any nonterminating, nonrepeating decimal represents an irrational number. We can say that the set of all nonterminating, nonrepeating decimals is the same as the set of all irrational numbers.

One example of an irrational number is the positive number whose square (whose product with itself) is equal to the integer 2. Algebraically, this irrational number is the positive solution of the equation $x \cdot x = 2$. It is denoted by the symbol $\sqrt{2}$. Another example of an irrational number is the number that first occurred in classical geometry and is commonly denoted by the Greek letter π. It is the ratio of the circumference to the diameter of any circle. Some other irrational numbers are $-\sqrt{2}$, $\sqrt{3}$, $-\sqrt{3}$, $\sqrt{5}$, and $-\sqrt{5}$. It can be shown that the sum or difference of any rational number and any irrational number is itself an irrational number. Likewise, the product or quotient of any nonzero rational number with any irrational number is itself irrational. Thus numbers such as $\frac{4}{5} + \sqrt{2}$, $7 - \sqrt{3}$, 2π, $3\sqrt{2}$, $\sqrt{2}/2$, and $2/\sqrt{3}$ are also irrational numbers.

Any irrational number can be approximated by a rational number. For example, a nine-decimal-place approximation to $\sqrt{2}$ is found on a hand calculator to be the rational number 1.414213562. Thus $\sqrt{2}$ is approximately equal to the three-place rational number 1.414.* Using the symbol $\simeq$ for "approximately equal to,"

$$\sqrt{2} \simeq 1.414.$$

Likewise, a fourteen-decimal-place approximation to π is the rational number 3.14159265358979. Thus π is approximately equal to the five-place rational number 3.14159.

$$\pi \simeq 3.14159$$

Note that the rational number $\frac{22}{7}$ is not equal to π. In fact, $\frac{22}{7}$ only agrees with π as far as the first two decimal places.

$$\frac{22}{7} = 3.\overline{142857}$$

An irrational number that is frequently encountered as a logarithm base in calculus and applications is denoted by the letter e. We have not developed sufficient vocabulary to define the number e precisely. Suffice it to say that e, as shown on a hand calculator, is approximately equal to the rational number 2.7182818284.

$$e \simeq 2.7182818284$$

The irrational numbers together with the rational numbers make up a set called the **real numbers.** Each real number is represented by exactly one point on a number line. Conversely, each point on a number line represents exactly one real number. Some real numbers are shown on the

---

*For one method of computing $\sqrt{2}$ to as many decimal places as you please, see Doris S. Stockton, *Essential Mathematics,* Scott, Foresman and Company, Glenview, Illinois, 1972, pp. 318–322.

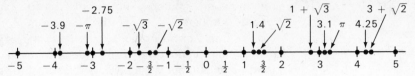

FIGURE 1.3 Real numbers on a number line

number line in Figure 1.3. Useful properties of the real numbers are listed in Appendix A.

Of any two real numbers, the one further to the right on a number line is said to be the greater. The set of all real numbers greater than 0 is called the set of **positive real numbers**. The set of all real numbers less than 0 is called the set of **negative real numbers.**

The symbol $>$ is used for "is greater than," and the symbol $<$ is used for "is less than." The symbols $\geq$ and $\leq$ are used for "is greater than or equal to" and "is less than or equal to," respectively. For example,

$$\pi > 3,$$
$$\pi < \tfrac{22}{7},$$

and
$$\sqrt{2} < \tfrac{3}{2}.$$

Also, if x is positive, then

$$x > 0,$$

and if x is nonnegative, then

$$x \geq 0.$$

It is sometimes convenient to denote sets of real numbers symbolically by using "set-builder notation." In this notation, the symbols

$$\{x \mid x \text{ is so and so}\}$$

are read

"the set of all x such that x is so and so,"

and unless otherwise specified, it is understood that x is a real number. Thus the set of all positive real numbers could be written $\{x \mid x > 0\}$, and these symbols could be read "the set of all x such that x is greater than zero." The set of all even integers could be written $\{x \mid x = 2k$ where k is an integer$\}$, and the set of all odd integers could be written $\{x \mid x = 2k + 1$ where k is an integer$\}$ or could be written $\{x \mid x = 2k - 1$ where k is an integer$\}$.

The graph of any set of real numbers is the set of all their corresponding points on a number line. The graphs of some common sets of real numbers are shown in Figure 1.4. Heavy dots are used to indicate endpoints that

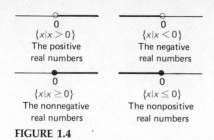

$\{x|x > 0\}$
The positive
real numbers

$\{x|x < 0\}$
The negative
real numbers

$\{x|x \geq 0\}$
The nonnegative
real numbers

$\{x|x \leq 0\}$
The nonpositive
real numbers

FIGURE 1.4

Exercise 1.1, 61–68

are to be included in the graph, and circles are used to indicate excluded endpoints.

One commonly encountered set of real numbers is a set of positive integers called *primes*. A **prime** is any integer greater than the integer 1 which is evenly divisible by no positive integer other than itself and 1. The first five primes are 2, 3, 5, 7, and 11. Some primes are represented by dots on a number line in Figure 1.5.

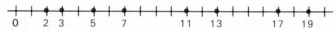

0   2 3   5   7   11   13   17   19

FIGURE 1.5   Primes on a number line

We have noted that associated with each positive integer p there is a negative integer called its negative and denoted by $-p$. In general, associated with each real number x there is a real number called its negative and denoted by $-x$. The symbol $-x$ is read "negative x." The numbers x and $-x$ are represented by points on a number line on opposite sides of the zero point and equally distant from it. If x is positive, $-x$ is negative. If x is negative, $-x$ is positive. Also, the negative of $-x$ is x itself.

$$-(-x) = x$$
$$-(-7) = 7$$

The number of units on a number line between the point representing any real number and the point representing 0 is called the *absolute value* of the real number. For example, the absolute value of 3 is 3, and the absolute value of -3 is also 3, because 3 and -3 are each represented by a point located 3 units from the point representing 0. The absolute value of any real number x is denoted by the symbol $|x|$ and is formally defined as follows:

$$|x| = x, \qquad \text{if } x \text{ is positive or zero,}$$
$$|x| = -x, \qquad \text{if } x \text{ is negative.}$$

For example,

$$|3| = 3,$$
$$|0| = 0,$$

Exercise 1.1, 69–85 and
$$|-3| = -(-3) = 3.$$

We have noted that each real number is represented by exactly one point on a number line and that each point on a number line represents exactly one real number. There do exist numbers other than real numbers that arise quite naturally in algebra. Examples of such numbers are the solutions of the equation $x \cdot x = -4$. We know that no real number x is a solution of $x \cdot x = -4$ because the product of any real number with itself is not negative. (A real number is either positive, negative, or zero. A positive times a positive is positive, a negative times a negative is positive, and zero times zero is zero.) The solutions of the equation $x \cdot x = -4$ are numbers called *imaginary numbers*. If we define a new number denoted by the letter i and satisfying the requirement that $i \times i = -1$, then the solutions of $x \cdot x = -4$ are the so-called imaginary numbers $2i$ and $-2i$. This can be checked by substitution in the equation $x \cdot x = -4$.

$$x \cdot x = -4 \qquad\qquad x \cdot x = -4$$
$$(2i)(2i) = 4(i \times i) \qquad (-2i)(-2i) = +4(i \times i)$$
$$= 4(-1) \qquad\qquad = 4(-1)$$
$$= -4 \qquad\qquad = -4$$

$2i$ is a solution $\qquad\qquad$ $-2i$ is a solution

In general, denoting $i \times i$ by the symbol i^2, an **imaginary number** is any number that can be written in the form $a + bi$ where b is a *nonzero* real number, a is any real number, and i is a number such that $i^2 = -1$.

$$\underset{\text{Real}}{a} + \underset{\text{Nonzero real}}{bi} \qquad\qquad \text{Imaginary number}$$

$$i^2 = -1$$

For example, $3 + 2i$, $3 + (-2)i = 3 - 2i$, $0 + 2i = 2i$, and $0 + (-2)i = -2i$ are imaginary numbers.

Any real number or any imaginary number is a special kind of a more general class of numbers called *complex numbers*. A **complex number** is any number that can be written in the form $a + bi$ where a and b are real numbers (with b not necessarily nonzero) and $i^2 = -1$. The a is called the *real part,* and the b is called the *imaginary part* of the complex number $a + bi$.

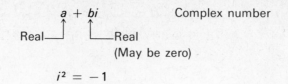

$$i^2 = -1$$

For example,

$$\tfrac{3}{5} - 7i, \qquad \sqrt{2} + 6i, \qquad -7 + \pi i, \quad \text{and} \quad 8 + 0i$$

are complex numbers. Any real number is a special complex number in which the imaginary part equals zero. Any imaginary number is a special complex number in which the imaginary part does *not* equal zero. The numbers

$$7 = 7 + 0i, \qquad \tfrac{-3}{2} = \tfrac{-3}{2} + 0i, \qquad \text{and} \qquad \sqrt{2} = \sqrt{2} + 0i$$

are real numbers and complex numbers. The numbers

$$2i = 0 + 2i, \qquad 3 + 5i, \quad \text{and} \quad 3 - 5i$$

are imaginary numbers and complex numbers.

Note that we may say that

$$0 = 0 + 0i$$

is a real number and a complex number, and that

$$i = 0 + (1)i$$

is an imaginary number and a complex number. It is true that

$$0 \times i = 0$$

and

$$1 \times i = i.$$

In Sections 1.4 and 1.6 it will be shown that sums, differences, products, and quotients of complex numbers can be written in the form $a + bi$ where a and b are real numbers.

A diagram showing relationships we have noted among common sets of numbers appears in Figure 1.6 (see page 10).

EXAMPLE 1

Let A be the set

$$\{-7,\ 0,\ 7,\ \tfrac{2}{3},\ \pi,\ \sqrt{2},\ -\sqrt{2},\ e,\ 3i,\ 2 + 5i,\ \tfrac{22}{7}\}.$$

List the integers, list the rational numbers, list the irrational numbers, and list the imaginary numbers in A.

Solution

The integers in A are -7, 0, and 7.

The rational numbers in A are -7, 0, 7, $\tfrac{2}{3}$, and $\tfrac{22}{7}$.

The irrational numbers in A are π, $\sqrt{2}$, $-\sqrt{2}$, and e.

The imaginary numbers in A are $3i$ and $2 + 5i$.

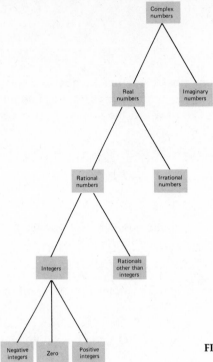

FIGURE 1.6   Relationships among common sets of numbers

EXAMPLE 2

List each of the following sets to which the number 3 belongs:

positive integers	real numbers
whole numbers	positive real numbers
negative integers	nonnegative real numbers
integers	negative real numbers
even integers	nonpositive real numbers
odd integers	primes
rational numbers	imaginary numbers
irrational numbers	complex numbers

Solution

Positive integers, whole numbers, integers, odd integers, rational numbers, real numbers, positive real numbers, nonnegative real numbers, primes, and complex numbers.

Exercise 1.1, 16–45

EXAMPLE 3

Classify each of the following as a terminating, repeating, or non-terminating and nonrepeating decimal, and state which kind of number, rational or irrational, it represents:

(a) $3.\overline{25}$ (b) 3.25 (c) $3.25225222522225\ldots$

Solution

(a) $3.\overline{25}$ is a repeating decimal. The pattern 25 repeats indefinitely. It represents a rational number.

(b) 3.25 is a terminating decimal. There are no nonzero digits beyond the second decimal place. It represents a rational number.

(c) 3.25225222522225 . . . is a nonterminating and nonrepeating decimal. After each digit 5 there appears one more digit 2 than appeared after the preceding 5, and this continues indefinitely. It represents an irrational number.

EXAMPLE 4

State which kind of decimal—terminating, repeating, or nonterminating and nonrepeating—represents each of the following:
(a) $\frac{3}{8}$ (b) $\sqrt{3}$ (c) $\frac{5}{3}$

Solution

(a) Formally dividing 8 into 3.000, we get 0.375. Thus $\frac{3}{8}$ is represented by a terminating decimal.

(b) $\sqrt{3}$ is an irrational number. It is represented by a nonterminating and nonrepeating decimal. (On a calculator, $\sqrt{3} \simeq 1.732050808$.)

(c) Formally dividing 3 into 5.000, we get 1.666 . . . $= 1.\overline{6}$. Thus $\frac{5}{3}$ is represented by a repeating decimal.

Exercise 1.1, 46–60

EXERCISE 1.1 COMMON SETS OF NUMBERS

▶ Let A be the set

$\{-5,\ -1,\ 0,\ 1,\ 4,\ \frac{3}{4},\ \pi,\ 7,\ 11,\ \sqrt{3},\ e,\ -3i,\ 4+3i,\ \frac{22}{7},\ -1.5,\ 1.\overline{5},$
$1.020020002 \ldots \}$.

In each of the following, list the specified members of A.

EXAMPLE

Nonnegative rational numbers

Solution

$0,\ 1,\ 4,\ \frac{3}{4},\ 7,\ 11,\ \frac{22}{7},\ 1.\overline{5}$

1. Nonnegative integers
2. Whole numbers
3. Negative integers
4. Integers
5. Even integers
6. Odd integers
7. Rational numbers
8. Irrational numbers
9. Real numbers
10. Nonpositive real numbers
11. Nonnegative real numbers
12. Primes
13. Imaginary numbers
14. Complex numbers
15. Nonpositive rational numbers

► List each of the following sets to which the given number belongs:

positive integers	real numbers
whole numbers	positive real numbers
negative integers	nonnegative real numbers
integers	negative real numbers
even integers	nonpositive real numbers
odd integers	primes
rational numbers	imaginary numbers
irrational numbers	complex numbers

EXAMPLE

$\sqrt{2}$

Solution

Irrational numbers, real numbers, positive real numbers, nonnegative real numbers, and complex numbers.

16. 7	**17.** $\frac{2}{3}$	**18.** 0	**19.** $\sqrt{3}$
20. 125	**21.** -3	**22.** $\frac{-4}{5}$	**23.** $\frac{22}{7}$
24. -4	**25.** 17π	**26.** 23	**27.** -23
28. $-\sqrt{2}$	**29.** $\sqrt{5}$	**30.** π	**31.** e
32. $0 + 0i$	**33.** $2 - 3i$	**34.** $5i$	**35.** i
36. $3 + \sqrt{2}$	**37.** $-\sqrt{3}$	**38.** $2 + \sqrt{3}$	**39.** $-4\sqrt{2}$
40. $7 + 9i$	**41.** $\sqrt{2} + 3i$	**42.** $\sqrt{2}i$	**43.** $-3i$
44. $-\pi$	**45.** $\frac{8}{3}$		

► Classify each of the following as a terminating, repeating, or nonterminating and nonrepeating decimal, and state which kind of number, rational or irrational, it represents.

EXAMPLE

$8.\overline{35}$

Solution

Repeating decimal; rational number

46. 1.25	**47.** $1.\overline{25}$	**48.** 4.37
49. $4.\overline{37}$	**50.** 4.373373337 . . .	**51.** 23.5
52. $6.\overline{41}$	**53.** 1.0001	**54.** $1.\overline{01}$
55. 1.001000100001 . . .	**56.** 1.414	**57.** 3.14159
58. 2.71828	**59.** 1.57	**60.** $5.\overline{9}$

► Use set-builder notation to denote each of the following sets, and make a graph of each set on a number line.

EXAMPLES

(a) All real numbers greater than or equal to 3

(b) All real numbers greater than 3 and less than 7

(c) All real numbers less than 3 or greater than 7

Solutions

(a) $\{x \mid x \geq 3\}$

(b) $\{x \mid 3 < x < 7\}$

(c) $\{x \mid x < 3 \quad \text{or} \quad x > 7\}$

61. All real numbers greater than 2

62. All real numbers less than 2

63. All real numbers greater than 2 and less than 5

64. All real numbers less than 2 or greater than 5

65. All real numbers less than or equal to 2 and greater than or equal to -2

66. All integers less than or equal to 2 and greater than or equal to -2

67. All nonnegative integers less than or equal to 9

68. All nonpositive integers greater than -5

69. All integers with absolute value less than 5

70. All real numbers with absolute value less than 5

71. All real numbers with absolute value greater than or equal to 5

72. All real numbers with absolute value equal to $\frac{3}{2}$

73. All nonnegative integers with absolute value less than 7

74. All nonpositive integers with absolute value less than 7 and greater than 2

75. All real numbers greater than or equal to -3

▶ Find the value of each of the following.

EXAMPLE

$-|2 - 7|$

Solution

$-|2 - 7| = -|-5| = -(5) = -5$

76. $|9 - 11|$ **77.** $|9 - 9|$ **78.** $-|3 - 7|$

79. $-|-8|$ **80.** $-(-8)$ **81.** $-|-1 - 1|$

82. $-|1 - 1|$ **83.** $|-8| - |2|$ **84.** $|-8| + |-2|$

85. $(-8 - 2)$

CHECK LIST FOR THE STUDENT

If you have learned the material in Section 1.1, you should be able to do the following:

1. Write an expression representing each of the following sets: the positive integers, the whole numbers, the negative integers, the integers, the even integers, and the odd integers.

2. Complete the statement: A rational number is any number that can be expressed in the form m/n where m and n are _____ and n is _____.

3. State the definition of a fraction, and state the definition of the numerator and denominator of a fraction.

4. Complete the statement: $PN/DN = P/D$ for _____ N and D.

5. Complete the statement: The set of all rational numbers is the same as the set of all _____ or _____ decimals.

6. Complete the statement: The set of all irrational numbers is the same as the set of all _____, _____ decimals.

7. Give three-decimal-place approximations to $\sqrt{2}$, π, and e.

8. Use set-builder notation and write an expression for each of the following: the positive real numbers, the negative real numbers, the nonnegative real numbers, the nonpositive real numbers.

9. State the definition of a prime.

10. Complete the statement: If x is negative, then $-x$ is _____; if x is positive, then $-x$ is _____.

11. State the definition of the absolute value of any real number x.

12. State a definition of an imaginary number.

13. State a definition of a complex number; define the real part and the imaginary part of a complex number.

14. Show the relationships among the following sets on a diagram (see Figure 1.6): negative integers, zero, positive integers, integers, rational numbers other than integers, rational numbers, irrational numbers, real numbers, imaginary numbers, complex numbers.

SECTION QUIZ 1.1

The names of some common sets of numbers are: negative integers, positive integers, integers, rational numbers, irrational numbers, real numbers, imaginary numbers, and complex numbers. To which of these sets do the numbers in each of the following sets belong?

1. (a) $\{1, 2, 3, 4, \ldots\}$
   (b) $\{0, 1, 2, 3, 4, \ldots\}$

2. (a) $\{-1, -2, -3, -4, \ldots\}$
   (b) $\{\ldots, -4, -3, -2, -1, 0, 1, 2, 3, 4, \ldots\}$

3. (a) $\{2, 4, 6, 8, \ldots\}$
   (b) $\{\frac{2}{3}, \frac{3}{4}, \frac{4}{3}, 5, \frac{-7}{8}, 2.5, 0\}$

4. (a) $\{\sqrt{2}, \sqrt{3}, \pi)$
   (b) $\{2 + 3i, 5i, 2 - 3i\}$

Let A be the set

$$\{-3, \ -2, \ 0, \ 2, \ \tfrac{3}{5}, \ 2.3, \ 2.\overline{3}, \ 9, \ \sqrt{3}, \ \pi, \ \tfrac{22}{7}, \ 3.14, \ \tfrac{-2}{3}, \ 1.732\}.$$

In each of the following, list the specified members of A.

5. All the rational numbers of A

6. All the irrational numbers of A

7. Classify each of the following as rational or irrational, and state which kind of decimal—terminating, repeating, or nonterminating and nonrepeating—represents it.
   (a) $\sqrt{2}$ (b) $\tfrac{3}{7}$ (c) $\tfrac{3}{5}$

Use set-builder notation to denote each of the following sets, and make a graph of each on a number line.

8. (a) All nonnegative real numbers
   (b) All negative real numbers

9. (a) All nonpositive real numbers
   (b) All positive real numbers

10. Find the value of $-|-7| + |-3|$.

1.2 Integral Exponents and Roots

The purpose of this section is to review the definitions of integral exponents and roots of real numbers and to review the laws of exponents.

Our first main objective is to define a^m for real number a and integer m so that a^m is a real number. We will accomplish this by first defining a^k for positive integral k and then defining a^0 and a^{-k} for nonzero a. We will then define roots of real numbers and review the laws of exponents.

Definition   If k is a positive integer and a is a real number, a^k represents the product of k a's.

$$a^5 = \underbrace{a \cdot a \cdot a \cdot a \cdot a}_{\text{Five } a\text{'s}}$$

$$a^2 = a \cdot a$$

$$a^1 = a$$

The number a^k is sometimes called the **kth power of a**. The a is called the **base,** and the k is called the **exponent** of a^k.

Note that if k is positive, $0^k = 0$.

Definition   If a is any nonzero real number, $a^0 = 1$.

$$5^0 = 1 \qquad (\sqrt{2})^0 = 1$$

$$(\tfrac{-2}{3})^0 = 1 \qquad 0^0 \text{ is not defined}$$

Definition   If k is a positive integer and a is any nonzero real number, $a^{-k} = 1/a^k$.

$$0^{-k} \text{ is not defined}$$

$$a^{-3} = 1/a^3, \qquad \text{if } a \text{ is not zero}$$

$$a^{-1} = 1/a, \qquad \text{if } a \text{ is not zero}$$

$$7^{-2} = 1/7^2 = \tfrac{1}{49} \qquad (\tfrac{3}{4})^{-2} = \frac{1}{(\tfrac{3}{4})^2} = \frac{1}{\tfrac{9}{16}} = \tfrac{16}{9}$$

$$7^{-1} = \tfrac{1}{7} \qquad (\tfrac{3}{4})^{-1} = \frac{1}{(\tfrac{3}{4})} = \tfrac{4}{3}$$

Definition   If n is a positive integer and b is any real number, **an nth root of b** is a number whose nth power equals b.

For example, a fourth root of 81 is 3 because $3^4 = 81$. Another fourth root of 81 is -3 because $(-3)^4 = 81$.

If n is odd, there is exactly one real nth root of any real number b. It is designated by the symbol $\sqrt[n]{b}$. If $n = 3$, the nth root is called a **cube root.**

$$n \text{ odd} \rightarrow \text{Exactly one real } n\text{th root of } b, \ \sqrt[n]{b}$$

$$\sqrt[3]{8} = 2 \qquad \text{The only real cube root of 8}$$

$$\sqrt[3]{-8} = -2 \qquad \text{The only real cube root of } -8$$

If n is even and b is positive, there are exactly two real nth roots of b. One of these roots is positive, one is negative, and they have the same absolute value. The *positive root* is designated by the symbol $\sqrt[n]{b}$. If $n = 2$, an nth root is called a **square root**. It is customary to write $\sqrt{b}$ for $\sqrt[2]{b}$ so that the positive square root of positive number b is written $\sqrt{b}$.

n even, and

b positive $\rightarrow$ Exactly two real nth roots of b, $\sqrt[n]{b}$ and $-\sqrt[n]{b}$

$\sqrt{25} = 5$ The positive square root of 25

$-\sqrt{25} = -5$ The negative square root of 25

$\sqrt[4]{81} = 3$ The positive fourth root of 81

$-\sqrt[4]{81} = -3$ The negative fourth root of 81

If n is even and b is negative, there is no real nth root of b. In this case the nth roots of b are imaginary numbers and $\sqrt[n]{b}$ does not represent a real number.

n even, and

b negative $\rightarrow$ No real nth root of b

$\sqrt[n]{b}$ is not a real number.

$\sqrt[4]{-81}$ is not a real number.

$\sqrt{-9}$ is not a real number.

If n is 2 and b is negative, say $b = -p$ where p is positive, then $\sqrt{b} = \sqrt{-p}$ is defined to be the imaginary number $\sqrt{p}\,i$ where $i^2 = -1$.

$$\sqrt{-p} = \sqrt{p}\,i \quad \text{for } p \text{ positive.}$$
$$\sqrt{-2} = \sqrt{2}\,i$$
$$\sqrt{-9} = \sqrt{9}\,i$$
$$= 3i$$

If n is even or odd, $\sqrt[n]{0} = 0$.

$$\sqrt{0} = 0 \qquad \sqrt[3]{0} = 0$$

In the symbol $\sqrt[n]{b}$, n is called the **index**. The symbol $\sqrt[n]{b}$ itself is called a **radical,** and the b is called its **radicand.**

Our symbolism has been defined so that for any real number x, $\sqrt{x^2}$ is nonnegative. Thus if x is itself negative, $\sqrt{x^2}$ must be $-x$. If x is positive or zero, $\sqrt{x^2} = x$. In other words, for any real number x

$$\sqrt{x^2} = |x|.$$
$$\sqrt{2^2} = |2| = 2$$
$$\sqrt{(-2)^2} = |-2| = -(-2) = 2$$
$$\sqrt{0^2} = 0$$

Likewise, if n is odd, $\sqrt[n]{x^n} = x$. If n is even, the symbol $\sqrt[n]{x^n}$ is defined to be nonnegative, so if n is even, $\sqrt[n]{x^n} = |x|$.

$$\sqrt[n]{x^n} = x, \qquad \text{if } n \text{ is odd}$$
$$\sqrt[n]{x^n} = |x|, \qquad \text{if } n \text{ is even.}$$
$$\sqrt[3]{5^3} = 5, \qquad \sqrt[3]{(-5)^3} = -5, \qquad \sqrt[3]{0^3} = 0$$
$$\sqrt[4]{5^4} = 5, \qquad \sqrt[4]{(-5)^4} = |-5| = 5, \qquad \sqrt[4]{0^4} = 0$$

EXAMPLE 1

Find the value of each of the following.

(a) $\left(\frac{2}{3}\right)^2$ (b) $\left(\frac{2}{3}\right)^0$ (c) $\left(\frac{2}{3}\right)^{-2}$ (d) $\left(\frac{2}{3}\right)^{-1}$

(e) 0^2 (f) $\sqrt[3]{64}$ (g) $\sqrt[3]{-64}$ (h) $\sqrt{64}$

(i) $\sqrt{3^2}$ (j) $\sqrt{(-3)^2}$ (k) $\sqrt[3]{4^3}$ (l) $\sqrt[3]{(-4)^3}$

(m) $\sqrt[6]{5^6}$ (n) $\sqrt[6]{(-5)^6}$

Solution

(a) $\left(\frac{2}{3}\right)^2 = \frac{2}{3} \times \frac{2}{3} = \frac{4}{9}$ (b) $\left(\frac{2}{3}\right)^0 = 1$

(c) $\left(\frac{2}{3}\right)^{-2} = \dfrac{1}{\left(\frac{2}{3}\right)^2} = \dfrac{1}{\frac{4}{9}} = \frac{9}{4}$ (d) $\left(\frac{2}{3}\right)^{-1} = \dfrac{1}{\left(\frac{2}{3}\right)} = \frac{3}{2}$

(e) $0^2 = 0 \times 0 = 0$ (f) $\sqrt[3]{64} = 4$

(g) $\sqrt[3]{-64} = -4$ (h) $\sqrt{64} = 8$

(i) $\sqrt{3^2} = 3$ (j) $\sqrt{(-3)^2} = |-3| = 3$

(k) $\sqrt[3]{4^3} = 4$ (l) $\sqrt[3]{(-4)^3} = -4$

Exercise 1.2, 1–75   (m) $\sqrt[6]{5^6} = 5$ (n) $\sqrt[6]{(-5)^6} = |-5| = 5$

If a^r and a^s are real numbers with real-number base a and integral exponents r and s, then the given definitions can be used to show that the following so-called *exponent laws* are true. Since we shall note in the next section that these laws are also true for rational-number exponents r and s, we shall state them here for that more general case.

Exponent Laws

If a^r, a^s, a^{rs}, and b^r are real numbers with a and b real and r and s rational, then

Law 1 $a^r a^s = a^{r+s}$

Law 2 $\dfrac{a^r}{a^s} = a^{r-s}, \quad a \neq 0$

Law 3 $(a^r)^s = a^{rs}$

Law 4 $(ab)^r = a^r b^r$

Law 5 $\left(\dfrac{a}{b}\right)^r = \dfrac{a^r}{b^r}, \qquad b \neq 0$

In Examples 2–6, let x and y represent real numbers.

EXAMPLE 2

(a) $x^3 x^4 = x^{3+4}$

$\quad = x^7$

Law 1: $a^r a^s = a^{r+s}$

(b) $10^3 \cdot 10^{-4} = 10^{3+(-4)}$

$\quad = 10^{-1}$

$\quad = \frac{1}{10}$

Law 1: $a^r a^s = a^{r+s}$

(c) $2^{-3} \cdot 2^{-4} = 2^{-3+(-4)}$

$\quad = 2^{-7}$

$\quad = 1/2^7$

Law 1: $a^r a^s = a^{r+s}$

(d) If $x \neq 0$,

$\quad x \cdot x^{-1} = x^{1+(-1)}$

$\quad = x^0$

$\quad = 1$

Law 1: $a^r a^s = a^{r+s}$

EXAMPLE 3

(a) If $x \neq 0$,

$\dfrac{x^5}{x^3} = x^{5-3}$

$\quad = x^2$

Law 2: $\dfrac{a^r}{a^s} = a^{r-s}$,

$a \neq 0$

(b) If $x \neq 0$,

$\dfrac{x^3}{x^5} = x^{3-5}$

$\quad = x^{-2}$

$\quad = 1/x^2$

Law 2: $\dfrac{a^r}{a^s} = a^{r-s}$,

$a \neq 0$

(c) If $x \neq 0$

$x^3/x^3 = x^{3-3}$

$\quad = x^0$

$\quad = 1$

Law 2: $\dfrac{a^r}{a^s} = a^{r-s}$,

$a \neq 0$

(d) $\dfrac{10^{-3}}{10^{-2}} = 10^{-3-(-2)}$

$\quad = 10^{-3+2}$

$\quad = 10^{-1}$

$\quad = \frac{1}{10}$

Law 2: $\dfrac{a^r}{a^s} = a^{r-s}$,

$a \neq 0$

EXAMPLE 4

(a) $(x^3)^2 = x^{3 \times 2}$

$\quad = x^6$

Law 3: $(a^r)^s = a^{rs}$

(b) If $x \neq 0$

$$(x^{-1})^2 = x^{(-1)\times(2)}$$

$$= x^{-2}$$

$$= \frac{1}{x^2}$$

Law 3: $(a^r)^s = a^{rs}$

(c) $(2^{-3})^{-2} = 2^{(-3)\times(-2)}$

$$= 2^6$$

$$= 64$$

Law 3: $(a^r)^s = a^{rs}$

EXAMPLE 5

(a) $(2x)^3 = 2^3 x^3$

$$= 8x^3$$

Law 4: $(ab)^r = a^r b^r$

(b) $(3\sqrt{2})^2 = 3^2 \times (\sqrt{2})^2$

$$= 9 \times 2$$

$$= 18$$

Law 4: $(ab)^r = a^r b^r$

(c) $(1.1 \times 10^{-3})^2 = (1.1)^2 \times (10^{-3})^2$

$$= 1.21 \times 10^{-6}$$

Law 4: $(ab)^r = a^r b^r$

Law 3: $(a^r)^s = a^{rs}$

EXAMPLE 6

(a) If $y \neq 0$

$$\left(\frac{x}{y}\right)^2 = \frac{x^2}{y^2}$$

Law 5: $\left(\frac{a}{b}\right)^r = \frac{a^r}{b^r}$,

$b \neq 0$

(b) $\left(\dfrac{\sqrt{2}}{2}\right)^2 = \dfrac{(\sqrt{2})^2}{2^2}$

$$= \tfrac{2}{4}$$

$$= \tfrac{1}{2}$$

Law 5: $\left(\frac{a}{b}\right)^r = \frac{a^r}{b^r}$,

$b \neq 0$

In the last part of Example 2 we saw that the product of nonzero x and x^{-1} is 1. The number $x^{-1} = 1/x$ is called the *reciprocal* of x. In general, each of two real numbers is called the **reciprocal** of the other provided that their product is 1.

EXAMPLE 7

Find the reciprocal of $\frac{2}{3}$.

Solution

The reciprocal of $\frac{2}{3}$ is $\left(\frac{2}{3}\right)^{-1}$.

$$\left(\frac{2}{3}\right)^{-1} = \frac{1}{\frac{2}{3}} = \frac{3}{2}$$

Exercise 1.2, 76–95

We have asserted that the exponent laws are true for rational exponents and real-number bases. These laws are also true for integral exponents and bases that are powers of the imaginary number i where $i^2 = -1$. The powers of i are defined as are the powers of the nonzero real number a.

PART ONE | BASIC ALGEBRA

For k a positive integer, i^k represents the product of k i's, $i^{-k} = 1/i^k$, and $i^0 = 1$.

Note that even powers of i are either -1 or $+1$, and odd powers of i are either $-i$ or $+i$. For example, let us use the exponent laws and the definition of i to compute a few positive integral powers of i.

$$i^2 = -1 \qquad i^3 = i^2 \cdot i$$
$$\qquad\qquad\qquad = (-1) \cdot i \qquad i^2 = -1$$
$$\qquad\qquad\qquad = -i \qquad\qquad\qquad\qquad i^3 = -i$$

$$i^4 = (i^2)^2 \qquad i^5 = i^4 \cdot i$$
$$\quad = (-1)^2 \qquad\quad = 1 \cdot i \qquad i^4 = 1$$
$$\quad = 1 \qquad\qquad\quad = i \qquad\qquad\qquad\qquad i^5 = i$$

$$i^6 = (i^2)^3 \qquad i^7 = i^6 \cdot i$$
$$\quad = (-1)^3 \qquad\quad = (-1) \cdot i \qquad i^6 = -1$$
$$\quad = -1 \qquad\qquad\; = -i \qquad\qquad\qquad\qquad i^7 = -i$$

Similarly, negative integral powers of i are either -1 or $+1$ or else are $-i$ or $+i$. For example:

$$i^{-2} = 1/i^2 \qquad i^{-1} = i^{-2} \cdot i$$
$$\qquad = \tfrac{1}{-1} \qquad\qquad = (-1) \cdot i \qquad i^{-2} = -1$$
$$\qquad = -1 \qquad\qquad\quad = -i \qquad\qquad\qquad\qquad i^{-1} = -i$$

Exercise 1.2, 96–110

EXERCISE 1.2 INTEGRAL EXPONENTS AND ROOTS

▶ Find the value of each of the following.

EXAMPLES

(a) $\sqrt[3]{\frac{-1}{8}}$ (b) $(\sqrt{3})^2$ (c) $(\sqrt{-3})^2$

Solutions

(a) $\sqrt[3]{\frac{-1}{8}} = -\frac{1}{2}$ (b) $(\sqrt{3})^2 = 3$

(c) $(\sqrt{-3})^2 = (\sqrt{3}i)^2$
$$= (\sqrt{3})^2 i^2$$
$$= 3(-1)$$
$$= -3$$

1. $(\frac{3}{5})^2$	**2.** $(\frac{3}{5})^0$	**3.** $(\frac{3}{5})^{-2}$	**4.** $(\frac{3}{5})^{-1}$
5. 0^3	**6.** $\sqrt[3]{27}$	**7.** $\sqrt[3]{-27}$	**8.** $\sqrt{25}$
9. $\sqrt{\frac{1}{25}}$	**10.** $\sqrt{17^2}$	**11.** $\sqrt{(-17)^2}$	**12.** $\sqrt[3]{8^3}$
13. $\sqrt[3]{(-8)^3}$	**14.** $\sqrt[6]{2^6}$	**15.** $\sqrt[6]{(-2)^6}$	**16.** $(\frac{-7}{8})^2$
17. $(\sqrt{2})^2$	**18.** $(\sqrt{2})^0$	**19.** $(\sqrt{2})^{-2}$	**20.** $(\frac{1}{10})^{-1}$

21. 0^{23}	**22.** $\sqrt[5]{32}$	**23.** $\sqrt[5]{-32}$	**24.** $\sqrt{16}$
25. $\sqrt{11^2}$	**26.** $\sqrt{(-11)^2}$	**27.** $\sqrt[5]{7^5}$	**28.** $\sqrt[5]{(-7)^5}$
29. $\sqrt[4]{7^4}$	**30.** $\sqrt[4]{(-7)^4}$	**31.** $\sqrt{4^2}$	**32.** $\sqrt{(-4)^2}$
33. $(\sqrt{-4})^2$	**34.** $\sqrt{9^2}$	**35.** $\sqrt{(-9)^2}$	**36.** $(\sqrt{-9})^2$
37. $\sqrt{7^2}$	**38.** $\sqrt{(-7)^2}$	**39.** $(\sqrt{7})^2$	**40.** $(\sqrt{-7})^2$

▶ In each of the following, if the expression is defined and is a real number, give its value. If the expression is not defined or is not real, say "not defined" or "not real."

EXAMPLES

(a) $\sqrt{0^3} = \sqrt{0} = 0$ (b) $3 \times 0 = 0$

(c) $3 + 0 = 3$ (d) $1/0^3 = 1/0$ is not defined

(e) $3^0 = 1$ (f) $(\sqrt{0})^0 = 0^0$ is not defined

(g) $\sqrt{-25}$ is not real   (h) $\sqrt[4]{-81}$ is not real

41. 0^3	**42.** $\sqrt[3]{0}$	**43.** 0^{-3}	**44.** 0^0
45. 0×0	**46.** $\sqrt{0}$	**47.** $(\sqrt{0})^0$	**48.** 1^0
49. 0^1	**50.** $\frac{1}{0}$	**51.** 0^{-1}	**52.** $\frac{0}{1}$
53. 1×0	**54.** $\frac{0}{0}$	**55.** $\frac{3}{0}$	**56.** $\frac{0}{3}$
57. $(0^2)^0$	**58.** $(3 \times 0)^2$	**59.** $(\frac{0}{1})^3$	**60.** $(\frac{0}{1})^4$
61. $\frac{-1}{0}$	**62.** 0^{-2}	**63.** $\frac{-2}{0}$	**64.** $\frac{0}{-2}$
65. $(-2) \times 0$	**66.** $\sqrt[3]{-1}$	**67.** $\sqrt{(-36)^2}$	**68.** $\sqrt{-36}$
69. $\sqrt{-2}$	**70.** $\sqrt[4]{16}$	**71.** $\sqrt[4]{-16}$	**72.** $\sqrt{-81}$
73. $\sqrt[4]{1}$	**74.** $\sqrt[3]{1}$	**75.** $\sqrt[3]{-1}$	

▶ Let x and y represent real numbers, and express each of the following in a simplified form containing no negative exponents. State any restrictions that must be placed on x and y so that the given expressions will be defined.

EXAMPLES

(a) $\dfrac{xy^{-2}}{x^2 y}$ (b) $(3x)^0$ (c) $(\sqrt{x})^2$

Solutions

(a) $\dfrac{1}{xy^3}$, $x \neq 0, y \neq 0$

(b) $(3x)^0 = 1$, $x \neq 0$

(c) $(\sqrt{x})^2 = x$

76. $\dfrac{3x^3 y^2}{7xy^3}$	**77.** $(x^3 y^2)(x^5 y^7)$	**78.** $\dfrac{x^3 y^{-5}}{x^5 y^2}$

79. $(\sqrt{2}xy)^2$ **80.** $(5x^2)^0$ **81.** $(\sqrt{2}x)^2$

82. $\sqrt{4x^2}$ **83.** $x^{-3}y^{-3}$ **84.** $(x^3y^3)(x^{-3}y^{-3})$

85. $\dfrac{y^{-2}}{y^{-1}}$ **86.** $(x^3y^2)^{-1}$ **87.** $(x^2y^5)^2$

88. $(xy)(xy)^{-1}$ **89.** $\left(\dfrac{5x^3y^{-7}}{\sqrt{2}x^{-1}y^{-1}}\right)^2$ **90.** $\left(\dfrac{7x^{-5}y^{-9}}{\sqrt{3}x^2y^{-2}}\right)^{-1}$

91. $\left(\dfrac{x}{y}\right)^{-1}$ **92.** $\left(\dfrac{3x^2}{2y^3}\right)^{-1}$ **93.** $\left(\dfrac{\sqrt{2}x^{-1}}{y^2}\right)^2$

94. $\left(\dfrac{x}{y}\right)^{-2}$ **95.** $\left(\dfrac{x}{\sqrt{3}}\right)^2$

▶ Express each of the following as either -1 or $+1$ or as $-i$ or $+i$.

96. i^8 **97.** i^9

98. i^{-3} **99.** i^{-4}

100. i^{10} **101.** i^{11}

102. i^{-5} **103.** i^{-6}

104. $i^6 \times i^2$ **105.** $i^{-3} \times i^7$

106. $(i^{-1})^2$ **107.** i^{4k+2} where k is an integer

108. i^{4k} where k is an integer **109.** i^{4k+1} where k is an integer

110. i^{4k+3} where k is an integer

CHECK LIST FOR THE STUDENT

If you have learned the material in Section 1.2, you should be able to do the following:

1. State the meaning of a^k where a is a real number and k is a positive integer.

2. State the meaning of a^0 where a is any nonzero real number.

3. State the meaning of a^{-k} where a is any nonzero real number and k is a positive integer.

4. State the meaning of an nth root of b where n is a positive integer and b is any real number.

5. Tell how many real nth roots of b there are if n is an odd integer and b is any real number.

6. Tell how many real nth roots of b there are if n is an even integer and b is any positive real number.

7. State the meaning of $\sqrt{b}$ where b is any positive real number.

8. Tell how many real nth roots of b there are if n is an even integer and b is any negative real number.

9. Give the names for n and b in the symbol $\sqrt[n]{b}$.

10. Complete the statement: For any real number x, $\sqrt{x^2} =$ _____; $\sqrt[n]{x^n} =$ _____ if n is odd; $\sqrt[n]{x^n} =$ _____ if n is even.

11. State and use the five exponent laws.

12. State the definition of the reciprocal of nonzero x.

13. Express any given integral power of i as either -1, $+1$, $-i$, or $+i$.

SECTION QUIZ 1.2

Find the value of each of the following.

1. (a) 5^3 (b) 5^1 (c) 5^{-3} (d) 5^0 (e) 0^5

2. (a) $\sqrt[3]{64}$ (b) $\sqrt[3]{-64}$ (c) $\sqrt{9}$ (d) $-\sqrt{9}$ (e) $\sqrt[3]{(-5)^3}$

3. (a) $(\sqrt{5})^2$ (b) $\sqrt{0}$ (c) $\sqrt{(3)^2}$ (d) $\sqrt{(-3)^2}$ (e) $(\frac{3}{5})^{-2}$

4. In each of the following, if the expression is defined and is a real number, give its value. If the expression is not defined or is not real, say "not defined" or "not real."
   (a) $5 + 0$ (b) 5×0 (c) $\frac{5}{0}$ (d) 0^0 (e) $\sqrt{-9}$

5. Let x and y represent real numbers, and give the integer that belongs in each of the following parentheses.

(a) $x^2 x^5 = x^{(\ \ )}$ (b) $\dfrac{x^5}{x^2} = x^{(\ \ )}$, $x \neq 0$,   (c) $(x^2)^5 = x^{(\ \ )}$

(d) $(xy)^5 = x^{(\ \ )} y^{(\ \ )}$,   (e) $\left(\dfrac{x}{y}\right)^5 = \dfrac{x^{(\ \ )}}{y^{(\ \ )}}$, $y \neq 0$

Let x and y represent real numbers, and express each of the following in a simplified form containing no negative exponents. State any restrictions that must be placed on x and y so that the given expressions will be defined.

6. $(x^3 y^{-2})(x^{-5} y^3)$ **7.** $\dfrac{xy^{-3}}{x^3 y^2}$ **8.** $\left(\dfrac{x^3}{y^2}\right)^{-1}$ **9.** $\left(\dfrac{3x^2 y^{-5}}{4x^{-1} y^{-1}}\right)^2$

10. Let $i^2 = -1$, and express each of the following as either -1, $+1$, $-i$, or $+i$:
(a) i^4 (b) i^3

1.3 Rational Exponents and Radicals

The purpose of this section is to review the definition of rational exponents, to continue to review the exponent laws, and to review properties of radicals.

Our primary objective is to extend the definition of a^r for real numbers a to include rational numbers r that are not necessarily integers and to state this definition so that a^r will be a real number and the exponent laws will hold true. To this end, let the rational number r be written in the form m/n where m and n are integers and n is positive. Then a^r is $a^{m/n}$, and we define

$$a^{m/n} = (\sqrt[n]{a})^m$$

where a and m are not both zero, a is nonzero if m is negative, and a is nonnegative if n is even. Note that a and m cannot both be zero because then $a^{m/n}$ would be 0^0, and 0^0 is not defined. Also, a must be nonzero if m is negative, because otherwise $(\sqrt[n]{a})^m$ would be of the form 0^{-k} where $-k$ is a negative integer, and such 0^{-k} is not defined. Finally, a must be nonnegative if n is even, because otherwise a^r might not be a real number. For example,

$$0^{0/2} \quad \text{is not defined}$$
$$0^{-3/2} \quad \text{is not defined}$$
$$(-16)^{3/2} \quad \text{is not defined.}$$

EXAMPLE 1   Find the values of $8^{2/3}$, $8^{-2/3}$, $(-8)^{2/3}$, $(16)^{3/2}$, and $0^{4/5}$.

Solution

$$8^{2/3} = (\sqrt[3]{8})^2 \qquad 8^{-2/3} = (\sqrt[3]{8})^{-2} \qquad (-8)^{2/3} = (\sqrt[3]{-8})^2$$
$$= 2^2 \qquad\qquad = (2)^{-2} \qquad\qquad = (-2)^2$$
$$= 4 \qquad\qquad = \tfrac{1}{4} \qquad\qquad = 4$$

$$16^{3/2} = (\sqrt{16})^3 \qquad 0^{4/5} = (\sqrt[5]{0})^4$$
$$= 4^3 \qquad\qquad = 0^4$$
$$= 64 \qquad\qquad = 0$$

Note that $0^{m/n} = 0$ provided that m and n are both positive integers. Note also that $b^{1/n} = \sqrt[n]{b}$ provided that $\sqrt[n]{b}$ is a real number.

$$0^{2/3} = 0$$
$$25^{1/2} = \sqrt{25} = 5$$
$$64^{1/3} = \sqrt[3]{64} = 4$$

In defining $a^{m/n}$, we have placed restrictions on m, n, and a so that $\sqrt[n]{a}$ is a real number and $a^{m/n} = (\sqrt[n]{a})^m$ is a real number. In that case,

$(\sqrt[n]{a})^m$ is equal to $\sqrt[n]{a^m}$ so that we can write

$$a^{m/n} = (\sqrt[n]{a})^m = \sqrt[n]{a^m}$$

provided that $\sqrt[n]{a}$ is a real number. For example:

$$
\begin{aligned}
8^{2/3} &= (\sqrt[3]{8})^2 & \text{and} && 8^{2/3} &= \sqrt[3]{8^2} \\
&= 2^2 & && &= \sqrt[3]{64} \\
&= 4 & && &= 4 \\[4pt]
4^{-3/2} &= (\sqrt{4})^{-3} & \text{and} && 4^{-3/2} &= \sqrt{4^{-3}} \\
&= (2)^{-3} & && &= \sqrt{1/4^3} \\
&= 1/2^3 & && &= \sqrt{\tfrac{1}{64}} \\
&= \tfrac{1}{8} & && &= \tfrac{1}{8}
\end{aligned}
$$

The question sometimes arises of whether or not it is permissible to reduce a rational exponent to lowest terms. If $mk/nk = m/n$, can we say that $a^{mk/nk} = a^{m/n}$? It is true that $\frac{2}{6} = \frac{1}{3}$, but is it true that $a^{2/6} = a^{1/3}$? Not if a is negative, because $a^{2/6}$ has not been defined for negative a. If n is even and a is negative, $a^{m/n}$ is not defined. If, however, $a^{m/n}$ and $a^{mk/nk}$ are both defined, then it is true that $a^{mk/nk} = a^{m/n}$ for any positive integer k. This says that

$$(64)^{2/6} = (64)^{1/3},$$

but it does not say that $(-64)^{2/6}$ is equal to $(-64)^{1/3}$, because $(-64)^{2/6}$ has not been defined. If the base a of exponent m/n is nonnegative and $a^{m/n}$ is defined, we can be sure that the value of $a^{m/n}$ does not depend on the form of the rational exponent. In particular, if a is positive,

$$a^{mk/nk} = a^{m/n}, \qquad m,\ n,\ \text{and } k \text{ are integers,}$$
$$n \text{ and } k \text{ are positive.}$$

This equation is also true if a is zero and both exponents are positive.

EXAMPLE 2   Find the value of $8^{14/21}$ and $(49)^{9/18}$.

Solution

$$8^{14/21} = 8^{2/3} = (\sqrt[3]{8})^2 = 2^2 = 4$$

Exercise 1.3, 1–50 $\quad (49)^{9/18} = (49)^{1/2} = \sqrt{49} = 7$

The given definitions can be used to show that the exponent laws stated in Section 1.2 are true for rational exponents. Use of these laws is demonstrated in the following examples.

EXAMPLE 3

$$
\begin{aligned}
8^2 \times 8^{1/3} &= 8^{2+(1/3)} \\
&= 8^{7/3} && \text{Law 1: } a^r a^s = a^{r+s} \\
&= 128
\end{aligned}
$$

EXAMPLE 4

$$\frac{8^{4/3}}{8^{1/3}} = 8^{(4/3)-(1/3)} \qquad \text{Law 2: } \frac{a^r}{a^s} = a^{r-s}, \quad a \neq 0$$

$$= 8^{3/3}$$

$$= 8$$

EXAMPLE 5

$$(27^{1/3})^2 = 27^{2/3} \qquad \text{Law 3: } (a^r)^s = a^{rs}$$

$$= 9$$

EXAMPLE 6

$$(3\sqrt{5})^2 = 3^2(\sqrt{5})^2 \qquad \text{Law 4: } (ab)^r = a^r b^r$$

$$= 9 \times 5$$

$$= 45$$

EXAMPLE 7

$$\left(\frac{4}{9}\right)^{3/2} = \frac{4^{3/2}}{9^{3/2}} \qquad \text{Law 5: } \left(\frac{a}{b}\right)^r = \frac{a^r}{b^r}, \quad b \neq 0$$

Exercise 1.3, 51–60

$$= \tfrac{8}{27}$$

EXAMPLE 8

Express the following in the form ax^r where a is a real number written with no negative or fractional exponents, and r is a rational number. Assume that x is a positive real number.

$$\tfrac{1}{2}(3x^5)^{(1/2)-1}(15x^4)$$

Solution

$$\tfrac{1}{2}(3x^5)^{(1/2)-1}(15x^4) = \tfrac{15}{2}(3x^5)^{-1/2}x^4$$

$$= \tfrac{15}{2}(3)^{-1/2}x^{-5/2}x^4$$

$$= \tfrac{15}{2} \times \tfrac{1}{\sqrt{3}}x^{4-(5/2)}$$

$$= \frac{15}{2\sqrt{3}}x^{3/2}$$

EXAMPLE 9

Express the following in the form ax^r where a is a real number, r is a rational number, and x is a positive real number.

$$\frac{1}{x^{2/3} + x^{1/3}x^{1/3} + x^{2/3}}$$

Solution

$$\frac{1}{x^{2/3} + x^{1/3}x^{1/3} + x^{2/3}} = \frac{1}{x^{2/3} + x^{2/3} + x^{2/3}} \qquad \text{Law 1: } a^r a^s = a^{rs}$$

$$= \frac{1}{3x^{2/3}}$$

$$= (3x^{2/3})^{-1}$$

$$= 3^{-1}x^{-2/3} \qquad \text{Law 4: } (ab)^r = a^r b^r$$

$$= \tfrac{1}{3}x^{-2/3}$$

Exercise 1.3, 61–70

Since a radical such as $\sqrt[n]{b}$ can be expressed in exponential form, radicals inherit properties from the exponent laws. Two of the most commonly used of these properties follow.

Properties of Radicals If $\sqrt[n]{a}$ and $\sqrt[n]{b}$ are real numbers, then:

1. $\sqrt[n]{ab} = \sqrt[n]{a}\,\sqrt[n]{b}$ (Derived from Law 4: $(ab)^r = a^r b^r$)

2. $\sqrt[n]{\dfrac{a}{b}} = \dfrac{\sqrt[n]{a}}{\sqrt[n]{b}},\ b \neq 0$ (Derived from Law 5: $\left(\dfrac{a}{b}\right)^r = \dfrac{a^r}{b^r},\ b \neq 0$)

EXAMPLE 10 Use the properties of radicals to find the values of $\sqrt[3]{64 \times 1000}$ and $\sqrt[3]{\frac{-27}{8}}$.

Solution
$$\sqrt[3]{64 \times 1000} = \sqrt[3]{64} \times \sqrt[3]{1000} \qquad \text{Property 1: } \sqrt[n]{ab} = \sqrt[n]{a}\,\sqrt[n]{b}$$
$$= 4 \times 10$$
$$= 40$$

$$\sqrt[3]{\frac{-27}{8}} = \frac{\sqrt[3]{-27}}{\sqrt[3]{8}} \qquad \text{Property 2: } \sqrt[n]{\frac{a}{b}} = \frac{\sqrt[n]{a}}{\sqrt[n]{b}},\ b \neq 0$$

Exercise 1.3, 71–80
$$= \frac{-3}{2}$$

In the following examples assume that the variables are all positive so that all the radicals represent real numbers.

EXAMPLE 11 Write $\sqrt[3]{5x^3 y^6 z^{10}}$ in a form whose radicand contains no factor with an exponent greater than or equal to the index 3.

Solution
$$\sqrt[3]{5x^3 y^6 z^{10}} = \sqrt[3]{5x^3 y^6 z^9 \cdot z}$$
$$= \sqrt[3]{5z}\,\sqrt[3]{x^3 y^6 z^9}$$
$$= \sqrt[3]{5z}\,\sqrt[3]{x^3}\,\sqrt[3]{y^6}\,\sqrt[3]{z^9}$$
$$= \sqrt[3]{5z} \cdot x \cdot y^2 \cdot z^3$$
$$= \sqrt[3]{5z}\, xy^2 z^3$$

Some radicals can be written in an equivalent form having a smaller index. For example,
$$\sqrt[6]{x^2} = x^{2/6} = x^{1/3} = \sqrt[3]{x}.$$

The next example demonstrates this with a more complicated radicand.

EXAMPLE 12 Write $\sqrt[10]{49x^4 y^6}$ in a form that cannot be rewritten with a smaller index.

Solution
$$\sqrt[10]{49x^4 y^6} = \sqrt[10]{(7x^2 y^3)^2}$$
$$= (7x^2 y^3)^{2/10}$$
$$= (7x^2 y^3)^{1/5}$$
$$= \sqrt[5]{7x^2 y^3}$$

If a radical has a fractional radicand, it can be written otherwise. For example, $\sqrt{2/3}$ can be written $\sqrt{6}/3$ with no fraction in the radicand.

$$\sqrt{\frac{2}{3}} = \sqrt{\frac{6}{9}}$$

$$= \frac{\sqrt{6}}{\sqrt{9}} \qquad \text{Property 2: } \sqrt[n]{\frac{a}{b}} = \frac{\sqrt[n]{a}}{\sqrt[n]{b}}, \ b \neq 0$$

$$= \frac{\sqrt{6}}{3}$$

EXAMPLE 13

Assume that the variables are positive and write $\sqrt[3]{7x/2y}$ in a form having no fraction in the radicand.

Solution

$$\sqrt[3]{\frac{7x}{2y}} = \sqrt[3]{\frac{(7x)(2y)^2}{(2y)(2y)^2}}$$

$$= \sqrt[3]{\frac{28xy^2}{(2y)^3}}$$

$$= \frac{\sqrt[3]{28xy^2}}{\sqrt[3]{(2y)^3}} \qquad \text{Property 2: } \sqrt[n]{\frac{a}{b}} = \frac{\sqrt[n]{a}}{\sqrt[n]{b}}, \ b \neq 0$$

$$= \frac{\sqrt[3]{28xy^2}}{2y}$$

To simplify a radical is to write it in a form

1. That cannot be rewritten with a smaller index, and

2. Whose radicand contains no factor with an exponent greater than or equal to the index.

EXAMPLE 14

Assume that the variables are all positive and simplify

$$\sqrt[4]{\frac{18^2 x^6 y^8}{25 z^2}}.$$

Solution

Step 1   If possible, rewrite the given radical with a smaller index.

$$\sqrt[4]{\frac{18^2 x^6 y^8}{25 z^2}} = \sqrt[4]{\left(\frac{18 x^3 y^4}{5z}\right)^2}$$

$$= \left(\frac{18 x^3 y^4}{5z}\right)^{2/4} = \left(\frac{18 x^3 y^4}{5z}\right)^{1/2}$$

$$= \sqrt{\frac{18 x^3 y^4}{5z}}$$

Step 2 If possible, remove factors from the radicand.

$$\sqrt[4]{\frac{18^2 x^6 y^8}{25z^2}} = \sqrt{\frac{18x^3 y^4}{5z}}$$

$$= \sqrt{\frac{9 \cdot 2 \cdot x^2 \cdot x \cdot y^4}{5z}}$$

$$= \sqrt{9x^2 y^4} \sqrt{\frac{2x}{5z}}$$

$$= 3xy^2 \sqrt{\frac{2x}{5z}}$$

Exercise 1.3, 81–100

EXERCISE 1.3 RATIONAL EXPONENTS AND RADICALS

▶ Express each of the following in two radical forms.

EXAMPLES

(a) $7^{3/5} = \sqrt[5]{7^3} = (\sqrt[5]{7})^3$ (b) $7^{-3/5} = \sqrt[5]{7^{-3}} = (\sqrt[5]{7})^{-3}$

1. $8^{4/5}$	**2.** $8^{-4/5}$	**3.** $5^{2/3}$	**4.** $5^{-2/3}$	**5.** $\pi^{-1/2}$
6. $\pi^{3/5}$	**7.** $2^{3/2}$	**8.** $2^{-3/2}$	**9.** $6^{7/8}$	**10.** $6^{-7/8}$

▶ Express each of the following in exponential form.

EXAMPLE

$$\sqrt{10^3} = 10^{3/2}$$

11. $\sqrt[3]{7^2}$	**12.** $(\sqrt[3]{23})^2$	**13.** $\sqrt{3}$	**14.** $\sqrt[3]{2}$
15. $\sqrt{7^3}$	**16.** $(\sqrt[3]{7})^2$	**17.** $\sqrt{5}$	**18.** $\sqrt[3]{5^2}$
19. $(\sqrt{2})^3$	**20.** $\sqrt[3]{2^2}$		

▶ Find the value of each of the following.

EXAMPLE

$$25^{3/2} = \sqrt{25^3} = (\sqrt{25})^3 = 5^3 = 125$$

21. $49^{3/2}$	**22.** $27^{2/3}$	**23.** $(-27)^{2/3}$
24. $0^{2/3}$	**25.** $27^{-2/3}$	**26.** $(-27)^{-2/3}$
27. $16^{-3/2}$	**28.** $100^{1/2}$	**29.** $(0.01)^{-1/2}$
30. $64^{1/3}$	**31.** $\left(\frac{8}{125}\right)^{1/3}$	**32.** $(1000)^{2/3}$
33. $(0.001)^{-2/3}$	**34.** $\left(\frac{9}{100}\right)^{1/2}$	**35.** $(-32)^{1/5}$

36. $(32)^{-3/5}$ **37.** $(144)^{-1/2}$ **38.** $(\sqrt{11})^2$

39. $(\sqrt{9})^2$ **40.** $(\frac{1}{36})^{3/2}$ **41.** $(0.09)^{1/2}$

42. $(0.0004)^{1/2}$ **43.** $(0.008)^{1/3}$ **44.** $(0.0016)^{3/4}$

45. $(0.00001)^{1/5}$ **46.** $8^{-6/9}$ **47.** $4^{4/8}$

48. $(\frac{1}{81})^{6/4}$ **49.** $(32)^{8/10}$ **50.** $(\frac{8}{27})^{2/6}$

▶ Using the exponent laws, express each of the following in a form containing no radicals or negative exponents. Assume all variables are positive.

EXAMPLE

$$\begin{aligned}(x^{3/2}y^{-3})(x^{1/2}y) &= x^{3/2+1/2}y^{-3+1}\\ &= x^2 y^{-2}\\ &= \frac{x^2}{y^2}\end{aligned}$$

51. $(x^{1/2}y^{3/2})(x^{-3/2}y^2)$ **52.** $(a^{3/2}b^{-1})(a^3b^{-3})$

53. $\dfrac{x^{1/2}y^{3/2}}{x^{-3/2}y^2}$ **54.** $\dfrac{a^{3/2}b^{-1}}{a^3b^{-3}}$

55. $\left(\dfrac{x^{5/2}}{y^{-1/2}}\right)^4$ **56.** $\left(\dfrac{x^{3/2}}{y^{1/2}}\right)^{-1/2}$

57. $(x^{-1}y^{-1})^{-1/2}(x^2y^3)^{1/2}$ **58.** $\left(\dfrac{a^3b^{-1/2}}{ab}\right)^{2/3}$

59. $(b^3a^{2/3})^{1/2}(ab)^2$ **60.** $\left(\dfrac{x^{-2/3}y^{2/3}}{x^{1/3}y}\right)^3$

▶ Express each of the following in the form ax^r where a is a real number written with no negative or fractional exponents, and r is a rational number. Assume that x is a positive real number.

EXAMPLE

$$\begin{aligned}\tfrac{1}{2}(5x^3)^{(1/2)-1}(15x^2) &= \tfrac{1}{2}(5x^3)^{-1/2}(15x^2)\\ &= \frac{15}{2\sqrt{5}}x^{-3/2}x^2\\ &= \frac{15}{2\sqrt{5}}x^{1/2}\end{aligned}$$

61. $\tfrac{1}{2}(7x^3)^{(1/2)-1}(21x^2)$ **62.** $\tfrac{1}{3}(5x^2)^{(1/3)-1}(10x)$

63. $\tfrac{1}{3}(2x^4)^{(1/3)-1}(8x^3)$ **64.** $(\tfrac{2}{3})(3)x^{(2/3)-1}$

65. $(\tfrac{-1}{2})(7x^3)^{(-1/2)-1}(21x^2)$ **66.** $(\tfrac{-1}{3})(2x^5)^{(-1/3)-1}(10x^4)$

67. $(\tfrac{-1}{2})(6x^5)^{(-1/2)-1}(30x^4)$ **68.** $\tfrac{1}{2}(3x^{2/3})^{-1/2}(3)(\tfrac{2}{3})x^{-1/3}$

69. $\frac{1}{2}(2^{1/2})(x^{1/3})^{(1/2)-1}(\frac{1}{3})x^{-2/3}$

70. $\dfrac{1}{x^{3/4} + x^{2/4} \cdot x^{1/4} + x^{1/4} \cdot x^{2/4} + x^{3/4}}$

▶ Use the properties of radicals to find the value of each of the following.

EXAMPLES

(a) $\sqrt[3]{125 \times 8} = \sqrt[3]{125} \times \sqrt[3]{8} = 5 \times 2 = 10$

(b) $\sqrt[3]{\dfrac{-1000}{8}} = \dfrac{\sqrt[3]{-1000}}{\sqrt[3]{8}} = \dfrac{-10}{2} = -5$

71. $\sqrt[3]{125 \times 1000}$ **72.** $\sqrt[3]{\frac{-27}{1000}}$ **73.** $\sqrt[5]{32 \times 243}$

74. $\sqrt[5]{\frac{243}{32}}$ **75.** $\sqrt{\frac{4}{8100}}$ **76.** $\sqrt{16 \times 2500}$

77. $\sqrt{16 \times 10^{10}}$ **78.** $\sqrt{\dfrac{1}{16 \times 10^{10}}}$ **79.** $\sqrt[3]{64 \times 125}$

80. $\sqrt[3]{\frac{-64}{125}}$

▶ Assume that all the variables are positive, and simplify each of the following.

EXAMPLE

$\sqrt[6]{64x^8y^2} = \sqrt[6]{(8x^4y)^2} = \sqrt[3]{8x^4y} = 2x\sqrt[3]{xy}$

81. $\sqrt[6]{64x^6y^8}$ **82.** $\sqrt[6]{64x^{15}y^9}$ **83.** $\sqrt{81x^6y^3}$

84. $\sqrt[3]{54x^5y^2}$ **85.** $\sqrt[3]{-16x^3y^7}$ **86.** $\sqrt[4]{81x^3y^8z^7}$

87. $\sqrt[5]{64x^{10}y^{11}}$ **88.** $\sqrt[6]{64x^2y^7}$ **89.** $\sqrt[5]{-81x^9y^3z^{12}}$

90. $\sqrt{36x^2y^9}$

▶ Assume that all the variables are positive, and write each of the following in a form with a nonfractional radicand.

EXAMPLE

$\sqrt{\dfrac{3}{2x}} = \sqrt{\dfrac{3(2x)}{4x^2}} = \dfrac{\sqrt{6x}}{2x}$

91. $\sqrt{\dfrac{2}{3x}}$ **92.** $\sqrt{\frac{1}{3}}$ **93.** $\sqrt[3]{\dfrac{5y}{2x}}$ **94.** $\sqrt[3]{\dfrac{2}{x^2y}}$ **95.** $\sqrt{\frac{1}{5}}$

96. $\sqrt[3]{\dfrac{7x^2}{2y}}$ **97.** $\sqrt[4]{\dfrac{3x}{y^2}}$ **98.** $\sqrt[5]{\dfrac{2xy}{y^3}}$ **99.** $\sqrt[3]{\frac{1}{2}}$ **100.** $\sqrt[3]{\dfrac{3x^2}{y^2}}$

CHECK LIST FOR THE STUDENT

If you have learned the material in Section 1.3, you should be able to do the following:

1. Complete the statement: Where m and n are integers and n is positive, we define $a^{m/n}$ to be _____ with the understanding that a and m are not both zero, that a is _____ if m is negative, and that a is _____ if n is even.

2. Write real number $a^{m/n}$ in two radical forms.

3. Write real number $(\sqrt[n]{a})^m$ and real number $\sqrt[n]{a^m}$ in exponential form.

4. Complete the statement: If $a^{m/n}$ and $a^{mk/nk}$ are both defined, then it is true that $a^{mk/nk} =$ _____ for positive integer k.

5. Use the exponent laws with rational exponents.

6. Complete the statements of the two main properties of radicals:

$$\sqrt[n]{ab} = \underline{\hspace{1cm}}, \qquad \sqrt[n]{\frac{a}{b}} = \underline{\hspace{1cm}}, \qquad b \neq 0.$$

7. Use the two main properties of radicals.

8. Simplify radicals.

SECTION QUIZ 1.3

1. (a) Express $5^{3/4}$ in two radical forms.
   (b) Express $\sqrt[3]{5^2}$ in exponential form.

2. Find the value of each of the following:
   (a) $64^{2/3}$ (b) $64^{-2/3}$ (c) $(-64)^{2/3}$ (d) $64^{1/2}$ (e) $0^{2/3}$

3. Using the exponent laws, express each of the following in a form containing no radicals or negative exponents. Assume all variables are positive.

   (a) $(x^{2/3}y^{4/5})(x^{-4/3}y^{6/5})$ (b) $\left(\dfrac{x^{1/2}y^{-2}}{xy^2}\right)^{1/2}$

4. Use the properties of radicals to find the value of each of the following:
   (a) $\sqrt[3]{64 \times 27}$ (b) $\sqrt[3]{\frac{27}{-125}}$

5. Assume that all variables are positive, and tell what belongs in the blank parentheses in each of the following:
   (a) $\sqrt{8x^2y^5} = (\ \ )\sqrt{2y}$
   (b) $\sqrt{5x/3y} = \sqrt{(\ \ )}/3y$
   (c) $\sqrt[6]{64x^4y^8} = (\ \ )\sqrt[3]{x^2y}$

1.4 Polynomials

The purpose of this section is to define polynomials, to review addition, subtraction, multiplication, and division of polynomials, and to show how to evaluate polynomials. Synthetic division, and addition, subtraction, and multiplication of complex numbers are also covered.

Our first objective is to introduce some preliminary definitions so that we can use them in stating the definition of a polynomial.

A **term** is a combination of numbers, variables, or parenthetical expressions formed by multiplication or division. A single number, variable, or parenthetical expression is also called a term. For example,

$$5x^3, \quad -2x^2y, \quad -x^3, \quad (x+h)^2x^3, \quad \frac{3x^2y}{2z}, \quad \sqrt{2}, \quad x, \quad \text{and} \quad (x+h)$$

are all terms. A term containing no variables is called a **constant term**. The numbers $\sqrt{2}$ and -6 are constant terms.

The numerical factor of a term is called its **coefficient**. (In any product ab, the a and the b are called *factors*.) The coefficient of $5x^3$ is 5, the coefficient of $-2x^2y$ is -2, the coefficient of $-x^3 = (-1)x^3$ is -1, and the coefficient of $3x^2y/2z$ is $\frac{3}{2}$. If two terms are exactly alike except for their coefficients, they are said to be **like terms**. The terms

$$2x^2y \quad \text{and} \quad \tfrac{7}{5}x^2y$$

Exercise 1.4, 1–10 are like terms.

A **polynomial in one variable** x is a sum of terms of the form ax^n where a is the coefficient and n is a nonnegative integer. If $n = 0$, the term ax^n simply consists of the coefficient a because $x^0 = 1$. A polynomial of just one term is called a **monomial,** a polynomial of exactly two terms is called a **binomial,** and a polynomial of exactly three terms is called a **trinomial.**

A polynomial *over the reals* is a polynomial whose coefficients are all real numbers. A polynomial *over the integers* is one whose coefficients are all integers. The expression

$$\sqrt{2}x^3 + \tfrac{3}{4}x^2 - 2x + 7$$

is a polynomial over the reals. It is the sum of the terms $\sqrt{2}x^3, \tfrac{3}{4}x^2, -2x,$ and 7. The expression

$$9x^3 - 3x^2 - 2x + 7$$

is a polynomial over the integers. It is the sum of the terms $9x^3, -3x^2, -2x,$ and 7. Note that since any integer is a real number, a polynomial over the integers is also a polynomial over the reals.

A polynomial in two variables x and y is a sum of terms of the form ax^ny^m where a is the coefficient and n and m are nonnegative integers. The expression

$$3x^2y^3 - 2xy + 7x - 8y + 5$$

is a polynomial in x and y. It is the sum of the terms $3x^2y^3$, $-2xy$, $7xy^0$, $-8x^0y$, and $5x^0y^0$. In general a polynomial in more than one variable is a sum of terms having no variables in their denominators and having only nonnegative exponents on their variables. The expression

$$3x^2y^3z - 5x^2y + \tfrac{7}{2}yz - 3x$$

is a polynomial in x, y, and z. The expression

$$3x^2y^3/z - 5x^2y + \tfrac{7}{2}yz - 3x$$

is not a polynomial. There is a variable in the denominator of the first term. The expression

$$3x^2y^3z - 5x^{-2}y + \tfrac{7}{2}y^{1/2}z - 3x$$

Exercise 1.4, 11–40

is not a polynomial. There is a negative exponent on x in the second term, and there is a nonintegral exponent on y in the third term.

The **degree of a term** is the sum of the exponents of its variables. The degree of a term *in one particular variable* is the exponent of that variable. The degree of $7x^2y^3$ is 5. The degree of $7x^2y^3$ in x is 2 and in y is 3. The degree of the monomial $7x^2$ is 2, and the degree of the monomial $8 = 8x^0$ is 0. (Any nonzero real number can be considered a monomial of zero degree. The number zero itself is not assigned a degree.)

The **degree of a polynomial** is the degree of its term having the highest degree. The degree of

$$7x^2y^3 + 5x^4 - 2x$$

is 5, and the degree of

$$5x^3 - 2x^2 + 7$$

is 3. The degree of a polynomial in one particular variable is the highest exponent appearing on that variable anywhere in the polynomial. The degree of $7x^2y^3 + 5x^4 - 2x$ in the variable x is 4 and in the variable y is 3.

A polynomial of third degree in x technically has four terms, even though some of the terms may not appear because their coefficients are zero. For example,

$$5x^3 - 2x^2 + 7$$

technically has the four terms $5x^3$, $-2x^2$, $0x$, and 7. Likewise, a polynomial of fourth degree in x technically has five terms. The fourth-degree polynomial

$$7x^4 - 2x^3 + 9x - 8$$

has the five terms $7x^4$, $-2x^3$, $0x^2$, $9x$, and -8. In general, an nth-degree polynomial in x has $n + 1$ terms. If we denote the $n + 1$ coefficients by the symbols

$$a_0, a_1, a_2, a_3, \ldots, a_{n-1}, a_n$$

while acknowledging that any of these might be zero except for the coefficient of x^n, then an nth-degree polynomial in x may be written in the form

$$a_0 x^n + a_1 x^{n-1} + a_2 x^{n-2} + a_3 x^{n-3} + \cdots + a_{n-1} x + a_n$$

where $a_0 \neq 0$ and n is a nonnegative integer. Unless otherwise specified, it is understood that the variable x may only represent real numbers and that the coefficients are all real numbers, so that unless otherwise specified, by *polynomial* we mean *polynomial over the reals*.

Exercise 1.4, 41–50

Polynomials in x are sometimes represented by symbols such as $P(x)$, $Q(x)$, $R(x)$, etc. These are read "P of x," "Q of x," "R of x," etc. If $P(x)$ represents a polynomial, $P(a)$ is said to be the *value of the polynomial at $x = a$*. For example, if

$$P(x) = 5x^3 - 2x^2 - 4x + 1,$$

then $P(-2)$ is the value of $P(x)$ at $x = -2$.

$$
\begin{aligned}
P(-2) &= 5(-2)^3 - 2(-2)^2 - 4(-2) + 1 \\
&= 5(-8) - 2(4) + 8 + 1 \\
&= -40 - 8 + 8 + 1 \\
&= -39
\end{aligned}
$$

To find $P(a)$, we merely go through the expression for $P(x)$ and replace each x by a.

$$P(a) = 5a^3 - 2a^2 - 4a + 1$$

To find $P(a + h)$, we go through the expression for $P(x)$ and replace each x by $a + h$.

$$P(x) = 5x^3 - 2x^2 - 4x + 1$$
$$P(a + h) = 5(a + h)^3 - 2(a + h)^2 - 4(a + h) + 1$$

To find $P(x + h)$, we replace each x in $P(x)$ by $x + h$.

$$P(x) = 5x^3 - 2x^2 - 4x + 1$$
$$P(x + h) = 5(x + h)^3 - 2(x + h)^2 - 4(x + h) + 1$$

EXAMPLE 1

If $P(x) = 2x^3 - 8x^2 + 7x$ and $Q(x) = 3x^2 - 4x$, find $P(2)$, $Q(2)$, $P(Q(2))$, and $Q(P(2))$.

Solution

$$P(x) = 2x^3 - 8x^2 + 7x \qquad Q(x) = 3x^2 - 4x$$
$$P(2) = 2(2)^3 - 8(2)^2 + 7(2) \qquad Q(2) = 3(2)^2 - 4(2)$$
$$= 16 - 32 + 14 \qquad\qquad\quad = 12 - 8$$
$$= -2 \qquad\qquad\qquad\qquad = 4$$

$$P(Q(2)) = P(4) \text{ since } Q(2) = 4.$$
$$P(4) = 2(4)^3 - 8(4)^2 + 7(4)$$
$$= 128 - 128 + 28$$
$$= 28$$

$$Q(P(2)) = Q(-2) \text{ since } P(2) = -2.$$
$$Q(-2) = 3(-2)^2 - 4(-2)$$
$$= 12 + 8$$

Exercise 1.4, 61–65

$$= 20$$

Like terms may be added or subtracted simply by adding or subtracting their coefficients and multiplying the result by the common variable part. For example,

$$3x^3 + 5x^3 = (3 + 5)x^3 = 8x^3$$

and

$$2x^2y - 7x^2y = (2 - 7)x^2y = -5x^2y.$$

Such operations are sometimes called *collecting like terms*. Collecting like terms is an application of the distributive property of the real numbers (Appendix A). To *simplify* a polynomial is to collect its like terms.

EXAMPLE 2

Simplify the polynomial

$$3x^3 + 9 - 2x + 5x^3 - 7x + 8x^2 - 7 - 9x^2.$$

Solution

Since the commutative property of addition (Appendix A) assures us that we may rearrange the order in which the terms are added, we may write the given polynomial in the form

$$3x^3 + 5x^3 + 8x^2 - 9x^2 - 2x - 7x + 9 - 7.$$

Collecting like terms, we have

Exercise 1.4, 51–60

$$8x^3 - x^2 - 9x + 2.$$

Polynomials may be added as demonstrated in the next example.

EXAMPLE 3

Simplify the sum of

$$3x^3 - 2x^2 + 5x - 8 \quad \text{and} \quad 2x - 7x^2 + 6x^3 - 1.$$

Solution

The work may be arranged either horizontally or vertically. Arranging it horizontally, we simply collect like terms.

$$3x^3 - 2x^2 + 5x - 8 + 2x - 7x^2 + 6x^3 - 1$$
$$= 3x^3 + 6x^3 - 2x^2 - 7x^2 + 5x + 2x - 8 - 1$$
$$= 9x^3 - 9x^2 + 7x - 9$$

Arranging the work vertically, we place like terms in the same column and then add.

$$\begin{array}{r} 3x^3 - 2x^2 + 5x - 8 \\ + 6x^3 - 7x^2 + 2x - 1 \\ \hline 9x^3 - 9x^2 + 7x - 9 \end{array}$$

If A and B are algebraic expressions, $A - B$ is the same as $A + (-B)$. Thus to subtract B from A, we simply change the sign of B and add. For example, $(5x^3 - 9x^2 + 8) - (2x^3 - 6x^2 - 3x + 1)$ is the sum of $(5x^3 - 9x^2 + 8)$ and $-(2x^3 - 6x^2 - 3x + 1)$.

A minus sign in front of a parenthesis may be read (-1) *times* the entire expression within the parentheses. It follows from the distributive property of the real numbers (Appendix A) that to remove parentheses with a minus sign in front, we must change the sign of every term inside. On the other hand, if a parenthesis is preceded by no sign or by a plus sign, the parentheses may simply be removed without changing the expression inside.

$$-(2x^3 - 6x^2 - 3x + 1) = -2x^3 + 6x^2 + 3x - 1$$
$$(5x^3 - 9x^2 + 8) = 5x^3 - 9x^2 + 8$$
$$+(5x^3 - 9x^2 + 8) = 5x^3 - 9x^2 + 8$$

The work performed in subtracting polynomials may be arranged horizontally or vertically. This is shown in the next example.

EXAMPLE 4

Simplify $(5x^3 - 9x^2 + 8) - (2x^3 - 6x^2 - 3x + 1)$.

Solution

Arranging the work horizontally, we remove parentheses and collect like terms.

$$(5x^3 - 9x^2 + 8) - (2x^3 - 6x^2 - 3x + 1)$$
$$= 5x^3 - 9x^2 + 8 - 2x^3 + 6x^2 + 3x - 1$$
$$= 5x^3 - 2x^3 - 9x^2 + 6x^2 + 3x + 8 - 1$$
$$= 3x^3 - 3x^2 + 3x + 7$$

Arranging the work vertically, we place like terms in the same column, and then change signs and add.

$$\begin{array}{r} 5x^3 - 9x^2 \qquad + 8 \\ \mp\ \underline{2x^3 \mp 6x^2 \mp 3x \pm 1} \\ 3x^3 - 3x^2 + 3x + 7 \end{array}$$

Exercise 1.4, 66–75

To multiply two polynomials, we take each term of one of them, multiply it by every term of the other one, and then add the individual products. This is illustrated in Example 5.

EXAMPLE 5 Simplify the product $(3x^2 + 4)(2x^3 - 5x^2 - 7)$.

Solution Arranging the work horizontally, we multiply each term of $(3x^2 + 4)$ by every term of $(2x^3 - 5x^2 - 7)$, add the products, and collect like terms.

$$(3x^2 + 4)(2x^3 - 5x^2 - 7)$$
$$= 3x^2(2x^3 - 5x^2 - 7) + 4(2x^3 - 5x^2 - 7)$$
$$= 6x^5 - 15x^4 - 21x^2 + 8x^3 - 20x^2 - 28$$
$$= 6x^5 - 15x^4 + 8x^3 - 21x^2 - 20x^2 - 28$$
$$= 6x^5 - 15x^4 + 8x^3 - 41x^2 - 28$$

Arranging the work vertically, we put like terms among the individual products in the same column, and then add.

$$2x^3 - 5x^2 - 7$$
$$3x^2 + 4$$
$$\overline{6x^5 - 15x^4 \qquad\quad - 21x^2} \longleftarrow 3x^2(2x^3 - 5x^2 - 7)$$
$$\underline{\qquad\qquad 8x^3 - 20x^2 - 28} \longleftarrow 4(2x^3 - 5x^2 - 7)$$
$$6x^5 - 15x^4 + 8x^3 - 41x^2 - 28 \leftarrow \text{Sum of individual products}$$

Exercise 1.4, 76–110

EXAMPLE 6 If $P(x) = 3x^2 - 5x + 1$, express $P(a + h)$ as a polynomial in a and h.

Solution
$$P(a + h) = 3(a + h)^2 - 5(a + h) + 1$$
$$(a + h)^2 = (a + h)(a + h)$$
$$= a^2 + ah + ha + h^2$$
$$= a^2 + ah + ah + h^2$$
$$= a^2 + 2ah + h^2$$
$$P(a + h) = 3(a^2 + 2ah + h^2) - 5(a + h) + 1$$
$$= 3a^2 + 6ah + 3h^2 - 5a - 5h + 1$$

EXAMPLE 7 If $P(x) = 3x^2 - 5x + 1$, express $P(x + h) - P(x)$ as a polynomial in x and h.

Solution
$$P(x + h) = 3(x + h)^2 - 5(x + h) + 1$$
$$P(x + h) - P(x) = 3(x + h)^2 - 5(x + h) + 1 - (3x^2 - 5x + 1)$$
$$= 3(x^2 + 2xh + h^2) - 5x - 5h + 1 - 3x^2$$
$$+ 5x - 1$$
$$= 3x^2 + 6xh + 3h^2 - 5x - 5h + 1 - 3x^2$$
$$+ 5x - 1$$
$$= 3x^2 - 3x^2 - 5x + 5x + 1 - 1 + 6xh + 3h^2$$
$$- 5h$$

Exercise 1.4, 111–140

$$= 6xh + 3h^2 - 5h$$

If p and d are positive integers, there exist nonnegative integers q and r such that

$$\frac{p}{d} = q + \frac{r}{d}$$

where r is less than d. Otherwise stated,

$$p = qd + r.$$

$$\frac{17}{3} = 5 + \frac{2}{3} \quad \text{or} \quad 17 = 5 \times 3 + 2$$
$$\frac{12}{4} = 3 + \frac{0}{4} \quad \text{or} \quad 12 = 3 \times 4 + 0$$
$$\frac{4}{7} = 0 + \frac{4}{7} \quad \text{or} \quad 4 = 0 \times 7 + 4$$

A similar situation exists for polynomials. If $P(x)$ and $D(x)$ are polynomials with $D(x)$ not zero, there exist polynomials $Q(x)$ and $R(x)$ such that

$$\frac{P(x)}{D(x)} = Q(x) + \frac{R(x)}{D(x)}$$

where $R(x)$ is of lower degree than $D(x)$ or $R(x) = 0$. That is,

$$P(x) = Q(x) \cdot D(x) + R(x).$$

Here the numerator $P(x)$ is called the **dividend**, the denominator $D(x)$ is called the **divisor**, the $Q(x)$ is called the **quotient**, and the $R(x)$ is called the **remainder**. If the division comes out even, $R(x)$ is zero. To find the quotient and the remainder, we divide the polynomial $P(x)$ by the polynomial $D(x)$ in much the same manner as we perform long division of integers. To illustrate, we will divide $x - 3x^2 + 2x^3 - 2$ by $x^2 - 4x + 5$. Note that before the process is begun, the dividend and divisor are arranged in descending powers of x, and note that the process is repeated until it produces a remainder that is zero or is of lower degree than the divisor.

$$\frac{\overset{\text{Dividend}}{2x^3 - 3x^2 + x - 2}}{\underset{\text{Divisor}}{x^2 - 4x + 5}} = \overset{\text{Quotient}}{2x + 5} + \frac{\overset{\text{Remainder}}{11x - 27}}{\underset{\text{Divisor}}{x^2 - 4x + 5}}$$

The student should check that

$$2x^3 - 3x^2 + x - 2 = (2x + 5)(x^2 - 4x + 5) + (11x - 27).$$
$$\text{Dividend} \quad = \text{(Quotient)(Divisor)} \quad + \text{Remainder.}$$

EXAMPLE 8

Find the quotient and remainder.

$$\frac{x^2 - 2x + 5}{x - 3}$$

Solution

$$
\begin{array}{r}
x + 1 \\
x - 3{\overline{\smash{\big)}\,x^2 - 2x + 5}} \\
\underline{x^2 - 3x} \\
x + 5 \\
\underline{x - 3} \\
8
\end{array}
$$

$$\frac{x^2 - 2x + 5}{x - 3} = x + 1 + \frac{8}{x - 3}$$

Check $(x + 1)(x - 3) + 8 = x^2 - 2x - 3 + 8$
$$= x^2 - 2x + 5$$

Exercise 1.4, 141–150   The quotient is $x + 1$. The remainder is 8.

In Example 8, the divisor is a polynomial of the form $x - c$ where c is the real number 3. Note that the remainder 8 happens to be the value of the dividend at $x = 3$. Indeed, if

$$P(x) = x^2 - 2x + 5,$$

then
$$P(3) = 3^2 - 2(3) + 5$$
$$= 9 - 6 + 5$$
$$= 8.$$

This is not just a coincidence. Whenever a polynomial $P(x)$ is divided by a polynomial of the form $x - c$, the remainder is equal to the value of $P(x)$ at $x = c$. This fact is known as the *remainder theorem*.

Remainder Theorem   If the polynomial $P(x)$ is divided by $x - c$, the remainder equals $P(c)$.

Proof   If $P(x)$ is divided by $x - c$, the remainder must be a number R because it must be zero or of lower degree than $x = c$. So if $Q(x)$ is the quotient,

$$P(x) = Q(x) \cdot (x - c) + R.$$

It follows that
$$P(c) = Q(c) \cdot (c - c) + R$$
$$= Q(c) \cdot 0 + R$$
$$= R.$$

Exercise 1.4, 151–160   That is, the remainder R equals $P(c)$.

Since there is a short method of dividing a polynomial by $x - c$, the remainder theorem can readily be used to evaluate a polynomial. This short method is called *synthetic division* or *synthetic substitution*. It amounts to leaving out all unnecessary symbols from the usual long-division process. For example, consider the division of $x^4 - 7x^3 + 3x - 5$ by $x - 2$.

$$
\begin{array}{r}
x^3 - 5x^2 - 10x - 17 \\
x - 2 \overline{)\, x^4 - 7x^3 \qquad\quad + 3x - 5} \\
\underline{x^4 - 2x^3} \\
-5x^3 \qquad\quad + 3x - 5 \\
\underline{-5x^3 + 10x^2} \\
-10x^2 + 3x - 5 \\
\underline{-10x^2 + 20x} \\
-17x - 5 \\
\underline{-17x + 34} \\
-39
\end{array}
$$

If we examine this work, we see that we have done an unnecessary amount of writing. The colored numbers and the coefficients of x^3 and x^4 in the dividend and quotient are really the only essential symbols. For example, we know that the dividend is of fourth degree and that the quotient is going to be a third-degree polynomial in x, so the only symbols that we need in order to identify dividend and quotient are their coefficients. We know that the divisor is of the form $x - c$, so the only identifying symbol we need for the divisor is the number $c = 2$. If we eliminate almost all nonessential symbols, insert ones for the coefficients of x^3 and x^4, and take care to acknowledge any zero coefficients (in the dividend, the x^2 term is missing so its coefficient is zero), we are left with the following array:

$$
\begin{array}{r}
1 \\
-2 \overline{)\, 1 - 7 \qquad 0 \qquad 3 - 5} \\
\underline{-2} \\
-5 \\
\underline{10} \\
-10 \\
\underline{20} \\
-17 \\
\underline{34} \\
-39
\end{array}
$$

The actual arithmetic computations that produced this array consisted of successive multiplications by -2 followed by subtractions. But if we work with 2 rather than -2, the signs of the products will be changed, and we can add rather than subtract. In that case, the array becomes

$$
\begin{array}{r}
1 \\
2{\overline{\smash{\big)}\,1 - 7 \quad\quad 0 \quad\quad 3 - 5}} \\
\underline{2} \\
- 5 \\
\underline{- 10} \\
- 10 \\
\underline{- 20} \\
- 17 \\
\underline{- 34} \\
- 39
\end{array}
$$

If we now compress this array to two lines, we have

$$
\begin{array}{r}
2{\overline{\smash{\big)}\,1 - 7 \quad\quad 0 \quad\quad 3 - 5}} \\
\underline{2 - 10 - 20 - 34} \\
1 - 5 - 10 - 17 - 39
\end{array}
$$

The remainder is -39, and the numbers 1, -5, -10, and -17 are the coefficients of the quotient. Since the quotient is a third-degree polynomial, it must be

$$x^3 - 5x^2 - 10x - 17.$$

Now the work and the arithmetic computations that produced the coefficients 1, -5, -10, and -17 and the remainder -39 can be outlined as follows. To divide $x^4 - 7x^3 + 3x - 5$ by $x - 2$ using synthetic division:

Step 1 Write down the 2, and write down all the coefficients (including any zeros) of the dividend.

Step 2 Bring down the first coefficient and multiply it by 2. $1 \times 2 = 2$

Step 3 Add. $-7 + 2 = -5$

Step 4 Multiply by 2. $-5 \times 2 = -10$

Step 5 Add. $0 + (-10) = -10$

Step 6 Multiply by 2. $-10 \times 2 = -20$

Step 7 Add. $3 + (-20) = -17$

Step 8 Multiply by 2. $-17 \times 2 = -34$

Step 9 Add. $-5 + (-34) = -39$

EXAMPLE 9   Use synthetic division to find the quotient and remainder.

$$\frac{3x^3 - 2x^2 - 8}{x + 5}$$

Solution   Here $x - c$ is $x + 5$, so $c = -5$. The quotient will be a second-degree polynomial.

$$
\begin{array}{r}
-5)\overline{3 \quad - \quad 2 \quad\quad 0 \, - \quad\quad 8} \\
\underline{- \, 15 \quad\quad 85 - 425} \\
3 - 17 \quad\quad 85 - 433
\end{array}
$$

The quotient is $3x^2 - 17x + 85$. The remainder is -433. (Note that the value of $3x^3 - 2x^2 - 8$ at $x = -5$ is indeed -433.)

     The synthetic division process will work *only* for divisors of the form $x - c$ where c may be positive or negative. It will not work for divisors of

Exercise 1.4, 161–180 any other form.

     A complex number was defined in Section 1.1 as a number that can be written in the form $a + bi$ where a and b are real numbers and $i^2 = -1$. Keeping in mind that $i^2 = -1$, we may add, subtract, or multiply complex numbers as if they were polynomials. These operations are demonstrated in the following examples.

EXAMPLE 10   Express the sum $(3 + 2i) + (5 - 7i)$ in the form $a + bi$ where a and b are real numbers.

Solution   Arranging the work horizontally,

$$
\begin{aligned}
(3 + 2i) + (5 - 7i) &= 3 + 2i + 5 - 7i \\
&= (3 + 5) + (2i - 7i) \\
&= 8 - 5i.
\end{aligned}
$$

Arranging the work vertically,

$$
\begin{array}{r}
3 + 2i \\
+ \, 5 - 7i \\
\hline
8 - 5i.
\end{array}
$$

EXAMPLE 11   Express the difference $(3 + 2i) - (5 - 7i)$ in the form $a + bi$ where a and b are real numbers.

Solution   Arranging the work horizontally,

$$
\begin{aligned}
(3 + 2i) - (5 - 7i) &= 3 + 2i - 5 + 7i \\
&= (3 - 5) + (2i + 7i) \\
&= -2 + 9i.
\end{aligned}
$$

Arranging the work vertically, we change signs and add.

$$3 + 2i$$
$$+ \underline{-5 + 7i}$$
$$-2 + 9i$$

EXAMPLE 12

Express the product $(3 + 2i)(5 - 7i)$ in the form $a + bi$ where a and b are real numbers.

Solution

Arranging the work horizontally, we multiply each term of $3 + 2i$ by every term of $5 - 7i$, add the products, and collect like terms.

$$(3 + 2i)(5 - 7i) = 15 - 21i + 10i - 14i^2$$
$$= 15 - 21i + 10i - 14(-1)$$
$$= (15 + 14) + (-21i + 10i)$$
$$= 29 - 11i$$

Arranging the work vertically,

$$5 - 7i$$
$$\underline{3 + 2i}$$
$$15 - 21i$$
$$\underline{ + 10i - 14i^2}$$
$$15 - 11i - 14i^2 = 15 - 11i - 14(-1)$$
$$= 29 - 11i.$$

Exercise 1.4, 181–200

(In Section 1.6 we shall see how to express a fraction whose numerator and denominator are complex numbers as a single complex number.)

EXERCISE 1.4 POLYNOMIALS

▶ Identify like terms in each of the following.

EXAMPLE

$5x^3y,\ 7xy^3,\ -3x^3y,\ \frac{5}{3}y^3,\ \frac{9}{2}xy^3,\ 7x^3$

Solution

$5x^3y,\ -3x^3y \qquad$ are like terms.

$7xy^3,\ \frac{9}{2}xy^3 \qquad$ are like terms.

1. $2x^2y,\ 3x,\ 7y,\ 5x^2y,\ -7,\ 8y,\ 6x^2,\ 8xy$

2. $\sqrt{2}x,\ 7x^3,\ \sqrt{2}y,\ 7x^3y,\ -x^3,\ 9x,\ 9y,\ 7xy^3$

3. $x^5y^2,\ -3x^4y^3,\ 2x^2y^5,\ -x^5y^2,\ -x^2y^5,\ x^2,\ y^5,\ x^5$

4. $y^3,\ -2y^2x,\ \frac{2}{5}yx^2,\ \frac{7}{8}x^3,\ -9y^2x,\ \sqrt{2}y^3,\ x^3y^3,\ xy^2$

5. $-x^5y,\ 7x^4y^2,\ \frac{4}{5}x^5,\ 3y,\ -8x^2y^4,\ 7x^5y,\ -x^4y^2,\ 19y$

6. $3yz,\ 3xz,\ 3xy,\ 3x,\ 3y,\ 3z,\ 2yz,\ 2xz,\ 2xy,\ 2x$

7. x^3y, xy^3, x^2y^2, x^3y^3, $2x^3y$, $2xy^3$, $2xy$, xy

8. $13x^3y$, $5x^5yz$, $6x^6yz$, $-13x^3y$, $7x^7yz$, $8x^8yz$

9. $\sqrt{2}x^2y$, $3x^3y^7$, $\sqrt{2}x^2y^2$, $3x^7y^3$, $\sqrt{2}xy$, $3xy$, $3x^2y$

10. $5x$, $5x^2y$, $5y$, $5yz$, $5z$, $3x^2y$, $-yz$, $5xyz$

▶ State whether or not each of the following is a polynomial. If it is a polynomial, state whether or not it is a polynomial over the integers.

EXAMPLES

(a) $5x^3 - 2x^{-4} + 7x - 8$
(b) $5x^3 - 2x^{3/2} + 7x - 8$
(c) $5x^3 - \frac{2}{3}x^4 + 7x - 8$
(d) $5x^3 - 2x^4 + 7x - 8$

Solutions

(a) Not a polynomial. (Not all exponents are nonnegative integers.)
(b) Not a polynomial. (Not all exponents are nonnegative integers.)
(c) A polynomial, but not over the integers.
(d) A polynomial over the integers.

11. $2x^3 - \frac{3}{2}x^2 + 7x - 8$ **12.** $2x^{3/2} - 3x^2 + 7x - 8$

13. $2x^3 - 5x^2 + 6x^{-1} - 8$ **14.** $\sqrt{2}x^2 - 3x + 7$

15. $x - 9$ **16.** $x^2 - 25$

17. $x^3 + 8$ **18.** $x^3 - 1$

19. $3x^2 - \sqrt{2}x^3 + 5x$ **20.** $3x^{-2} - \sqrt{2}x^3 + 5x$

21. $\frac{2}{3}x^2 - 2x^3 + 5x$ **22.** $3x^2 - 2x^3 + 5x$

23. $3x^{2/3} - 2x^3 + 5x$ **24.** $3x^{-1} - 2x^3 + 5x$

25. $3x + 7$ **26.** $-x^2 - 2x + 4$

27. $2x - \frac{3}{5}$ **28.** $x^2 - 2x^{-1} + 4$

29. $\sqrt{x} - 2$ **30.** $\sqrt{x} + 3$

▶ State whether or not each of the following is a polynomial.

EXAMPLES

(a) $2x^2y^3z^2 - 5xy^3 + \frac{7}{2}y^2$
(b) $2x^2y^3z - 5xy^{3/2} + \frac{7}{2}y^{-2}$
(c) $\dfrac{2x^2y^3}{z} - 5xy^3 + \frac{7}{2}y^2$

Solutions

(a) A polynomial.
(b) Not a polynomial. (Not all exponents are nonnegative integers.)
(c) Not a polynomial. (There is a variable in the denominator of $2x^2y^3/z$.)

31. $5x^3y^2 - 2xy^3 + 7x - 9$ **32.** $5x^{-3}y^2 - 2xy^3 + 7x - 9$

33. $5x^{3/2}y^2 - 2xy^3 + 7x - 9$ **34.** $\dfrac{5x^3}{y^2} - 2xy^3 + 7x - 9$

35. $x + y - z$ **36.** $x - y$

37. $x^2 - xy + y^2$ **38.** $2xy^3 - 5xy^{-1} + 1$

39. $2xy^3 - 5xy^{1/2} + 1$ **40.** $2xy^3 - \dfrac{5}{xy} + 1$

▶ State the degree of each of the following, and then state its degree in each variable.

EXAMPLE

$3x^2y^4 - 5x^3y^5 + 2xy^6$

Solution

Eighth-degree polynomial. In x, third degree. In y, sixth degree.

41. $2x^3y^5 - 7x^4y^6 + 3x^2y^7$ **42.** $3xy - 2x + 7y - 2$

43. $7xy^2 - 3x^3y - 8x^5$ **44.** $2x^3yz - 5x^4 + 3x^2z - 2yz^2$

45. $3x + 5y - 7$ **46.** $y - 3x - 6$

47. $y - 3x^2 + 2x - 5$ **48.** $x^2 + xy + y^2$

49. $x - y$ **50.** $6x^4yz - 7z^5 + 2x^3y^2 - 3zy^3$

▶ Simplify each of the following (collect like terms).

EXAMPLE

$5x^3 - 3x^2 + 7x^3 - 2x^2 + 4x$

Solution

$5x^3 + 7x^3 - 3x^2 - 2x^2 + 4x = 12x^3 - 5x^2 + 4x$

51. $6x^3 - 2x^4 + 5x^3 - 3x^4 - 7$

52. $3x^2y - 5xy + 7x^2y - 3xy$

53. $x^2 - 2xy + y^2 - 3xy + 7$

54. $x^2 + 2xh + h^2 + 3x + 3h - 5 - x^2 - 3x + 5$

55. $3x + 3h + 7 - 3x - 7$

56. $3x^2 + 6xh + 3h^2 - 2x - 2h + 1 - 3x^2 + 2x - 1$

57. $5x + 5h - 8 - 5x + 8$

58. $2x^2 + 4xh + 2h^2 - 3x - 3h + 4 - 2x^2 + 3x - 4$

59. $x^3 + 3x^2h + 3xh^2 + h^3 - x^3$

60. $5x^2 + 10xh + 5h^2 - 3x - 3h - 5x^2 + 3x$

EXAMPLE

If $P(x) = 5x^3$ and $Q(x) = 7x - 5$, find $P(Q(1))$.

Solution

$Q(1) = 7(1) - 5 = 2$
$P(Q(1)) = P(2) = 5 \times (2)^3$
$\qquad\qquad\quad = 5 \times (8) = 40$

61. If $P(x) = 2x^3 - 5x^2$ and $Q(x) = x^4 - 3x$, find $P(1), Q(1), P(-1), Q(-1)$, $P(Q(1))$, and $Q(P(1))$.

62. If $P(x) = 3x^2$ and $Q(x) = 6x + 4$, find $P(0), Q(0), P(1), Q(1), P(-1)$, $Q(-1), P(Q(0))$, and $Q(P(0))$.

63. If $P(x) = x^2 - 3x$ and $Q(x) = 3x^2 - 2x$, find $P(3), Q(3), P(-3), Q(-3)$, $P(Q(-3))$, and $Q(P(-3))$.

64. If $P(x) = 3x - 5$ and $Q(x) = x^4 - 6$, find $P(2), Q(2), P(-2), Q(-2)$, $P(Q(2))$, and $Q(P(2))$.

65. If $P(x) = 3x^2 - 2x$ and $Q(x) = 2x + 7$, find $P(a), Q(a), P(Q(a))$, and $Q(P(a))$.

▶ Simplify each of the following.

66. $(3x^3 - 5x^2 + 7) + (2x^3 - 6x^2 + 8)$

67. $(3x^3 - 5x^2 + 7) - (2x^3 - 6x^2 + 8)$

68. $(6x^4 - 5x^3 + 2x^2 - 8) + (2x^4 - 3x^3 - 8x^2 + x - 16)$

69. $(6x^4 - 5x^3 + 2x^2 - 8) - (2x^4 - 3x^3 - 8x^2 + x - 16)$

70. $(x^2 - x + 1) + (x^2 - x + 1)$

71. $(x^2 - x + 1) - (x^2 - x + 1)$

72. $(3x^5 - 2x^4 + 3x^2 - 7) + (8x^5 - 7x^3 - 7x^4 - 3x^2 + 9)$

73. $(3x^5 - 2x^4 + 3x^2 - 7) - (8x^5 - 7x^3 - 7x^4 - 3x^2 + 9)$

74. $(2x^3 + 3x^2 - 5x - 9) + (-x^3 - 2x^2 + 7x - 8)$

75. $(2x^3 + 3x^2 - 5x - 9) - (-x^3 - 2x^2 + 7x - 8)$

▶ Simplify each of the following products.

76. $(2x^2 + 3)(5x^3 - 2x^2 + 4)$ **77.** $(8x^2 - 2)(6x^3 - 5x^2 + 1)$

78. $(3x + 7)(x^2 - 5x + 6)$

79. $(5x^2 - 2x)(x^3 - 3x^2 + 2x + 1)$

80. $(x^2 - 4)(x^2 + 4)$ **81.** $(3x + 7)(3x - 8)$

82. $(2x - 4)(5x + 6)$ **83.** $(x - 7)(x + 4)$

84. $(4x - 3)(2x + 5)$ **85.** $(x - 2)(x + 2)$

86. $(x - 2)(x^2 + 2x + 4)$ **87.** $(x + 2)(x^2 - 2x + 4)$

88. $(5x - 1)(3x + 7)$ **89.** $(x - y)(x^2 + xy + y^2)$

90. $(x^2 - y^2)(x^2 + y^2)$ **91.** $(x + y)(x^2 - xy + y^2)$

92. $(x + 5)^2$ **93.** $(x - 5)^2$

94. $(a + b)^2$ **95.** $(a - b)^2$

96. $(a - b)^3$ **97.** $(3 - \sqrt{2})(3 + \sqrt{2})$

98. $(2 - \sqrt{3})(2 + \sqrt{3})$ **99.** $(\sqrt{2} - \sqrt{3})(\sqrt{2} + \sqrt{3})$

100. $(\sqrt{5} - \sqrt{7})(\sqrt{5} + \sqrt{7})$

▶   Assume that all the radicands are positive, and write out each of the following products.

EXAMPLE

$(\sqrt{x} - 5)^2$

Solution

$x - 10\sqrt{x} + 25$

101. $(\sqrt{x} - 3)^2$ **102.** $(\sqrt{x} + 3)^2$

103. $(\sqrt{x} - \sqrt{y})^2$ **104.** $(\sqrt{x} - \sqrt{y})(\sqrt{x} + \sqrt{y})$

105. $(\sqrt{x - 2} + 2)^2$ **106.** $(\sqrt{x + 5} - 3)(\sqrt{x + 5} + 3)$

107. $(\sqrt{x - 3} + 2\sqrt{x})^2$ **108.** $(\sqrt{x - 3} - 2\sqrt{x})^2$

109. $(7 - \sqrt{x + 4})^2$ **110.** $(7 + \sqrt{x + 4})^2$

▶   In each of the following, express $P(a + h)$ as a polynomial in a and h and then express $P(x + h)$ as a polynomial in x and h.

EXAMPLE

$P(x) = 3x^2 - 2x + 7$

Solution

$$P(a + h) = 3(a + h)^2 - 2(a + h) + 7$$
$$= 3a^2 + 6ah + 3h^2 - 2a - 2h + 7$$
$$P(x + h) = 3(x + h)^2 - 2(x + h) + 7$$
$$= 3x^2 + 6xh + 3h^2 - 2x - 2h + 7$$

111. $P(x) = 2x^2 - 3x + 5$ **112.** $P(x) = 5x^2 - 7x + 6$

113. $P(x) = x^2 + 5x - 3$ **114.** $P(x) = 2x^3 - 5x^2 + 1$

115. $P(x) = x^3 - x^2 + 3x + 2$ **116.** $P(x) = 3x - 2$

117. $P(x) = 7x + 15$ **118.** $P(x) = 6x^2 - 3x$

119. $P(x) = 9x^2 + 1$ **120.** $P(x) = 8x^3 - 1$

▶   In each of the following, express $P(x + h) - P(x)$ as a polynomial in x and h.

EXAMPLE

$P(x) = 3x^2 - 4x + 5$

Solution

$$P(x + h) = 3(x + h)^2 - 4(x + h) + 5$$
$$P(x + h) - P(x) = 3(x + h)^2 - 4(x + h) + 5 - (3x^2 - 4x + 5)$$
$$= 3x^2 + 6xh + 3h^2 - 4x - 4h + 5 - 3x^2 + 4x - 5$$
$$= 6xh + 3h^2 - 4h$$

121. $P(x) = 3x^2 - 7x + 2$ **122.** $P(x) = -5x^2 + 3x + 1$

123. $P(x) = 2x^3 - x^2$ **124.** $P(x) = 6x - 5$

125. $P(x) = 4x^3$ **126.** $P(x) = 2x^2 - 5x + 3$

127. $P(x) = -x^2 + x - 1$ **128.** $P(x) = 7x - 4$

129. $P(x) = 2x^3 - 5x$ **130.** $P(x) = -3x^2 - 2x + 4$

▶ In each of the following, express $P(Q(x))$ as a polynomial.

EXAMPLE

$P(x) = 3x^2 + 4x$

$Q(x) = 2x + 5$

Solution

$$P(Q(x)) = 3(2x + 5)^2 + 4(2x + 5)$$
$$= 12x^2 + 30x + 75 + 8x + 20$$
$$= 12x^2 + 38x + 95$$

131. $P(x) = 5x^2 + 3x,\ Q(x) = 3x + 2$

132. $P(x) = 3x + 7,\ Q(x) = 3x^2 - 2x$

133. $P(x) = x^3,\ Q(x) = x + 2$

134. $P(x) = x^3,\ Q(x) = x^2$

135. $P(x) = x^2 - 3x,\ Q(x) = 3x^2 - 2$

136. $P(x) = 2x - 5,\ Q(x) = 6x^5 - 3x^4 + 2$

137. $P(x) = 3,\ Q(x) = 2x^2 + 5x$

138. $P(x) = 7,\ Q(x) = 8x^3 - 5x^2$

139. $P(x) = 2x^2 - 3x + 1,\ Q(x) = x + 1$

140. $P(x) = x^5 + x,\ Q(x) = 2x^3$

▶ In each of the following, find the quotient and the remainder.

141. $\dfrac{5x^3 - 2x^2 + x - 2}{x^2 - 3x + 6}$ **142.** $\dfrac{2x^3 - 5x^2 + 7}{x^2 - 2x}$

143. $\dfrac{4x^3 - 6x^2 + 2x - 1}{2x^2 + x + 3}$ **144.** $\dfrac{x^4 - 3x^2 + 2x - 1}{x^2 - 2x + 3}$

145. $\dfrac{2x^4 - x^3 + 3x^2 - 5}{x^3 - 5x^2 + 6}$ **146.** $\dfrac{x^3 - 2x + 8}{x - 2}$

147. $\dfrac{6x^2 + 17x - 45}{3x - 5}$ **148.** $\dfrac{10x^2 - 27x - 28}{2x - 7}$

149. $\dfrac{x^2 - 9}{x - 3}$ **150.** $\dfrac{12x^2 - 64x + 50}{6x - 5}$

▶ In each of the following, evaluate $P(c)$ by substitution, and then divide $P(x)$ by $x - c$ and find the remainder. Compare the remainder with $P(c)$.

EXAMPLE

$P(x) = 3x^2 - 2x - 3, \ c = 2$

Solution

$P(2) = 3(2)^2 - 2(2) - 3 = 12 - 4 - 3 = 5$

$$
\begin{array}{r}
3x \ + 4 \\
x - 2 \overline{) 3x^2 - 2x - 3} \\
\underline{3x^2 - 6x} \\
4x - 3 \\
\underline{4x - 8} \\
5 \leftarrow \text{Remainder}
\end{array}
$$

151. $P(x) = 2x^2 - x - 9, \ c = 3$

152. $P(x) = 2x^2 + 11x + 20, \ c = -3$

153. $P(x) = 3x^3 - x^2 - 3x - 7, \ c = 1$

154. $P(x) = 6x^2 + 7x - 8, \ c = -2$

155. $P(x) = 2x^3 - x^2 - 7x - 4, \ c = 2$

156. $P(x) = x^5 - 3x^3 - x^2 + x^4 + 2x + 3, \ c = -1$

157. $P(x) = 5x^3 + 17x^2 - 12x - 10, \ c = -4$

158. $P(x) = x^4 - 2x^2 + 17x - 7x^3 - 23, \ c = 7$

159. $P(x) = 7x^3 + 33x^2 - 7x + 11, \ c = -5$

160. $P(x) = x^7 + 3x^4 - x^6 - 3x^3 - 2x^2 + 9x - 9, \ c = 1$

▶ In each of the following, use synthetic division to find the quotient and remainder.

161. $\dfrac{x^5 - 2x^4 + 3x^3 - x^2 + 2x + 1}{x - 2}$

162. $\dfrac{x^5 - 2x^4 + 3x^2 - 1}{x + 2}$

163. $\dfrac{3x^4 - 2x^3 + 7x - 4}{x + 3}$

164. $\dfrac{3x^4 - 2x^3 + 7x - 4}{x - 3}$

165. $\dfrac{5x^2 - 2x + 7}{x - 4}$

166. $\dfrac{5x^2 - 2x + 7}{x + 4}$

167. $\dfrac{2x^5 - 3x^3 + 5x^2 - 2}{x - 1}$

168. $\dfrac{2x^5 - 3x^3 + 5x^2 - 2}{x + 1}$

169. $\dfrac{5x^4 - 3x^2 - 4x + 6}{x - 7}$

170. $\dfrac{5x^4 - 3x^2 - 4x + 6}{x + 7}$

▶ In each of the following, use synthetic division to find $P(c)$.

EXAMPLE

$P(x) = 2x^3 - 5x^2 + 7, \ c = 3$

Solution

$$3\overline{)\begin{array}{cccc} 2 & -5 & 0 & 7 \\ & 6 & 3 & 9 \end{array}}$$
$$\begin{array}{cccc} 2 & 1 & 3 & 16 \end{array} \quad P(3) = 16$$

171. $P(x) = 5x^3 - 4x^2 - 2x + 8, \ c = 2$

172. $P(x) = 2x^4 - 5x^3 + 3x, \ c = 3$

173. $P(x) = x^5 - 4x^4 + 5x^2 - 11, \ c = 2$

174. $P(x) = 2x^3 - 5x^2 + 7x - 18, \ c = -3$

175. $P(x) = 8x^5 - 3x^3 + 2x - 1, \ c = -2$

176. $P(x) = 7x^4 - 2x^2 + 3x - 11, \ c = -1$

177. $P(x) = 6x^3 - 5x^2 + 3x - 12, \ c = 4$

178. $P(x) = 11x^5 - 3x^4 + 2x^3 - 8, \ c = 6$

179. $P(x) = x^3 - 18, \ c = -6$

180. $P(x) = 2x^6 - 5x^4 + 3x^3 - 2x^2 + 5x - 8, \ c = 2$

▶ Express each of the following in the form $a + bi$ where a and b are real numbers and $i^2 = -1$.

EXAMPLES

$(2 + 5i) + (6 - 3i) = 8 + 2i$

$(2 + 5i) - (6 - 3i) = -4 + 8i$

$(2 + 5i)(6 - 3i) = 12 + 30i - 6i - 15i^2$
$$= 27 + 24i$$

$(2 + 5i)^2 = 4 + 20i + 25i^2 = -21 + 20i$

181. $(3 - 7i) + (4 + 6i)$

182. $(3 - 7i) - (4 + 6i)$

183. $(3 - 7i)(4 + 6i)$

184. $(3 - 7i)^2$

185. $5(3 - 7i)$

186. $5 + (3 - 7i)$

187. $(3 - 7i) - 5$

188. $(2 + 3i)^2 - 2(2 + 3i) + 4$

189. $3(4 - i)^2 + 2(4 - i) + 6$

190. $(\sqrt{2} + 3i)^2 - 4(\sqrt{2} + 3i) + 1$

191. $(\sqrt{2} + \sqrt{3}i)(\sqrt{2} - \sqrt{3}i)$

192. $(5 - 2i)(5 + 2i)$

193. $(2 + \sqrt{3}i)^2 - 4(2 + \sqrt{3}i) + 7$

194. $(4 + 3i)(4 - 3i)$

195. $(\frac{3}{5} + i)(\frac{3}{5} - i)$

▶ In each of the following, find the value of the polynomial for each of the given values of x.

EXAMPLE

Solution

$x^2 - 4x + 13; \; x = 2 + 3i, \; x = 2 - 3i$

$$(2 + 3i)^2 - 4(2 + 3i) + 13 = 4 + 12i + 9i^2 - 8 - 12i + 13$$
$$= 9 + 9i^2 + 12i - 12i$$
$$= 9 - 9$$
$$= 0$$
$$(2 - 3i)^2 - 4(2 - 3i) + 13 = 4 - 12i + 9i^2 - 8 + 12i + 13$$
$$= 9 + 9i^2 - 12i + 12i$$
$$= 0$$

196. $x^2 - 8x + 41; \; x = 4 + 5i, \; x = 4 - 5i$

197. $x^2 - 10x + 74; \; x = 5 + 7i, \; x = 5 - 7i$

198. $x^2 + 4x + 31; \; x = 2 + 5i, \; x = 2 - 5i$

199. $x^2 - 2x + 7; \; x = 1 + 2i, \; x = 1 - 2i$

200. $x^2 - 6x + 10; \; x = 3 + i, \; x = 3 - i$

CHECK LIST FOR THE STUDENT

If you have learned the material in Section 1.4, you should be able to do the following:

1. Give examples of terms, constant terms, coefficients, factors, and like terms.

2. Define polynomial, monomial, binomial, and trinomial.

3. Define polynomial over the reals and polynomial over the integers.

4. Give an example of a polynomial in two variables x and y.

5. Define the degree of a term and the degree of a polynomial.

6. Write an expression for an nth-degree polynomial in x.

7. Use the notation $P(x)$ for a polynomial in x, and find $P(a)$ for some specific real number a and a given polynomial $P(x)$.

8. Simplify a polynomial.

9. Remove parentheses with either a plus sign or a minus sign in front.

10. Add, subtract, multiply, and divide polynomials.

11. Express $P(x + h) - P(x)$ as a polynomial in x and h.

12. State and verify the remainder theorem.

13. Use synthetic division or synthetic substitution.

14. Add, subtract, and multiply complex numbers.

SECTION QUIZ 1.4

1. State whether or not each of the following is a polynomial.
   (a) $3x^4 - 2x^{-3} + 7x^2 - 9$
   (b) $3x^4 - 2x^3 + 7x^{3/5} - 9$
   (c) $3x^4 - 5x^3 + 7x - 9$
   (d) $3x^4 - 5x^3 + 2x^2 + 7x - 9$
   (e) $3x^4 - 5x^3 + \frac{4}{5}x^2 + 7x - 9$

2. (a) State the degree of $5x^3 - 2x^4 + 3x^2 + 9x - 8$.
   (b) Simplify $2x^3 - 3x^2 - 7x^3 + x^2 - 3x + 7$ by collecting like terms.

3. Let $P(x) = x^3 - 5x^2 + 2x - 1$ and $Q(x) = x^2 + 3$.
   (a) Find $P(2)$.   (b) Find $Q(1)$.   (c) Find $P(Q(1))$.

Let $P(x) = 2x^3 + 3x^2 - 5x - 6$ and $Q(x) = x^2 - 4x - 7$.

4. Find $P(x) + Q(x)$.

5. Find $P(x) - Q(x)$.

6. Find the product $P(x) \times Q(x)$.

7. Find the quotient and remainder.

$$\frac{2x^3 + 3x^2 - 5x - 6}{x^2 - 4x - 7}$$

8. Let $P(x) = 3x^2 - 2x - 5$, and express $P(x + h)$ as a polynomial in x and h.

9. Use synthetic division to find the quotient and remainder.

$$\frac{2x^3 + 3x^2 - 5x - 6}{x - 1}$$

10. Express each of the following in the form $a + bi$ where a and b are real numbers and $i^2 = -1$.

(a) $(2 + 5i) + (3 - 7i)$

(b) $(2 + 5i) - (3 - 7i)$

(c) $(2 + 5i) \times (3 - 7i)$

1.5 Factoring

The purpose of this section is to review factoring of polynomials with integral coefficients.

As we said in Section 1.4, in any product ab, the a and the b are called **factors.** Thus in the product 3×4, the 3 and the 4 are factors, and in the product $(x - 3)(x + 3)$, the $(x - 3)$ and the $(x + 3)$ are factors. To factor an integer or polynomial is to express it as a product of factors.

It can be shown that (except for the order of the factors) any positive integer that is not itself a prime can be factored into *primes* (Figure 1.5, Section 1.1) in exactly one way. For example, except for order, the only way to express 12 as a product of primes is to write

$$12 = 3 \times 2 \times 2$$
$$= 3 \times 2^2,$$

and the only way to express 72 as a product of primes is to write

$$72 = 2 \times 2 \times 2 \times 3 \times 3$$
$$= 2^3 \times 3^2.$$

Exercise 1.5, 1–10

A polynomial over the integers is said to be a **prime polynomial over the integers** if it is evenly divisible (the remainder is zero) by no polynomial with integral coefficients other than itself and $+1$ or -1. For example,

$$x - 7, \quad 2x + 3, \quad \text{and} \quad x^2 + 5$$

are prime polynomials over the integers.

It can be shown that (except for the order of the factors) any polynomial over the integers that is not itself a prime can be factored into prime factors with integral coefficients in exactly one way. For example, except for order, the only way to express $x^2 - 9$ as a product of such prime factors is to write

$$x^2 - 9 = (x - 3)(x + 3),$$

and the only way to express $3x^2 + 13x - 10$ as a product of prime polynomials with integral coefficients is to write

$$3x^2 + 13x - 10 = (3x - 2)(x + 5).$$

Expressing a polynomial over the integers as a product of prime polynomials with integral coefficients is sometimes called **factoring completely.** If a polynomial over the integers *can* be factored into prime polynomials with integral coefficients, then it is said to be **factorable.** Otherwise it is said to be **not factorable over the integers.**

Some types of factorable polynomials occur more frequently than others. Before listing some of these more common types, let us first list the special products giving rise to them.

SPECIAL PRODUCTS

1. $a(b + c) = ab + ac$
2. $(a - b)(a + b) = a^2 - b^2$
3. $(a + b)^2 = a^2 + 2ab + b^2$ }
   $(a - b)^2 = a^2 - 2ab + b^2$ }   Square of a binomial
4. $(a + b)(a^2 - ab + b^2) = a^3 + b^3$
   $(a - b)(a^2 + ab + b^2) = a^3 - b^3$

Note particularly that the square of a binomial is the sum of *three* terms. We caution all students to avoid the very common error of overlooking the "middle term." In other words, $(a + b)^2$ is *not* equal to just $a^2 + b^2$, but $(a + b)^2$ is equal to a^2 plus $2ab$ plus b^2.

$$(a + b)^2 = a^2 + 2ab + b^2$$

Likewise, $(a - b)^2$ is *not* equal to $a^2 - b^2$, as so many students seem to believe; $(a - b)^2$ is equal to a^2 minus $2ab$ plus b^2.

$$(a - b)^2 = a^2 - 2ab + b^2$$

EXAMPLE 1   Write out the product $(x + 7)^2$ and the product $(x - 7)^2$.

Solution
$(x + 7)^2 = x^2 + 14x + 49$
$(a + b)^2 = a^2 + 2ab + b^2$
$(x - 7)^2 = x^2 - 14x + 49$
$(a - b)^2 = a^2 - 2ab + b^2$

The common types of factorable polynomials arising in our list of special products are factored in the following list. Students should verify each of the examples by multiplying out the factors.

Exercise 1.5, 11–70

COMMON TYPES OF FACTORABLE POLYNOMIALS

I. $ab + ac = a(b + c)$ Common monomial
   $15x - 6 = 3(5x - 2)$ factor
   $15x^3 - 6x^2 = 3x^2(5x - 2)$

II. $a^2 - b^2 = (a - b)(a + b)$ Difference of two
   $9x^2 - 4 = (3x - 2)(3x + 2)$ squares

III. $a^2 + 2ab + b^2 = (a + b)^2$
  $a^2 - 2ab + b^2 = (a - b)^2$ Perfect-square
  $x^2 + 6x + 9 = (x + 3)^2$ trinomials
  $x^2 - 6x + 9 = (x - 3)^2$

IV. $a^3 + b^3 = (a + b)(a^2 - ab + b^2)$ Sum of two cubes
  $a^3 - b^3 = (a - b)(a^2 + ab + b^2)$ Difference of two cubes
  $x^3 + 125 = (x + 5)(x^2 - 5x + 25)$
  $x^3 - 125 = (x - 5)(x^2 + 5x + 25)$

EXAMPLE 2

Factor each of the following completely.
(a) $5x^3 - 5x$ (b) $x^4 - y^4$
(c) $9x^2 + 6xy + y^2$ (d) $9x^2 - 6xy + y^2$
(e) $8x^3 + 27$ (f) $8x^3 - 27$

Solution

We are to express each of the given polynomials as a product of prime factors with integral coefficients. In each case, if there is a common monomial factor, we should remove that (factor that out) first.

(a) $5x^3 - 5x = 5x(x^2 - 1)$ Common monomial factor

$\qquad\qquad = 5x(x - 1)(x + 1)$ Difference of two squares

(b) $x^4 - y^4 = (x^2 - y^2)(x^2 + y^2)$ Difference of two squares

$\qquad\qquad = (x - y)(x + y)(x^2 + y^2)$ Difference of two squares

(c) $9x^2 + 6xy + y^2 = (3x + y)^2$ Perfect-square trinomial

(d) $9x^2 - 6xy + y^2 = (3x - y)^2$ Perfect-square trinomial

(e) $8x^3 + 27 = (2x + 3)(4x^2 - 6x + 9)$ Sum of two cubes

(f) $8x^3 - 27 = (2x - 3)(4x^2 + 6x + 9)$ Difference of two cubes

Type I of the common types we have listed arises from the distributive law (Appendix A). An extension of the distributive law is used in reverse to remove any monomial that is common to all the terms of any polynomial. For example, the terms of the polynomial

$$6x^5 - 14x^4 - 8x^3$$

all have the factor $2x^3$ in common, so we can factor out $2x^3$ (divide each term by $2x^3$) and write

$$6x^5 - 14x^4 - 8x^3 = 2x^3(3x^2 - 7x - 4).$$

EXAMPLE 3	Factor $3x^3 - 30x^2 + 75x$ completely.
Solution	$3x^3 - 30x^2 + 75x = 3x(x^2 - 10x + 25)$
	$\qquad\qquad\qquad\quad = 3x(x - 5)^2$

Type II, the difference of two squares, is easy to recognize and usually causes no problems.

Polynomials of Type III, perfect-square trinomials, are also relatively easy to recognize. Each is the sum of *three* terms; the first and last terms of each are perfect squares; the "middle term" of each is either plus or minus twice the product of the numbers whose squares constitute the first and last terms. Each perfect-square trinomial is the square of a binomial.

Perhaps it will help to emphasize that the square of a binomial, a perfect-square trinomial, is the sum of *three* terms, if we note the following.

Where k is a real number, the expression $x^2 + kx$ becomes a perfect-square trinomial if the number $(k/2)^2$ is added to it.

In particular, $x^2 + 6x$ becomes a perfect square if the number $(6/2)^2 = (3)^2$ is added to it.

$$x^2 + 6x + (3)^2 = (x + 3)^2$$

The process of determining what should be added to an expression like $x^2 + kx$ in order to make it become a perfect square trinomial is called *completing the square*. We will use it in Section 1.7.

EXAMPLE 4

Tell what number should be put in the parentheses in order to make $x^2 - 18x + (\ \ )$ become a perfect-square trinomial, and then factor that perfect-square trinomial.

Solution

The number that should go in the parentheses is the square of half the coefficient of x. Half the coefficient of x is $\frac{-18}{2} = -9$, and its square is $(\frac{-18}{2})^2 = (-9)^2 = 81$.

$$x^2 - 18x + (81) = (x - 9)^2$$

Exercise 1.5, 116–125 This can be checked by finding the product $(x - 9)^2$.

Type IV is another source of many common errors. Note that $a^3 + b^3$ is *not equal to* $(a + b)^3$. In fact,

$$a^3 + b^3 = (a + b)(a^2 - ab + b^2),$$

whereas

$$(a + b)^3 = (a + b)(a + b)^2$$
$$= (a + b)(a^2 + 2ab + b^2).$$

Similarly, note that $a^3 - b^3$ is *not equal to* $(a - b)^3$;

$$a^3 - b^3 = (a - b)(a^2 + ab + b^2),$$

whereas
$$(a - b)^3 = (a - b)(a - b)^2$$
$$= (a - b)(a^2 - 2ab + b^2).$$

EXAMPLE 5

(a) Factor $x^3 + 8$ completely, and factor $x^3 - 8$ completely.

(b) Write out the product $(x + 2)^3$ and the product $(x - 2)^3$.

Solution

(a) $x^3 + 8 = (x + 2)(x^2 - 2x + 4)$
$x^3 - 8 = (x - 2)(x^2 + 2x + 4)$

(b) $(x + 2)^3 = (x + 2)(x + 2)^2$
$\qquad = (x + 2)(x^2 + 4x + 4)$
$\qquad = x^3 + 6x^2 + 12x + 8$
$(x - 2)^3 = (x - 2)(x - 2)^2$
$\qquad = (x - 2)(x^2 - 4x + 4)$
$\qquad = x^3 - 6x^2 + 12x - 8$

In addition to the common types of factorable polynomials we have listed, factorable second-degree trinomials that are not perfect squares are frequently encountered. For example,

$$15x^2 + 2x - 8$$

is such a trinomial. Each such trinomial is the product of two prime binomial factors of first degree. For example, $15x^2 + 2x - 8$ is the product of $(3x - 2)$ and $(5x + 4)$. Any factorable trinomial of the form $ax^2 + bx + c$ where a, b, and c are integers is the product of two binomials of the form $(px + q)$ and $(rx + s)$ where p, q, r, and s are integers.

$$(px + q)(rx + s) = \underbrace{(pr)x^2}_{\substack{\text{Product} \\ \text{of the} \\ \text{first terms}}} \quad \underbrace{(ps)x + (qr)x}_{\substack{\text{Sum of the} \\ \text{``cross products''}}} \quad \underbrace{(qs)}_{\substack{\text{Product} \\ \text{of the} \\ \text{second terms}}}$$

$$(3x - 2)(5x + 4) = 15x^2 \qquad 12x - 10x \qquad (-2)(4)$$

Where a, b, and c are integers, to factor $ax^2 + bx + c$ so that

$$ax^2 + bx + c = (px + q)(rx + s),$$

we systematically look for a pair of binomial factors $(px + q)$ and $(rx + s)$ with integral coefficients such that

1. The product of the first terms is ax^2,

$$pr = a$$

2. The product of the second terms is c,

$$qs = c$$

3. The sum of the "cross products" (the sum of the products of the first and last terms) is bx.

$$(ps) + (qr) = b$$

For example, to factor $15x^2 + 2x - 8$, we look for factors of the form $(px + q)(rx + s)$ with integral coefficients such that

1. $pr = 15$
2. $qs = -8$
3. $ps + qr = 2$

We can do this by first listing all possibilities for integers p, q, r, and s such that $pr = 15$ and $qs = -8$ and then selecting a combination for which $ps + qr = 2$. Such a list would reveal that if $p = 3$, $q = -2$, $r = 5$, and $s = 4$, $ps + qr = 2$ so that

$$15x^2 + 2x - 8 = (3x - 2)(5x + 4).$$

EXAMPLE 6

Factor $10x^3 + 26x^2 - 12x$ completely.

Solution

The monomial factor $2x$ is common to all three terms so we factor $2x$ out first.

$$10x^3 + 26x^2 - 12x = 2x(5x^2 + 13x - 6)$$
$$= 2x(5x - 2)(x + 3)$$

This can be checked by multiplying out the factors.

EXAMPLE 7

Factor $x^2 + 3x + 7$ completely, if possible. If not possible, write "not factorable over the integers."

Solution

There is no monomial factor (other than 1) common to all three terms of $x^2 + 3x + 7$. If $x^2 + 3x + 7$ is factorable, there must exist integers p, q, r, and s such that $(px + q)(rx + s)$ are the factors. That is, there must exist integers p, q, r, and s such that

$$pr = 1, \qquad qs = 7,$$

and $ps + qr = 3$. The only possibilities for p, r, q, and s such that $pr = 1$ and $qs = 7$ are:

p	r	q	s	$ps + qr$
1	1	1	7	8
1	1	7	1	8
1	1	-1	-7	-8
1	1	-7	-1	-8
-1	-1	1	7	-8
-1	-1	7	1	-8
-1	-1	-1	-7	8
-1	-1	-7	-1	8

But in none of these possibilities is the sum of the cross products, $ps + qr$, equal to 3. It is not possible to factor $x^2 + 3x + 7$ into a product of polynomials with integral coefficients that are prime polynomials. The polynomial $x^2 + 3x + 7$ is itself a prime polynomial. We say that $x^2 + 3x + 7$ is *not factorable over the integers.*

Exercise 1.5, 71–115

We have seen that an extension of the distributive law can be used in reverse to factor any existing common *monomial* factors out of a polynomial. Sometimes the terms of a polynomial can be grouped in such a way that an extension of the distributive law can be used in reverse to factor out common *binomial* factors. For example, consider the polynomial

$$x^4 - 3x^3 - 2x^2 + 6x + 7x - 21.$$

We can group and factor the terms so that the new expression has the binomial factor $(x - 3)$ common to all the terms. We can then factor out that common binomial factor. Indeed,

$$\underbrace{x^4 - 3x^3}\ \underbrace{-\ 2x^2 + 6x}\ \underbrace{+\ 7x - 21} = \underbrace{x^3(x - 3)}\ \underbrace{-\ 2x(x - 3)}\ +\ \underbrace{7(x - 3)}$$
$$= (x - 3)(x^3 - 2x \qquad + 7 \qquad)$$

EXAMPLE 8

Group the terms of

$$2x^5 - 3x^4 - 10x^3 + 15x^2 + 18x - 27$$

in pairs and then factor out a common binomial factor.

Solution

$$\underbrace{2x^5 - 3x^4}\ \underbrace{-\ 10x^3 + 15x^2}\ \underbrace{+\ 18x - 27} = \underbrace{x^4(2x - 3)}\ \underbrace{-\ 5x^2(2x - 3)}\ +\ \underbrace{9(2x - 3)}$$

Exercise 1.5, 126–130

$$= (2x - 3)(x^4 - 5x^2 + 9)$$

Earlier in this section we saw that

$$a^2 - b^2 = (a - b)(a + b)$$

and

$$a^3 - b^3 = (a - b)(a^2 + ab + b^2).$$

By multiplying out the factors, it can also be shown that

$$a^4 - b^4 = (a - b)(a^3 + a^2b + ab^2 + b^3)$$

and

$$a^5 - b^5 = (a - b)(a^4 + a^3b + a^2b^2 + ab^3 + b^4).$$

In general, if n is any positive integer, the "formula" for factoring $(a - b)$ out of $a^n - b^n$ is

$$a^n - b^n = (a - b)(a^{n-1} + a^{n-2}b + a^{n-3}b^2 + \cdots + ab^{n-2} + b^{n-1}).$$

Note the pattern of the terms in the second factor. Reading left to right, the first term is a^{n-1}. Each term thereafter contains a raised to an exponent that is one less than the exponent of a in the preceding term. The next to the last term contains $a = a^1$, and the last term technically contains $a^{1-1} = a^0 = 1$ as a factor. The b's start appearing in the second term, and each term contains b raised to an exponent that is one more than the exponent of b in the preceding term. Note that the degree of every single term is $n - 1$. If we use this formula to factor $a^7 - b^7$, we get

$$a^7 - b^7 = (a - b)(a^6 + a^5b + a^4b^2 + a^3b^3 + a^2b^4 + ab^5 + b^6).$$

Note that the degree of every single term in the second factor is 6. The terms are formed by starting with $a^6 = a^6b^0$ and then successively dropping the exponent on the a by 1 while raising the exponent on the b by 1 until $b^6 = b^6a^0$ is reached. The factorization can be checked by multiplying out the factors.

EXERCISE 1.5 FACTORING

▶ Factor each of the following into a product of primes.

EXAMPLE
$180 = 5 \times 36 = 5 \times 6 \times 6 = 5 \times 2 \times 3 \times 2 \times 3 = 5 \times 2^2 \times 3^2$

1. 225	**2.** 1800	**3.** 1400	**4.** 1350	**5.** 338
6. 3388	**7.** 714	**8.** 539	**9.** 9020	**10.** 1287

▶ Write out each of the following products.

EXAMPLE
$(3x - 2)(3x + 2) = 9x^2 - 4$

11. $(a - b)(a + b)$ **12.** $(a + b)^2$

13. $(a - b)^2$ **14.** $(a + b)(a^2 - ab + b^2)$

15. $(a - b)(a^2 + ab + b^2)$ **16.** $(a - b)(a^3 + a^2b + ab^2 + b^3)$

17. $(a - b)(a^4 + a^3b + a^2b^2 + ab^3 + b^4)$

18. $(x - 7)(x + 8)$ **19.** $(x + 3)(x - 5)$

20. $(x - 2)(x + 9)$ **21.** $(x + 6)(x - 4)$

22. $(x + 8)(x - 2)$ **23.** $(3x - 7)(2x + 1)$

24. $(2x - 5)(3x + 8)$ **25.** $(6x - 4)(2x + 3)$

26. $(5x - 3)(3x - 2)$ **27.** $(7x - 2)(3x + 4)$

28. $2x^3(7x - 5)$ **29.** $3x^5(7x^2 - 2x + 9)$

30. $2x^7(3x^3 - 2x^2 + 5x - 1)$ **31.** $x(5x^4 - 4x^2 + 2x - 7)$

32. $-x(x^3 - x^2 + x - 1)$ **33.** $-5x^2(x^4 + 3x^2 - 2x - 9)$

34. $(2x - 3)(x^3 - 2x^2 + 7)$ **35.** $(x - 8)(3x^5 - 2x^3 + x^2 - 1)$

36. $(x - 7)(x^3 - 2x^2 + 3x - 1)$ **37.** $(3x + 7)(x^7 - 2x^5 + 6)$

38. $(3x - 4)(3x + 4)$ **39.** $(x - 2y)(x + 2y)$

40. $(3x + 4)^2$ **41.** $(3x - 4)^2$

42. $(x - 2y)^2$ **43.** $(x + 2y)^2$

44. $(x^2 - y^2)(x^2 + y^2)$ **45.** $(x^3 - y^3)(x^3 + y^3)$

46. $(x^2 - y^2)^2$ **47.** $(x^3 + y^3)^2$

48. $(2 - 3x)^2$ **49.** $(-x + 2)^2$

50. $(-x - 1)^2$ **51.** $(3x + 2)(3x - 2)$

52. $(3x + 2)^2$ **53.** $(x^3 - 2)(x^3 + 2)$

54. $(x^3 - 2)^2$ **55.** $(x + h)^2$

56. $(a - 2)(a^2 + 2a + 4)$ **57.** $(x - 1)^2$

58. $(3x + 7)^2$ **59.** $(3x + 7)(3x - 7)$

60. $(3x - 7)^2$ **61.** $(3x - 7)(3x + 1)$

62. $(x + 1)(x^2 - x + 1)$ **63.** $(x - 1)(x^2 + x + 1)$

64. $(3x - 7)(9x^2 + 21x + 49)$ **65.** $(3x + 7)(9x^2 - 21x + 49)$

66. $(x - 2y)(x^2 + 2xy + 4y^2)$ **67.** $(x + 2y)(x^2 - 2xy + 4y^2)$

68. $(x + 2)(x^2 - 2x + 4)$

69. $(2 - 3x)(8 + 12x + 18x^2 + 27x^3)$

70. $(2 - 3x)(16 + 24x + 36x^2 + 54x^3 + 81x^4)$

▶   If possible, express each of the following as a product of prime polynomials with integral coefficients. If such a factorization is not possible, write "not factorable."

71. $25x^2 - 16$ **72.** $25x^2 - 40x + 16$

73. $3x^2yz - 2x^5y^2 + x^3y^2z$ **74.** $8w^3 - 27z^3$

75. $8w^3 + 27z^3$ **76.** $x^2 - 5x - 36$

77. $9x^2 - 49y^2$ **78.** $9x^4 - 42x^3y + 49x^2y^2$

79. $27x^4y - 64xy^4$ **80.** $3x^2 - 19x + 20$

81. $x^8 - 2x^4y^4 + y^8$ **82.** $x^8 - y^8$

83. $x^9 - y^9$ **84.** $81x^2 - 100y^2$

85. $x^2 - 7x - 44$ **86.** $6x^2 - 13x - 5$

87. $64x^6 - y^6$ **88.** $x^3y^6 + x^6y^3$

89. $2x^2 - 3x - 20$ **90.** $x^2 + 5x + 3$

91. $36x^2 - 4y^2$ **92.** $36x^2 - 24xy + 4y^2$

93. $2a^4b^2 + 4a^3b^3 + 2a^2b^4$ **94.** $x^2 - 2x + 2$

95. $121a^2 - 169y^2$ **96.** $12x^3 - 34x^2 + 14x$

97. $9w^2 + 48wz + 64z^2$ **98.** $9w^2 - 64z^2$

99. $9w^2 + 64z^2$ **100.** $14x^2 + 33x - 5$

101. $27a^3b + 8b^4$ **102.** $15x^2 - 8x + 1$

103. $16x^3y^2 - 25xz^2$ **104.** $6x^4 + 19x^2 - 20x^2$

105. $x^4y - x^3y - 56x^2y$ **106.** $2a^2 - 2a - 40$

107. $x^2 + y^2$ **108.** $a^6 + 2a^3b^3 + b^6$

109. $6x^2 - 6x - 12$ **110.** $21a^2 - 13a + 2$

111. $x^4 - 16y^4$ **112.** $32a^5 - b^5$

113. $a^{10} - b^{10}$ **114.** $x^{11} - y^{11}$

115. $x^{12} - y^{12}$

▶ In each of the following, tell what number should be put in the parentheses in order to make the expression become a perfect-square trinomial, and then write that trinomial as the square of a binomial with rational coefficients.

EXAMPLE $x^2 - 5x + (\ \ )$

Solution $(\frac{-5}{2})^2 = \frac{25}{4}; \ x^2 - 5x + (\frac{25}{4}) = (x - \frac{5}{2})^2$

116. $x^2 + 5x + (\ \ )$ **117.** $x^2 + 7x + (\ \ )$ **118.** $x^2 - 3x + (\ \ )$

119. $x^2 - 9x + (\ \ )$ **120.** $x^2 + x + (\ \ )$ **121.** $x^2 - 2x + (\ \ )$

122. $x^2 - 4x + (\ \ )$ **123.** $x^2 + 2x + (\ \ )$ **124.** $x^2 + 4x + (\ \ )$

125. If a and b are real numbers and $a \neq 0$, tell what should be put in the parentheses in order to make the expression

$$x^2 + \frac{b}{a}x + (\ \ )$$

become a perfect-square trinomial, and then write that trinomial as the square of a binomial.

▶ In each of the following, group the terms in pairs and then factor out a common binomial factor.

EXAMPLE

$$x^4 - 7x^3 - 2x^2 + 14x + 6x - 42 = (x^4 - 7x^3) - (2x^2 - 14x)$$
$$+ (6x - 42)$$
$$= x^3(x - 7) - 2x(x - 7) + 6(x - 7)$$
$$= (x - 7)(x^3 - 2x + 6)$$

$$xy^5 - 2x + 3y^5 - 6 = (xy^5 + 3y^5) - (2x + 6)$$
$$= y^5(x + 3) - 2(x + 3)$$
$$= (x + 3)(y^5 - 2)$$

126. $x^4 - 5x^3 - 4x^2 + 20x + 8x - 40$

127. $xy^7 - 3x + 2y^7 - 6$

128. $a^5 - 3a^4b - ab^2 + 3b^3 + ab - 3b^2$

129. $3a^6 - 2a^4 - 3a^5 + 2a^3$

130. $x^3 + yx^2 - 2x^2y - 2xy^2 + 3x + 3y$

CHECK LIST FOR THE STUDENT

If you have learned the material in Section 1.5, you should be able to do the following:

1. Define prime polynomial over the integers.

2. Tell how many ways (except for the order of the factors) any polynomial over the integers that is not itself a prime can be factored into prime factors with integral coefficients.

3. Tell what is meant by factoring a polynomial over the integers completely.

4. Factor out a common monomial factor.

5. Factor the difference of two squares.

6. Factor perfect-square trinomials.

7. Tell what should be added to $x^2 + kx$ in order to make it become a perfect-square trinomial.

8. Factor the sum of two cubes.

9. Factor the difference of two cubes.

10. Factor any factorable second-degree trinomial over the integers.

11. Group terms and factor out a common binomial factor.

12. Factor $a^n - b^n$.

SECTION QUIZ 1.5

1. Write out each of the following products.
   (a) $(x - y)(x + y)$ (b) $(x + y)^2$
   (c) $(x - y)(x^2 + xy + y^2)$ (d) $(3x + 2)(x - 7)$
   (e) $-2x(3x^2 - x + 1)$

If possible, express each of the following as a product of prime polynomials with integral coefficients. If such a factorization is not possible, write "not factorable."

2. $9x^2 - 12x + 4$ 3. $9x^2 - 4$

4. $27x^3 + 8$ 5. $x^2 + 6x + 9$

6. $x^2 + 6x + 1$ 7. $9x^2 + 6x - 8$

8. $2a + 2b - 3ab - 3a^2$ 9. $2a^5b - 6a^3b^2 + 14a^2b$

10. Tell what number should be put in the parentheses in order to make the following expression become a perfect-square trinomial, and then factor that trinomial.

$$x^2 - 12x + (\ \ )$$

1.6 Fractions; Rational Expressions

The purpose of this section is to review manipulations with fractions, particularly those fractions in which the numerator and denominator are polynomials. Some fractions with radicals in their denominators and fractions with imaginary-number denominators are also considered.

In Section 1.1, we saw that where P and nonzero D and N are mathematical expressions,

$$\frac{PN}{DN} = \frac{P}{D}. \qquad \text{Fundamental principle of fractions}$$

It is likewise true that for nonzero D,

$$\frac{P}{D} = \frac{-P}{-D} = -\frac{P}{-D} = -\frac{-P}{D}.$$

For example,

$$\frac{3}{4} = \frac{-3}{-4} = -\frac{3}{-4} = -\frac{-3}{4}$$

$$\frac{x+2}{5x} = \frac{-(x+2)}{-5x} = -\frac{x+2}{-5x} = -\frac{-(x+2)}{5x}, \qquad \text{if } x \neq 0.$$

Note that any fraction may be thought of as having three signs associated with it: the sign in front, the sign of the numerator, and the sign of the denominator. (If the sign in front is plus, it is not usually written.) We may change exactly two of these three signs without changing the value of the fraction.

EXAMPLE 1

State which sign belongs in the parentheses in each of the following, assuming that $x + y \neq 0$.

(a) $\dfrac{x-y}{x+y} = (\ \ ) \dfrac{y-x}{-x-y}$

(b) $\dfrac{x-y}{x+y} = (\ \ ) \dfrac{x-y}{-x-y}$

(c) $\dfrac{x-y}{x+y} = (\ \ ) \dfrac{y-x}{x+y}$

Solution

(a) $\dfrac{x-y}{x+y} = (\ \ ) \dfrac{y-x}{-x-y}$

Since $y - x = -(x - y)$ and $-x - y = -(x + y)$, the signs of both the numerator and the denominator have been changed. The sign in front stays the same.

$$\frac{x - y}{x + y} = (+)\ \frac{y - x}{-x - y} \qquad \frac{P}{D} = \frac{-P}{-D}$$

(b) $\dfrac{x - y}{x + y} = (\quad)\ \dfrac{x - y}{-x - y}$

Only the sign of the denominator has been changed, and since the numerator is unchanged, the sign in the front must be changed.

$$\frac{x - y}{x + y} = (-)\ \frac{x - y}{-x - y} \qquad \frac{P}{D} = -\ \frac{P}{-D}$$

(c) $\dfrac{x - y}{x + y} = (\quad)\ \dfrac{y - x}{x + y}$

Only the sign of the numerator has been changed, and since the denominator is unchanged, the sign in front must be changed.

Exercise 1.6, 1–15

$$\frac{x - y}{x + y} = (-)\ \frac{y - x}{x + y} \qquad \frac{P}{D} = -\ \frac{-P}{D}$$

If variables occur in a denominator, let it be understood that they are not permitted to take on values for which the denominator is zero. A fraction P/D may be considered to be an indicated division $P \div D$, and division by zero is not defined.

Numerical fractions—that is, fractions in which the numerator and denominator are real numbers—can sometimes be simplified or "reduced to lowest terms" by using the fundamental principle of fractions. A fraction in which the numerator and nonzero denominator are integers (that is, a fraction that is a rational number) is said to be in **lowest terms** if the numerator and denominator have no common integral factors other than $+1$ or -1. A fraction in which the numerator and nonzero denominator are polynomials with integral coefficients is said to be in lowest terms if the numerator and denominator have no common factors with integral coefficients other than $+1$ or -1. The fractions

$$\frac{6}{8} \quad \text{and} \quad \frac{x^2 - 9}{x^2 + 5x + 6}$$

are not in lowest terms but can be reduced to lowest terms by factoring their numerators and denominators and using the fundamental principle of fractions.

PART ONE | BASIC ALGEBRA

$$\frac{6}{8} = \frac{3 \times 2}{4 \times 2} = \frac{3}{4}$$

$$\frac{x^2 - 9}{x^2 + 5x + 6} = \frac{(x - 3)(x + 3)}{(x + 2)(x + 3)}$$

$$= \frac{x - 3}{x + 2}, \qquad \text{if } x + 3 \neq 0 \quad \text{and} \quad x + 2 \neq 0.$$

To reduce a fraction such as these to lowest terms, factor its numerator and denominator completely and then divide the numerator and denominator by their common factors. Dividing the numerator and denominator of a fraction by a common factor is sometimes called *canceling* that common factor.

EXAMPLE 2

Reduce each of the following to lowest terms, if possible.

(a) $\dfrac{(x - 3) \times 5(x - 2)}{(x - 3)(x + 7)}$ (b) $\dfrac{(x - 3) + 5(x - 2)}{(x - 3)(x + 7)}.$

Solution

(a) $\dfrac{(x - 3) \times 5(x - 2)}{(x - 3)(x + 7)}$

The factor $(x - 3)$ is common to both numerator and denominator. Dividing both numerator and denominator by $(x - 3)$ and indicating this division by slash or "cancellation" marks, we have

$$\frac{(x - 3) \times 5(x - 2)}{(x - 3)(x + 7)} = \frac{\cancel{(x - 3)} \times 5(x - 2)}{\cancel{(x - 3)}(x + 7)}$$

$$= \frac{5(x - 2)}{x + 7}$$

$$= \frac{5x - 10}{x + 7}$$

(b) $\dfrac{(x - 3) + 5(x - 2)}{(x - 3)(x + 7)}$

Here $(x - 3)$ is not a factor of the numerator. The numerator and denominator do not have any common factors. The given fraction is already in lowest terms. Note that the term $(x - 3)$ cannot be canceled here. Students should beware of trying to cancel terms simply because they appear on the same page or in the same problem. In any fraction, only terms that are *factors* of *both* numerator and denominator can be canceled.

$$\frac{(x - 3) + 5(x - 2)}{(x - 3)(x + 7)} = \frac{x - 3 + 5x - 10}{(x - 3)(x + 7)}$$

$$= \frac{6x - 13}{(x - 3)(x + 7)}$$

Exercise 1.6, 16–40

Note that this result is *not* what we would get if we erroneously crossed out the $(x - 3)$ simply because we saw it in two places in the same fraction.

The fundamental principle of fractions can also be used to express a fraction in an equivalent form with a larger or higher-degree denominator. For example, to express $\frac{3}{4}$ in a form with denominator 36, multiply numerator and denominator by 9.

$$\frac{3}{4} = \frac{3 \times 9}{4 \times 9} = \frac{27}{36}$$

To express $(x - 3)/(x + 2)$ in a form with denominator $(x + 2)(x - 7)$, multiply numerator and denominator by $(x - 7)$.

$$\frac{x - 3}{x + 2} = \frac{(x - 3)(x - 7)}{(x + 2)(x - 7)} = \frac{x^2 - 10x + 21}{x^2 - 5x - 14}$$

Exercise 1.6, 41–50

We must perform such manipulations when it is necessary to express two or more fractions in forms having the same denominator.

A fraction in which the numerator and nonzero denominator are polynomials in one variable is called a **rational expression**. The fractions

$$\frac{x^2 - 9}{x^2 + 5x + 6}, \quad \frac{x^2 + 5x + 6}{1}, \quad \frac{3x^5}{7}, \quad \frac{\sqrt{2}x^3 - 5x^2 + 7}{\frac{3}{8}x^4 - \sqrt{3}x + 6}$$

are all rational expressions, even though the last one has irrational coefficients. The numerators and denominators are all polynomials (1, 7, and $3x^5$ are special polynomials of one term each) over the real numbers. Most of the rational expressions with which we shall work, however, have numerators and denominators that are polynomials with integral coefficients.

Fractions are multiplied or divided as follows:

$$\frac{A}{B} \times \frac{C}{D} = \frac{AC}{BD}, \qquad\qquad B \neq 0, \quad D \neq 0$$

$$\frac{A}{B} \div \frac{C}{D} = \frac{A}{B} \times \frac{D}{C} = \frac{AD}{BC}, \qquad B \neq 0, \quad D \neq 0, \quad C \neq 0$$

EXAMPLE 3

Find the product in each of the following.

(a) $\dfrac{3}{5} \times \dfrac{2}{7}$ (b) $\dfrac{x - 2}{x + 5} \times \dfrac{x + 2}{x - 3}$

Solution

(a) $\dfrac{3}{5} \times \dfrac{2}{7} = \dfrac{3 \times 2}{5 \times 7} = \dfrac{6}{35}$

(b). $\dfrac{x-2}{x+5} \times \dfrac{x+2}{x-3} = \dfrac{(x-2)(x+2)}{(x+5)(x-3)}$

$$= \dfrac{x^2 - 4}{x^2 + 2x - 15}$$

EXAMPLE 4

Perform the division in each of the following.

(a) $\dfrac{3}{5} \div \dfrac{8}{7}$

(b) $\dfrac{x-3}{x+2} \div \dfrac{x^2+5}{4x}$

Solution

(a) $\dfrac{3}{5} \div \dfrac{8}{7} = \dfrac{3}{5} \times \dfrac{7}{8}$

(b) $\dfrac{x-3}{x+2} \div \dfrac{x^2+5}{4x} = \dfrac{x-3}{x+2} \times \dfrac{4x}{x^2+5}$

$$= \dfrac{21}{40}$$

$$= \dfrac{(x-3)(4x)}{(x+2)(x^2+5)}$$

EXAMPLE 5

Divide $\sqrt{2}/2$ by $\tfrac{1}{3}$.

Solution

$$\dfrac{\sqrt{2}}{2} \div \dfrac{1}{3} = \dfrac{\sqrt{2}}{2} \times \dfrac{3}{1}$$

$$= \dfrac{3\sqrt{2}}{2}$$

Fractions with common nonzero denominators are added or subtracted as follows:

$$\dfrac{A}{B} + \dfrac{C}{B} = \dfrac{A+C}{B}, \quad B \neq 0$$

$$\dfrac{A}{B} - \dfrac{C}{B} = \dfrac{A-C}{B}, \quad B \neq 0$$

EXAMPLE 6

Find the sum in each of the following.

(a) $\dfrac{8}{13} + \dfrac{17}{13}$

(b) $\dfrac{x^2+2x+9}{x+5} + \dfrac{3x^2-8x+4}{x+5}$

Solution

(a) $\dfrac{8}{13} + \dfrac{17}{13} = \dfrac{8+17}{13}$

$$= \dfrac{25}{13}$$

(b) $\dfrac{x^2+2x+9}{x+5} + \dfrac{3x^2-8x+4}{x+5} = \dfrac{x^2+2x+9+3x^2-8x+4}{x+5}$

$$= \dfrac{4x^2-6x+13}{x+5}$$

EXAMPLE 7   Find the difference in each of the following.

(a) $\dfrac{8}{13} - \dfrac{17}{13}$ (b) $\dfrac{x^2 + 2x + 9}{x + 5} - \dfrac{3x^2 - 8x + 4}{x + 5}$

Solution   (a) $\dfrac{8}{13} - \dfrac{17}{13} = \dfrac{8 - 17}{13}$ (b) $\dfrac{x^2 + 2x + 9}{x + 5} - \dfrac{3x^2 - 8x + 4}{x + 5}$

$\qquad\qquad = \dfrac{-9}{13}$ $\qquad\qquad = \dfrac{x^2 + 2x + 9 - (3x^2 - 8x + 4)}{x + 5}$

$\qquad\qquad\qquad\qquad\qquad = \dfrac{x^2 + 2x + 9 - 3x^2 + 8x - 4}{x + 5}$

$\qquad\qquad\qquad\qquad\qquad = \dfrac{-2x^2 + 10x + 5}{x + 5}$

Fractions not having common denominators may be added or subtracted by first writing them in equivalent forms having common denominators.

EXAMPLE 8   Find the sum in each of the following.

(a) $\dfrac{5}{7} + \dfrac{3}{14}$ (b) $\dfrac{x^2 + 5}{(x + 2)(x - 7)} + \dfrac{x - 3}{x + 2}$

Solution   (a) $\dfrac{5}{7} + \dfrac{3}{14} = \dfrac{10}{14} + \dfrac{3}{14}$

$\qquad\qquad = \dfrac{10 + 3}{14}$

$\qquad\qquad = \dfrac{13}{14}$

(b) $\dfrac{x^2 + 5}{(x + 2)(x - 7)} + \dfrac{x - 3}{x + 2} = \dfrac{x^2 + 5}{(x + 2)(x - 7)} + \dfrac{(x - 3)(x - 7)}{(x + 2)(x - 7)}$

$\qquad\qquad\qquad\qquad\qquad = \dfrac{x^2 + 5 + (x - 3)(x - 7)}{(x + 2)(x - 7)}$

$\qquad\qquad\qquad\qquad\qquad = \dfrac{x^2 + 5 + (x^2 - 10x + 21)}{(x + 2)(x - 7)}$

$\qquad\qquad\qquad\qquad\qquad = \dfrac{x^2 + 5 + x^2 - 10x + 21}{(x + 2)(x - 7)}$

$\qquad\qquad\qquad\qquad\qquad = \dfrac{2x^2 - 10x + 26}{(x + 2)(x - 7)}$

When more than one operation is indicated in any one example, work from left to right, performing the multiplications and divisions first. If grouping symbols such as parentheses or brackets are involved, work from inside out and then from left to right.

EXAMPLE 9   Perform the indicated operations in each of the following.

(a) $\frac{2}{5} \times \frac{3}{7} + \frac{3}{10} \div \frac{7}{3}$ (b) $\left(\frac{3x}{x+2} + \frac{7}{x}\right) \times \frac{x^4}{3} \div \frac{x^3}{x+2}$

Solution

(a) $\frac{2}{5} \times \frac{3}{7} + \frac{3}{10} \div \frac{7}{3} = \frac{6}{35} + \frac{3}{10} \times \frac{3}{7}$

$= \frac{6}{35} + \frac{9}{70}$

$= \frac{12}{70} + \frac{9}{70}$

$= \frac{21}{70}$

$= \frac{3}{10}$

(b) $\left(\frac{3x}{x+2} + \frac{7}{x}\right) \times \frac{x^4}{3} \div \frac{x^3}{x+2}$

$= \left(\frac{3x^2}{x(x+2)} + \frac{7(x+2)}{x(x+2)}\right) \times \frac{x^4}{3} \div \frac{x^3}{x+2}$

$= \frac{3x^2 + 7x + 14}{x(x+2)} \times \frac{x^4}{3} \div \frac{x^3}{x+2}$

$= \frac{x^4(3x^2 + 7x + 14)}{3x(x+2)} \times \frac{x+2}{x^3}$

$= \frac{x^4(3x^2 + 7x + 14)(x+2)}{3x^4(x+2)}$

$= \frac{3x^2 + 7x + 14}{3}$

In expressing fractions in equivalent forms so that they will have common denominators, any common denominator is acceptable. Computations are usually simpler, however, if the smallest positive or lowest-degree common denominator is used. The **lowest common denominator (least common multiple)** of two or more positive integral denominators is the smallest positive integer that is evenly divisible by each of them. The **lowest common denominator (least common multiple)** of two or more denominators that are polynomials with integral coefficients is the lowest-degree polynomial with integral coefficients which is evenly divisible by each of them. For example, the lowest common denominator of $\frac{7}{24}$, $\frac{19}{30}$, and $\frac{13}{18}$ is 360, because 360 is the smallest positive integer which is evenly divisible by 24, by 18, and also by 30. The lowest common denominator of

$$\frac{7}{(x-1)^3(x+5)}, \quad \frac{19}{(x-1)(x+5)(x-4)}, \quad \text{and} \quad \frac{13}{(x-1)(x+5)^2}$$

is $(x-1)^3(x+5)^2(x-4)$, because this product is the lowest-degree polynomial over the integers which is evenly divisible by each of the given denominators.

To find the lowest common denominator (LCD) of a set of denominators that are integers or polynomials over the integers, factor each denominator completely and then form the product of the prime factors, taking each prime factor the largest number of times it appears in any individual factorization. For example, to find the LCD of $\frac{7}{24}$, $\frac{19}{30}$, and $\frac{13}{18}$, first factor 24, 30, and 18 completely, and then form the product of prime factors as follows:

$$\begin{aligned} 24 &= 2^3 \times 3 & \text{LCD} &= 2^3 \times 3^2 \times 5 \\ 30 &= 2 \times 3 \times 5 & &= 8 \times 9 \times 5 \\ 18 &= 3^2 \times 2 & &= 360 \end{aligned}$$

To find the LCD of

$$\frac{7}{(x-1)^3(x+5)}, \quad \frac{19}{(x-1)(x+5)(x-4)}, \quad \text{and} \quad \frac{13}{(x-1)(x+5)^2},$$

first make sure each denominator is factored completely, and then form the product of prime factors as follows:

$$(x-1)^3(x+5) \qquad \text{is factored completely.}$$
$$(x-1)(x+5)(x-4) \qquad \text{is factored completely.}$$
$$(x-1)(x+5)^2 \qquad \text{is factored completely.}$$
$$\text{LCD} = (x-1)^3(x+5)^2(x-4)$$

Exercise 1.6, 51–60

EXAMPLE 10

Perform the indicated operations.

$$\frac{3}{x^2-4} + \frac{5x}{x^2-4x+4} - \frac{9}{(x+1)(x^2+2x+1)}$$

Solution

First factor each denominator completely, and then find the LCD.

$$x^2-4 = (x-2)(x+2)$$
$$x^2-4x+4 = (x-2)^2$$
$$(x+1)(x^2+2x+1) = (x+1)(x+1)^2 = (x+1)^3$$
$$\text{LCD} = (x-2)^2(x+2)(x+1)^3$$

Next write each of the individual fractions in equivalent form having the LCD as denominator.

$$\frac{3}{x^2 - 4} = \frac{3}{(x - 2)(x + 2)} = \frac{3(x - 2)(x + 1)^3}{(x - 2)(x + 2)(x - 2)(x + 1)^3}$$

$$= \frac{3(x - 2)(x + 1)^3}{(x - 2)^2(x + 2)(x + 1)^3}$$

$$\frac{5x}{x^2 - 4x + 4} = \frac{5x}{(x - 2)^2} = \frac{5x(x + 2)(x + 1)^3}{(x - 2)^2(x + 2)(x + 1)^3}$$

$$\frac{9}{(x + 1)(x^2 + 2x + 1)} = \frac{9}{(x + 1)^3} = \frac{9(x - 2)^2(x + 2)}{(x + 1)^3(x - 2)^2(x + 2)}$$

$$= \frac{9(x - 2)^2(x + 2)}{(x - 2)^2(x + 2)(x + 1)^3}$$

Now perform the indicated operations.

$$\frac{3}{x^2 - 4} + \frac{5x}{x^2 - 4x + 4} - \frac{9}{(x + 1)(x^2 + 2x + 1)}$$

$$= \frac{3(x - 2)(x + 1)^3}{(x - 2)^2(x + 2)(x + 1)^3} + \frac{5x(x + 2)(x + 1)^3}{(x - 2)^2(x + 2)(x + 1)^3}$$

$$- \frac{9(x - 2)^2(x + 2)}{(x - 2)^2(x + 2)(x + 1)^3}$$

$$= \frac{3(x - 2)(x + 1)^3 + 5x(x + 2)(x + 1)^3 - 9(x - 2)^2(x + 2)}{(x - 2)^2(x + 2)(x + 1)^3}$$

We may stop here because we have not been instructed to express the numerator and denominator as polynomials.

EXAMPLE 11   Find the sum of $5/x$, $3/(x - 1)$, and $2/(x + 1)$; express its numerator and denominator as polynomials.

Solution $\text{LCD} = x(x - 1)(x + 1)$

$$= x(x^2 - 1)$$

$$\frac{5}{x} + \frac{3}{x - 1} + \frac{2}{x + 1} = \frac{5(x^2 - 1)}{x(x^2 - 1)} + \frac{3x(x + 1)}{x(x^2 - 1)} + \frac{2x(x - 1)}{x(x^2 - 1)}$$

$$= \frac{5(x^2 - 1) + 3x(x + 1) + 2x(x - 1)}{x(x^2 - 1)}$$

$$= \frac{5x^2 - 5 + 3x^2 + 3x + 2x^2 - 2x}{x^3 - x}$$

$$= \frac{10x^2 + x - 5}{x^3 - x}$$

Exercise 1.6, 61–100

We have now seen how to express the sum of rational expressions as a single rational expression. Sometimes it is necessary to reverse this procedure and decompose single rational expressions into sums of fractions with lower-degree denominators. For example, by actually performing the addition, we can verify that

$$\frac{9x^2 + 8x - 12}{x^3 - 4x}$$

can be decomposed into the sum of

$$\frac{3}{x}, \quad \frac{5}{x-2}, \quad \text{and} \quad \frac{1}{x+2}.$$

The fractions in such a decomposition are commonly called *partial fractions*. We shall work some with partial fractions in Section 1.10, and we shall see how to decompose certain rational expressions into sums of partial fractions in Section 3.5.

Exercise 1.6, 101–110

Fractions in which the numerator and denominator contain fractions are called **complex fractions**. They can be simplified by multiplying numerator and denominator through by the LCD of all the fractions appearing in the numerator and denominator. This is illustrated in the next example.

EXAMPLE 12

Express the following as a fraction having no fractions in the numerator or denominator.

$$\frac{\dfrac{y}{3x^2} - \dfrac{1}{6x}}{\dfrac{y^2}{x} + \dfrac{5y}{12}}$$

Solution

The LCD of the fractions $y/3x^2$, $1/6x$, y^2/x, and $5y/12$ is $12x^2$. Multiplying numerator and denominator through by $12x^2$, we have

$$\frac{12x^2\left(\dfrac{y}{3x^2} - \dfrac{1}{6x}\right)}{12x^2\left(\dfrac{y^2}{x} + \dfrac{5y}{12}\right)} = \frac{12x^2\left(\dfrac{y}{3x^2}\right) - 12x^2\left(\dfrac{1}{6x}\right)}{12x^2\left(\dfrac{y^2}{x}\right) + 12x^2\left(\dfrac{5y}{12}\right)}$$

$$= \frac{4y - 2x}{12xy^2 + 5x^2y}.$$

Note that *every term* of the numerator and denominator must be multiplied by the LCD.

Exercise 1.6, 111–115

Some fractions with radicals in their denominators may be written in equivalent forms having rational denominators, as illustrated in the next example.

EXAMPLE 13

Express $7/(\sqrt{5} - \sqrt{2})$ in a form having no irrational number in the denominator.

Solution

We may use the fundamental principle of fractions and multiply numerator and denominator by $\sqrt{5} + \sqrt{2}$.

$$\frac{7}{\sqrt{5} - \sqrt{2}} = \frac{7(\sqrt{5} + \sqrt{2})}{(\sqrt{5} - \sqrt{2})(\sqrt{5} + \sqrt{2})}$$

$$= \frac{7\sqrt{5} + 7\sqrt{2}}{(\sqrt{5})^2 - (\sqrt{2})^2}$$

$$= \frac{7\sqrt{5} + 7\sqrt{2}}{5 - 2}$$

$$= \frac{7\sqrt{5} + 7\sqrt{2}}{3}$$

Any fraction of the form

$$\frac{c}{\sqrt{a} - \sqrt{b}} \quad \text{or} \quad \frac{c}{\sqrt{a} + \sqrt{b}}$$

where c and positive a and b are real numbers may be expressed in an equivalent form having no radical in the denominator by multiplying numerator and denominator by $\sqrt{a} + \sqrt{b}$ or by $\sqrt{a} - \sqrt{b}$, respectively. This process is sometimes called *rationalizing the denominator*.

$$\frac{c}{\sqrt{a} - \sqrt{b}} = \frac{c(\sqrt{a} + \sqrt{b})}{(\sqrt{a} - \sqrt{b})(\sqrt{a} + \sqrt{b})}$$

$$= \frac{c\sqrt{a} + c\sqrt{b}}{(\sqrt{a})^2 - (\sqrt{b})^2}$$

$$= \frac{c\sqrt{a} + c\sqrt{b}}{a - b}$$

$$\frac{c}{\sqrt{a} + \sqrt{b}} = \frac{c(\sqrt{a} - \sqrt{b})}{(\sqrt{a} + \sqrt{b})(\sqrt{a} - \sqrt{b})}$$

$$= \frac{c\sqrt{a} - c\sqrt{b}}{(\sqrt{a})^2 - (\sqrt{b})^2}$$

$$= \frac{c\sqrt{a} - c\sqrt{b}}{a - b}$$

Exercise 1.6, 116–120

As the next example illustrates, any fraction with an imaginary-number denominator may be written in an equivalent form having a real-number denominator.

EXAMPLE 14   Express

$$\frac{3 + 2i}{5 - 7i}$$

in a form having a real-number denominator, and express the fraction as a single complex number.

Solution   We may use the fundamental principle of fractions and multiply numerator and denominator by $5 + 7i$.

$$\frac{3 + 2i}{5 - 7i} = \frac{(3 + 2i)(5 + 7i)}{(5 - 7i)(5 + 7i)}$$

$$= \frac{15 + 21i + 10i + 14i^2}{25 - 49i^2}$$

$$= \frac{15 + 31i + 14(-1)}{25 - 49(-1)}$$

$$= \frac{15 + 31i - 14}{25 + 49}$$

$$= \frac{1 + 31i}{74}$$

Now use the definition of addition of fractions with common denominators.

$$\frac{3 + 2i}{5 - 7i} = \frac{1 + 31i}{74} = \frac{1}{74} + \frac{31}{74}i$$

We have shown in the last example that the fraction $\frac{3 + 2i}{5 - 7i}$ is equal to the single complex number whose real part is $\frac{1}{74}$ and whose imaginary part is $\frac{31}{74}$. Any fraction in which the numerator and nonzero denominator are complex numbers can be written as a single complex number. If the denominator is an imaginary number, $a - bi$, multiply numerator and denominator by $a + bi$; if the denominator is an imaginary number, $a + bi$, multiply numerator and denominator by $a - bi$. Each of the two numbers $a - bi$ and $a + bi$ is called the **conjugate** of the other.

EXAMPLE 15   Express

$$\frac{2 + 5i}{4 + 7i}$$

as a single complex number.

Solution

We multiply numerator and denominator by the conjugate of $4 + 7i$.

$$\frac{2 + 5i}{4 + 7i} = \frac{(2 + 5i)(4 - 7i)}{(4 + 7i)(4 - 7i)}$$

$$= \frac{8 - 14i + 20i - 35i^2}{16 - 49i^2}$$

$$= \frac{8 + 6i + 35}{16 + 49}$$

$$= \frac{43 + 6i}{65}$$

$$= \frac{43}{65} + \frac{6}{65}i$$

Exercise 1.6, 121–125

EXERCISE 1.6 FRACTIONS; RATIONAL EXPRESSIONS

▶ State which sign belongs in the parentheses in each of the following.

EXAMPLES

(a) $\dfrac{x - y}{y - x} = (\quad) \dfrac{x - y}{x - y}$ (b) $\dfrac{x - y}{y - x} = (\quad) \dfrac{y - x}{x - y}$

Solutions

(a) Minus   (b) Plus

1. $\dfrac{c + d}{c - d} = (\quad) \dfrac{d + c}{d - c}$ 2. $\dfrac{c + d}{c - d} = (\quad) \dfrac{-c - d}{c - d}$

3. $\dfrac{c + d}{c - d} = (\quad) \dfrac{-c - d}{d - c}$ 4. $-\dfrac{x - y}{y - x} = (\quad) \dfrac{y - x}{x - y}$

5. $-\dfrac{x - y}{y - x} = (\quad) \dfrac{x - y}{x - y}$ 6. $-\dfrac{x - y}{y - x} = (\quad) \dfrac{y - x}{y - x}$

7. $\dfrac{a^2 - 2ab + b^2}{a - b} = (\quad) \dfrac{-a^2 + 2ab - b^2}{b - a}$

8. $\dfrac{a^2 - 2ab + b^2}{a - b} = (\quad) \dfrac{a^2 - 2ab + b^2}{b - a}$

9. $\dfrac{a^2 - 2ab + b^2}{a - b} = (\quad) \dfrac{-a^2 + 2ab - b^2}{a - b}$

10. $\dfrac{(a - b)^2}{a + b} = (\quad) \dfrac{(b - a)^2}{a + b}$ 11. $\dfrac{(a - b)^2}{a + b} = (\quad) \dfrac{(a - b)^2}{-a - b}$

12. $\dfrac{(a - b)^2}{a + b} = (\quad) \dfrac{(b - a)^2}{-a - b}$

13. $\dfrac{(x-y)(x+y)}{x(y-x)} = (\ \ ) \dfrac{(y-x)(x+y)}{-x(x-y)}$

14. $\dfrac{(x-y)(x+y)}{x(y-x)} = (\ \ ) \dfrac{(y-x)(-x-y)}{x(x-y)}$

15. $\dfrac{(x-y)(x+y)}{x(y-x)} = (\ \ ) \dfrac{(x-y)(x+y)}{-x(x-y)}$

▶ If possible, reduce each of the following to lowest terms. If the given fraction is already in lowest terms, simply write "in lowest terms."

EXAMPLES

(a) $\dfrac{75}{90} = \dfrac{5^2 \cdot 3}{5 \cdot 3^2 \cdot 2} = \dfrac{5}{3 \cdot 2} = \dfrac{5}{6}$

(b) $\dfrac{3(x-1) + 2x(x-1)}{(x-1)(x+8)} = \dfrac{(x-1)(3+2x)}{(x-1)(x+8)} = \dfrac{3+2x}{x+8}$

16. $\frac{36}{108}$ **17.** $\frac{72}{324}$ **18.** $\frac{45}{56}$ **19.** $\frac{392}{224}$

20. $\frac{1092}{735}$ **21.** $\frac{65}{24}$ **22.** $\frac{360}{900}$

23. $\dfrac{x^2 - 4}{x^2 + 5x - 6}$

24. $\dfrac{x^3 - 25x}{x^5 - 4x^4 - 5x^3}$

25. $\dfrac{x^2 - 5x - 6}{x^2 - 5x + 6}$

26. $\dfrac{3(x-2)(x+5)}{(x-2)(x-7)}$

27. $\dfrac{3(x-2) + (x+5)}{(x-2)(x-7)}$

28. $\dfrac{3(x-2) + (x+5)(x-2)}{(x-2)(x-7)}$

29. $\dfrac{6(x+1)(x-9)}{(x-9)(x+4)}$

30. $\dfrac{6 + (x+1)(x-9)}{(x-9)(x+4)}$

31. $\dfrac{6(x-9) + (x+1)(x-9)}{(x-9)(x+4)}$

32. $\dfrac{2x^3(x-5) - x(x-5)}{x^2(x-5)}$

33. $\dfrac{2x^3 - x(x-5)}{x^2(x-5)}$

34. $\dfrac{2x^3(x-5) - x(x-5)}{x^2}$

35. $\dfrac{17(x^2 - y^2) + y^2}{17xy}$

36. $\dfrac{29x^2 - (x^2 + y^2)}{29x^2}$

37. $\dfrac{x^2 - 3x - 10}{x^2 - 5x - 14}$

38. $\dfrac{3x^2 - 5x - 2}{3x^2 - 7x + 2}$

39. $\dfrac{x^3 - 1}{x^2 - 2x + 1}$

40. $\dfrac{x^5 y^2 + x^2 y^5}{x^3 y + 2x^2 y^2 + xy^3}$

Find the missing numerator in each of the following.

EXAMPLES

(a) $\dfrac{3}{7} = \dfrac{}{112}$ (b) $\dfrac{x - 2}{x + 5} = \dfrac{}{x^2 + 4x - 5}$

Solutions

(a) $112 \div 7 = 16, \quad 112 = 7 \times 16$ (b) $x^2 + 4x - 5 = (x + 5)(x - 1)$

$\dfrac{3}{7} = \dfrac{3 \times 16}{7 \times 16} = \dfrac{48}{112}$

$\dfrac{x - 2}{x + 5} = \dfrac{(x - 2)(x - 1)}{(x + 5)(x - 1)}$

$= \dfrac{x^2 - 3x + 2}{x^2 + 4x - 5}$

41. $\dfrac{2}{5} = \dfrac{}{275}$

42. $\dfrac{x - 3}{x + 7} = \dfrac{}{x^2 + 5x - 14}$

43. $\dfrac{3}{8} = \dfrac{}{200}$

44. $\dfrac{7}{x - 9} = \dfrac{}{x^2 - 81}$

45. $\dfrac{19}{x^2 + x - 6} = \dfrac{}{(x^2 - 4)(x + 3)}$

46. $\dfrac{7}{90} = \dfrac{}{15660}$

47. $\dfrac{-5}{3 - x} = \dfrac{}{x^2 - x - 6}$

48. $\dfrac{3x}{x^2 + x - 6} = \dfrac{}{(x^2 - 4)(x^2 - 9)}$

49. $\dfrac{1}{x^5(x - 2)} = \dfrac{}{x^{10} - 8x^7}$

50. $\dfrac{7x}{x^2 - 3x + 9} = \dfrac{}{x^3 + 27}$

Find the LCD in each of the following.

EXAMPLES

(a) $\dfrac{3}{40}, \dfrac{7}{10}, \dfrac{9}{16}$

(b) $\dfrac{3}{x^2 - 16}, \dfrac{5x}{(x - 4)^2}, \dfrac{7x^4}{x^2 - x - 12}$

Solutions

(a) $40 = 2^3 \cdot 5$

$10 = 2 \cdot 5$

$16 = 2^4$

$\text{LCD} = 2^4 \cdot 5 = 80$

(b) $x^2 - 16 = (x - 4)(x + 4)$

$(x - 4)^2 = (x - 4)^2$

$x^2 - x - 12 = (x - 4)(x + 3)$

$\text{LCD} = (x - 4)^2(x + 4)(x + 3)$

51. $\frac{7}{12}, \frac{8}{30}, \frac{11}{36}, \frac{4}{15}$ **52.** $\frac{1}{40}, \frac{19}{315}, \frac{8}{210}, \frac{3}{380}$

53. $\dfrac{5}{63x^3}, \dfrac{-7}{27x^5}, \dfrac{4}{81x}$

54. $\dfrac{3x}{x-2}, \dfrac{2x^3}{(x-2)^3(x+1)}, \dfrac{5x^7}{(x-2)(x+1)^2}$

55. $\dfrac{-9}{x^2-2x+1}, \dfrac{5}{(x-1)^3}, \dfrac{7x^3}{x^3-1}$

56. $\dfrac{2x^7}{x^2-4x+4}, \dfrac{7x^3}{x^2-3x+2}, \dfrac{-8}{x^3-8}$

57. $\dfrac{5}{(x-2)^5(x+1)^3}, \dfrac{-6}{x^3(x-2)^7(x+1)}, \dfrac{3}{x(x-2)(x-1)}$

58. $\dfrac{1}{x}, \dfrac{1}{x+1}, \dfrac{1}{x-1}$

59. $\dfrac{1}{x^2-y^2}, \dfrac{1}{x+y}, \dfrac{1}{(x-y)^2}$

60. $\dfrac{1}{x^3}, \dfrac{-1}{y^3}, \dfrac{1}{x^3-y^3}$

▶ Perform the indicated operations in each of the following. If possible, reduce the answer to lowest terms.

EXAMPLES

(a) $\frac{7}{8} \times \frac{3}{5} = \frac{21}{40}$

(b) $\dfrac{3x}{x-1} \times \dfrac{7x^2}{x^2-1} = \dfrac{21x^3}{(x-1)(x^2-1)}$

(c) $\frac{7}{8} \div \frac{3}{5} = \frac{7}{8} \times \frac{5}{3} = \frac{35}{24}$

(d) $\dfrac{3x}{x-1} \div \dfrac{7x^2}{x^2-1} = \dfrac{3x}{x-1} \times \dfrac{x^2-1}{7x^2}$

$$= \dfrac{3x\cancel{(x-1)}(x+1)}{\cancel{(x-1)} \cdot 7x^2}$$

$$= \dfrac{3x+3}{7x}$$

(e) $\frac{7}{8} + \frac{3}{5} = \frac{35}{40} + \frac{24}{40}$

$$= \frac{59}{40}$$

(f) $\dfrac{3x}{x-1} + \dfrac{7x^2}{x^2-1} = \dfrac{3x(x+1)}{(x-1)(x+1)}$

$$+ \dfrac{7x^2}{(x-1)(x+1)}$$

$$= \dfrac{3x(x+1)+7x^2}{(x-1)(x+1)}$$

$$= \dfrac{10x^2+3x}{x^2-1}$$

61. $\frac{5}{8} \times \frac{7}{11}$ **62.** $\frac{5}{8} \div \frac{7}{11}$ **63.** $\frac{5}{8} + \frac{7}{11}$ **64.** $\frac{5}{8} - \frac{7}{11}$

65. $\frac{8}{9} \div \frac{5}{27}$ **66.** $\frac{7}{36} + \frac{9}{48} - \frac{11}{60}$

67. $\frac{5}{42} - \frac{13}{45} + \frac{1}{105}$ **68.** $\frac{1}{20} + \frac{5}{21} - \frac{8}{45} + \frac{7}{30}$

69. $\frac{3}{8} \times \frac{2}{7} + \frac{6}{5} \div \frac{9}{4}$ **70.** $(\frac{3}{8} + \frac{2}{7}) \times \frac{6}{5} - \frac{9}{4}$

71. $(\frac{3}{8} \div \frac{2}{7}) \times \frac{6}{5} + \frac{9}{4}$ **72.** $[3 + (2 - \frac{4}{5})] \div \frac{7}{8}$

73. $(\frac{13}{14} - \frac{5}{21}) + \frac{11}{28} \div \frac{31}{32}$ **74.** $(\frac{23}{20} + \frac{7}{45}) \div [\frac{2}{3}(\frac{17}{12} - \frac{7}{30})]$

75. $\frac{13}{24} + \frac{13}{54} - \frac{5}{20} \times \frac{36}{13}$ **76.** $\frac{\sqrt{2}}{2} \times \frac{1}{2}$

77. $\frac{1}{\sqrt{3}} \times \frac{\sqrt{3}}{3}$ **78.** $\frac{\sqrt{3}}{2} \div \frac{1}{2}$

79. $\frac{1}{2} \div \frac{\sqrt{3}}{2}$ **80.** $1 \div \frac{\sqrt{3}}{3}$

81. $\frac{2x}{x-3} \times \frac{5x^2}{x^2-9}$ **82.** $\frac{2x}{x-3} \div \frac{5x^2}{x^2-9}$

83. $\frac{x^2+5x+6}{x^2+x-6} \times \frac{(x+1)^2}{x^2-9}$ **84.** $\frac{3xy^3}{25a^2b} \times \frac{10a^5b}{6x^3y}$

85. $\frac{2a^4b^4}{3x^2y^3} \div \frac{a^2b^2}{8xy^4}$ **86.** $\left(\frac{9x^2y^3}{5a^2b^3} \div \frac{xy}{7a^3b^2}\right) \times \frac{10ab^3}{3xy^4}$

87. $\left(\frac{4x^6y^3z}{6a^2b^2c^2} \times \frac{7a^3b^3c^3}{6x^2y^2z^2}\right) \div \frac{42x^2y^2z^2}{24a^2b^2c^2}$

88. $\frac{4x-8}{x^5} \times \frac{x^3}{x-2}$ **89.** $\frac{7x+21}{x-2} \times \frac{x^2-4}{x^2-9}$

90. $\frac{2x-10}{x+2} \div \frac{3x-15}{x^2-4}$ **91.** $\frac{3x}{x^2-9} + \frac{5x^2}{x^2-9}$

92. $\frac{7x^3}{(x-1)^3} - \frac{8x^2}{(x-1)^3}$ **93.** $\frac{2x}{x-3} + \frac{5x^2}{x^2-9}$

94. $\frac{8}{x-5} + \frac{7}{x+3}$ **95.** $\frac{2}{x-2} + \frac{-7}{x^2-10x+25}$

96. $\frac{3x}{(x-2)(x+1)} - \frac{2x^2}{x^2+2x+1}$ **97.** $\frac{x+1}{x-1} + \frac{x-2}{x+2}$

98. $\frac{5}{x} - \frac{7}{x^2-1} + \frac{8}{x^4-2x^3+x^2}$ **99.** $\left(\frac{7x}{x+3} + \frac{3}{x}\right) \times \left(\frac{x^5}{5} \div \frac{x^4}{x+3}\right)$

100. $\left(\frac{2x^3}{x-2} + \frac{5}{x^2}\right) \div \left(\frac{2}{3} \times \frac{x-2}{x^5}\right)$

In each of the following, verify that the given rational expression has been properly decomposed into partial fractions.

EXAMPLE

$$\frac{14x^2 - 5x - 5}{x^3 - x} = \frac{5}{x} + \frac{2}{x - 1} + \frac{7}{x + 1}$$

Solution

$$\frac{5}{x} + \frac{2}{x - 1} + \frac{7}{x + 1} = \frac{5(x - 1)(x + 1) + 2x(x + 1) + 7x(x - 1)}{x(x - 1)(x + 1)}$$

$$= \frac{5x^2 - 5 + 2x^2 + 2x + 7x^2 - 7x}{x(x^2 - 1)}$$

$$= \frac{14x^2 - 5x - 5}{x^3 - x}$$

101. $\dfrac{17x - 1}{3x^2 + 5x - 2} = \dfrac{2}{3x - 1} + \dfrac{5}{x + 2}$

102. $\dfrac{16x^2 + x - 27}{(x + 2)(x^2 - 2x - 3)} = \dfrac{7}{x + 2} + \dfrac{3}{x + 1} + \dfrac{6}{x - 3}$

103. $\dfrac{11x^2 - 15x - 50}{x^3 - 25x} = \dfrac{3}{x - 5} + \dfrac{6}{x + 5} + \dfrac{2}{x}$

104. $\dfrac{3x + 2}{x^2 - 2x + 1} = \dfrac{3}{x - 1} + \dfrac{5}{(x - 1)^2}$

105. $\dfrac{7x + 22}{x^2 + 6x + 9} = \dfrac{7}{x + 3} + \dfrac{1}{(x + 3)^2}$

106. $\dfrac{x^2 + 7x + 15}{(x + 2)^3} = \dfrac{1}{x + 2} + \dfrac{3}{(x + 2)^2} + \dfrac{5}{(x + 2)^3}$

107. $\dfrac{4x^2 - 1}{x^3 - 1} = \dfrac{3x + 2}{x^2 + x + 1} + \dfrac{1}{x - 1}$

108. $\dfrac{2x^2 + 3x - 2}{x^3 - 8} = \dfrac{x + 3}{x^2 + 2x + 4} + \dfrac{1}{x - 2}$

109. $\dfrac{8x^2 + 49x + 63}{x^3 + 6x^2 + 9x} = \dfrac{7}{x} + \dfrac{1}{x + 3} + \dfrac{4}{(x + 3)^2}$

110. $\dfrac{2x^4 + x^3 + 4x^2 - 2x + 2}{(x^2 - x + 1)(x^3 + 1)} = \dfrac{x + 3}{x^2 - x + 1} + \dfrac{x - 2}{(x^2 - x + 1)^2}$

$$+ \frac{1}{x + 1}$$

Express each of the following as a fraction having no fractions in the numerator or denominator.

EXAMPLE

$$\frac{\dfrac{4}{x^3} - \dfrac{2}{x}}{\dfrac{7}{x^5} - \dfrac{1}{3x^2}} = \frac{3x^5\left(\dfrac{4}{x^3} - \dfrac{2}{x}\right)}{3x^5\left(\dfrac{7}{x^5} - \dfrac{1}{3x^2}\right)} = \frac{12x^2 - 6x^4}{21 - x^3}$$

111. $\dfrac{\dfrac{5}{x^4} - \dfrac{3}{x^7}}{\dfrac{3}{x} + \dfrac{1}{2x^3}}$

112. $\dfrac{\dfrac{6}{x-3} + \dfrac{1}{x+3}}{x - \dfrac{4}{x-3}}$

113. $\dfrac{\dfrac{5}{x+2} - \dfrac{2}{x-2}}{\dfrac{3}{x+2} + \dfrac{1}{x}}$

114. $\dfrac{\dfrac{6}{x} + \dfrac{5}{x^2}}{\dfrac{7}{x^3} - \dfrac{3}{x^4}}$

115. $\dfrac{\dfrac{1}{x-1} - \dfrac{x}{x+1}}{\dfrac{x^2}{x^2-1} + 3}$

Express each of the following in a form having no irrational number in the denominator.

EXAMPLE

$$\frac{5}{7 + \sqrt{3}} = \frac{5(7 - \sqrt{3})}{(7 + \sqrt{3})(7 - \sqrt{3})} = \frac{35 - 5\sqrt{3}}{49 - 3} = \frac{35 - 5\sqrt{3}}{46}$$

116. $\dfrac{4}{5 - \sqrt{2}}$

117. $\dfrac{4}{5 + \sqrt{2}}$

118. $\dfrac{2}{\sqrt{3} - \sqrt{5}}$

119. $\dfrac{2}{\sqrt{3} + \sqrt{5}}$

120. $\dfrac{-3}{4 - \sqrt{7}}$

Express each of the following as a single complex number.

EXAMPLE

$$\frac{2 + 3i}{5 - 4i} = \frac{(2 + 3i)(5 + 4i)}{(5 - 4i)(5 + 4i)} = \frac{-2 + 23i}{25 + 16} = \frac{-2}{41} + \frac{23}{41}i$$

121. $\dfrac{2 + 3i}{5 + 4i}$

122. $\dfrac{3 + 7i}{2 - 4i}$

123. $\dfrac{3 + 7i}{2 + 4i}$

124. $\dfrac{5 + 8i}{7 + i}$

125. $\dfrac{7 + 9i}{i}$

CHECK LIST FOR THE STUDENT

If you have learned the material in Section 1.6, you should be able to do the following:

1. Write a fraction in three equivalent forms by changing signs.

2. Reduce to lowest terms a fraction whose numerator and nonzero denominator are integers or are polynomials with integral coefficients.

3. Express a fraction whose numerator and nonzero denominator are integers, or are polynomials with integral coefficients, in an equivalent form with a larger or higher-degree denominator.

4. Multiply, divide, add, and subtract fractions with nonzero denominators.

5. Perform more than one of the operations addition, subtraction, multiplication, and division in the same example.

6. Find the lowest common denominator (least common multiple) of two or more fractions whose numerators and nonzero denominators are integers or are polynomials with integral coefficients.

7. Express a complex fraction in a form having no fractions in the numerator or denominator.

8. Express any fraction of the form

$$\frac{c}{\sqrt{a} - \sqrt{b}} \quad \text{or} \quad \frac{c}{\sqrt{a} + \sqrt{b}},$$

where c and positive a and positive b are real numbers, in an equivalent form having no radical in the denominator.

9. Express any fraction in which the numerator and nonzero denominator are complex numbers as a single complex number.

SECTION QUIZ 1.6

1. Find what belongs in the parentheses in each of the following.

(a) $-\dfrac{x - y}{y - x} = (\quad) \dfrac{y - x}{x - y}$

(b) $\dfrac{a - 3}{a + 4} = \dfrac{(\qquad\qquad)}{a^2 + 3a - 4}$

2. Reduce the following to lowest terms.

$$\frac{2x^2 - 18}{4x^2 - 4x - 24}$$

3. Find the lowest common denominator.

$$\frac{4}{x^2 - 10x + 25}, \quad \frac{3x}{x^2 - 25}, \quad \frac{x^2}{(x - 5)^3}, \quad \frac{6x^4}{(x^2 + 10x + 25)(x + 1)}$$

Perform the indicated operations in each of the following. If possible, reduce the answer to lowest terms.

4. $\dfrac{5x}{x-2} \times \dfrac{x^2-4}{7x^3}$

5. $\dfrac{x^2-6x+9}{7(x^2-1)} \div \dfrac{x-3}{x-1}$

6. $\dfrac{5x^2}{x^2+2x-3} + \dfrac{2x}{x^2+5x+6}$

7. $\dfrac{3}{x-1} - \dfrac{2}{x^2+2x-3}$

8. $\left(\dfrac{3}{x} - \dfrac{2}{x^2}\right) \div \dfrac{1}{x^5}$

9. Express the following as a fraction having no fractions in the numerator or denominator.

$$\dfrac{\dfrac{3}{x^2} - \dfrac{5}{x^3}}{\dfrac{7}{x} - \dfrac{1}{x^4}}$$

10. (a) Express the following in a form having no irrational number in the denominator.

$$\dfrac{3}{\sqrt{5}-1}$$

(b) Express the following in the form $a + bi$ where a and b are real numbers and $i^2 = -1$.

$$\dfrac{3+5i}{2-7i}$$

1.7 Linear and Quadratic Equations

The purpose of this section is to review properties of equations, to review the solution of linear equations, to review the solution of quadratic equations by factoring and by the quadratic formula, and to review completing the square. A derivation of the quadratic formula is included at the end of the section.

An **equation** is a statement of equality. The expression on the left of the equals sign is called its *left side;* the expression on the right is called its *right side.* A **solution** or **root** of an equation in one variable is a number such that the equation becomes a true statement when the variable is replaced by that number. A solution of an equation is said to *satisfy* the equation. The set of all solutions of an equation is sometimes called the **solution set** of the equation. To *solve* an equation is to find all its solutions or to find its solution set. If an equation has no solutions, we say that its solution set is *empty* or that its solution set is the *empty set* or the *null* set. Two equations are said to be **equivalent** if they have exactly the same solutions, that is, if they have the same solution set. If two equations are equivalent, each is said to be an *equivalent form* of the other.

Some properties of equations are as follows:

Property 1 If the same number or polynomial in the variable(s) of an equation is added to or subtracted from both sides of an equation, the result is an equation with exactly the same solution(s) as the original equation. See Examples 1(a) and 1(b).

Property 2 If both sides of an equation are multiplied or divided by the same nonzero number, the result is an equation with exactly the same solution(s) as the original equation. See Example 1(c).

Property 3 If both sides of an equation in one variable are multiplied by a nonzero polynomial in that variable, the result is an equation such that any solution of the original equation is also a solution of the new equation, but the new equation may have additional solutions that are not solutions of the original equation. See Example 1(d).

Property 4 If both sides of an equation in one variable are divided by a nonzero polynomial in that variable, any solution of the new equation is also a solution of the original equation, but the original equation may have additional solutions that are not solutions of the new equation. Solutions of the original equation may have been "lost." See Example 1(e).

Property 5 If both sides of an equation in one variable are raised to the same positive integral power, any solution of the original equation is also a solution of the new equation, but the new equation may have additional solutions that are not solutions of the original equation. See Example 1(f).

Note: When both sides of an equation in one variable are multiplied by a polynomial in that variable or are raised to the same positive integral power, the solutions of the new equation must be considered candidates for solutions of the original equation, and each must be checked by substitution in the original equation to ascertain whether or not it actually satisfies the original equation. When both sides of an equation in one variable are divided by a polynomial in that variable, solutions of the original equation might be lost, in the sense that the new equation might not have as solutions all the solutions of the original equation.

Whenever Property 1 or Property 2 is applied, the resulting equation is equivalent to the original and is said to be the original equation written in an equivalent or another form. These and the other properties of equations on our list are illustrated in the following example.

EXAMPLE 1

In each of the following, state the relationship between the solutions of the original equation (Equation (1)) and the new equation (Equation (2)).

(a) (1) $x - 3 = 7$
    (2) $x = 7 + 3$

(b) (1) $10 = -x^2 + 7x$
    (2) $x^2 - 7x + 10 = 0$

(c) (1) $3x = 7$
    (2) $x = \frac{7}{3}$

(d) (1) $15 = 5x$
    (2) $15(x - 2) = 5x(x - 2)$

(e) (1) $3(x - 2)(x - 5) = 6(x - 5)$
    (2) $3(x - 2) = 6$

(f) (1) $2x + 1 = 3x$
    (2) $4x^2 + 4x + 1 = 9x^2$

Solution

(a) (1) $x - 3 = 7$
    (2) $x = 7 + 3$

The number 3 was added to both sides of Equation (1) to produce Equation (2). Equations (1) and (2) have exactly the same solution(s). Property 1

(b) (1) $10 = -x^2 + 7x$
    (2) $x^2 - 7x + 10 = 0$

The polynomial $x^2 - 7x$ was added to both sides of Equation (1) to produce Equation (2). Equations (1) and (2) have exactly the same solution(s). Property 1

(c) (1) $3x = 7$
    (2) $x = \frac{7}{3}$

Both sides of Equation (1) were divided by the number 3 to produce Equation (2). Equations (1) and (2) have exactly the same solution(s). Property 2

(d) (1) $15 = 5x$
    (2) $15(x - 2) = 5x(x - 2)$

Both sides of Equation (1) were multiplied by the polynomial $x - 2$ to produce Equation (2). Any solution of Equation (1) is also a solution of Equation (2), but Equation (2) may have additional solutions that are not solutions of Equation (1). Property 3

In this particular example, Equation (2) has one more solution than Equation (1). It can be verified by substitution that both 2 and 3 are solutions of Equation (2), but of these only 3 is a solution of Equation (1).

(e) (1) $3(x - 2)(x - 5) = 6(x - 5)$
    (2) $3(x - 2) = 6$

Both sides of Equation (1) were divided by the polynomial $x - 5$ to produce Equation (2). Any solution of Equation (2) is also a solution of Equation (1), but Equation (1) may have additional solutions that are not solutions of Equation (2). Property 4

In this example, Equation (1) has one more solution than Equation (2). It can be verified by substitution in Equation (1) that both 5 and 4 are solutions of it, but of these only 4 is a solution of Equation (2).

(f) (1) $2x + 1 = 3x$
    (2) $4x^2 + 4x + 1 = 9x^2$

Both sides of Equation (1) were squared to produce Equation (2). Any solution of Equation (1) is also a solution of Equation (2), but Equation (2) may have additional solutions that are not solutions of Equation (1). Property 5

In this particular example, Equation (2) has one more solution than Equation (1). It can be verified by substitution that both 1 and $-\frac{1}{5}$ are solutions of $4x^2 + 4x + 1 = 9x^2$, but of these only 1 is a solution of the original equation $2x + 1 = 3x$.

An **nth-degree polynomial equation** is an equation of the form $P = Q$ where P and Q are polynomials and the polynomial $P - Q$ is of degree n. For example,

$$3x - 2 = 7x + 5$$

is a first-degree polynomial equation, and

$$3x^2 + 13x = 10$$

is a second-degree polynomial equation.

A **linear** or **first-degree equation in one variable**, say x, is a first-degree polynomial equation in one variable that can be written in the form

$$ax + b = 0$$

where a and b are specific numbers and $a \neq 0$. Any linear equation in one variable has exactly one solution. The solution of $ax + b = 0$ is

$$x = \frac{-b}{a}.$$

As illustrated in the next example, the properties of equations can be used to find the solution of a linear equation in one variable.

EXAMPLE 2

Find the solution of $x - 2 = 4 - 2x$.

Solution

We shall use the properties of equations to isolate the terms involving x on one side of the equation and hence produce an equivalent equation whose solution is obvious.

$$x - 2 = 4 - 2x$$

$x + 2x - 2 = 4$ Adding $2x$ to both sides. Property 1

$3x = 6$ Adding 2 to both sides. Property 1

$x = 2$ Dividing both sides by 3. Property 2

Check

Show that the original equation becomes a true statement when x is replaced by 2.

$$x - 2 = 4 - 2x$$
$$2 - 2 = 4 - 2(2)$$
$$0 = 4 - 4$$ True. The solution is 2.

Exercise 1.7, 11–30

A **quadratic** or **second-degree equation** is a second-degree polynomial equation in one variable, say x, that can be written in the ''standard'' form

$$ax^2 + bx + c = 0$$

where a, b, and c are specific real numbers and $a \neq 0$. Any quadratic equation has either

1. Exactly one real-number solution,
2. Exactly two distinct real-number solutions, or
3. Exactly two distinct imaginary-number solutions.

The solution(s) of $ax^2 + bx + c = 0$ is (are) given by the *quadratic formula*

$$x = \frac{-b \pm \sqrt{b^2 - 4ac}}{2a}.$$ Quadratic formula

If $b^2 - 4ac = 0$, this formula gives the single real-number solution $x = -b/2a$, but if $b^2 - 4ac$ is not zero, the formula gives the two distinct solutions

$$x = \frac{-b + \sqrt{b^2 - 4ac}}{2a} \quad \text{and} \quad x = \frac{-b - \sqrt{b^2 - 4ac}}{2a}.$$

A derivation of the quadratic formula appears near the end of this section.

We see from the quadratic formula that the nature of the solutions of a quadratic equation depends on the value of the expression $b^2 - 4ac$. In particular,

1. If $b^2 - 4ac = 0$, $ax^2 + bx + c = 0$ has exactly one real solution.
2. If $b^2 - 4ac > 0$, $ax^2 + bx + c = 0$ has exactly two distinct real solutions.
3. If $b^2 - 4ac < 0$, $ax^2 + bx + c = 0$ has exactly two distinct imaginary solutions. (The two imaginary solutions are conjugates.)

EXAMPLE 3

In each of the following determine the nature of the solutions without actually finding them.
(a) $4x^2 = 12x - 9$ (b) $4x^2 - 4x = 3$ (c) $2x^2 - 2x + 5 = 0$

Solution

(a) To determine a, b, and c, we must first write the equation in standard form.

$$4x^2 - 12x + 9 = 0$$

Since $a = 4$, $b = -12$, and $c = 9$,

$$b^2 - 4ac = (-12)^2 - 4(4)(9)$$
$$= 144 - 144$$
$$= 0 \quad b^2 - 4ac \text{ is zero.}$$

There is exactly one real-number solution of $4x^2 = 12x - 9$.

(b) Written in standard form, the given equation is

$$4x^2 - 4x - 3 = 0.$$

Here $a = 4$, $b = -4$, and $c = -3$.

$$b^2 - 4ac = (-4)^2 - 4(4)(-3)$$
$$= 16 + 48$$
$$= 64 \quad b^2 - 4ac \text{ is positive.}$$

There are exactly two distinct real-number solutions of $4x^2 - 4x = 3$.

(c) The equation

$$2x^2 - 2x + 5 = 0$$

is already written in standard form. Here $a = 2$, $b = -2$, and $c = 5$.

$$b^2 - 4ac = (-2)^2 - 4(2)(5)$$
$$= 4 - 40$$
$$= -36 \qquad b^2 - 4ac \text{ is negative.}$$

Exercise 1.7, 31–40

There are exactly two distinct imaginary-number solutions of $2x^2 - 2x + 5 = 0$.

If the coefficients of a quadratic equation are integers, and if the left side of the standard form can be readily factored into first-degree polynomials with integral coefficients, probably the easiest way to find the solutions is to utilize the *zero factor law.*

Zero Factor Law   If the product of two factors is equal to zero, then one or the other or both of the factors is zero.

For example, if

$$(x - 3)(x + 7) = 0,$$

then $\qquad\qquad x - 3 = 0 \quad \text{or} \quad x + 7 = 0.$

If the product of more than two factors is equal to zero, an extension of the zero factor law is applicable. For example, if

$$(2x - 3)(x - 7)(x^2 - 6x + 3) = 0,$$

Exercise 1.7, 41–50 then $\quad 2x - 3 = 0, \quad \text{or} \quad x - 7 = 0, \quad \text{or} \quad x^2 - 6x + 3 = 0.$

Using the zero factor law to solve a quadratic equation by factoring, we do the following:

1. Make sure the equation is in standard form (0 is isolated on the right side).

2. Factor the left side.

3. Equate each factor to zero, and solve for the variable.

Check by substituting back in the original equation.

EXAMPLE 4   Solve $3x^2 + 13x = 10$ by factoring.

Solution   1. Write the equation in standard form.

$$3x^2 + 13x - 10 = 0$$

2. Factor the left side.

$$(3x - 2)(x + 5) = 0$$

3. Equate each factor to zero, and solve for the variable.

$$3x - 2 = 0 \qquad x + 5 = 0$$
$$x = \tfrac{2}{3} \qquad\qquad x = -5$$

Check by showing that $\frac{2}{3}$ and -5 both satisfy the original equation.

$$3x^2 + 13x = 10$$

$$3(\tfrac{2}{3})^2 + 13(\tfrac{2}{3}) = 10 \qquad\qquad 3(-5)^2 + 13(-5) = 10$$
$$3(\tfrac{4}{9}) + \tfrac{26}{3} \quad\;\; = 10 \qquad\qquad\quad 3(25) - 65 \quad\;\; = 10$$
$$\tfrac{4}{3} + \tfrac{26}{3} \quad\;\; = 10 \qquad\qquad\qquad\; 75 - 65 \quad\;\; = 10 \quad \text{True}$$
$$\tfrac{30}{3} \quad\;\; = 10 \quad \text{True}$$

The solutions of $3x^2 + 13x = 10$ are $\frac{2}{3}$ and -5.

EXAMPLE 5

Solve $4x^2 - 12x = -9$ by factoring.

Solution

1. Write the equation in standard form.

$$4x^2 - 12x + 9 = 0$$

2. Factor the left side.

$$(2x - 3)^2 = 0$$

3. Equate each factor to zero, and solve for the variable.

$$2x - 3 = 0 \qquad 2x - 3 = 0$$
$$x = \tfrac{3}{2}$$

Check

$$4x^2 - 12x \;\; = -9$$
$$4(\tfrac{3}{2})^2 - 12(\tfrac{3}{2}) = -9$$
$$4(\tfrac{9}{4}) - 18 \quad\; = -9$$
$$9 - 18 \quad\; = -9 \qquad \text{True}$$

Exercise 1.7, 51–60 The solution is $\frac{3}{2}$.

If a quadratic equation cannot easily be solved by factoring, it can always be solved by using the quadratic formula.

EXAMPLE 6

Solve $3x^2 + 13x = 10$ by using the quadratic formula.

Solution

$$3x^2 + 13x - 10 = 0$$
$$a = 3, \qquad b = 13, \qquad c = -10$$

$$x = \frac{-b \pm \sqrt{b^2 - 4ac}}{2a}$$

$$x = \frac{-13 \pm \sqrt{(13)^2 - 4(3)(-10)}}{6}$$

$$= \frac{-13 \pm \sqrt{169 + 120}}{6}$$

$$= \frac{-13 \pm \sqrt{289}}{6} = \frac{-13 \pm 17}{6}$$

PART ONE | BASIC ALGEBRA

$$x = \frac{-13 + 17}{6} \quad \text{or} \quad x = \frac{-13 - 17}{6}$$

$$= \frac{4}{6} \qquad\qquad = \frac{-30}{6}$$

$$= \frac{2}{3} \qquad\qquad = -5$$

The solutions of $3x^2 + 13x = 10$ are $\frac{2}{3}$ and -5. These were checked in Example 4, where this same equation was solved by factoring.

EXAMPLE 7

Solve $2x^2 - 2x + 5 = 0$.

Solution

The equation $2x^2 - 2x + 5 = 0$ is already in standard form, and the left side cannot be factored into first-degree polynomials with integral coefficients. Here

$$b^2 - 4ac = (-2)^2 - 4(2)(5)$$
$$= -36.$$

There are two distinct imaginary-number solutions. They can be found by using the quadratic formula.

$$a = 2, \qquad b = -2, \qquad c = 5$$

$$x = \frac{-(-2) \pm \sqrt{(-2)^2 - 4(2)(5)}}{4}$$

$$= \frac{2 \pm \sqrt{-36}}{4}$$

$$= \frac{2 \pm 6i}{4}$$

$$x = \frac{2 + 6i}{4} \quad \text{or} \quad x = \frac{2 - 6i}{4}$$

$$= \frac{2}{4} + \frac{6}{4}i \qquad\qquad = \frac{2}{4} - \frac{6}{4}i$$

$$= \frac{1}{2} + \frac{3}{2}i \qquad\qquad = \frac{1}{2} - \frac{3}{2}i$$

Check

$$2x^2 - 2x + 5 = 0$$

$$2(\tfrac{1}{2} + \tfrac{3}{2}i)^2 - 2(\tfrac{1}{2} + \tfrac{3}{2}i) + 5 = 0$$

$$2(\tfrac{1}{4} + \tfrac{3}{2}i + \tfrac{9}{4}i^2) - (1 + 3i) + 5 = 0$$

$$\tfrac{1}{2} + 3i - \tfrac{9}{2} - 1 - 3i + 5 = 0$$

$$\tfrac{-8}{2} + 4 = 0$$

$$-4 + 4 = 0 \qquad \text{True}$$

Exercise 1.7, 61–70
The student should show by similar substitution that $\frac{1}{2} - \frac{3}{2}i$ also checks. The solutions of $2x^2 - 2x + 5 = 0$ are $\frac{1}{2} + \frac{3}{2}i$ and $\frac{1}{2} - \frac{3}{2}i$.

The quadratic formula is derived by employing the process called *completing the square* that was first introduced in Section 1.5. The technique itself provides an additional method for solving a quadratic equation. As a method of solution of a quadratic equation, it tends to be impractical, but completing the square is extremely useful in other contexts. For example, in some applications of the calculus, completing the square is needed to write equations in forms from which it is easy to identify various features of their graphs.

The process of completing the square hinges on the fact that where k is a specific real number, the expression $x^2 + kx$ becomes a perfect-square trinomial when the number $(k/2)^2$ is added to it. Solution of a quadratic by completing the square will be illustrated in the next example, and then the technique will be used to derive the quadratic formula. Other uses of completing the square will be illustrated in the exercises.

EXAMPLE 8

Solve $3x^2 + 13x - 10 = 0$ by completing the square.

Solution

The idea is to write the equation in the form $(x - p)^2 = q$ so that we can conclude that $x - p = +\sqrt{q}$ or $x - p = -\sqrt{q}$. To do this, we proceed as follows:

1. Divide the equation through by the coefficient of x^2.

$$x^2 + \tfrac{13}{3}x - \tfrac{10}{3} = 0$$

2. Write the equation in an equivalent form with the nonvariable term isolated on the right side.

$$x^2 + \tfrac{13}{3}x = \tfrac{10}{3}$$

3. To both sides of the equation add the square of half the coefficient of x. (This "completes the square" on the left side and makes the left side become a perfect-square trinomial.)

Half the coefficient of x is $\frac{1}{2} \times \frac{13}{3} = \frac{13}{6}$.
The square of half the coefficient of x is $\left(\frac{13}{6}\right)^2$.

$$x^2 + \tfrac{13}{3}x + \left(\tfrac{13}{6}\right)^2 = \tfrac{10}{3} + \left(\tfrac{13}{6}\right)^2$$

4. Write the equation in the form $(x - p)^2 = q$.

$$(x + \tfrac{13}{6})^2 = \tfrac{10}{3} + \tfrac{169}{36}$$
$$(x + \tfrac{13}{6})^2 = \tfrac{120}{36} + \tfrac{169}{36}$$
$$= \tfrac{289}{36}$$

5. Conclude that $x - p = +\sqrt{q}$ or $x - p = -\sqrt{q}$.

$$x + \tfrac{13}{6} = +\sqrt{\tfrac{289}{36}} \quad \text{or} \quad x + \tfrac{13}{6} = -\sqrt{\tfrac{289}{36}}$$
$$= \tfrac{17}{6} \qquad\qquad\qquad = -\tfrac{17}{6}$$

$$x = \tfrac{17}{6} - \tfrac{13}{6} \quad \text{or} \quad x = \tfrac{-17}{6} - \tfrac{13}{6}$$
$$= \tfrac{4}{6} \qquad\qquad\qquad = \tfrac{-30}{6}$$
$$= \tfrac{2}{3} \qquad\qquad\qquad = -5$$

The solutions of $3x^2 + 13x - 10 = 0$ are $\tfrac{2}{3}$ and -5. These same solutions were found by factoring in Example 4 and by the quadratic formula in Example 6. They were checked in Example 4.

Exercise 1.7, 71–80

The quadratic formula will now be derived by completing the square. Given the equation

$$ax^2 + bx + c = 0$$

where $a \neq 0$.

1. Divide through by the coefficient of x^2.

$$x^2 + \frac{b}{a}x + \frac{c}{a} = 0$$

2. Write the equation in an equivalent form with the nonvariable term isolated on the right side.

$$x^2 + \frac{b}{a}x = \frac{-c}{a}$$

3. To both sides of the equation add the square of

$$\frac{1}{2} \times \frac{b}{a} = \frac{b}{2a}.$$

$$x^2 + \frac{b}{a}x + \left(\frac{b}{2a}\right)^2 = \frac{-c}{a} + \left(\frac{b}{2a}\right)^2$$

4. Write the equation in the form $(x - p)^2 = q$.

$$\left(x + \frac{b}{2a}\right)^2 = \frac{-c}{a} + \frac{b^2}{4a^2}$$

$$= \frac{-4ac + b^2}{4a^2}$$

$$= \frac{b^2 - 4ac}{4a^2}$$

5. Conclude that $x - p = +\sqrt{q}$ or $x - p = -\sqrt{q}$.

$$x + \frac{b}{2a} = +\sqrt{\frac{b^2 - 4ac}{4a^2}} \quad \text{or} \quad x + \frac{b}{2a} = -\sqrt{\frac{b^2 - 4ac}{4a^2}}$$

$$= \frac{\sqrt{b^2 - 4ac}}{|2a|} \qquad\qquad = \frac{-\sqrt{b^2 - 4ac}}{|2a|}$$

Since $|2a|$ is either *plus* $2a$ or *minus* $2a$, $x + b/2a$ is either plus or minus $\sqrt{b^2 - 4ac}/2a$. In other words,

$$x + \frac{b}{2a} = \frac{\pm\sqrt{b^2 - 4ac}}{2a}$$

$$x = \frac{-b}{2a} \frac{\pm\sqrt{b^2 - 4ac}}{2a}$$

$$x = \frac{-b \pm \sqrt{b^2 - 4ac}}{2a}.$$

Additional exercises, Exercise 1.7, 81–112

EXERCISE 1.7 LINEAR AND QUADRATIC EQUATIONS

▶ In each of the following, state the relationship between the solutions of Equation (1) and Equation (2) without actually finding those solutions.

EXAMPLES

(a) (1) $x - 1 = 4$
    (2) $x^2 + 4x - 5 = 4x + 20$

(b) (1) $x^2 - 4 = 3x + 6$
    (2) $x - 2 = 3$

Solutions

(a) Both sides of (1) were multiplied by $(x + 5)$. Any solution of (1) is also a solution of (2), but (2) may have additional solutions. (The solutions of (2) are 5 and -5. The only solution of (1) is 5.)

(b) Both sides of (1) were divided by $(x + 2)$. Any solution of (2) is also a solution of (1), but (1) may have additional solutions. (The only solution of (2) is 5, but the solutions of (1) are 5 and -2.)

1. (1) $x - 2 = 5$
    (2) $x^2 + x - 6 = 5x + 15$

2. (1) $x^2 - 9 = 2x - 6$
    (2) $x + 3 = 2$

3. (1) $x - 2 = 5$
    (2) $x - 2 + x + 3 = 5 + x + 3$

4. (1) $7x^2 - 14x = 8$
    (2) $x^2 - 2x = \frac{8}{7}$

5. (1) $x + 3 = 2x$
    (2) $x^2 + 6x + 9 = 4x^2$

6. (1) $5x^2 + 2x = 11$
    (2) $x^2 + \frac{2}{5}x + \frac{1}{25} = \frac{11}{5} + \frac{1}{25}$

7. (1) $x + 4 = 7$
    (2) $x^2 + 3x - 4 = 7x - 7$

8. (1) $x^2 + x - 6 = 5x - 10$
    (2) $x + 3 = 5$

9. (1) $2x - 1 = 5x$
    (2) $4x^2 - 4x + 1 = 25x^2$

10. (1) $3x^2 + 6x = 9$
    (2) $x^2 + 2x + 1 = 4$

▶ Solve each of the following.

11. $3x - 7 = -2x + 8$

12. $5x + 3 = 4 - x$

13. $3(x - 2) = 2(x + 1)$

14. $0 = 8 - 2x$

15. $5x^2 - 3x + 7 = 8x + 5x^2$

16. $\frac{2}{3}x - \frac{4}{5} = x$

17. $x + \dfrac{2}{3} = \dfrac{x}{2}$ **18.** $2x + 16 - x^2 = 8 - x^2$

19. $\frac{2}{3}(x + 4) = \frac{1}{2}(x - 1)$ **20.** $5x - \frac{4}{5} = \frac{2}{5}x + \frac{3}{5}$

▶ Solve each of the following for the letter indicated.

EXAMPLE $l = a + (n - 1)d, \quad n \neq 1; \quad$ solve for d.

Solution $l - a = (n - 1)d$

$$\frac{l - a}{n - 1} = d$$

21. $V = 2\pi rh, \quad r \neq 0; \quad$ solve for h.

22. $A = \frac{1}{2}bh, \quad b \neq 0; \quad$ solve for h.

23. $V = \frac{1}{3}\pi r^2 h, \quad r \neq 0; \quad$ solve for h.

24. $r = \dfrac{1}{1 + i}, \quad r \neq 0; \quad$ solve for i.

25. $A = P + Pni, \quad ni \neq -1; \quad$ solve for P.

26. $S = \dfrac{a}{1 - r}, \quad r \neq 1, \, S \neq 0; \quad$ solve for r.

▶ Solve each of the following for y', and make note of any restrictions that must be put on the variables.

EXAMPLE $4x + 6xy + 3x^2 y' + 3y^2 y' = 0$

Solution $3x^2 y' + 3y^2 y' = -4x - 6xy$

$(3x^2 + 3y^2)y' = -4x - 6xy$

$$y' = \frac{-4x - 6xy}{3x^2 + 3y^2}, \quad 3x^2 + 3y^2 \neq 0$$

27. $2x + xy' + y + 2y' = 0$ **28.** $6x + 7xy' + 7y + 6yy' = 0$

29. $2x + 2yy' = 0$ **30.** $6x + 5x^2 y' + 10xy - 8y^2 y' = 0$

▶ In each of the following, determine the nature of the solutions without actually finding them.

EXAMPLE $3x^2 - 2x = 5$

Solution $3x^2 - 2x - 5 = 0, \quad b^2 - 4ac = 4 - 4(3)(-5) = 4 + 60 > 0$

There are two distinct real-number solutions.

31. $3x^2 = 11x - 8$ **32.** $2x^2 - 7x + 10 = 0$

33. $x^2 - 6x = -7$ **34.** $x^2 - 6x = -9$ **35.** $x^2 - 6x = -10$

36. $3x^2 = 2x - 1$ **37.** $4x^2 = -28x - 49$

38. $3x^2 = 20x + 7$ **39.** $3x^2 = 5x - 4$ **40.** $x^2 - 17 = 0$

▶ Use the zero factor law to solve each of the following.

EXAMPLES

Solutions

(a) $(x - 5)(x^2 - 16) = 0$ (b) $x(x + 1) = 6$

(a) $(x - 5)(x - 4)(x + 4) = 0$

    $x - 5 = 0$ $x - 4 = 0$ $x + 4 = 0$

        $x = 5$ $x = 4$ $x = -4$

(b) $x^2 + x - 6 = 0$

    $(x - 2)(x + 3) = 0$

    $x - 2 = 0$ $x + 3 = 0$

       $x = 2$ $x = -3$

41. $(x - 8)(x + 4)(x - 7) = 0$

42. $(x^2 - 1)(x^2 - 4)(x^2 - 9) = 0$

43. $(x + 4)(x^2 + x - 6) = 0$ **44.** $x(x + 1) = 0$

45. $x(x + 1) = 12$ **46.** $x^3 - 5x^2 - 14x = 0$

47. $x(x - 3) = 0$ **48.** $x(x - 3) = 28$

49. $(x - 4)^2(x^2 - 4) = 0$ **50.** $x(x + 7)(x - 3)(x^2 - 9) = 0$

▶ Solve each of the following by factoring.

51. $x^2 - 7x - 18 = 0$ **52.** $x^2 + 5x - 24 = 0$

53. $x^2 - 2x = 35$ **54.** $x^2 = 12 - 4x$

55. $6x^2 + x - 1 = 0$ **56.** $15x^2 = 2 - 7x$

57. $56x^2 = x + 1$ **58.** $9x^3 = 4x^2$

59. $9x^2 + 30x = -25$ **60.** $x^3 = 4x^2 - 4x$

▶ Use the quadratic formula to solve each of the following.

61. $3x^2 + 5x - 2 = 0$ **62.** $x^2 = 2x - 10$

63. $x^2 = 2x + 10$ **64.** $4x^2 - 20x = -25$

65. $5x^2 = 16x - 20$ **66.** $x^2 + 9 = 0$

67. $x^2 - 9 = 0$ **68.** $5x^2 + 4x - 1 = 0$

69. $2x^2 + 2x + 13 = 0$ **70.** $8x^2 = 2x + 1$

▶ Solve each of the following by completing the square.

EXAMPLE

$2x^2 - x - 3 \quad = 0$

Solution

$x^2 - \tfrac{1}{2}x - \tfrac{3}{2} \quad = 0$

$x^2 - \tfrac{1}{2}x \quad = \tfrac{3}{2}$

$x^2 - \tfrac{1}{2}x + (\tfrac{-1}{4})^2 = \tfrac{3}{2} + (\tfrac{-1}{4})^2$

$(x - \tfrac{1}{4})^2 = \tfrac{25}{16}$

$x - \tfrac{1}{4} = \tfrac{5}{4}$ or $x - \tfrac{1}{4} = \tfrac{-5}{4}$

$x = \tfrac{6}{4} \qquad\qquad x = \tfrac{-4}{4}$

$= \tfrac{3}{2} \qquad\qquad = -1$

71. $6x^2 - 7x - 5 = 0$ **72.** $2x^2 - 7x + 2 = 0$

73. $15x^2 + 2x - 1 = 0$ **74.** $x^2 + 2x - 35 = 0$

75. $x^2 - 10x + 25 = 0$ **76.** $10x^2 + 3x - 1 = 0$

77. $3x^2 - 22x + 7 = 0$ **78.** $2x^2 + 11x + 12 = 0$

79. $2x^2 + 3x - 5 = 0$ **80.** $9x^2 + 3x - 20 = 0$

▶ Solve each of the following equations.

EXAMPLE

$x^4 - 13x^2 + 36 = 0$

Solution

This may be treated as a quadratic in x^2. Let $x^2 = y$. The equation becomes

$$y^2 - 13y + 36 = 0.$$

We can solve for y by factoring.

$$(y - 4)(y - 9) = 0$$
$$y = 4 \quad\text{or}\quad y = 9$$

But $y = x^2$, so $x^2 = 4$ or $x^2 = 9$.

$$x = 2 \text{ or } -2 \quad\text{or}\quad x = 3 \text{ or } -3$$

81. $x^4 + 15x^2 - 16 = 0$ **82.** $x^4 - 17x^2 + 16 = 0$

83. $x^4 + 5x^2 - 36 = 0$ **84.** $x^6 - 9x^3 + 8 = 0$

85. $x^6 - 28x^3 + 27 = 0$

▶ An equation of the form $Ax^2 + By^2 + Cx + Dy + E = 0$, where A, B, C, D, and E are all real numbers and A and B are not both zero, represents a graph called a conic section. In order to identify features of the graph it is frequently convenient to write the equation in a "standard" form. In each of the following, write the given equation in the indicated form, and state the values of h, k, r, p, a, or b where these letters represent real numbers and a, b, and r are positive. (For the student's information, the name of each conic section is given in parentheses. Some conic sections are treated in detail in Part 2 of this text.)

EXAMPLE

$$4x^2 + 9y^2 - 8x + 36y + 4 = 0; \quad \frac{(x - h)^2}{a^2} + \frac{(y - k)^2}{b^2} = 1 \quad \text{(Ellipse)}$$

Solution

$$4x^2 - 8x + 9y^2 + 36y = -4$$
$$4(x^2 - 2x) + 9(y^2 + 4y) = -4$$

Completing the square in both parentheses by adding appropriate numbers to both sides of the equation, we have

$$4(x^2 - 2x + 1) + 9(y^2 + 4y + 4) = -4 + 4 + 36.$$
$$4(x - 1)^2 + 9(y + 2)^2 = 36$$
$$\frac{(x - 1)^2}{9} + \frac{(y + 2)^2}{4} = 1$$
$$h = 1, \quad k = -2, \quad a = 3, \quad b = 2$$

86. $x^2 + y^2 - 4x + 6y + 9 = 0;$ $(x - h)^2 + (y - k)^2 = r^2$ (Circle)

87. $4x^2 - 9y^2 - 8x - 36y - 68 = 0;$ $\dfrac{(x - h)^2}{a^2} - \dfrac{(y - k)^2}{b^2} = 1$
(Hyperbola)

88. $x^2 - 2x - 6y - 11 = 0;$ $(x - h)^2 = 2p(y - k)$ (Parabola)

89. $y^2 - 8x - 6y + 1 = 0;$ $(y - k)^2 = 2p(x - h)$ (Parabola)

90. $16x^2 + 25y^2 - 32x - 150y - 159 = 0;$ $\dfrac{(x - h)^2}{a^2} + \dfrac{(y - k)^2}{b^2} = 1$
(Ellipse)

91. $-4x^2 + 9y^2 - 8x - 54y + 41 = 0;$ $\dfrac{(y - k)^2}{a^2} - \dfrac{(x - h)^2}{b^2} = 1$
(Hyperbola)

92. $3x^2 + 3y^2 - 6x + 12y - 12 = 0;$ $(x - h)^2 + (y - k)^2 = r^2$
(Circle)

93. $x^2 + 4y^2 + 10x - 32y + 85 = 0;$ $\dfrac{(x - h)^2}{a^2} + \dfrac{(y - k)^2}{b^2} = 1$
(Ellipse)

94. $4x^2 - 16y^2 - 40x + 36 = 0;$ $\dfrac{(x - h)^2}{a^2} - \dfrac{(y - k)^2}{b^2} = 1$
(Hyperbola)

95. $4x^2 + 16y^2 - 24x + 32y - 12 = 0$; $\dfrac{(x - h)^2}{a^2} + \dfrac{(y - k)^2}{b^2} = 1$

(Ellipse)

▶ Where h and positive a and k are real numbers, write each of the following in the form indicated and state the values of h and k.

EXAMPLE $\qquad \sqrt{4x^2 - 24x + 20}$; $\sqrt{a}\,\sqrt{(x - h)^2 - k^2}$

Solution $\qquad \sqrt{4x^2 - 24x + 20} = \sqrt{4}\,\sqrt{x^2 - 6x + 5}$
$$= 2\sqrt{(x^2 - 6x \quad\;\;) + 5}$$

Completing the square in the parentheses by adding and subtracting the appropriate number, we have

$$\sqrt{4x^2 - 24x + 20} = 2\sqrt{(x^2 - 6x + 9) + 5 - 9}$$
$$= 2\sqrt{(x - 3)^2 - 4} \qquad\qquad h = 3, k = 2$$

96. $\sqrt{9x^2 - 36x - 45}$; $\sqrt{a}\,\sqrt{(x - h)^2 - k^2}$

97. $\sqrt{-9x^2 + 36x + 45}$; $\sqrt{a}\,\sqrt{k^2 - (x - h)^2}$

98. $\sqrt{4x^2 - 8x + 13}$; $\sqrt{a}\,\sqrt{(x - h)^2 + k^2}$

99. $\sqrt{4x^2 - 12x - 135}$; $\sqrt{a}\,\sqrt{(x - h)^2 - k^2}$

100. $\sqrt{-x^2 + 16x - 15}$; $\sqrt{a}\,\sqrt{k^2 - (x - h)^2}$

101. $\sqrt{x^2 + 10x + 34}$; $\sqrt{a}\,\sqrt{(x - h)^2 + k^2}$

102. $\sqrt{9x^2 - 72x + 135}$; $\sqrt{a}\,\sqrt{(x - h)^2 - k^2}$

103. $\sqrt{4x^2 - 20x - 11}$; $\sqrt{a}\,\sqrt{(x - h)^2 - k^2}$

104. $\sqrt{-4x^2 + 4x + 3}$; $\sqrt{a}\,\sqrt{k^2 - (x - h)^2}$

105. $\sqrt{4x^2 + 4x + 17}$; $\sqrt{a}\,\sqrt{(x - h)^2 + k^2}$

106. In one season, Charlie was at bat 560 times and his batting average (his number of hits divided by his number of times at bat) was 0.125. How many hits did Charlie make that season?

107. The sales tax on a car was $162.50, and this was 5% of the list price. What was the list price?

108. A professor received $200 for a review on which she worked 26 hours. If her federal, state, and self-employment taxes took 48% of her salary, what was her "take-home" pay per hour?

109. If a rocket is fired from the Earth with an initial velocity of 98 meters per second, its distance s (in meters) above the Earth t seconds later is given approximately by the equation $s = -4.9t^2 + 98t$. Using this equation, how long will it take for the rocket to reach the height of 490 meters? How long will the rocket be in flight?

110. If a rocket is fired from the Earth with an initial velocity of 320 feet per second, its distance s (in feet) above the Earth t seconds later is given approximately by the equation $s = -16t^2 + 320t$. Using this equation, at how many seconds after fire is the rocket 1536 feet above the Earth?

111. A chemist needs to make a cylindrical container 2 meters high out of a special piece of material. He finds he must make the container with a radius r meters so that $2\pi r^2 + 4\pi r = 7.5$. If he uses 3 as a rough estimate for π, what will he obtain for r?

112. A steel beam 4 meters long, $\frac{1}{6}$ meters wide, and h meters deep can safely support a load of $6{,}500{,}000\, h^2$ newtons at its center. (A newton is a unit of force; it is the unit of load in this problem.) What is the depth of a steel beam 4 meters long and $\frac{1}{6}$ meters wide that can safely support a load of $1{,}625{,}000$ newtons at its center?

CHECK LIST FOR THE STUDENT

If you have learned the material in Section 1.7, you should be able to do the following:

1. Give the meaning of a solution or root of an equation in one variable.

2. Tell what is meant by equivalent equations and by an equivalent form of an equation.

3. Tell how the solutions of a given equation are related to the solutions of a new equation obtained from the given equation by doing each of the following:

(a) Adding or subtracting the same polynomial in its variable(s) to or from both sides,

(b) Multiplying or dividing both sides by the same nonzero number.

4. Tell how the solutions of a given equation in one variable are related to the solutions of a new equation obtained from the given equation by doing each of the following:

(a) Multiplying both sides by the same nonzero polynomial in the one variable,

(b) Dividing both sides by the same nonzero polynomial in the one variable,

(c) Raising both sides to the same positive integral power.

5. Give an example of a linear or first-degree equation in one variable.

6. Tell the number of solutions of any linear equation in one variable; solve any linear equation in one variable.

7. Give an example of a quadratic or second-degree equation in one variable.

8. List the three possibilities as far as the number and nature of the solutions of any quadratic equation are concerned.

9. State the quadratic formula for the solutions of $ax^2 + bx + c = 0$.

10. Tell how the number and nature of the solutions of $ax^2 + bx + c = 0$ are related to the value of $b^2 - 4ac$.

11. State and use the zero factor law.

12. Solve a quadratic equation by factoring.

13. Solve a quadratic equation using the quadratic formula.

14. Solve a quadratic equation by completing the square.

15. Complete squares and write certain equations or expressions in prescribed forms.

SECTION QUIZ 1.7

Solve each of the following equations.

1. $3x - 2 = -5x + 7$

2. $\frac{2}{5}(x - 1) = \frac{-1}{4}(x + 2)$

3. $x(x - 1)(x + 3) = 0$

4. Solve the equation

$$C = \frac{A}{A + B}$$

for B. (Assume that $A + B \neq 0$.)

In each of the following determine the number of distinct real-number solutions without actually finding them.

5. (a) $x^2 - 2x - 2 = 0$
    (b) $x^2 - 2x + 2 = 0$

6. (a) $x^2 - 2x + 1 = 0$
    (b) $x^2 - 2x = 0$

7. Solve each of the following by factoring.
    (a) $x^2 - 3x = 0$
    (b) $x^2 - 3x = 10$

8. Use the quadratic formula to solve the following equation.

$$9x^2 - 12x - 1 = 0$$

9. Solve the following equation by completing the square.

$$5x^2 - 9x - 2 = 0$$

10. Where h and positive a and k are real numbers, write $\sqrt{9x^2 - 18x - 27}$ in the form $\sqrt{a}\ \sqrt{(x - h)^2 - k^2}$, and state the values of h and k.

1.8 Radical, Fractional, and Absolute Value Equations

The purpose of this section is to consider some radical equations, fractional equations, and some equations involving absolute values.

A **radical equation** is an equation containing one or more radicals with one or more variables in their radicands. We shall consider radical equations in which the radicals are square roots. An example of such an equation in one variable is

$$\sqrt{2x - 4} - \sqrt{x - 1} = 1.$$

To solve such an equation in one variable, we do the following:

1. Isolate a radical on one side.

2. Square both sides.

3. Repeat this procedure until the equation contains no square roots of variable expressions.

4. Solve the latest equation for the variable.

5. Substitute in the original equation to determine which solutions of the latest equation are solutions of the original. (This step is necessary because the latest equation may have some solutions that are not solutions of the original equation.)

EXAMPLE 1   Solve the equation $\sqrt{2x - 4} - \sqrt{x - 1} = 1$.

Solution   1. Isolate a radical on one side.

$$\sqrt{2x - 4} = 1 + \sqrt{x - 1}$$

2. Square both sides.

$$2x - 4 = 1 + 2\sqrt{x - 1} + (x - 1)$$

3. Repeat this procedure: Isolate a radical and square both sides.

$$2x - x - 4 = 2\sqrt{x - 1}$$
$$x - 4 = 2\sqrt{x - 1}$$
$$x^2 - 8x + 16 = 4(x - 1)$$

4. Solve the latest equation.

$$x^2 - 8x + 16 = 4x - 4$$
$$x^2 - 12x + 20 = 0$$
$$(x - 2)(x - 10) = 0$$
$$x = 2 \quad \text{or} \quad x = 10$$

The numbers 2 and 10 are solutions of $x^2 - 12x + 20 = 0$; they are candidates for solutions of the original equation.

5. Substitute 2 and 10 for x in the original equation to determine whether or not they are solutions of it.

$$\sqrt{2x - 4} - \sqrt{x - 1} = 1$$
$$x = 2: \quad \sqrt{4 - 4} - \sqrt{2 - 1} = 1$$
$$0 - 1 \qquad\quad = 1 \qquad \text{False}$$

The number 2 is not a solution of the original equation.

$$x = 10: \quad \sqrt{20 - 4} - \sqrt{10 - 1} = 1$$
$$\sqrt{16} - \sqrt{9} \qquad = 1$$
$$4 - 3 \qquad\quad = 1 \qquad \text{True}$$

Exercise 1.8, 1–10

The number 10 is the only solution of the equation $\sqrt{2x - 4} - \sqrt{x - 1} = 1$.

It can be shown that certain equations containing square roots of expressions in two variables represent certain graphs. For example, the equation

$$\sqrt{(x - 3)^2 + y^2} + \sqrt{(x + 3)^2 + y^2} = 10$$

represents a graph called an ellipse. In applications, it is frequently necessary to write such equations in standard form without radicals. This is illustrated in the next example.

EXAMPLE 2

Write the equation $\sqrt{(x - 3)^2 + y^2} + \sqrt{(x + 3)^2 + y^2} = 10$ in the form

$$\frac{x^2}{a^2} + \frac{y^2}{b^2} = 1$$

where a and b are positive real numbers. Give the values of a and b.

Solution

To write the equation in a form containing no radicals, we proceed as we did in Example 1.

1. Isolate a radical on one side.

2. Square both sides.

3. Repeat this procedure.

First we isolate a radical.

$$\sqrt{(x - 3)^2 + y^2} = 10 - \sqrt{(x + 3)^2 + y^2}$$

Then we square both sides.

$$(x - 3)^2 + y^2 = 100 - 20\sqrt{(x + 3)^2 + y^2} + (x + 3)^2 + y^2$$

Collecting like terms before repeating the procedure will make the algebraic manipulations much less cumbersome.

$$\cancel{x^2} - 6x + \cancel{9} + \cancel{y^2} = 100 - 20\sqrt{(x + 3)^2 + y^2} + \cancel{x^2} + 6x + \cancel{9} + \cancel{y^2}$$

The slashes indicate that $x^2 + 9 + y^2$ was subtracted from both sides of the equation. Also subtracting $6x$ from both sides, we have:

$$-12x = 100 - 20\sqrt{(x + 3)^2 + y^2}$$

Dividing both sides through by the common factor -4 further simplifies the equation.

$$3x = -25 + 5\sqrt{(x + 3)^2 + y^2}$$

Now we begin to repeat the procedure by isolating the term containing the radical.

$$3x + 25 = 5\sqrt{(x + 3)^2 + y^2}$$

Squaring both sides eliminates the remaining radical.

$$(3x + 25)^2 = 25[(x + 3)^2 + y^2]$$

Since our aim is to write this in the form

$$\frac{x^2}{a^2} + \frac{y^2}{b^2} = 1,$$

we multiply out the indicated products and collect like terms.

$$9x^2 + 150x + 625 = 25[x^2 + 6x + 9 + y^2]$$
$$9x^2 + 150x + 625 = 25x^2 + 150x + 225 + 25y^2$$
$$-16x^2 - 25y^2 = 225 - 625$$
$$-16x^2 - 25y^2 = -400$$

To put this in the required form, we divide both sides through by -400.

$$\frac{16x^2}{400} + \frac{25y^2}{400} = 1$$

Since $400 = 16 \times 25$, we finally obtain the required form.

$$\frac{x^2}{25} + \frac{y^2}{16} = 1$$

Comparing this with

$$\frac{x^2}{a^2} + \frac{y^2}{b^2} = 1,$$

we see that $a^2 = 25$ and $b^2 = 16$. Since a and b are positive,

Exercise 1.8, 11–25

$$a = 5 \quad \text{and} \quad b = 4.$$

A **fractional equation** is an equation containing at least one fraction with a variable expression for denominator. An example of a fractional equation in one variable is

$$\frac{x-4}{x-2} - \frac{1}{x} = \frac{-4}{x(x-2)}.$$

To solve such an equation we do the following:

1. Multiply both sides through by the LCD of all the fractions.

2. Solve the new equation.

3. Substitute in the original equation to determine which solutions of the new equation are also solutions of the original. (The new equation may have some solutions that are not solutions of the original equation.)

EXAMPLE 3

Solve the equation

$$\frac{x-4}{x-2} - \frac{1}{x} = \frac{-4}{x(x-2)}.$$

Solution

1. Multiply both sides through by $x(x-2)$, the LCD of all the fractions.

$$\frac{\cancel{x(x-2)}(x-4)}{\cancel{x-2}} - \frac{x\cancel{(x-2)}}{\cancel{x}} = \frac{\cancel{x(x-2)}(-4)}{\cancel{x(x-2)}}$$

$$x(x-4) - (x-2) = -4$$

2. Solve the new equation.

$$x^2 - 4x - x + 2 = -4$$
$$x^2 - 5x + 6 = 0$$
$$(x-2)(x-3) = 0$$
$$x = 2 \quad \text{or} \quad x = 3$$

3. Substitute 2 and 3 for x in the original equation to determine whether or not they are solutions of it.

$x = 2$: The number 2 is not a solution of the original equation. Two terms in the original equation are not defined when $x = 2$ because their denominators are zero for $x = 2$.

$x = 3$: $\dfrac{3-4}{3-2} - \dfrac{1}{3} = \dfrac{-4}{3(3-2)}$

$\qquad -1 - \dfrac{1}{3} = \dfrac{-4}{3} \qquad$ True

The number 3 is the only solution of the original equation.

EXAMPLE 4   Solve the equation

$$\frac{x-1}{x+1} + \frac{6}{x-1} = \frac{x^2+11}{x^2-1}.$$

Solution   The LCD of all the fractions is $x^2 - 1 = (x-1)(x+1)$.

1. Multiply through by $x^2 - 1 = (x-1)(x+1)$.

$$\frac{(x-1)\cancel{(x+1)}(x-1)}{\cancel{x+1}} + \frac{6\cancel{(x-1)}(x+1)}{\cancel{x-1}} = \frac{(x^2+11)\cancel{(x^2-1)}}{\cancel{x^2-1}}$$

$$(x-1)(x-1) + 6(x+1) \qquad = x^2 + 11$$

$$x^2 - 2x + 1 + 6x + 6 \qquad = x^2 + 11$$

$$4x + 7 \qquad = 11$$

2. Solve the new equation.

$$4x = 4$$
$$x = 1$$

3. The number 1 is the only candidate for a solution of the original equation, but it is *not* a solution. Two terms of the original equation are not defined for $x = 1$. The original equation has no solutions; its solution set Exercise 1.8, 26–35 is empty.

Certain equations and inequalities involving absolute value appear frequently in mathematics. Inequalities will be treated in the next section. An example of an equation involving an absolute value is

$$|x - 5| = 2.$$

We use the term *absolute value equation* for equations involving absolute values of variable expressions.

To work with equations such as $|x - 5| = 2$, we need to be familiar with the definition of absolute value. In Section 1.1 we defined the absolute value of real number x as follows:

$$|x| = x, \text{ if } x \text{ is positive or zero,}$$
$$|x| = -x, \text{ if } x \text{ is negative.}$$

We noted that $|x|$ represents the number of units between the point representing x on a number line and the point representing zero. It is likewise true that for real number a, $|x - a|$ represents the number of units between the point representing x on a number line and the point representing a. Furthermore,

$$|x - a| = x - a, \qquad \text{if } x - a \text{ is positive or zero} \quad (\text{if } x - a \geq 0),$$
$$|x - a| = -(x - a), \qquad \text{if } x - a \text{ is negative} \quad (\text{if } x - a < 0).$$

Specifically,

$$|7 - 3| = 7 - 3 = 4,$$
$$|3 - 7| = -(3 - 7) = 4.$$

Here 4 is the number of units between the point representing 7 and the point representing 3.

We can solve an equation such as $|x - 5| = 2$ by "inspection" or by employing the definition of absolute value and then solving the resulting two linear equations. In particular, since the equation $|x - 5| = 2$ states that the point for x must be located 2 units from the point for 5, by inspection we see that x must be $5 + 2 = 7$ or $5 - 2 = 3$. (See Figure 1.7) We get the same solutions if we employ the definition of absolute value. Indeed, to say that

$$|x - 5| = 2$$

means that

$$x - 5 = 2 \quad \text{or} \quad -(x - 5) = 2.$$

Solving these two equations, we have

$$x = 7 \quad \text{or} \quad -x + 5 = 2$$
$$-x = -3$$
$$x = 3.$$

The numbers 7 and 3 are the solutions of $|x - 5| = 2$. Each of the numbers 7 and 3 represents a point on the number line located 2 units from the point representing 5.

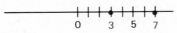

FIGURE 1.7   The solutions of $|x - 5| = 2$

The following example utilizes the fact that the absolute value of a product of real numbers is the product of their absolute values. In other words, if a and b are real numbers,

$$|ab| = |a|\,|b|.$$

EXAMPLE 5   Solve $|3x - 5| = 6$, and graph the set of solutions on a number line.

Solution

Method 1   In order to solve the given equation by inspection, we must write it in the form $|x - a| = c$ where a and c are real numbers. To do this, we write $3x - 5 = 3(x - \frac{5}{3})$ and proceed as follows:

$$|3x - 5| = 6$$
$$|3(x - \tfrac{5}{3})| = 6$$
$$|3|\,|x - \tfrac{5}{3}| = 6 \qquad |ab| = |a|\,|b|$$
$$3|x - \tfrac{5}{3}| = 6$$
$$|x - \tfrac{5}{3}| = 2$$

The equation $|x - \frac{5}{3}| = 2$ states that the point for x must be located 2 units from the point for $\frac{5}{3}$. It follows that x must be $\frac{5}{3} + 2 = \frac{11}{3}$ or $\frac{5}{3} - 2 = \frac{-1}{3}$.

Method 2   To say that

$$|3x - 5| = 6$$

means that $3x - 5 = 6$ or $-(3x - 5) = 6$.

Solving these, we have

$$3x = 11 \qquad -3x + 5 = 6$$
$$x = \tfrac{11}{3} \qquad\quad -3x = 1$$
$$x = \tfrac{-1}{3}.$$

The numbers $\frac{11}{3}$ and $\frac{-1}{3}$ are the solutions of the given equation. They are represented by points on a number line as shown in Figure 1.8.

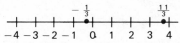

FIGURE 1.8   The solutions of $|3x - 5| = 6$

In the second method that was demonstrated in Example 5, we had to solve *two* linear equations; there were *two* cases to consider. In general, in using the second method to solve $|ax + b| = c$ where a, b, and positive c are specific real numbers and $a \neq 0$, we have to solve two linear equations because $|ax + b|$ is equal to either $ax + b$ or $-(ax + b)$.

Exercise 1.8, 36–45   If $|ax + b| = c$, then $ax + b = c$ or $ax + b = -c$.

Likewise, if $P(x)$ is any polynomial in x with real-number coefficients and if c is a specific positive real number, to solve $|P(x)| = c$ we must consider two equations, because $|P(x)|$ is equal to either $P(x)$ or $-P(x)$.

If $|P(x)| = c$, then $P(x) = c$ or $P(x) = -c$.

EXAMPLE 6   Solve $|x^2 - 5| = 4$, and graph the set of solutions on a number line.

Solution   To say that $|x^2 - 5| = 4$ means that

$$(1)\ x^2 - 5 = 4 \quad \text{or} \quad (2)\ x^2 - 5 = -4.$$

Exercise 1.8, 46–50
Additional exercises,
Exercise 1.8, 51–55 In Case (1), $x^2 = 9$ so $x = 3$ or $x = -3$. In Case (2), $x^2 = 1$ so $x = 1$ or $x = -1$. The solutions of $|x^2 - 5| = 4$ are 3, -3, 1, and -1. They are graphed in Figure 1.9.

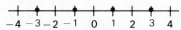

FIGURE 1.9   The solutions of $|x^2 - 5| = 4$

EXERCISE 1.8 RADICAL, FRACTIONAL, AND ABSOLUTE VALUE EQUATIONS

▶ Solve each of the following.

EXAMPLE

$x = 5 + \sqrt{x - 3}$

Solution

$\begin{aligned} x - 5 &= \sqrt{x - 3} \\ x^2 - 10x + 25 &= x - 3 \\ x^2 - 11x + 28 &= 0 \\ (x - 4)(x - 7) &= 0 \\ x = 4, \quad x &= 7 \end{aligned}$

Substitute in original equation.

$x = 4$: $4 = 5 + \sqrt{4 - 3}$ False

$x = 7$: $7 = 5 + \sqrt{7 - 3}$ True

The only solution is 7.

1. $x = 3 - 3\sqrt{x - 3}$ **2.** $\sqrt{5x - 19} - 2\sqrt{x - 4} = 1$

3. $\sqrt{8}\sqrt{x - 2} - x = 2$

4. $\sqrt{28x - 59} = 1 + 2\sqrt{6x - 9}$

5. $x = 1 + \sqrt{x - 19}$

6. $\sqrt{65 - 16x} - 2\sqrt{14 - 5x} = 1$

7. $x = 8 + \sqrt{8x - 16}$ **8.** $\sqrt{12x - 15} = 1 + 2\sqrt{2x}$

9. $x - 2 = 3\sqrt{x - \frac{8}{9}}$ **10.** $x = 2\sqrt{6 - x} - 2$

▶ Write each of the following in the form

$$\frac{x^2}{a^2} + \frac{y^2}{b^2} = 1$$

where a and b are positive real numbers. Give the values of a and b.

11. $\sqrt{(x - 4)^2 + y^2} + \sqrt{(x + 4)^2 + y^2} = 10$

12. $\sqrt{(x - 5)^2 + y^2} + \sqrt{(x + 5)^2 + y^2} = 26$

13. $\sqrt{(x - 12)^2 + y^2} + \sqrt{(x + 12)^2 + y^2} = 26$

14. $\sqrt{(x - 8)^2 + y^2} + \sqrt{(x + 8)^2 + y^2} = 20$

15. $\sqrt{(x - 6)^2 + y^2} + \sqrt{(x + 6)^2 + y^2} = 20$

▶ Each of the following represents a graph called a hyperbola. Write each in the form

$$\frac{x^2}{a^2} - \frac{y^2}{b^2} = 1$$

where a and b are positive real numbers. Give the values of a and b.

16. $\sqrt{(x + 5)^2 + y^2} - \sqrt{(x - 5)^2 + y^2} = 6$

17. $\sqrt{(x + 13)^2 + y^2} - \sqrt{(x - 13)^2 + y^2} = 24$

18. $\sqrt{(x + 5)^2 + y^2} - \sqrt{(x - 5)^2 + y^2} = 8$

19. $\sqrt{(x + 13)^2 + y^2} - \sqrt{(x - \sqrt{13})^2 + y^2} = 4$

20. $\sqrt{(x + \sqrt{5})^2 + y^2} - \sqrt{(x - \sqrt{5})^2 + y^2} = 4$

▶ Each of the following represents a graph called a parabola. Where p is a real number, write each in the form $y^2 = 2px$ or the form $x^2 = 2py$, as indicated. Give the value of $p/2$.

EXAMPLE

$\sqrt{(x - 3)^2 + y^2} = x + 3; \quad y^2 = 2px$

Solution

$$(x - 3)^2 + y^2 = (x + 3)^2$$

$$x^2 - 6x + 9 + y^2 = x^2 + 6x + 9$$

$$y^2 = 12x$$

$$2p = 12$$

$$p = 6, \quad \frac{p}{2} = 3$$

21. $\sqrt{(x - 5)^2 + y^2} = x + 5; \quad y^2 = 2px$

22. $\sqrt{x^2 + (y - 3)^2} = y + 3; \quad x^2 = 2py$

23. $\sqrt{x^2 + (y + 5)^2} = y - 5; \quad x^2 = 2py$

24. $\sqrt{(x - 7)^2 + y^2} = x + 7; \quad y^2 = 2px$

25. $\sqrt{(x + 2)^2 + y^2} = x - 2; \quad y^2 = 2px$

▶ Solve each of the following.

EXAMPLE

$$\frac{x - 2}{x + 2} + \frac{5}{x - 2} = \frac{20}{x^2 - 4}$$

Solution

The LCD is $x^2 - 4 = (x - 2)(x + 2)$.

$$\frac{(x - 2)(x + 2)(x - 2)}{x + 2} + \frac{5(x - 2)(x + 2)}{x - 2} = \frac{20(x^2 - 4)}{x^2 - 4}$$

$$x^2 - 4x + 4 + 5x + 10 = 20 \qquad x = 2\text{: The number 2 is not a solution of}$$

$$x^2 + x - 6 = 0 \qquad\qquad\qquad \text{the original equation.}$$

$$(x - 2)(x + 3) = 0 \qquad x = -3\text{:} \quad \frac{-5}{-1} + \frac{5}{-5} = \frac{20}{9 - 4}$$

$$x = 2, \qquad x = -3$$

$$5 - 1 = 4 \qquad \text{True}$$

The only solution is -3.

26. $\dfrac{x - 1}{x + 1} + \dfrac{3}{x - 1} = \dfrac{6}{x^2 - 1}$

27. $\dfrac{x - 1}{x - 3} = \dfrac{x + 3}{x - 3} - \dfrac{4}{x - 1}$

28. $1 + \dfrac{1}{x^2 - 4} + \dfrac{1}{x + 2} + \dfrac{1}{x - 2} = 0$

29. $\dfrac{x - 2}{x(x + 2)} - \dfrac{2}{x(x - 2)} + \dfrac{8}{x^3 - 4x} = 0$

30. $\dfrac{x + 3}{x - 3} - \dfrac{x - 3}{x + 3} = 1 - \dfrac{x^2}{x^2 - 9}$

31. $1 + \dfrac{1}{x} - \dfrac{8}{x(x + 3)} = 0$

32. $\dfrac{x - 3}{x - 1} + \dfrac{1}{x - 3} = \dfrac{2}{x^2 - 4x + 3}$

33. $2 + \dfrac{x - 2}{x - 1} = \dfrac{2}{x - 1}$

34. $\dfrac{x - 2}{x + 2} - \dfrac{1}{x + 2} + \dfrac{8}{x^2 + 5x + 6} = 0$

35. $\dfrac{x - 2}{x + 1} - \dfrac{3(x - 2)}{x^2 + 3x + 2} = \dfrac{6}{(x + 1)(x + 2)}$

▶ Solve each of the following, and graph the set of solutions on a number line.

EXAMPLE

$|x + 1| = 3$

Solution

$x + 1 = 3$ or $-(x + 1) = 3$

$\qquad x = 2 \qquad\qquad -x - 1 = 3$

$\qquad\qquad\qquad\qquad\qquad -x = 4$

$\qquad\qquad\qquad\qquad\qquad x = -4$

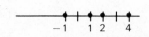

36. $|x + 1| = 5$ **37.** $|2x + 2| = 5$ **38.** $|x - 5| = 8$

39. $|3x - 4| = 6$ **40.** $|x + 9| = 11$ **41.** $|2x + 9| = 11$

42. $|3x - 7| = 4$ **43.** $|2x + 8| = 6$ **44.** $|3x - 10| = 0$

45. $|6x| = 12$

▶ Find the set of real-number solutions of each of the following, and graph it on a number line.

EXAMPLE

$|x^2 - 3x - 1| = 3$

Solution

$x^2 - 3x - 1 = 3$ or $x^2 - 3x - 1 = -3$

$x^2 - 3x - 4 = 0 \qquad\qquad x^2 - 3x + 2 = 0$

$(x - 4)(x + 1) = 0 \qquad\quad (x - 1)(x - 2) = 0$

$x = 4$ or $x = -1 \qquad\quad x = 1$ or $x = 2$

46. $|x^2 - 13| = 3$ **47.** $|x^2 + 3x - 10| = 12$

48. $|x^2 - 7| = 2$ **49.** $|x^2 - 6| = 3$

50. $|x^2 + 2x - 1| = 2$

▶ In each of the following, h, k, a, and b are real numbers, and a and b are positive. Put the given equation in the form indicated and give the values of h, k, a, and b.

51. $\sqrt{(x + 1)^2 + (y - 1)^2} + \sqrt{(x - 5)^2 + (y - 1)^2} = 10;$

$$\frac{(x - h)^2}{a^2} + \frac{(y - k)^2}{b^2} = 1$$

52. $\sqrt{(x + 6)^2 + (y - 1)^2} - \sqrt{(x - 4)^2 + (y - 1)^2} = 6;$

$$\frac{(x - h)^2}{a^2} - \frac{(y - k)^2}{b^2} = 1$$

53. $\sqrt{(x + 6)^2 + (y + 2)^2} - \sqrt{(x - 4)^2 + (y + 2)^2} = 8;$

$$\frac{(x - h)^2}{a^2} - \frac{(y - k)^2}{b^2} = 1$$

54. $\sqrt{(x - 3)^2 + (y - 1)^2} + \sqrt{(x + 5)^2 + (y - 1)^2} = 12;$

$$\frac{(x - h)^2}{a^2} + \frac{(y - k)^2}{b^2} = 1$$

55. Write the equation $\sqrt{(x - 4)^2 + (y - 2)^2} = x + 4$ in the form $(y - k)^2 = 2p(x - h)$ where h, k, and p are real numbers. Give the value of $p/2$.

CHECK LIST FOR THE STUDENT

If you have learned the material in Section 1.8, you should be able to do the following:

1. Tell what is meant by a radical equation.

2. Solve one-variable radical equations in which the radicals are square roots.

3. Write certain radical equations in prescribed standard forms without radicals.

4. Tell what is meant by a fractional equation.

5. Solve fractional equations in one variable.

6. Tell what is meant by an absolute value equation.

7. Solve any equation of the form $|ax + b| = c$ where nonzero a, b, and positive c are specific real numbers; graph the set of solutions on a number line.

8. Solve equations of the form $|P(x)| = c$ where $P(x)$ is a polynomial in x with real-number coefficients and c is a specific positive real number; graph the set of solutions on a number line.

SECTION QUIZ 1.8

1. Solve the following equation.

$$x + \sqrt{x + 28} - 2 = 0$$

2. Write the following equation in the form

$$\frac{x^2}{a^2} + \frac{y^2}{b^2} = 1$$

where a and b are positive real numbers. Give the values of a and b.

$$\sqrt{(x - \sqrt{3})^2 + y^2} + \sqrt{(x + \sqrt{3})^2 + y^2} = 4$$

3. Solve the following equation.

$$\frac{x - 1}{x + 2} - \frac{2(x - 1)}{(x + 2)^2} - \frac{6}{(x + 2)^2} = 0$$

Solve each of the following equations, and graph its set of solutions on a number line.

4. $|3x - 7| = 8$

5. (a) $|x^2 - 2| = 3$
   (b) $|x^2 + x - 2| = 10$

1.9 Linear, Quadratic, and Absolute Value Inequalities

The purpose of this section is to review properties of inequalities and to consider the solution of linear inequalities in one variable, the solution of quadratic inequalities, and the solution of certain inequalities involving absolute values.

A statement that one expression is less than ($<$), greater than ($>$), less than or equal to ($\leq$), or greater than or equal to ($\geq$) another expression is called an **inequality.** Just as in the case for equations, the expression on the left of the inequality symbol is called its *left side;* the expression on the right is called its *right side.* A solution of an inequality in one variable is a real number such that the inequality becomes a true statement when the variable is replaced by that number. A solution of an inequality is said to *satisfy* the inequality. The set of all solutions of an inequality is sometimes called its *solution set.* To *solve* an inequality is to find its solution set. Two inequalities are said to be **equivalent** and each is said to be an *equivalent form* of the other if they have exactly the same solution set.

Inequalities have properties that are very similar to some of the properties of equations. These properties are listed in the following paragraphs. The major difference between the properties of inequalities and the similar properties of equations is that in an inequality, if both sides are multiplied or divided by the same *negative* number, then the sense of the inequality is reversed (changed from $<$ to $>$, etc.).

Some properties of inequalities are as follows:

Property 1   If the same number or polynomial in the variable(s) of an inequality is added to or subtracted from both sides of an inequality, the result is an inequality with exactly the same solution set as the original inequality.

Example: $3x + 4 < 7 - 5x$

has the same solution set as

$$3x + 5x < 7 - 4 \qquad \text{Adding } 5x - 4 \text{ to both sides}$$

Property 2   If both sides of an inequality are multiplied or divided by the same *positive* number, the result is an inequality with exactly the same solution set as the original inequality.

Example: $3x < 6$

has the same solution set as

$$x < 2 \qquad \text{Dividing both sides by the same positive number}$$

Property 3   If both sides of an inequality are multiplied or divided by the same *negative* number *and the sense of the inequality is reversed,* the result is an inequality with exactly the same solution set as the original inequality.

Example: $-3x < 6$

has the same solution set as

$x > -2$ Dividing both sides by the same negative number and reversing the sense of the inequality

A **linear inequality in one variable** is an inequality in which the one variable occurs to the first power and no other power. It can be written in one of the forms

$$ax + b < 0 \qquad ax + b > 0$$
$$ax + b \leq 0 \qquad ax + b \geq 0$$

where $a \neq 0$ and a and b are real numbers. Such inequalities may be solved by using the properties of inequalities, as illustrated in the following example.

EXAMPLE 1

Solution

Solve $-2x - 2 \leq 4 + x$, and graph the set of solutions on a number line.

$-2x - 2 \leq 4 + x$

$-2x - x \leq 4 + 2$ Adding $2 - x$ to both sides

$-3x \leq 6$

$x \geq -2$ Dividing both sides by the same negative number and reversing the sense of the inequality

The graph is shown in Figure 1.10. As in Section 1.1, a heavy dot has been used to show that the endpoint -2 is included on the graph.

Exercise 1.9, 1–10

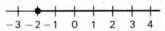

FIGURE 1.10   The solutions of $-2x - 2 \leq 4 + x$; $\{x | x \geq -2\}$

A **quadratic inequality** is an inequality in one variable in which the variable occurs to the second power and to no higher power. A quadratic inequality can be written in one of the "standard" forms

$$ax^2 + bx + c < 0 \qquad ax^2 + bx + c > 0$$
$$ax^2 + bx + c \leq 0 \qquad ax^2 + bx + c \geq 0$$

where a, b, and c are real numbers and $a \neq 0$. If $ax^2 + bx + c$ factors into first-degree factors with integral coefficients, then the quadratic inequality can be solved as illustrated in the next example.

EXAMPLE 2   Solve $x^2 + 2x > 3$, and graph the set of solutions on a number line.

Solution   First write the inequality in standard form.

$$x^2 + 2x - 3 > 0$$

Factor the left side.

$$(x - 1)(x + 3) > 0$$

The inequality now states that the product of the factors $(x - 1)$ and $(x + 3)$ is positive. That means that the two factors must be of the same sign. We may use either one of the following two methods to find the values of x for which this is true.

Method 1   Indicate the signs of each of the factors on a number line, and read off the values of x for which the factors have the same sign. To do this, note that

$$x - 1 \text{ is zero, } x - 1 = 0, \text{ for } x = 1;$$
$$x - 1 \text{ is positive, } x - 1 > 0, \text{ for } x > 1;$$
$$x - 1 \text{ is negative, } x - 1 < 0, \text{ for } x < 1.$$

Also,

$$x + 3 \text{ is zero, } x + 3 = 0, \text{ for } x = -3;$$
$$x + 3 \text{ is positive, } x + 3 > 0, \text{ for } x > -3;$$
$$x + 3 \text{ is negative, } x + 3 < 0, \text{ for } x < -3.$$

These facts are shown on the number line in Figure 1.11.

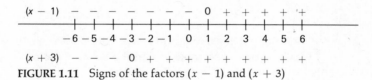

FIGURE 1.11   Signs of the factors $(x - 1)$ and $(x + 3)$

In Figure 1.11 we see that the two factors have the same signs for $x > 1$ or for $x < -3$. The solution set is $\{x \mid x > 1 \text{ or } x < -3\}$. It is graphed in Figure 1.12. As agreed in Section 1.1, open circles are used at the endpoints to show that they are not included on the graph.

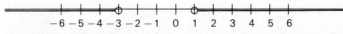

FIGURE 1.12   The solutions of $x^2 + 2x > 3$; $\{x \mid x > 1 \text{ or } x < -3\}$

Method 2   Proceed to analyze the problem algebraically. We seek the values of x for which the two factors $(x - 1)$ and $(x + 3)$ have the same sign. In other words, either

(1) $x - 1$ is positive   or   (2) $x - 1$ is negative
and   and
$x + 3$ is positive $x + 3$ is negative.

In Case 1, x must be such that

$$x - 1 > 0 \quad \text{and} \quad x + 3 > 0.$$

That is,

$$x > 1 \quad \text{and} \quad x > -3.$$

The only numbers x both greater than 1 and also greater than -3 are those x greater than 1. Thus the given inequality is satisfied at least for all x such that

$$x > 1.$$

In Case 2, x must be such that

$$x - 1 < 0 \quad \text{and} \quad x + 3 < 0.$$

That is,

$$x < 1 \quad \text{and} \quad x < -3.$$

The only numbers x both less than 1 and also less than -3 are those x less than -3. Thus the given inequality is satisfied at least for all x such that

$$x < -3.$$

Combining the findings in Cases 1 and 2, we conclude, in agreement with the results of Method 1, that the solution set of the given inequality is

$$\{x \mid x > 1 \text{ or } x < -3\}.$$

If $ax^2 + bx + c$ cannot be factored into first-degree linear factors with integral coefficients as was true in the case illustrated in Example 2, the quadratic inequality can still be solved using the techniques of Example 2 provided that the equation $ax^2 + bx + c = 0$ has real roots. In such instances we use the fact that if r and s are roots of $ax^2 + bx + c = 0$, then $(x - r)$ and $(x - s)$ are factors of $ax^2 + bx + c$ so that

$$ax^2 + bx + c = a(x - r)(x - s).$$

This is a special case of the so-called *factor theorem*. This special case is utilized in the next example.

EXAMPLE 3   Solve $2x^2 - 12x \le -14$, and graph the set of solutions on a number line.

Solution   First write the given inequality in standard form.

$$2x^2 - 12x + 14 \le 0$$
$$2(x^2 - 6x + 7) \le 0$$

Note that $x^2 - 6x + 7$, and hence $2x^2 - 12x + 14$, cannot be factored into linear factors with integral coefficients, but the equation $2x^2 - 12x + 14 = 0$ does have real roots because

$$b^2 - 4ac = 144 - 4(2)(14)$$
$$= 32 \text{ is positive.}$$

The roots of $2x^2 - 12x + 14 = 0$ can be found by using the quadratic formula, and these roots are

$$x = \frac{12 \pm \sqrt{144 - 4(2)(14)}}{4}$$

$$= \frac{12 \pm \sqrt{32}}{4}$$

$$= \frac{12 \pm 4\sqrt{2}}{4}$$

$$= 3 \pm \sqrt{2}.$$

Since $3 + \sqrt{2}$ and $3 - \sqrt{2}$ are roots of $2x^2 - 12x + 14 = 0$,

$$(x - (3 + \sqrt{2})) \quad \text{and} \quad (x - (3 - \sqrt{2}))$$

are factors of $2x^2 - 12x + 14$ so that

$$2x^2 - 12x + 14 = 2(x - 3 - \sqrt{2})(x - 3 + \sqrt{2}).$$

(It would be good practice for the student to verify this by multiplying out the factors.) The given quadratic inequality can now be written

$$2(x - 3 - \sqrt{2})(x - 3 + \sqrt{2}) \le 0,$$

and we can proceed to solve it using the techniques of Example 2.

Method 1   We want to find the values of x for which the two factors $(x - 3 - \sqrt{2})$ and $(x - 3 + \sqrt{2})$ have opposite signs or one of them is zero so that their product will be negative or zero. The factor

$(x - 3 - \sqrt{2})$ is zero, $x - 3 - \sqrt{2} = 0$, for $x = 3 + \sqrt{2}$;
$(x - 3 - \sqrt{2})$ is positive, $x - 3 - \sqrt{2} > 0$, for $x > 3 + \sqrt{2}$;
$(x - 3 - \sqrt{2})$ is negative, $x - 3 - \sqrt{2} < 0$, for $x < 3 + \sqrt{2}$.

The factor

$$(x - 3 + \sqrt{2}) \text{ is zero, } x - 3 + \sqrt{2} = 0, \text{ for } x = 3 - \sqrt{2};$$
$$(x - 3 + \sqrt{2}) \text{ is positive, } x - 3 + \sqrt{2} > 0, \text{ for } x > 3 - \sqrt{2};$$
$$(x - 3 + \sqrt{2}) \text{ is negative, } x - 3 + \sqrt{2} < 0, \text{ for } x < 3 - \sqrt{2}.$$

These facts are shown on the number line in Figure 1.13

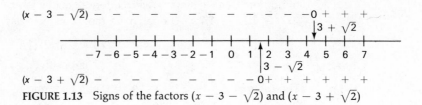

FIGURE 1.13   Signs of the factors $(x - 3 - \sqrt{2})$ and $(x - 3 + \sqrt{2})$

We can see from Figure 1.13 that one of the factors is zero or the factors are of opposite sign for all numbers x between and including $3 - \sqrt{2}$ and $3 + \sqrt{2}$. The solution set of the given inequality is

$$\{x \mid 3 - \sqrt{2} \leq x \leq 3 + \sqrt{2}\}.$$

It is graphed in Figure 1.14.

FIGURE 1.14   The solutions of $2x^2 - 12x \leq -14; \{x \mid 3 - \sqrt{2} \leq x \leq 3 + \sqrt{2}\}$

Method 2   We seek the values of x for which the product

$$(x - 3 - \sqrt{2})(x - 3 + \sqrt{2})$$

is negative or zero. This will be true whenever the factors are of opposite sign or one factor is zero, that is whenever

(1) $(x - 3 - \sqrt{2} \geq 0$ *and* $x - 3 + \sqrt{2} \leq 0$

or

(2) $x - 3 - \sqrt{2} \leq 0$ *and* $x - 3 + \sqrt{2} \geq 0.$

In Case 1, x must be such that

$$x \geq 3 + \sqrt{2} \quad \text{and} \quad x \leq 3 - \sqrt{2}.$$

But there are no numbers x which are both greater than or equal to $3 + \sqrt{2}$ and also less than or equal to $3 - \sqrt{2}$. Case 1 yields no solutions. In Case 2, x must be such that

$$x \leq 3 + \sqrt{2} \quad and \quad x \geq 3 - \sqrt{2}.$$

Written otherwise,

$$3 - \sqrt{2} \leq x \leq 3 + \sqrt{2}.$$

Exercise 1.9, 11–30 In agreement with the results of Method 1, the solution set is $\{x \mid 3 - \sqrt{2} \leq x \leq 3 + \sqrt{2}\}$.

In mathematics we frequently encounter certain inequalities and equations involving absolute values. Some absolute value equations were considered in Section 1.8. Here we shall consider some *absolute value inequalities*—that is, inequalities involving absolute values of variable expressions.

In Section 1.1 we noted that for any real number x, $|x|$ represents the number of units between the point representing x on a number line and the point representing zero. Thus to say that $|x| \leq c$ where c is some specific positive real number is to say that x is located no more than c units from the zero point. In other words, x must be between c and $-c$ or

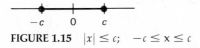

FIGURE 1.15 $|x| \leq c; \quad -c \leq x \leq c$

at most *at* c or $-c$. Otherwise stated, if $|x| \leq c$, then $-c \leq x \leq c$.

$$|x| \leq c \quad \text{provided that} \quad -c \leq x \leq c$$

This is shown in Figure 1.15.

To say that $|x| \geq c$ where c is positive is to say that x is located more than c units from the zero point or possibly exactly c units from the zero point. In other words, x must be to the right of c, or to the left of $-c$, or possibly at c or $-c$. Otherwise stated, if $|x| \geq c$, then $x \geq c$ or $x \leq -c$.

$$|x| \geq c \quad \text{provided that} \quad x \geq c \quad \text{or} \quad x \leq -c$$

This is shown in Figure 1.16.

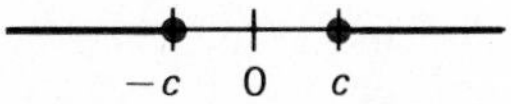

FIGURE 1.16 $|x| \geq c; \quad x \geq c \text{ or } x \leq -c$

EXAMPLE 4

Solve $|x| \leq 3$ and $|x| \geq 3$, and graph the set of solutions of each of them on a separate number line.

Solution

If $|x| \leq 3$, then $-3 \leq x \leq 3$. The graph is shown in Figure 1.17.

If $|x| \geq 3$, then $x \geq 3$ or $x \leq -3$. The graph is shown in Figure 1.18

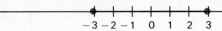

FIGURE 1.17 $|x| \leq 3$

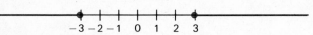

FIGURE 1.18 $|x| \geq 3$

If a, b, and c are specific real numbers, $a \neq 0$, and c is positive, then $|ax + b| \leq c$ means that $(ax + b)$ is between c and $-c$ on a number line or at most at c or $-c$.

$$|ax + b| \leq c \quad \text{provided that} \quad -c \leq ax + b \leq c$$

In other words, $|ax + b| \leq c$ means that

$$ax + b \leq c \quad \text{and also} \quad ax + b \geq -c.$$

Since $|ax + b|$ equals either $ax + b$ or $-(ax + b)$, to say that $|ax + b| \geq c$ means that $ax + b \geq c$ or else $-(ax + b) \geq c$ so that $(ax + b) \leq -c$.

$$|ax + b| \geq c \quad \text{provided that} \quad ax + b \geq c \text{ or } ax + b \leq -c$$

In other words, $|ax + b| \geq c$ means that *either*

$$ax + b \geq c \quad \text{or else} \quad ax + b \leq -c.$$

EXAMPLE 5

Solve $|3x - 2| \leq 4$, and graph the set of solutions on a number line.

Solution

If $|3x - 2| \leq 4$, then

$$-4 \leq 3x - 2 \leq 4.$$

In other words,

$$3x - 2 \leq 4 \quad \text{and also} \quad 3x - 2 \geq -4$$
$$3x \leq 6 \qquad\qquad\qquad 3x \geq -2$$
$$x \leq 2 \qquad\qquad\qquad x \geq \tfrac{-2}{3}$$

The solutions are those numbers x such that $x \leq 2$ and also $x \geq \tfrac{-2}{3}$. Otherwise written,

$$\tfrac{-2}{3} \leq x \leq 2.$$

PART ONE | BASIC ALGEBRA

The set of solutions is graphed in Figure 1.19.

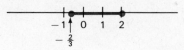

FIGURE 1.19   The solutions of $|3x - 2| \leq 4$; $-\frac{2}{3} \leq x \leq 2$

In this example, the solution set could also have been found by treating the two inequalities simultaneously, as follows:

$$-4 \leq 3x - 2 \leq 4$$

$$-4 + 2 \leq 3x \leq 4 + 2 \qquad \text{Adding 2 to all sides.}$$

$$-2 \leq 3x \leq 6$$

$$\frac{-2}{3} \leq x \leq 2 \qquad \text{Dividing all sides by 3.}$$

EXAMPLE 6

Solve $|3x - 2| \geq 4$, and graph the set of solutions on a number line.

Solution

If $|3x - 2| \geq 4$, then *either*

$$3x - 2 \geq 4 \quad \text{or else} \quad 3x - 2 \leq -4.$$

$$3x \geq 6 \qquad\qquad\qquad 3x \leq -2$$

$$x \geq 2 \qquad\qquad\qquad x \leq \frac{-2}{3}$$

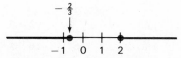

FIGURE 1.20   The solutions of $|3x - 2| \geq 4$; $x \geq 2$ or $x \leq -\frac{2}{3}$

Exercise 1.9, 31–40
Additional Exercises
Exercise 1.9, 41–45
Verbal Problems
Exercise 1.9, 46–55

The solutions are those numbers x such that $x \geq 2$ or else $x \leq \frac{-2}{3}$. This set of solutions is graphed in Figure 1.20.

EXERCISE 1.9 LINEAR, QUADRATIC, AND ABSOLUTE VALUE INEQUALITIES

▶ Solve each of the following, and graph the set of solutions on a number line.

EXAMPLE

$$-5x - 10 \leq 14 + 3x$$

Solution

$$-8x \leq 24$$

$$x \geq -3$$

1. $-3x - 3 \le 5 + 2x$ 2. $2x - 6 \ge 7x + 8$

3. $5x + 5 \le 4x + 5$ 4. $5x + 8 > 4x + 7$

5. $3x + 1 \ge 4x - 4$ 6. $6 + x < 8 + 2x$

7. $2x - 5 < 3x - 4$ 8. $3x + 1 \ge 2x - 2$

9. $8x - 1 \ge 7x + 2$ 10. $4x - 3 \le 3x - 5$

EXAMPLE $x^2 + x \le 6$

Solution $x^2 + x - 6 \le 0$

$(x + 3)(x - 2) \le 0$

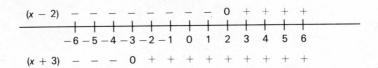

The two factors have opposite signs or one of them is zero for all numbers between and including -3 and 2. The solution set is $\{x \mid -3 \le x \le 2\}$.

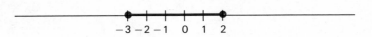

11. $x^2 + x \ge 6$ 12. $x^2 + 2x > 8$ 13. $x^2 + 2x \le 3$

14. $x^2 + 3x - 4 > 0$ 15. $x^2 \le x + 6$ 16. $0 \le x^2 + 3x - 10$

17. $x^2 < 28 - 3x$ 18. $3x + 4 \le x^2$ 19. $x^2 > -3x - 2$

20. $x^2 - 7x + 12 \le 0$

EXAMPLE $x^2 - 2x \ge 1$

Solution   The roots of $x^2 - 2x - 1 = 0$ are $x = \dfrac{2 \pm \sqrt{4 + 4}}{2}$ or $x = 1 \pm \sqrt{2}$. Factors

of $x^2 - 2x - 1$ are $(x - 1 - \sqrt{2})$ and $(x - 1 + \sqrt{2})$.

$$x^2 - 2x - 1 \ge 0$$

$$(x - 1 - \sqrt{2})(x - 1 + \sqrt{2}) \ge 0$$

$(x - 1 - \sqrt{2})$ $-$ $-$ $-$ $-$ $-$ $-$ $-$ $-$ -0 $+$ $+$ $+$

$$\big|1 + \sqrt{2}$$

$$-5 \ -4 \ -3 \ -2 \ -1 \ \Big| \ 0 \ \ 1 \ \ 2 \ \ 3 \ \ 4 \ \ 5$$

$$1 - \sqrt{2}$$

$(x - 1 + \sqrt{2})$ $-$ $-$ $-$ $-$ $-0+$ $+$ $+$ $+$ $+$ $+$

The two factors have the same sign or one of them is zero for all numbers greater than or equal to $1 + \sqrt{2}$ or for all numbers less than or equal to $1 - \sqrt{2}$. The solution set is $\{x \mid x \geq 1 + \sqrt{2} \text{ or } x \leq 1 - \sqrt{2}\}$.

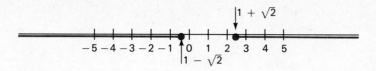

21. $x^2 - 2x \leq 1$ 22. $x^2 - 4x \leq -1$ 23. $x^2 - 2x > 2$

24. $x^2 \geq 6x - 7$ 25. $x^2 \leq -2x + 1$ 26. $x^2 + 4x \geq -1$

27. $x^2 - 2x - 4 \leq 0$ 28. $0 \leq x^2 - 10x + 23$ 29. $x^2 > -2x + 2$

30. $x^2 < 4x + 1$

EXAMPLE

$|3x + 7| \leq 2$

Solution

$-2 \leq 3x + 7 \leq 2$

$-9 \leq 3x \quad\quad \leq -5$

$-3 \leq x \quad\quad \leq -\frac{5}{3}$

31. $|3x + 7| \geq 2$ 32. $|2x - 1| \leq 3$ 33. $|2x - 1| \geq 3$

34. $|4x| > 8$ 35. $|3x| < 6$ 36. $|4x + 3| \geq 1$

37. $|4x + 3| \leq 1$ 38. $|2x - 5| \leq 3$ 39. $|2x - 5| \geq 3$

40. $|5x - 1| \leq 11$

► Solve each of the following.

EXAMPLE

$\dfrac{x - 5}{x + 3} < 8$

Solution

Case 1 If $x + 3 > 0$, $x > -3$ and $x - 5 < 8(x + 3)$.

Case 2 If $x + 3 < 0$, $x < -3$ and $x - 5 > 8(x + 3)$.

Case 1 $x > -3$ and $x < 8x + 29$

$\quad\quad\quad -7x < 29$

$\quad\quad\quad\quad x > -4\frac{1}{7}$

$x > -3$ and $x > -4\frac{1}{7}$ so $x > -3$

Case 2 $x < -3$ and $x > 8x + 29$

$\quad\quad\quad -7x > 29$

$\quad\quad\quad\quad x < -4\frac{1}{7}$

$x < -3$ and $x < -4\frac{1}{7}$ so $x < -4\frac{1}{7}$.

The solutions are those x for which $x > -3$ or $x < -4\frac{1}{7}$.

EXAMPLE $\dfrac{x + 2}{x - 4} \geq 6$

Solution

Case 1   If $x - 4 > 0$, $x > 4$ and $x + 2 \geq 6(x - 4)$.

Case 2   If $x - 4 < 0$, $x < 4$ and $x + 2 \leq 6(x - 4)$.

Case 1 $x > 4$ and $x \geq 6x - 26$

$$-5x \geq -26$$

$$x \leq 5\tfrac{1}{5}$$

$x > 4$ and $x \leq 5\tfrac{1}{5}$ so $4 < x \leq 5\tfrac{1}{5}$.

Case 2 $x < 4$ and $x \leq 6x - 26$

$$-5x \leq -26$$

$$x \geq 5\tfrac{1}{5}$$

$x < 4$ and $x \geq 5\tfrac{1}{5}$ is impossible.

No solutions

The solutions are those x for which $4 < x \leq 5\tfrac{1}{5}$.

41. $\dfrac{x - 6}{x + 2} < 9$ **42.** $\dfrac{x + 3}{x - 5} > 7$

43. $\dfrac{x + 1}{x - 5} \leq 2$ **44.** $\dfrac{x + 5}{x + 2} < 0$

45. $\dfrac{x}{x - 3} > 0$

46. If a family can save $1.20 a month by using a Cycle Defrost refrigerator rather than a No Frost refrigerator, in how many full months can they save at least $50?

47. If a truck gets 3.6 kilometers to a liter of gas, how many full liters of gas will be needed for the truck to go at least 320 kilometers?

48. George is making bread for a church supper. Each batch takes $4\tfrac{1}{2}$ cups of flour. He has only 23 cups of flour. How many batches of bread can he make?

49. Hank programmed a computer to locate a date for him for a dance. He specified that the absolute value of the difference in their ages must be no more than 5. If Hank is 31, what are possible ages for a date located for him by his program?

50. A college graduate wants at least 8 weeks to take a trip, and he is interviewing for a job with a company that has the following vacation policy. If an employee works for one full year, he gets one week vacation. For each five full years he works after that he gets an additional week of vacation. How many full years would an employee have to work before he would be getting at least 8 weeks vacation?

51. Delilah has $300 in an account that pays her 6% of $300 in interest every full year. At the end of each full year she withdraws just the interest. In how many full years will she have withdrawn at least as much as the $300 she keeps in the account?

52. Marian needs at least $8000 a year to put her nephew through college. She can get $6\frac{3}{4}$% of $1000 interest on a $1000 bond each full year. What is the least number of such bonds that will produce annual interest exceeding $8000?

53. Jane can afford to pay no more than $2\frac{1}{2}$ times her annual salary for a house. What is the smallest annual salary that will permit her to afford a $60,000 house?

54. The annual finance charge on a charge account is 18% of the price of the purchases. If Ginger can afford to pay no more than $450 per year in finance charges, what is the largest purchase price she can afford?

55. It will cost a company $125,000 to set up production for a new camera. Once the camera is in production, it will cost $25 to make and distribute each camera. If the cameras sell for $65 each, how many must the company sell in order for their profit to exceed $500,000?

CHECK LIST FOR THE STUDENT

If you have learned the material in Section 1.9, you should be able to do the following:

1. Define inequality, a solution of an inequality, equivalent inequalities, and an equivalent form of an inequality.

2. Tell how the solutions of a given inequality are related to the solutions of a new inequality obtained from the given inequality by doing each of the following:

(a) Adding or subtracting the same polynomial in its variable(s) to or from both sides.

(b) Multiplying or dividing both sides by the same positive number.

(c) Multiplying or dividing both sides by the same negative number and reversing the sense of the inequality.

3. Give an example of a linear inequality in one variable; solve any linear inequality in one variable, and graph the set of solutions on a number line.

4. Give an example of a quadratic inequality; solve quadratic inequalities in which the quadratic is factorable, and graph the set of solutions on a number line.

5. Use the factor theorem and solve quadratic inequalities.

6. Tell what is meant by an absolute value inequality.

7. Solve inequalities of the form $|ax + b| \leq c$ or of the form $|ax + b| \geq c$ where nonzero a, b, and positive c are specific real numbers; graph the set of solutions on a number line.

SECTION QUIZ 1.9

Solve each of the following, and graph its set of solutions on a number line.

1. $-9x - 2 \leq 6 - 5x$ 2. $x^2 - 3x - 4 \leq 0$ 3. $x^2 + 1 > 4x$

4. $|2x - 5| < 7$ 5. $|3x - 7| \geq 2$

1.10 Simultaneous Linear Equations

The purpose of this section is to show how to solve two simultaneous linear equations in two variables and to demonstrate the solution of n simultaneous linear equations in n variables for $n > 2$.

A **linear equation in two variables** is a first-degree polynomial equation containing exactly two variables. The equation

$$3x + 2y + 5 = 0$$

is a linear equation in x and y. The equation

$$5x - 6y = 0$$

is also a linear equation in x and y.

A **linear equation in n variables** is a first-degree polynomial equation containing exactly n variables. The equation

$$7x - 8y + 4z - 12 = 0$$

is a linear equation in x, y, and z. The equation

$$7A + 5B - C + D - 3E = -11$$

is a linear equation in A, B, C, D, and E.

A **solution** of a linear equation in n variables is a set of values for the n variables that makes the equation a true statement. A solution of a linear equation is said to *satisfy* the equation. For example, $x = 1$, $y = -4$ is a solution of

$$3x + 2y + 5 = 0$$

because
$$3(1) + 2(-4) + 5 = 0$$

is a true statement. Also, $A = 1$, $B = -2$, $C = 3$, $D = 1$, and $E = 2$ is a solution of

$$7A + 5B - C + D - 3E = -11$$

because
$$7(1) + 5(-2) - 3 + 1 - 3(2) = -11$$

is a true statement.

Any single linear equation in n variables with $n > 1$ has an unlimited number of solutions. A solution may be found simply by assigning values to $(n - 1)$ of the variables and solving the resulting linear equation in one variable for that variable. For example, we may find a solution of

$$7x - 8y + 4z - 12 = 0$$

by assigning values to x and y and then solving for z. In particular, if we let $x = 0$ and $y = 0$, we have

$$0 - 0 + 4z - 12 = 0$$
$$4z = 12$$
$$z = 3.$$

Thus $x = 0$, $y = 0$, $z = 3$ is one solution of $7x - 8y + 4z - 12 = 0$. Again, if we let $x = 0$ and $y = 1$, we have

$$0 - 8 + 4z - 12 = 0$$
$$4z = 20$$
$$z = 5.$$

The values $x = 0$, $y = 1$, and $z = 5$ constitute another solution of $7x - 8y + 4z - 12 = 0$.

By a *system of n simultaneous linear equations in n variables* with $n > 1$, we mean a set of n linear equations collectively containing terms of first degree in each of the n variables and containing no other variable terms. The pair of equations

$$\begin{cases} 3x + 2y + 5 = 0 \\ 5x - 4y + 12 = 0 \end{cases}$$

is an example of a system of two simultaneous equations in the two variables x and y. The pair of equations

$$\begin{cases} 3x + 2y + 5 = 0 \\ 8x \qquad - 6 = 0 \end{cases}$$

is also an example of a system of two simultaneous equations in the two variables x and y. The four equations

$$\begin{cases} 3A - 2B + C - D = 7 \\ \qquad\quad B + 3C - 4D = 5 \\ A - 3B \qquad\qquad = 4 \\ 4A + 2B \qquad - 5D = 2 \end{cases}$$

make up a system of four simultaneous linear equations in the four variables A, B, C, and D.

In working with simultaneous linear equations, it is customary to call the variables "unknowns." We may say that the system of four equations in the previous paragraph is a system of four simultaneous linear equations in the four *unknowns A, B, C,* and *D*. Similarly, we may say that the system

$$\begin{cases} 3x + 2y + 5 = 0 \\ 5x - 4y + 12 = 0 \end{cases}$$

is a system of two simultaneous linear equations in the two *unknowns x* and *y*.

A solution of a system of *n* simultaneous linear equations in *n* unknowns with $n > 1$ is a set of values of the unknowns that satisfies each and every one of the *n* equations. For example, $x = -2$, $y = \frac{1}{2}$ is a solution of the system

$$\begin{cases} 3x + 2y + 5 = 0 \\ 5x - 4y + 12 = 0 \end{cases}$$

because $x = -2$, $y = \frac{1}{2}$ satisfies both equations.

$$\begin{cases} 3(-2) + 2(\tfrac{1}{2}) + 5 = 0 \quad \text{True} \\ 5(-2) - 4(\tfrac{1}{2}) + 12 = 0 \quad \text{True} \end{cases}$$

Also, $A = 1$, $B = -1$, $C = 2$, $D = 0$ is a solution of the system

$$\begin{cases} 3A - 2B + C - D = 7 \\ B + 3C - 4D = 5 \\ A - 3B = 4 \\ 4A + 2B - 5D = 2 \end{cases}$$

because these values satisfy each and every equation of the system.

$$\begin{cases} 3(1) - 2(-1) + 2 - 0 = 7 \quad \text{True} \\ (-1) + 3(2) - 4(0) = 5 \quad \text{True} \\ 1 - 3(-1) = 4 \quad \text{True} \\ 4(1) + 2(-1) - 5(0) = 2 \quad \text{True} \end{cases}$$

It can be shown that a system of *n* simultaneous linear equations in *n* unknowns has either

1. Exactly one solution,

2. No solution, or

3. An unlimited number of solutions.

Criteria for determining which of these three applies to any given system of *n* simultaneous linear equations can be developed. In this section we shall consider systems having exactly one solution, and we shall show how to solve them.

First let us review two methods of solving a system of two simultaneous linear equations in two unknowns. The idea in each of the two methods is to eliminate one of the variables and thereby reduce the problem to that of solving a linear equation in one variable.

Method 1 *Substitution*

Step 1 Solve either one of the equations for one unknown in terms of the other.

Step 2 Substitute for that one unknown in the other equation.

Step 3   Solve the resulting linear equation in one variable.

Step 4   Substitute the value obtained in Step 3 in either one of the given equations, and solve for the remaining unknown.

To illustrate, let us solve the following system by substitution.

$$\begin{cases} 4x + 3y + 7 = 0 \\ 5x - 6y + 12 = 0 \end{cases}$$

Step 1   Suppose that we solve the first equation for y in terms of x.

$$3y = -4x - 7$$
$$y = \tfrac{-4}{3}x - \tfrac{7}{3}$$

Step 2   Substitute $\tfrac{-4}{3}x - \tfrac{7}{3}$ for y in the second equation.

$$5x - 6(\tfrac{-4}{3}x - \tfrac{7}{3}) + 12 = 0$$

Step 3   Solve for x.

$$5x - 6(\tfrac{-4}{3})x - 6(\tfrac{-7}{3}) + 12 = 0$$
$$5x + 8x + 14 + 12 = 0$$
$$13x = -26$$
$$x = -2$$

Step 4   Substitute -2 for x in the first equation of the given system, and solve for y.

$$4(-2) + 3y + 7 = 0$$
$$-8 + 3y + 7 = 0$$
$$3y = 1$$
$$y = \tfrac{1}{3}$$

The solution is $x = -2$, $y = \tfrac{1}{3}$. The solution should be checked by substituting it in both the given equations.

Check $\begin{cases} 4x + 3y + 7 = 0 \\ 5x - 6y + 12 = 0 \end{cases}$ $\begin{array}{l} 4(-2) + 3(\tfrac{1}{3}) + 7 = 0 \\ 5(-2) - 6(\tfrac{1}{3}) + 12 = 0 \end{array}$ $\begin{array}{l} \text{True} \\ \text{True} \end{array}$

Method 2   *Choice of Multiplier and Addition*

Step 1   Decide which unknown is to be eliminated.

Step 2   Choose a multiplier for each equation so that after each equation is multiplied through by its multiplier, the coefficients of the unknown to be eliminated will be the same except for sign.

Step 3   Multiply the equations through by their multipliers, and add. (This will eliminate the unknown to be eliminated.)

Step 4   Solve for the remaining variable.

Step 5   Substitute the value obtained in Step 4 in either of the given equations and solve for the other variable.

For example, let us solve the system

$$\begin{cases} 4x + 3y + \ 7 = 0 \\ 5x - 6y + 12 = 0 \end{cases}$$

by choice of multiplier and addition.

Step 1   Suppose that we decide to eliminate y.

Step 2   Multipliers 2 (for the first equation) and 1 (for the second equation) will make the coefficients of y become 6 and -6, the same except for sign. (Other multipliers will also make the coefficients of y become the same except for sign. For example, multipliers 4 for the first equation and 2 for the second equation will make the coefficients of y become 12 and -12. We may use any convenient multipliers that make the coefficients of the unknown to be eliminated become the same except for sign.)

Step 3

$$\begin{array}{ll} 2 \begin{vmatrix} 4x + 3y + \ 7 = 0 \end{vmatrix} & \quad 8x + 6y + 14 = 0 \\ 1 \begin{vmatrix} 5x - 6y + 12 = 0 \end{vmatrix} & \quad \underline{5x - 6y + 12 = 0} \\ & \quad 13x \qquad\ \ + 26 = 0 \end{array}$$

Step 4

$$13x + 26 = 0$$
$$13x = -26$$
$$x = -2$$

Step 5   Substitute -2 for x in the first of the given equations, and solve for y.

$$4(-2) + 3y + 7 = 0$$
$$-8 + 3y + 7 = 0$$
$$3y = 1$$
$$y = \tfrac{1}{3}$$

The solution is $x = -2$, $y = \tfrac{1}{3}$.

Exercise 1.10, 1–12

This agrees with the solution found by using Method 1. It was checked by substituting it in both the given equations.

If a system of three simultaneous linear equations in three unknowns has exactly one solution, it can be solved by adapting the elimination methods just used and systematically eliminating one of the three unknowns in order to reduce the problem to that of solving a pair of simultaneous linear equations in two unknowns. This is illustrated in the next example.

EXAMPLE 1   Find the solution of the following system.

$$\begin{cases} 3x - 2y + 6z = -6 \\ -3x + 10y + 11z = 13 \\ x - 2y - z = -3 \end{cases}$$

Solution   Suppose that we decide to eliminate z from the given system and reduce the problem to that of solving a pair of linear equations in x and y.

Step 1   Decide to eliminate z. We may now use either the method of substitution or the method of choice of multiplier and addition to eliminate z from the first and second, or from the first and third, or from the second and third of the given equations. We only need to eliminate z from two of these three pairs, and we will have a new pair of linear equations in the two unknowns x and y.

Step 2   Choose two pairs of the given equations, and eliminate z from each pair. Suppose we choose the first and third and also the second and third, and eliminate z from each of these two pairs.

$$\begin{array}{r|l} 1 & 3x - 2y + 6z = -6 \\ 6 & x - 2y - z = -3 \end{array} \qquad \begin{array}{r|l} 1 & -3x + 10y + 11z = 13 \\ 11 & x - 2y - z = -3 \end{array}$$

$$\begin{array}{r} 3x - 2y + 6z = -6 \\ 6x - 12y - 6z = -18 \\ \hline 9x - 14y = -24 \end{array} \qquad \begin{array}{r} -3x + 10y + 11z = 13 \\ 11x - 22y - 11z = -33 \\ \hline 8x - 12y = -20 \end{array}$$

Step 3   Solve the resulting pair of linear equations in x and y.

$$\begin{cases} 9x - 14y = -24 \\ 8x - 12y = -20 \end{cases}$$

1. Eliminate x; solve for y.

$$\begin{array}{r|l} 8 & 9x - 14y = -24 \\ -9 & 8x - 12y = -20 \end{array} \qquad \begin{array}{r} 72x - 112y = -192 \\ -72x + 108y = 180 \\ \hline -4y = -12 \\ y = 3 \end{array}$$

2. Substitute 3 for y in $9x - 14y = -24$, and solve for x.

$$\begin{aligned} 9x - 14(3) &= -24 \\ 9x - 42 &= -24 \\ 9x &= 18 \\ x &= 2 \end{aligned}$$

The solution of the resulting pair of equations in x and y is $x = 2$, $y = 3$.

Check $\begin{cases} 9x - 14y = -24 \\ 8x - 12y = -20 \end{cases}$ $\begin{aligned} 9(2) - 14(3) &= -24 \\ 8(2) - 12(3) &= -20 \end{aligned}$ True
    True

Step 4 Substitute the values found for x and y in any one of the original three equations, and solve for z.

$$3x - 2y + 6z = -6$$
$$3(2) - 2(3) + 6z = -6$$
$$6 - 6 + 6z = -6$$
$$z = -1$$

The solution of the given system of three equations in three unknowns is $x = 2$, $y = 3$, $z = -1$.

Check

$3x - 2y + 6z = -6 \qquad 3(2) - 2(3) + 6(-1) = -6$ True
$-3x + 10y + 11z = 13 \qquad -3(2) + 10(3) + 11(-1) = 13$ True
$x - 2y - z = -3 \qquad (2) - 2(3) - (-1) = -3$ True

Exercise 1.10, 13–18

If a system of n simultaneous linear equations in n unknowns has exactly one solution, even when n is greater than three the system can be solved by adapting the elimination methods just presented. This can be done by systematically eliminating one of the unknowns in order to reduce the problem to that of solving $(n - 1)$ simultaneous linear equations in $(n - 1)$ unknowns. This process can be repeated until the problem is reduced to that of solving a pair of linear equations in two unknowns. In Part 3 of this text, more sophisticated and usually more direct methods for solving n equations in n unknowns for $n > 2$ will be presented. The elimination methods presented in this section are apt to be quite tedious for any large n. If some of the unknowns do not appear in some of the given equations, however, these elimination methods might produce the solution quite readily. This is illustrated for $n = 5$ in the next example.

EXAMPLE 2

Find the solution of the following system.

$$\begin{cases} A + C = 4 \\ 3A + B + 2C + D = 6 \\ 3A + 3B + C + D + E = 3 \\ A + 3B = -5 \\ B = -2 \end{cases}$$

Solution

We can solve this system by repeatedly using the substitution method. From the last equation we see that $B = -2$. Substituting -2 for B in the fourth equation and solving for A, we have

$$A + 3(-2) = -5$$
$$A - 6 = -5$$
$$A = 1.$$

Substituting 1 for A in the first equation and solving for C, we have

$$1 + \quad C = 4$$
$$C = 3.$$

Now substituting $A = 1$, $B = -2$, $C = 3$ in the second equation and solving for D, we have

$$3(1) + (-2) + 2(3) + D = 6$$
$$3 - \quad 2 + \quad 6 \quad + D = 6$$
$$D = -1.$$

Finally, substituting $A = 1$, $B = -2$, $C = 3$, and $D = -1$ in the third equation, we can solve for the remaining unknown E.

$$3(1) + 3(-2) + 3 + (-1) + E = 3$$
$$3 - \quad 6 \quad + 3 \quad -1 + E = 3$$
$$E = 4$$

The solution of the given system is $A = 1$, $B = -2$, $C = 3$, $D = -1$, $E = 4$. This can be checked by substituting in each of the given equations.

Check

$A \quad + C \quad = 4$	$1 \quad + 3 \quad = 4$	True
$3A + B + 2C + D = 6$	$3(1) + (-2) + 2(3) + (-1) = 6$	True
$3A + 3B + C + D + E = 3$	$3(1) + 3(-2) + 3 + (-1) + 4 = 3$	True
$A + 3B \quad = -5$	$1 + 3(-2) \quad = -5$	True
$B \quad = -2$	$-2 \quad = -2$	True

Exercise 1.10, 19–25

Systems of n simultaneous linear equations in n unknowns such as the system presented in the previous example arise in many different contexts. For instance, such a system may arise in decomposing a given rational expression into a sum of partial fractions, that is fractions with lower degree denominators than the given rational expression. For example, the so-called *partial fraction theorem*, to be presented in Section 3.5, assures us that the expression

$$\frac{6x^2 + 7x - 4}{x(x - 1)(x + 2)}$$

can be expressed as a sum of partial fractions of the form

$$\frac{A}{x} + \frac{B}{x - 1} + \frac{C}{x + 2}$$

where A, B, and C represent particular real numbers. The actual values of

A, B, and C may be found by solving a system of three simultaneous linear equations in three unknowns. This is illustrated in the next example.

EXAMPLE 3 Assume that

$$\frac{6x^2 + 7x - 4}{x(x - 1)(x + 2)} = \frac{A}{x} + \frac{B}{x - 1} + \frac{C}{x + 2},$$

and find the values of A, B, and C.

Solution Combining the fractions on the right side of the given equation, we have

$$\frac{6x^2 + 7x - 4}{x(x - 1)(x + 2)} = \frac{A(x - 1)(x + 2) + Bx(x + 2) + Cx(x - 1)}{x(x - 1)(x + 2)}.$$

This equation states that two fractions with the same denominator are equal. In order for this equality to be true, we must have the two numerators equal. Equating the numerators, we have

$$6x^2 + 7x - 4 = A(x - 1)(x + 2) + Bx(x + 2) + Cx(x - 1).$$

We may now determine the values of A, B, and C by either one of two methods. The method to be presented first happens to be the longer method for this particular problem, but it is the method that applies to more general situations than the situation dictated by the denominator in this particular problem.

Method 1 *Equating Coefficients of Like Powers*
First multiply out the right side of

$$6x^2 + 7x - 4 = A(x - 1)(x + 2) + Bx(x + 2) + Cx(x - 1)$$

and collect like terms.

$$6x^2 + 7x - 4 = A(x^2 + x - 2) + Bx^2 + 2Bx + Cx^2 - Cx$$
$$= (A + B + C)x^2 + (A + 2B - C)x - 2A$$

In order for this to be true, the numbers A, B, and C must be such that the coefficients of like powers are equal. Equating the coefficients of x^2, equating the coefficients of x, and equating the constant terms, we get a system of three simultaneous linear equations in the three unknowns A, B, and C.

$$\begin{cases} A + B + C = 6 \\ A + 2B - C = 7 \\ \qquad\quad -2A = -4 \end{cases}$$

Now we must solve this system for A, B, and C. From the third equation we get

$$A = 2.$$

Substituting 2 for A in the first two equations, we get a pair of linear equations in the two unknowns B and C.

$$\begin{cases} B + C = 4 \\ 2B - C = 5 \end{cases}$$

Solving these by choice of multiplier and addition, we have

$$\begin{array}{r|l} 1 & B + C = 4 \\ 1 & 2B - C = 5 \\ \hline & 3B \quad\;\; = 9 \\ & B \quad\;\;\; = 3 \\ \\ & 3 + C = 4 \\ & C = 1. \end{array}$$

It follows that $A = 2$, $B = 3$, and $C = 1$ so that

$$\frac{6x^2 + 7x - 4}{x(x - 1)(x + 2)} = \frac{2}{x} + \frac{3}{x - 1} + \frac{1}{x + 2}.$$

Method 2   *Substituting Convenient Values*
In the equation

$$6x^2 + 7x - 4 = A(x - 1)(x + 2) + Bx(x + 2) + Cx(x - 1)$$

that we obtained when we first equated the numerators, if we successively let x take on values that make each of the three distinct linear factors x, $(x - 1)$, and $(x + 2)$ become zero, we will be able to compute A, B, and C rather easily. This process will work easily as long as the denominator in the original fraction that is to be decomposed into partial fractions has only unrepeated linear factors, as it does in this example. To illustrate, first let $x = 0$, then let $x = 1$, then let $x = -2$.

$x = 0$:

$$6(0)^2 + 7(0) - 4 = A(-1)(2) + B(0)(2) + C(0)(-1)$$
$$-4 = -2A$$
$$2 = A$$

$x = 1$:

$$6(1)^2 + 7(1) - 4 = A(0)(3) + B(1)(3) + C(1)(0)$$
$$9 = 3B$$
$$3 = B$$

$x = -2$:

$$6(-2)^2 + 7(-2) - 4 = A(-3)(0) + B(-2)(0) + C(-2)(-3)$$
$$24 \quad - \quad 14 \quad - 4 = 6C$$
$$6 = 6C$$
$$1 = C$$

Thus we find that $A = 2$, $B = 3$, and $C = 1$ in agreement with the results of Method 1. As was stated earlier, although Method 1 is longer for this particular example, it can be used in more general situations than Method 2.

Exercise 1.10, 26–35

EXAMPLE 4

A man is on a diet that requires him to consume exactly 1500 calories per day in exactly three meals. The number of calories he consumes at dinner must equal the total number of calories he consumes at breakfast and lunch. The number of calories he consumes at lunch must be 150 more than twice the number he consumes at breakfast. How many calories must he consume at each meal?

Solution

Let x, y, and z be the number of calories he consumes daily at breakfast, lunch, and dinner, respectively. According to the requirements of the diet,

$$x + y + z = 1500$$
$$z = x + y$$
$$y = 2x + 150.$$

Rewriting these equations, we have the following system of three linear equations in the three unknowns x, y, and z.

$$\left\{ \begin{array}{l} x + y + z = 1500 \\ x + y - z = 0 \\ -2x + y = 150 \end{array} \right.$$

We must solve this system for x, y, and z. Suppose that we decide to eliminate z. Adding the first two equations, we have

$$2x + 2y = 1500.$$

This equation and the third equation of the system, $-2x + y = 150$, make up a system of two linear equations in the two unknowns x and y.

$$\left\{ \begin{array}{l} 2x + 2y = 1500 \\ -2x + y = 150 \end{array} \right.$$

Eliminating x by adding, we have

$$3y = 1650$$
$$y = 550.$$

Substituting this value of y in the equation $-2x + y = 150$ and solving for x, we have

$$-2x + 550 = 150$$
$$-2x = -400$$
$$x = 200.$$

Substituting $x = 200$ and $y = 550$ in the second equation of the original system and solving for z, we have

$$200 + 550 - z = 0$$
$$z = 750.$$

The man must consume 200 calories at breakfast, 550 calories at lunch, and 750 calories at dinner.

Exercise 1.10, 36–45

EXERCISE 1.10 SIMULTANEOUS LINEAR EQUATIONS

▶ Find the solution of each of the following systems.

EXAMPLE

$$\begin{cases} A + B + C = 3 \\ 2A - 3B + 4C = 31 \\ 6A = 6 \end{cases}$$

Solution

$$\begin{aligned} 6A &= 6 \\ A &= 1 \end{aligned} \quad \begin{cases} 1 + B + C = 3 \\ 2 - 3B + 4C = 31 \end{cases} \quad \begin{cases} B + C = 2 \\ -3B + 4C = 29 \end{cases} \begin{array}{c} 3 \\ 1 \end{array} \left| \begin{array}{l} B + C = 2 \\ -3B + 4C = 29 \end{array} \right.$$

$$\begin{array}{r} 3B + 3C = 6 \\ -3B + 4C = 29 \\ \hline 7C = 35 \\ C = 5 \end{array}$$

$$A + B + C = 3$$
$$1 + B + 5 = 3$$
$$B = -3 \qquad \text{The solution is } A = 1,\ B = -3,\ C = 5.$$

1. $\begin{cases} 2x - 7y = 16 \\ 3x + 4y = -5 \end{cases}$ **2.** $\begin{cases} 2A + 3B = -2 \\ -A - 2B = 2 \end{cases}$

3. $\begin{cases} 4A - 6B = -36 \\ 3A + 8B = 24 \end{cases}$ **4.** $\begin{cases} x - 6y - 20 = 0 \\ 2x + 5y = -11 \end{cases}$

5. $\begin{cases} 2A + 4B - 2 = 0 \\ 3A + 9B - 6 = 0 \end{cases}$ **6.** $\begin{cases} 5x + 4y = -6 \\ -2x + 3y - 7 = 0 \end{cases}$

7. $\begin{cases} x + y = -1 \\ x - y = 17 \end{cases}$ **8.** $\begin{cases} 7A - 8B = -5 \\ -5A + 4B = 1 \end{cases}$

9. $\begin{cases} 8x - 5y - 17 = 0 \\ 4x + 3y - 3 = 0 \end{cases}$

10. $\begin{cases} 2A + B - 8 = 0 \\ 3A = 23 + 4B \end{cases}$

11. $\begin{cases} 11A + 8B + 52 = 0 \\ 9A = B - 35 \end{cases}$

12. $\begin{cases} 15A - 27B + 3 = 0 \\ 3A + 9B = 9 \end{cases}$

13. $\begin{cases} 3x - 2y + z = -7 \\ 4x + 5y + 2z = 8 \\ x + 3y - 4z = 19 \end{cases}$

14. $\begin{cases} A - 2B + 3C = -4 \\ 4A + 3B - 2C = 20 \\ A + 3B - 5C = 14 \end{cases}$

15. $\begin{cases} A + B + C = 1 \\ 2A - 3B - 1 = 0 \\ A + 4B - C = -8 \end{cases}$

16. $\begin{cases} 2x - 5y + 3z = 18 \\ x + 7y - 2z = -5 \\ -x + 4y - 3z = -15 \end{cases}$

17. $\begin{cases} 3x + 2y - 5z + 20 = 0 \\ -x - y + 4z = 15 \\ 2x + 3y = 5 \end{cases}$

18. $\begin{cases} x + y + a = 5 \\ x + y = 0 \\ y + z = 1 \end{cases}$

19. $\begin{cases} x + y + 2z - w = 4 \\ -x + 2y + z + w = 2 \\ 2x - y + 3z + 2w = 16 \\ 2x - 3y + 2z + 2w = 15 \end{cases}$

20. $\begin{cases} A + B = 4 \\ B + 3C + D = 15 \\ 2A + 3B - C - D = 3 \\ C + 4D = 12 \end{cases}$

21. $\begin{cases} A + B + C + D = 18 \\ B + C + D = 11 \\ A + B + C = 16 \\ A - 2B = -1 \end{cases}$

22. $\begin{cases} 3x - 2y + z + 2w = 23 \\ 3y - z - w = -6 \\ z + w = 9 \\ 2x + 4y - 3z - 2w = -10 \end{cases}$

23. $\begin{cases} 3A + B = 2 \\ A - 2B + 3C - D + 2E = 17 \\ B - 3C + 2D + 4E = 1 \\ A + 4C = 9 \\ 3C = 6 \end{cases}$

24. $\begin{cases} A - B + C - D + E = 0 \\ 2E = 2 \\ -C + D - E = 4 \\ B - 2C = 6 \\ A + 3C = -5 \end{cases}$

25. $\begin{cases} x + y + z + w + q = 0 \\ y + z + w + q = -1 \\ z + w + q = -1 \\ w + q = 0 \\ q = -2 \end{cases}$

In each of the following, find the values of the capital letters by two different methods.

EXAMPLE

$$\frac{x + 21}{(3x - 1)(x + 5)} = \frac{A}{3x - 1} + \frac{B}{x + 5}$$

Solution

Method 1 $x + 21 = A(x + 5) + B(3x - 1)$

$$= (A + 3B)x + (5A - B)$$

$$
\begin{array}{c|l}
1 & A + 3B = 1 \\
3 & 5A - B = 21
\end{array}
\qquad
\begin{array}{r}
A + 3\cancel{B} = 1 \\
15A - \cancel{3B} = 63 \\
\hline
16A = 64 \\
A = 4
\end{array}
\qquad
\begin{array}{r}
4 + 3B = 1 \\
B = -1
\end{array}
$$

Method 2 $x + 21 = A(x + 5) + B(3x - 1)$

$x = -5$: $16 = 0 + B(-16)$

$-1 = B$

$x = \frac{1}{3}$: $21\frac{1}{3} = A(5\frac{1}{3}) + 0$

$\frac{64}{3} = \frac{16}{3}A$

$4 = A$

26. $\dfrac{8x + 29}{(2x - 1)(3x + 4)} = \dfrac{A}{2x - 1} + \dfrac{B}{3x + 4}$

27. $\dfrac{-30x + 8}{(3x - 7)(5x + 9)} = \dfrac{A}{3x - 7} + \dfrac{B}{5x + 9}$

28. $\dfrac{2x^2 + 16x - 12}{x(x + 3)(x - 4)} = \dfrac{A}{x} + \dfrac{B}{x + 3} + \dfrac{C}{x - 4}$

29. $\dfrac{5x^2 - 30x - 30}{x(x - 2)(x + 5)} = \dfrac{A}{x} + \dfrac{B}{x - 2} + \dfrac{C}{x + 5}$

30. $\dfrac{-6x^2 - 17x + 70}{(x - 2)(x + 2)(x - 3)} = \dfrac{A}{x - 2} + \dfrac{B}{x + 2} + \dfrac{C}{x - 3}$

31. $\dfrac{4x^2 + 6x + 26}{(x - 1)(x + 1)(x + 5)} = \dfrac{A}{x - 1} + \dfrac{B}{x + 1} + \dfrac{C}{x + 5}$

32. $\dfrac{4x^2 - 16x + 24}{(x - 1)(x - 2)(x - 3)} = \dfrac{A}{x - 1} + \dfrac{B}{x - 2} + \dfrac{C}{x - 3}$

33. $\dfrac{6x^3 + 17x^2 + x - 4}{x(x - 1)(x + 1)(x + 4)} = \dfrac{A}{x} + \dfrac{B}{x - 1} + \dfrac{C}{x + 1} + \dfrac{D}{x + 4}$

34. $\dfrac{2x^3 - 34x^2 + 28x + 40}{x(x - 2)(x + 2)(x - 5)} = \dfrac{A}{x} + \dfrac{B}{x - 2} + \dfrac{C}{x + 2} + \dfrac{D}{x - 5}$

35. $\dfrac{7x^4 + x^3 - 82x^2 + 6x + 36}{x(x - 2)(x + 2)(x - 3)(x + 3)} = \dfrac{A}{x} + \dfrac{B}{x - 2} + \dfrac{C}{x + 2} + \dfrac{D}{x - 3}$

$$+ \dfrac{E}{x + 3}$$

36. One serving of a cereal contains one less than 12 times as many grams of carbohydrate as of protein. If one serving contains 25 grams of protein and carbohydrate, how many grams of each kind does it contain?

37. A rice cereal is fortified with three more essential vitamins than a corn cereal. The number of vitamins in the rice cereal is two less than twice the number in the corn cereal. How many vitamins are there in each cereal?

38. A dance record has 16 different numbers consisting of Latins and non-Latins. There are two more non-Latins than Latins. How many dances of each kind are there on the record?

39. One year Professor Smiley gave 50 more than 10 times the number of lectures given by Professor Carberry. If Carberry had given 10 more lectures than he did give, Smiley would have given 5 times as many lectures as Carberry. How many lectures were given that year by each of these professors?

40. One serving of a rice cereal contains the same number of calories as one serving of a corn cereal and contains 20 less calories than one serving of a wheat cereal. If there are 310 calories in a bowl of cereal consisting of one serving of each kind, how many calories are there in one serving of each kind of cereal?

41. Professor Douglas spends 2 hours per class-hour preparing his lectures. Each week he spends 2 hours more in advising students than he spends in class, and each week he works 10 additional hours studying, etc. If he works 60 hours per week in class, preparing lectures, advising, studying, etc., how many hours per week does he spend in class?

42. During the first 56 years of their marriage, Mary and Roland purchased 3 more major appliances than cars. If they had purchased 3 more cars or major appliances during that time, they would have purchased either a car or a major appliance for each 2 years of their marriage. How many cars and how many major appliances did they purchase in that 56 years?

43. Fred plays the tuba or piccolo at hockey and football games, and he does not switch instruments at a game. One season there were 39 games. If he had played the tuba at 3 fewer games, he would have played the tuba at 5 times as many games as he played the piccolo. At how many games did he play each instrument that season?

44. Tom plays the French horn and piano and takes a course in French. Each week he spends 22 hours practicing horn and piano and studying French. He spends 3 times as much time studying French as practicing the piano, and each week he practices the horn 7 hours longer than the piano. How many hours does he spend each week practicing each instrument and studying French?

45. Marianne had wheat-ear pennies, new pennies, and silver bicentennial coins. She had twice as many pennies as silver coins, and she had 1 more new penny than silver coins. If she gave away half of her wheat-ears, she would have 18 coins left. How many coins of each kind did she have originally?

CHECK LIST FOR THE STUDENT

If you have learned the material in Section 1.10, you should be able to do the following:

1. Give an example of a linear equation in the two variables x and y and an example of a linear equation in the four variables A, B, C, and D.

2. Define the solution of a linear equation in n variables where n is an integer greater than 1.

3. Find several solutions of a single linear equation in two or more variables.

4. Give an example of a system of two simultaneous linear equations in the two variables x and y; give an example of a system of four simultaneous linear equations in the four variables A, B, C, and D.

5. Define the solution of a system of n simultaneous linear equations in n unknowns where n is an integer greater than 1.

6. State the three possibilities as far as the number of solutions of a system of n simultaneous linear equations in n unknowns is concerned.

7. Solve a system of two simultaneous linear equations in two unknowns by substitution and also by choice of multiplier and addition.

8. Solve a system of n simultaneous linear equations in n unknowns where $n \geq 3$ by systematically eliminating unknowns.

9. Find the numerators in a given partial-fraction decomposition of a given rational expression by equating coefficients of like powers and solving a system of simultaneous equations or by substituting convenient values.

10. Solve some verbal problems leading to systems of simultaneous linear equations.

SECTION QUIZ 1.10

1. Find the four solutions of the equation $3x - 2y + 4 = 0$ for which

$$x = 0, \ y = 0, \ x = 1, \text{ and } y = 1.$$

2. Solve the following system of equations by substitution.

$$3x - 4y = -2$$
$$x + 2y = -4$$

3. Solve the following system of equations by choice of multiplier and addition.

$$5x - 3y = 3$$
$$4x + 2y = 20$$

4. Solve the following system of equations.

$$3x - 4y + z + 8 = 0$$
$$2x + y - z + 3 = 0$$
$$x - 2y + 3z - 4 = 0$$

5. Assume that

$$\frac{7x^2 - 11x - 6}{x(x + 1)(x - 2)} = \frac{A}{x} + \frac{B}{x + 1} + \frac{C}{x - 2},$$

and find the values of A, B, and C by two different methods.

Part 1 Review Exercises

1. Let A be the set

   $\{-6, 0, 1, 5, \frac{2}{7}, 7, \frac{1}{3}, \frac{-3}{4}, 2.\overline{7}, \pi, \frac{22}{7}, 3.14, \sqrt{2}, \sqrt{7}, 3 + 2i\}.$

   In each of the following, list the members of A that belong to the specified set.
   (a) The rational numbers   (b) The irrational numbers
   (c) The real numbers   (d) The integers
   (e) The nonnegative integers   (f) The even integers
   (g) The primes

2. Classify each of the following as rational or irrational, and state which kind of decimal, terminating, repeating, or nonterminating and nonrepeating, represents it.
   (a) $\sqrt{3}$ (b) π (c) $\frac{22}{7}$ (d) $\frac{7}{4}$

Find the value of each of the following.

3. (a) $|8|$ (b) $|-8|$ (c) $|0|$ (d) $-|-6|$

4. (a) 6^3 (b) 6^{-1} (c) 6^0 (d) $(-6)^0$

5. (a) $(\sqrt{6})^2$ (b) $(-\sqrt{6})^2$ (c) $-(\sqrt{6})^2$ (d) $\sqrt[5]{2^5}$

6. (a) $(\frac{3}{5})^{-1}$ (b) $(\frac{3}{5})^0$ (c) $(\frac{1}{8}) \times 0$ (d) $1/8^0$

7. (a) $(\frac{1}{8})^{-1/3}$ (b) $(\frac{1}{8})^{2/3}$ (c) $(49)^{1/2}$ (d) $0^{1/5}$

8. Express $23^{4/7}$ in two radical forms.

9. Express $\sqrt[5]{2^3}$ in exponential form.

Using the exponent laws, express each of the following in a form containing no radicals or negative exponents. Assume all variables are positive.

10. $a^3 a^7$ 11. a^3 / a^7 12. $(a^3)^{-7}$

13. $\left(\dfrac{x^2 y^7}{y^3 x^{-1}}\right)^{-3}$ 14. $(x^{3/5} y^{2/5})(x^{2/5} y^{3/5})^{-1}$

15. Let $i^2 = -1$, and express each of the following as either -1, $+1$, $-i$, or $+i$.
    (a) i^9 (b) i^{10}

16. Use the properties of radicals to find the value of each of the following.
    (a) $\sqrt[3]{125 \times (-8)}$ (b) $\sqrt[3]{\dfrac{64}{27}}$

17. Assume that all variables are positive, and tell what belongs in the blank parentheses in each of the following.
    (a) $\sqrt{18x^4 y^7} = (\quad) \sqrt{2y}$

    (b) $\sqrt{\dfrac{4x^3}{7y}} = (\quad) \sqrt{7xy}$

    (c) $\sqrt[9]{8x^6 y^{12}} = (\quad) \sqrt[3]{2x^2 y}$

18. Let $P(x) = x^4 - 3x^3 + 2x - 1$ and $Q(x) = 3x^2 + 1$.
    (a) Find $P(-1)$.   (b) Find $Q(P(1))$.

19. Let $P(x) = 2x^3 - x^2 + 4x - 5$ and $Q(x) = x^2 - x - 1$.

(a) Find $P(x) + Q(x)$.

(b) Find $P(x) - Q(x)$.

(c) Find the product $P(x) \times Q(x)$.

(d) Perform the division $P(x)/Q(x)$ and find the quotient and remainder.

20. Let $P(x) = 5x^2 - 3x - 7$. (a) Express $P(x + h)$ as a polynomial in x and h. (b) Express $P(x + h) - P(x)$ as a polynomial in x and h.

21. Use synthetic division to find the quotient and remainder:

$$\frac{x^3 - 3x^2 + 4x - 2}{x + 2}$$

22. Express each of the following in the form $a + bi$, where a and b are real numbers and $i^2 = -1$.

(a) $(3 + 7i) + (-1 + 6i)$

(b) $(3 + 7i) - (-1 + 6i)$

(c) $(3 + 7i) \times (-1 + 6i)$

If possible, express each of the following as a product of prime polynomials with integral coefficients. If such a factorization is not possible, write "not factorable."

23. $25a^2 - 16b^2$　　　**24.** $x^3 - 27y^3$　　　**25.** $9x^2 - 30x + 25$

26. $9x^2 - 9x - 10$　　　**27.** $xy + y^2 + 3x + 3y$　**28.** $x^3 + (a + b)^3$

29. $9x^2 + 4y^2$

30. Tell what number should be put in the parentheses in order to make the following expression become a perfect square trinomial, and then factor that trinomial.

$$x^2 - 15x + (\quad)$$

31. Find what belongs in the parentheses in each of the following.

(a) $\dfrac{x - 1}{1 - x} = (\quad)\dfrac{-x + 1}{x - 1}$　　(b) $\dfrac{x - 2}{x + 2} = \dfrac{(\quad)}{x^2 + 4x + 4}$

32. Reduce the following to lowest terms.

$$\frac{(3x^2 - 3x - 18)(x + 3)}{(x^2 - 6x + 9)(x^2 - 9)}$$

Perform the indicated operations in each of the following. If possible, reduce the answer to lowest terms.

33. $\dfrac{2x}{x^2 + 5x + 6} \times \dfrac{x^2 - 9}{x^3}$

34. $\dfrac{5x^3}{x^2 - 4x + 4} \div \dfrac{10x^7}{x^2 + x - 6}$

35. $\dfrac{3}{x} - \dfrac{2}{(x - 1)^2} + \dfrac{5}{x^2 - x}$

36. Express the following as a fraction having no fractions in the numerator or denominator.

$$\frac{\dfrac{2}{x^3y} - \dfrac{3}{x^4y^2}}{\dfrac{y}{x} + \dfrac{5y^3}{x^5}}$$

37. Express the following in a form having no irrational number in the denominator.

$$\frac{5}{\sqrt{5} + \sqrt{3}}$$

38. Express the following in the form $a + bi$, where a and b are real numbers and $i^2 = -1$.

$$\frac{2 + 11i}{3 - 2i}$$

39. Solve the equation $PV = P_1V_1$ for P_1.

Solve each of the following equations.

40. $\frac{3}{8}(x - 2) = \frac{-1}{5}(x + 4)$ **41.** $x^2(x + 4)(x - 7) = 0$

42. $x^2 = 7x$ **43.** $x^2 - 6x - 7 = 0$

44. $x^2 - 14x + 47 = 0$ **45.** $|x^2 - 5| = 4$

46. $|x^2 - 3x + 1| = 1$ **47.** $|x^2 + 6x - 3| = 4$

48. $x - 3 + \sqrt{2x + 18} = 0$ **49.** $3\sqrt{x} - \sqrt{x + 16} = 4$

50. $\sqrt{x + 9} - x + 3 = 0$ **51.** $\dfrac{x - 1}{x - 2} - \dfrac{2}{x - 2} - \dfrac{6}{x^2 - 4} = 0$

52. $\dfrac{x - 1}{x - 4} - \dfrac{4}{x - 4} - \dfrac{20}{x^2 - 16} = 0$ **53.** $\dfrac{x(x + 4)}{x - 1} - \dfrac{x(x + 9)}{(x - 1)(x - 3)} = 0$

54. $\dfrac{5}{x + 4} = \dfrac{2}{x - 3}$ **55.** $\dfrac{5}{x^2 + 9} = \dfrac{4}{x^2 + 4}$

56. $|2x - 3| = 5$ **57.** $|5x - 8| = 9$

Solve each of the following, and graph its set of solutions on a number line.

58. $-9x - 4 \geq 2 - 5x$ **59.** $x^2 - 7x - 8 \leq 0$

60. $x^2 + 4x - 45 > 0$ **61.** $|7x - 2| \leq 12$

62. $|3x + 5| \geq 16$

63. Solve the following system of equations.

$$7x - 5y = -14$$
$$14x + 4y = 14$$

Where a and b represent positive real numbers, write each of the following in the prescribed form, and give the values of a and b.

64. $\sqrt{(x + \sqrt{3})^2 + y^2} - \sqrt{(x - \sqrt{3})^2 + y^2} = 2,$ $\quad \dfrac{x^2}{a^2} - \dfrac{y^2}{b^2} = 1$

65. $\sqrt{(x + \sqrt{10})^2 + y^2} - \sqrt{(x - \sqrt{10})^2 + y^2} = 6,$ $\quad \dfrac{x^2}{a^2} - \dfrac{y^2}{b^2} = 1$

66. $\sqrt{(x - \sqrt{5})^2 + y^2} + \sqrt{(x + \sqrt{5})^2 + y^2} = 6,$ $\quad \dfrac{x^2}{a^2} + \dfrac{y^2}{b^2} = 1$

67. $\sqrt{(x - \sqrt{3})^2 + y^2} + \sqrt{(x + \sqrt{3})^2 + y^2} = 4,$ $\quad \dfrac{x^2}{a^2} + \dfrac{y^2}{b^2} = 1$

Where p represents a real number, write each of the following in the prescribed form, and give the value of p.

68. $\sqrt{(x - 4)^2 + y^2} = x + 4,$ $\quad y^2 = 2px$

69. $\sqrt{(y - 6)^2 + x^2} = y + 6,$ $\quad x^2 = 2py$

70. $\sqrt{(x - 2)^2 + y^2} = -x - 2,$ $\quad y^2 = 2px$

Part 1 Final Test

1. (a) Use braces and write an expression for the set of nonnegative integers.
   (b) Use set-builder notation and write an expression for the set of all positive real numbers. Make a graph of this set on a number line.

2. Let A be the set

   $$\{-5,\ 0,\ |-5|,\ 0.5,\ \tfrac{1}{5},\ \sqrt{5},\ 2.\overline{51},\ \pi,\ \tfrac{22}{7},\ 3.14,\ \sqrt{2},\ \tfrac{-4}{5}\}.$$

   (a) List all the rational numbers of A.
   (b) List all the irrational numbers of A.

3. Find the value of each of the following.
   (a) 3^5 (b) 3^1 (c) 3^0 (d) 3^{-5} (e) 0^3

4. In each of the following, if the expression is defined and is a real number, give its value. If the expression is not defined or is not real, write "not defined" or "not real."
   (a) $(\sqrt{3})^2$ (b) $\tfrac{3}{0}$ (c) $\sqrt{-16}$ (d) $\sqrt{(-5)^2}$ (e) $(\tfrac{2}{3})^{-1}$

5. Let x and y represent real numbers, and express each of the following in a simplified form containing no negative exponents. State any restrictions that must be placed on x and y so that the given expression will be defined.
   (a) $(x^4 y^{-2})(x^2 y^{-5})$ (b) $\left(\dfrac{3x^{-1}y^3}{x^{-3}y^5}\right)^2$

6. Find the value of each of the following.
   (a) $(-8)^{4/3}$ (b) $25^{1/2}$ (c) $8^{-1/3}$ (d) $\sqrt[3]{(27)^2}$ (e) $0^{1/3}$

7. Using the exponent laws, express each of the following in a form containing no radicals or negative exponents. Assume all variables are positive.
   (a) $(x^{6/5}y^{-3/10})^{1/3}(x^{-2/5}y^{11/10})$ (b) $\left(\dfrac{x^{-2}y^{-2}}{x^2 y^2}\right)^{-1/4}$

8. Assume that all variables are positive, and tell what belongs in the blank parentheses in each of the following.
   (a) $\sqrt{12x^7 y^3} = (\quad)\sqrt{3xy}$ (b) $\sqrt{\dfrac{3x^5}{2y^3}} = (\quad)\sqrt{6xy}$

9. (a) Let $P(x) = 3x^3 - 2x^2 + x - 7$ and
   $Q(x) = 5x^3 + 2x^2 - 2x - 3$.   Find $P(x) - Q(x)$.
   (b) Let $P(x) = 3x^2 - 2x + 1$ and $Q(x) = x - 7$.
   Find the product $P(x) \times Q(x)$.

10. Find the quotient and remainder.

    $$\dfrac{4x^3 - 2x^2 + x - 5}{x + 3}$$

11. Let $P(x) = 5x^2 - 3x - 7$. Express $P(x + h)$ as a polynomial in x and h.

12. If possible, express each of the following as a product of prime polynomials with integral coefficients. If such a factorization is not possible, write "not factorable."

(a) $25x^2 - 36$ (b) $25x^2 + 20x + 4$
(c) $27x^3 - 8$ (d) $25x^2 - 10x - 8$

13. Tell what number should be put in the parentheses to make the following expression become a perfect-square trinomial, and then factor that trinomial.

$$x^2 - 26x + (\quad)$$

14. (a) Find what belongs in the blank parentheses.

$$\frac{x - 2}{x + 3} = \frac{(\qquad)}{x^2 - 9}$$

(b) Reduce the following to lowest terms.

$$\frac{3x^3 - 12x}{x^4 - x^3 - 6x^2}$$

15. Express each of the following as a single fraction; if possible, reduce to lowest terms.

(a) $\dfrac{1}{x} - \dfrac{3}{x - 1} + \dfrac{5}{(x - 1)(x + 3)}$ (b) $\left(\dfrac{x^2 - 16}{x^2 + 3x - 4}\right) \div \left(\dfrac{x + 4}{x^2 - 2x + 1}\right)$

16. Express the following as a fraction having no fractions in the numerator or denominator.

$$\frac{\dfrac{3}{\sqrt{2x}} - \dfrac{\sqrt{2}}{x^2}}{\dfrac{\sqrt{2}}{x^3} + \dfrac{1}{\sqrt{2}}}$$

17. Solve each of the following equations.
(a) $2x - 5 = -7x + 2$ (b) $x(x + 2)(x - 5) = 0$

18. (a) Solve the equation $3x^2 - 5x = 0$ by factoring.
(b) Solve the equation $9x^2 - 30x + 23 = 0$.

Solve each of the following equations.
19. $x + \sqrt{3x + 15} - 1 = 0$

20. $\dfrac{x - 1}{x + 4} - \dfrac{4(x - 1)}{(x + 4)^2} - \dfrac{20}{(x + 4)^2} = 0$

Solve each of the following, and graph its set of solutions on a number line.
21. (a) $|5x - 6| = 9$ (b) $|x^2 - 3x + 1| = 1$

22. $-11x - 3 \geq 5 - 2x$

23. $x^2 - 3x - 10 \leq 0$

24. $|8x - 3| \leq 4$

25. Solve the following system of equations.

$$5x - 7y = -15$$
$$10x + 3y = \quad 4$$

Part 2 Functions and Graphs

The concept of a function is one of the most basic and universally employed concepts in all of mathematics and its applications. Part 2 of this text provides an introduction to the function concept. It demonstrates the graphs of frequently encountered functions and equations.

2.1 Functions, Relations, Cartesian Coordinates, and Graphs

The purpose of this section is to introduce functions and relations, to introduce Cartesian coordinates, and to define and illustrate the graphs of functions, relations, or equations.

Any particular function is defined provided that the following two things are given:

1. A set called the **domain**

2. A *rule* assigning a single thing (usually a number) to each element in the domain

The set of all things assigned to the domain elements by the rule of assignment is called the **range** of the function. Functions are commonly denoted by lower-case letters such as f, g, h, and k.

EXAMPLE 1

Let f be the function such that

1. The domain is $\{-2, -1, 0, 1, 2, 3\}$.

2. The rule assigns to each domain element the square of that element.

(a) What number is assigned to each of the domain elements?

(b) What is the range of f?

Solution

(a) To -2 the rule assigns $(-2)^2 = 4$.
To -1 the rule assigns $(-1)^2 = 1$.
To 0 the rule assigns $0^2 = 0$.
To 1 the rule assigns $1^2 = 1$.
To 2 the rule assigns $2^2 = 4$.
To 3 the rule assigns $3^2 = 9$.

(b) The range of f is the set $\{4, 1, 0, 9\}$. Note that, as required of a function, the rule of this f assigns a single thing to each element in the domain, but it is permissible and possible for the rule of a function to assign the same single thing to two different domain elements. In this example the rule assigns the single number 4 to both -2 and 2, and it also assigns the single number 1 to both -1 and 1. In stating the range, however, it is unnecessary for us to list 4 or 1 or any assignment more than once.

There are several common ways of giving the rule of assignment of any function. One way is to state the rule in words, as was done in Example 1. Another way is to use arrows to indicate the assignments. For example, the rule of f in Example 1 could be written

Rule of f:

$$-2 \rightarrow 4$$
$$-1 \rightarrow 1$$
$$0 \rightarrow 0$$
$$1 \rightarrow 1$$
$$2 \rightarrow 4$$
$$3 \rightarrow 9$$

If there are more than just a few elements in the domain, this way of giving the rule becomes cumbersome and impractical. In many cases it can be abbreviated by letting the letter x represent any element in the domain. To illustrate, the rule of f in Example 1 can be written

Rule of f:

$$x \rightarrow x^2$$

This can be condensed still more by writing

$$f: \qquad x \rightarrow x^2.$$

The most common and usually the most useful way of giving the rule of assignment of a function is to let the letter x represent any element in the domain and to let the symbol $f(x)$, read "f of x," represent the single thing assigned to x by the rule of f. Thus the rule of f in Example 1 could be written

$$f(x) = x^2.$$

Note that the symbol $f(x)$ does *not* mean that f is in some way multiplied by x. The symbol $f(x)$ stands for the single number assigned to x by the rule of f. The statement $f(x) = x^2$ can be read "f of x equals x squared" or "the number that f assigns to x is x squared."

EXAMPLE 2

Let g be the function such that

1. The domain is the set of all nonnegative real numbers.

2. $g(x) = x^2$.

Find $g(0)$, $g(5)$, $g(\frac{1}{2})$, $g(\sqrt{3})$, $g(a)$, and $g(a + h)$ where a is any nonnegative real number and h is positive.

Solution

Since $g(x) = x^2$, we can find $g(0)$ by replacing x by 0, we can find $g(5)$ by replacing x by 5, etc.

$$g(0) = 0^2 = 0$$
$$g(5) = 5^2 = 25$$
$$g(\tfrac{1}{2}) = (\tfrac{1}{2})^2 = \tfrac{1}{4}$$
$$g(\sqrt{3}) = (\sqrt{3})^2 = 3$$
$$g(a) = a^2$$

Exercise 2.1, 6–20

$$g(a + h) = (a + h)^2 = a^2 + 2ah + h^2$$

Let us look again at the function f of Example 1. The domain of that f is the set $\{-2, -1, 0, 1, 2, 3\}$, and the rule of f may be written

$$-2 \to 4$$
$$-1 \to 1$$
$$0 \to 0$$
$$1 \to 1$$
$$2 \to 4$$
$$3 \to 9.$$

Note that associated with this f there is a set of *ordered pairs,* namely the set $\{(-2, 4), (-1, 1), (0, 0), (1, 1), (2, 4), (3, 9)\}$. An **ordered pair** is a pair in which order of listing makes a difference so that the pair $(2, 4)$, for example, is not the same as the pair $(4, 2)$. The first member of each ordered pair is a domain element, and the second member is that single thing that is assigned to the first member by the rule of assignment of the function. The set of all first members is the domain, and the set of all second members is the range of the function. Note that since the rule of assignment of a function must assign exactly one single thing to each element in the domain, no two different ordered pairs have the same first element.

Just as there is a certain set of ordered pairs associated with the function f of Example 1, there is such a set of ordered pairs associated with any function. For this reason, many authors like to define a function as

A set of ordered pairs such that no two different ordered pairs have the same first member

or

A set of ordered pairs such that to each first member there corresponds a unique second member.

Certainly, if we are given the set of ordered pairs associated with some function, we are actually given the two essential things: domain and rule of assignment. The domain is simply the set of all first members, and the rule of assignment assigns to each domain element the second member of its ordered pair.

EXAMPLE 3

Does the set $\{(-1, -1), (1, -1), (4, -2), (9, -3)\}$ represent a function? If so, what is the domain, how can the rule of assignment be written with arrows, and what is the range?

Solution

The given set does represent a function; no two of its ordered pairs have the same first member.

The domain is the set $\{-1, 1, 4, 9\}$.

Using arrows, the rule of assignment can be written

$$-1 \rightarrow -1$$
$$1 \rightarrow -1$$
$$4 \rightarrow -2$$
$$9 \rightarrow -3.$$

Exercise 2.1, 21–25 The range is $\{-1, -2, -3\}$. (It is unnecessary to list the -1 twice.)

We have seen that every function must have a rule of assignment. For this reason, some authors prefer to define a function as

A correspondence or rule that associates exactly one thing with each element of a set called the domain.

But whether a function is defined as a certain set of ordered pairs or as a certain correspondence, the two essential ingredients, domain and rule of assignment, must be given.

Sometimes it is convenient not to state the domain explicitly but rather to give the domain according to the following convention.

Domain Convention

If the domain of a function is not explicitly stated, it is understood to be the set of all real numbers except those to which the rule of assignment would not assign a real number.

EXAMPLE 4

According to the domain convention, what is the domain of h such that $h(x) = \sqrt{x}$?

Solution

The domain of h is the set of all real numbers x except those for which $\sqrt{x}$ is not a real number. Since $\sqrt{x}$ is real for all real numbers x except negative numbers x, the domain of h is the set of all nonnegative numbers x. Otherwise written,

Domain of $h = \{x | x$ is a nonnegative real$\}$

or   Domain of $h = \{x | x \geq 0\}$.

EXAMPLE 5

According to the domain convention, what is the domain of k such that $k(x) = 1/(x - 5)$?

Solution

When $x = 5$, $1/(x - 5)$ is not defined so it is not a real number. Since $1/(x - 5)$ is a real number for all other real numbers x, the domain of k is the set of all real numbers except 5.

Exercise 2.1, 26–30

$$\text{Domain of } k = \{x \mid x \neq 5\}$$

Functions encountered in undergraduate mathematics usually have ranges that are sets of real numbers. Such functions are called *real-valued* functions. Functions whose domains *and* ranges are both sets of real numbers are called **real** functions. All the functions in our Examples 1–5 are real functions. We shall limit our discussion almost exclusively to real or real-valued functions.

Real functions can be represented geometrically on a *Cartesian coordinate* system. The configuration in Figure 2.1 is a Cartesian coordinate system. It consists of two mutually perpendicular number lines intersecting at their zero points, with the positive numbers increasing up and to the right. The common zero point is called the **origin**. The horizontal number line is usually called the **x-axis**, and the vertical number line is usually called the **y-axis**. The arrows indicate the directions in which the positive numbers increase. The two axes split the plane into four parts called **quadrants** labeled counterclockwise I, II, III, and IV, as shown in Figure 2.1. The axes themselves do not belong to any quadrant.

By means of a Cartesian coordinate system, any point in the plane is represented by exactly one ordered pair of real numbers, and conversely, any ordered pair of real numbers represents exactly one point in the plane. For example, the point located 2 units to the *right* of the y-axis and 3 units *below* the x-axis is represented by the pair $(2, -3)$. The pair $(-2, 3)$ represents the point located by beginning at the origin and moving 2 units *left* (the first member is negative) and 3 units *up* (the second member is positive). The points $(2, -3)$, $(-2, 3)$, and others are shown in Figure 2.1. To *plot* a point is to locate it by using its associated ordered pair of real numbers. To plot a point, begin at the origin and proceed right (if the first member is positive) or left (if the first member is negative) and then

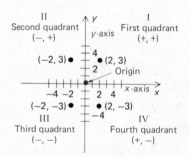

FIGURE 2.1 A Cartesian coordinate system

up (if the second member is positive) or down (if the second member is negative).

The members of the ordered pair of real numbers representing a point are called the **coordinates** of the point. The first member is called the **first coordinate**, the **x-coordinate**, or the **abscissa**. The second member is called the **second coordinate**, the **y-coordinate**, or the **ordinate**. The signs of the coordinates of all points in each of the four quadrants are shown in Figure 2.1.

Exercise 2.1, 31–40

A real function is represented on a Cartesian coordinate system by means of a *graph*. The **graph** of a function f is the set of all points (x, y) such that x is in the domain of f and $y = f(x)$. In other words, it is the set of all points associated with the ordered pairs of the function.

EXAMPLE 6

Let f be the function such that

1. The domain is $\{-2, -1, 0, 1, 2, 3\}$.
2. $f(x) = x^2$.

Sketch the graph of f. (Note that this is precisely the function f of Example 1.)

Solution

The graph of f will consist of the points $(-2, 4)$, $(-1, 1)$, $(0, 0)$, $(1, 1)$, $(2, 4)$, and $(3, 9)$. They are shown in Figure 2.2. Note that we do *not* connect these points because the graph of f consists *only* of these six points. Since there are only six elements in the domain, there are only six points on the graph.

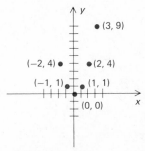

FIGURE 2.2 Graph of f: Domain of $f = \{-2, -1, 0, 1, 2, 3\}$
$$f(x) = x^2$$
(Graph of $y = x^2$ where $x = -2, -1, 0, 1, 2,$ or 3)

EXAMPLE 7

Let g be the function such that

1. The domain is $\{x \mid x \geq 0\}$.
2. $g(x) = x^2$.

Sketch the graph of *g*. (Note that this is precisely the function *g* of Example 2.)

Solution

Here the domain is the set of all nonnegative real numbers, so the graph consists of an unlimited number of points—one point for each nonnegative real number. Since we cannot possibly plot an unlimited number of points, we will just plot enough points so that our intuition will enable us to sketch the graph. Let us begin by making a "table of values" that lists the coordinates of points associated with several domain elements. *Any* nonnegative real number *x* will give us a point (x, y) on the graph, where $y = g(x)$. It will probably be easier for us to guess what the graph looks like, however, if we start with 0 and then take a few convenient successively larger nonnegative numbers.

x	0	$\frac{1}{2}$	1	$\frac{3}{2}$	2	$\frac{5}{2}$	3
$y = g(x) = x^2$	0	$\frac{1}{4}$	1	$\frac{9}{4}$	4	$\frac{25}{4}$	9

These points are plotted in Figure 2.3. If we were to plot more and more points on this graph, we would see that the points all appear to be strung together like beads on the curve shown in Figure 2.3. The curve terminates at (0, 0), but it continues on indefinitely beyond (3, 9). We shall see in Section 2.5 that our graph is the right half of a curve called a parabola.

Note that function *g* of this example has the same rule of assignment as function *f* of Example 6, but the two functions are different functions because their domains are different.

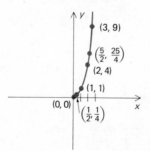

FIGURE 2.3   Graph of *g*:   Domain of $g = \{x|x \geq 0\}$
$$g(x) = x^2$$
(Graph of $y = x^2$ where $x \geq 0$)

EXAMPLE 8

Let *h* be the function such that

1. The domain is $\{x|x \geq 0\}$.
2. $h(x) = -\sqrt{x}$.

Sketch the graph of *h*.

Solution

Since the domain consists of an unlimited number of elements, there will be an unlimited number of points on the graph. We will plot a few of them and then sketch the graph. First let us make a table of values for h.

x	0	$\frac{1}{4}$	1	$\frac{9}{4}$	4	9
$y = h(x) = -\sqrt{x}$	0	$-\frac{1}{2}$	-1	$-\frac{3}{2}$	-2	-3

The points whose coordinates are listed in our table of values are plotted in Figure 2.4. They all appear to be strung together like beads on the curve shown in that figure. The curve terminates at (0, 0) and continues on indefinitely beyond (9, −3). We shall see in Section 2.5 that the graph of h is the lower half of a parabola that opens to the right.

Note that the function h of Example 8 has the same domain as the function g of Example 7, but the functions h and g are different functions because their rules of assignment are different. Two functions are the same or equal if and only if both their domains and their rules of assignment are the same.

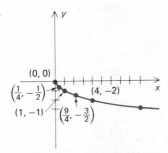

FIGURE 2.4 Graph of h: Domain of $h = \{x \mid x \geq 0\}$
$$h(x) = -\sqrt{x}$$
(Graph of $y = -\sqrt{x}$ where $x \geq 0$)

EXAMPLE 9

Let k be the function such that

1. The domain is the set of all real numbers.
2. $k(x) = x^2$.

Sketch the graph of k.

Let us make a table of values for k.

Solution

x	−3	−2	−1	0	2	3
$y = k(x) = x^2$	9	4	1	0	4	9

The points whose coordinates are listed in this table of values are plotted in Figure 2.5. They all appear to be strung together like beads on the

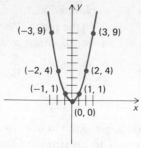

FIGURE 2.5 Graph of k: Domain of $k = \{x \mid x \text{ is a real number}\}$
$$k(x) = x^2$$
(Graph of $y = x^2$)

curve shown in Figure 2.5. In Section 2.5 we shall see that the graph of k is a parabola.

Compare the graphs of function f in Example 6, function g in Example 7, and function k in Example 9. The graph of k is the whole parabola; the graph of g is the right half of the parabola; the graph of f consists of exactly six points on the parabola.

The **graph of an equation in two variables,** say x and y, is the set of all points (x, y) whose coordinates satisfy the equation. Thus the graph of any function f is that part of the graph of the equation $y = f(x)$ for which x is in the domain of f. The graph of function f in Example 6 consists of that part of the graph of $y = x^2$ for which $x = -2, -1, 0, 1, 2,$ or 3; the graph of function g in Example 7 consists of that part of the graph of $y = x^2$ for which $x \geq 0$; the graph of function k in Example 9 is all of the graph of $y = x^2$. If the graph of an equation in x and y is the graph of a function, we say that the equation *defines* that function. The equation $y = x^2$ defines the function k in Example 9. In such a case the domain of the function is determined by the domain convention.

Exercise 2.1, 41–60

Any function is a special kind of a more general concept called a *relation.* A relation is defined provided that a set called the domain is given and provided that there is given a rule of assignment that assigns one *or more* things to each element in the domain. Thus a function is a special relation in which the rule of assignment assigns exactly *one* thing to each element in the domain. A **relation,** then, may be defined as any set of ordered pairs—without any restriction on the first members. Unless otherwise specified, we shall restrict our attention to relations in which the ordered pairs are pairs of real numbers.

The **graph of a relation** is the set of all points associated with its ordered pairs. The graph of any equation in two variables is the graph of a relation, so any equation in two variables may be thought of as representing a relation. For example, the graph of $y^2 = x$ is the graph of a relation. It is shown in Figure 2.6.

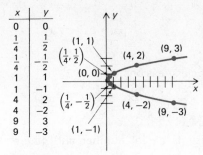

x	y
0	0
$\frac{1}{4}$	$\frac{1}{2}$
$\frac{1}{4}$	$-\frac{1}{2}$
1	1
1	-1
4	2
4	-2
9	3
9	-3

FIGURE 2.6 Graph of $y^2 = x$

Note that although the graph of $y^2 = x$ is the graph of a relation, it is not the graph of a function. Indeed, there exist distinct points on the graph with the same first coordinates. For example, the points (4, 2) and (4, -2) are both on the graph. Both of these points could not possibly be on the graph of a function, because the rule of assignment of a function assigns only a single number to each element in the domain; the rule of a function could not assign both 2 and -2 to the domain element 4.

There is an easy way to tell from looking at the graph whether or not the graph represents a function. Think of sweeping a vertical line across the graph. If the sweeping vertical line never intersects the graph in more than one point at a time, then the graph is the graph of a function. In other words, if no vertical line ever intersects the graph in more than one point, then the graph is the graph of a function. Applying the ''sweeping vertical line'' test in Figure 2.7, we see that the graph of $y^2 = x$ is not the graph of a function, while the graph of $y = x^2$ *is* the graph of a function.

Exercise 2.1, 61–80

The notation $f(x)$ that we have been using is used extensively in mathematics and its applications. For instance, it is used in expressions

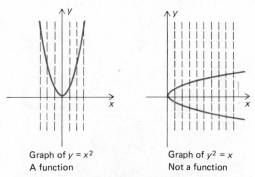

Graph of $y = x^2$
A function

Graph of $y^2 = x$
Not a function

FIGURE 2.7 Sweeping vertical line test

such as

$$\frac{f(x + h) - f(x)}{h}$$

for nonzero h, or expressions such as

$$\frac{f(x_1 + \Delta x) - f(x_1)}{\Delta x}$$

where x_1 represents a real number and where the symobl Δx, read "delta x," represents a nonzero real number. These expressions have many different physical, geometrical, and theoretical interpretations. For example, they may be interpreted as the average speed of a moving object, the slope of a line segment, or the density of a piece of wire. The next example provides practice with the notation.

EXAMPLE 10   (a) If $f(x) = x^2$ and h is nonzero, simplify the expression

$$\frac{f(x + h) - f(x)}{h}.$$

(b) If $f(x) = x^2 - 3x + 4$, if x_1 represents a real number in the domain of f, and if the symbol Δx represents a nonzero real number, simplify the expression

$$\frac{f(x_1 + \Delta x) - f(x_1)}{\Delta x}.$$

Solution   (a) $f(x) = x^2$

$$f(x + h) = (x + h)^2$$

$$\frac{f(x + h) - f(x)}{h} = \frac{(x + h)^2 - x^2}{h}$$

$$= \frac{x^2 + 2xh + h^2 - x^2}{h}$$

$$= \frac{2xh + h^2}{h}$$

$$= 2x + h$$

(b) $f(x) = x^2 - 3x + 4$

$$f(x_1 + \Delta x) = (x_1 + \Delta x)^2 - 3(x_1 + \Delta x) + 4$$

(The symbol Δx is printed in color to emphasize that it represents one single number.)

$$f(x_1) = x_1^2 - 3x_1 + 4$$

$$\frac{f(x_1 + \Delta x) - f(x_1)}{\Delta x} = \frac{(x_1 + \Delta x)^2 - 3(x_1 + \Delta x) + 4 - (x_1^2 - 3x_1 + 4)}{\Delta x}$$

$$= \frac{x_1^2 + 2x_1\,\Delta x + (\Delta x)^2 - 3x_1 - 3\,\Delta x + 4 - x_1^2 + 3x_1 - 4}{\Delta x}$$

$$= \frac{2x_1\,\Delta x + (\Delta x)^2 - 3\,\Delta x}{\Delta x}$$

$$= 2x_1 + \Delta x - 3$$

Exercise 2.1, 81–90
Verbal Problems
Exercise 2.1, 91–94

EXERCISE 2.1 FUNCTIONS, RELATIONS, CARTESIAN COORDINATES, AND GRAPHS

▶ In each of the following, tell what number is assigned to each of the domain elements, and give the range of the function.

EXAMPLE

Domain of $f = \{-2, 1, \frac{1}{3}\}$

$$f(x) = \frac{1}{x}$$

Solution

To -2 the rule assigns $\frac{-1}{2}$.

To 1 the rule assigns $\frac{1}{1} = 1$.

To $\frac{1}{3}$ the rule assigns $\dfrac{1}{\frac{1}{3}} = 3$.

The range is $\{\frac{-1}{2}, 1, 3\}$.

1. Domain of $f = \{-3, \frac{-1}{2}, 0, \frac{1}{2}, 3\}$
   $f(x) = x^2 + 1$

2. Domain of $g = \{-3, \frac{-1}{2}, 0, \frac{1}{2}, 3\}$
   $g(x) = x^2 + 2x - 1$

3. Domain of $h = \{0, 1, \frac{9}{4}, 4, 9\}$
   $h(x) = \sqrt{x}$

4. Domain of $k = \{-3, -2, -1, 0, 1, 2\}$
   $k(x) = x^3$

5. Domain of $f = \{\frac{-1}{8}, -1, 0, 1, 8\}$
   $f(x) = \sqrt[3]{x}$

▶ In each of the following, let a and $a + h$ be in the domain of f.

EXAMPLE

If $f(x) = 3x^2 - 2x + 1$, find $f(0)$, $f(-1)$, $f(4)$, $f(a)$, and $f(a + h)$.

Solution

$$f(0) = 3(0)^2 - 2(0) + 1 = 1$$
$$f(-1) = 3(-1)^2 - 2(-1) + 1 = 3 + 2 + 1 = 6$$
$$f(4) = 3(4)^2 - 2(4) + 1 = 48 - 8 + 1 = 41$$
$$f(a) = 3a^2 - 2a + 1$$
$$f(a + h) = 3(a + h)^2 - 2(a + h) + 1$$
$$= 3(a^2 + 2ah + h^2) - 2a - 2h + 1$$
$$= 3a^2 + 6ah + 3h^2 - 2a - 2h + 1$$

6. If $f(x) = 2x^2 - 4x + 5$, find $f(0)$, $f(-1)$, $f(3)$, $f(a)$, and $f(a + h)$.

7. If $f(x) = x^3 - x^2 + 1$, find $f(0)$, $f(-1)$, $f(1)$, $f(a)$, and $f(a + h)$.

8. If $f(x) = \sqrt{2x} + 3$, find $f(2)$, $f(8)$, $f(0)$, $f(a)$, and $f(a + h)$.

9. If $f(x) = \dfrac{1}{2x} + 1$, find $f(-1)$, $f(3)$, $f(\frac{1}{2})$, $f(a)$, and $f(a + h)$.

10. If $f(x) = x^{1/3}$, find $f(0)$, $f(\frac{1}{27})$, $f(-8)$, $f(a)$, and $f(a + h)$.

▶ In each of the following let Δx represent a single number, let x_1 represent a number, and let $x_1 + \Delta x$ be a number in the domain of f. Express $f(x_1 + \Delta x)$ as a polynomial in x_1 and Δx.

EXAMPLE

$$f(x) = 3x^2 - 2x + 1$$

Solution

$$f(x_1 + \Delta x) = 3(x_1 + \Delta x)^2 - 2(x_1 + \Delta x) + 1$$
$$= 3(x_1^2 + 2x_1 \Delta x + (\Delta x)^2) - 2x_1 - 2 \Delta x + 1$$
$$= 3x_1^2 + 6x_1 \Delta x + 3(\Delta x)^2 - 2x_1 - 2 \Delta x + 1$$

11. $f(x) = x^2 - 5x + 3$ 12. $f(x) = 2x^2 + 3x - 4$

13. $f(x) = 4x^2 - 2x + 1$ 14. $f(x) = 5x^2 - x + 7$

15. $f(x) = 2x^3$

▶ In each of the following let $x + h$ be in the domain of f, and express $f(x + h)$ as a polynomial in x and h.

EXAMPLE

$$f(x) = 3x^2 - 2x + 1$$

Solution

$$f(x + h) = 3(x + h)^2 - 2(x + h) + 1$$
$$= 3(x^2 + 2xh + h^2) - 2x - 2h + 1$$
$$= 3x^2 + 6xh + 3h^2 - 2x - 2h + 1$$

16. $f(x) = x^2 - 5x + 3$ 17. $f(x) = 2x^2 + 3x - 4$

18. $f(x) = 4x^2 - 2x + 1$ **19.** $f(x) = 5x^2 - x + 7$

20. $f(x) = 2x^3$

▶ In each of the following, state whether or not the given set represents a function. If it does represent a function, give the domain, show how the rule of assignment can be written with arrows, and give the range.

EXAMPLE $\{(\frac{-1}{2}, 0), (\frac{1}{2}, 1), (\frac{1}{3}, 1), (\frac{3}{2}, 2)\}$

Solution   The set represents a function.

Domain $= \{\frac{-1}{2}, \frac{1}{2}, \frac{1}{3}, \frac{3}{2}\}$

Rule $\frac{-1}{2} \to 0$

$\qquad \frac{1}{2} \to 1$

$\qquad \frac{1}{3} \to 1$

$\qquad \frac{3}{2} \to 2$

Range $= \{0, 1, 2\}$

21. $\{(\frac{-1}{2}, -1), (\frac{1}{2}, 0), (\frac{1}{3}, 0), (\frac{3}{2}, 1)\}$

22. $\{(-1, \frac{-1}{2}), (0, \frac{1}{2}), (0, \frac{1}{3}), (1, \frac{3}{2})\}$

23. $\{(-2, -8), (0, 0), (-1, -1), (2, 8)\}$

24. $\{(-2, 2), (2, 2), (-3, 3), (3, 3)\}$

25. $\{(16, 4), (16, -4), (9, 3), (25, 5)\}$

▶ According to the domain convention, what is the domain of the function in each of the following?

EXAMPLES   (a) $f(x) = 1/x$ (b) $g(x) = \sqrt{x - 3}$

Solutions   (a) Domain of $f = \{x \mid x \neq 0\}$ (b) Domain of $g = \{x \mid x \geq 3\}$

26. $f(x) = \dfrac{1}{x - 1}$ **27.** $g(x) = \sqrt{x - 1}$

28. $f(x) = \dfrac{1}{x^2}$ **29.** $g(x) = \sqrt{x + 2}$

30. $f(x) = \dfrac{1}{x - 2} + \sqrt{x + 1}$

▶ In each of the following, plot and label the points on a Cartesian coordinate system.

EXAMPLE $(-2, 1), (2, -1), (2, 1), (-2, -1)$

Solution

(-2, 1)   (2, 1)
  ●   ●

(-2, -1)   (2, -1)
  ●   ●

31. $(3, 5), (3, -5), (-3, 5), (-3, -5), (0, 0)$

32. $(0, 1), (0, -1), (1, 0), (-1, 0)$

33. $(0, 0), (1, 1), (2, 2), (3, 3), (-1, -1), (-2, -2), (-3, -3)$

34. $(0, 3), (1, 3), (2, 3), (3, 3), (-1, 3), (-2, 3)$

35. $(0, 0), (1, 1), (2, 2), (3, 3), (-1, 1), (-2, 2), (-3, 3)$

▶ In each of the following give the coordinates of the point which is located as described.

EXAMPLE   3 units left of y-axis, 4 units above x-axis

Solution $(-3, 4)$

36. 3 units right of y-axis, 4 units below x-axis

37. 5 units left of y-axis, 2 units below x-axis

38. 4 units right of y-axis, 2 units above x-axis

39. 2 units left of y-axis, on x-axis

40. on y-axis, 3 units below x-axis

▶ In each of the following, sketch the graph of the function.

EXAMPLE   (1) Domain of $f = \{x \mid x > -2\}$
              (2) $f(x) = x^2$

Solution

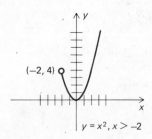

$y = x^2, x > -2$

The graph of f is the graph of $y = x^2$ where $x > -2$. An open circle is placed at the endpoint $(-2, 4)$ to show that it is not included on the graph.

41. (1) Domain of $f = \{-3, -2, -1, 0, 1\}$
    (2) $f(x = x^2$

42. (1) Domain of $f = \{x \mid x \leq 0\}$
    (2) $f(x) = x^2$

43. (1) Domain of $f = \{x \mid x < 0\}$
    (2) $f(x) = x^2$

44. (1) Domain of $f = \{x \mid x > 0\}$
    (2) $f(x) = x^2$

45. (1) Domain of $f = \{x \mid x \geq -1\}$
    (2) $f(x) = x^2$

46. (1) Domain of $f = \{x \mid x < -1\}$
    (2) $f(x) = x^2$

47. (1) Domain of $f = \{x \mid -1 \leq x \leq 2\}$
    (2) $f(x) = x^2$

48. (1) Domain of $f = \{x \mid -2 \leq x < 3\}$
    (2) $f(x) = x^2$

49. (1) Domain of $f = \{0, \frac{9}{4}, 4\}$
    (2) $f(x) = \sqrt{x}$

50. (1) Domain of $f = \{x \mid x \geq 0\}$
    (2) $f(x) = \sqrt{x}$

51. (1) Domain of $f = \{x \mid x \geq 1\}$
    (2) $f(x) = -\sqrt{x}$

52. (1) Domain of $f = \{x \mid 0 \leq x \leq 9\}$
    (2) $f(x) = -\sqrt{x}$

53. (1) Domain of $f = \{x \mid 0 \leq x \leq 9\}$
    (2) $f(x) = \sqrt{x}$

54. (1) Domain of $f = \{x \mid 1 \leq x \leq 9\}$
    (2) $f(x) = \sqrt{x}$

55. (1) Domain of $f = \{x \mid 1 \leq x \leq 4\}$
    (2) $f(x) = -\sqrt{x}$

In each of the following, complete the table of values and plot the points on a Cartesian coordinate system.

EXAMPLE

x	y = f(x) = |x|
−2	
−1	
0	
1	
2	

Solution

x	y = f(x) = |x|
−2	2
−1	1
0	0
1	1
2	2

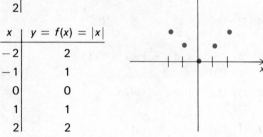

56.

x	y = f(x) = |x − 1|
−1	
0	
1	
2	
3	

57.

x	y = f(x) = |x + 1|
−3	
−2	
−1	
0	
1	

58.

x	y = f(x) = x
−3	
−2	
−1	
0	
1	
2	
3	

59.

x	y = f(x) = x + 2
−3	
−2	
−1	
0	
1	
2	
3	

60.

x	y = f(x) = x³
−3	
−2	
−1	
0	
1	
2	
3	

▶ Sketch the graph of each of the following relations and state whether or not the relation is a function.

EXAMPLE

$\{(x, y) \mid y^2 = x \text{ and } 0 \le x \le 4\}$

Solution

Not a function.

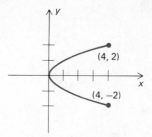

(4, 2)

(4, −2)

61. $\{(x, y) \mid y^2 = x \text{ and } 0 \le x \le 9\}$

62. $\{(x, y) \mid y^2 = x \text{ and } 0 \le x < 1\}$

63. $\{(x, y) \mid y = x^2 \text{ and } 0 \le x \le 1\}$

64. $\{(x, y) \mid y = x^2 \text{ and } -1 \le x \le 1\}$

65. $\{(9, 3), (9, -3), (4, 2), (1, 1)\}$

▶ In each of the following, tell whether or not the graph represents a function.

EXAMPLE

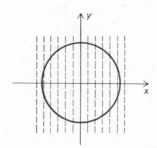

Solution

Not a function. (There exists a vertical line that intersects the graph in more than one point. In fact, there exist an unlimited number of such vertical lines.)

66.

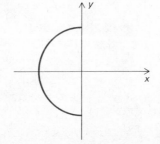

67.

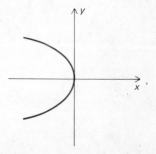

68.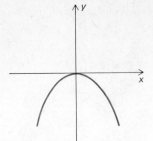

69.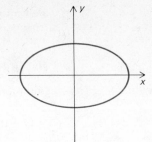

70.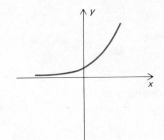

71.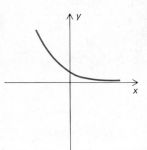

72.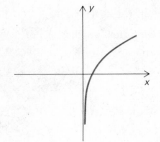

73.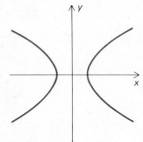

74.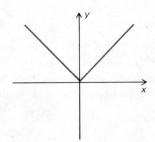

75.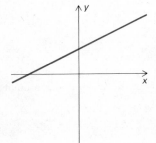

76.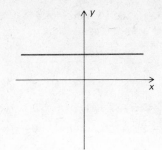

77.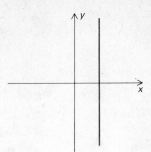

78.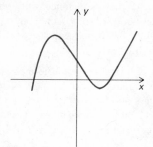

79.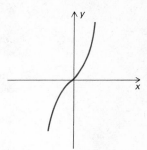

80.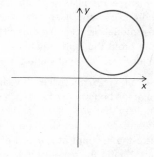

▶ In each of the following, let h be nonzero and write the expression

$$\frac{f(x + h) - f(x)}{h}$$

as a polynomial in x and h.

81. $f(x) = 3x^2 - 2x + 7$ **82.** $f(x) = x^3$

83. $f(x) = 4x^2 + 3x - 8$ **84.** $f(x) = x^3 - 2x$

85. $f(x) = 7x^2 - 4x + 1$

▶ In each of the following, let x_1 represent a real number in the domain of f, let Δx be nonzero, and write the expression

$$\frac{f(x_1 + \Delta x) - f(x_1)}{\Delta x}$$

as a polynomial in x_1 and Δx.

86. $f(x) = 3x^2 - 2x + 7$ **87.** $f(x) = x^3$

88. $f(x) = 4x^2 + 3x - 8$ **89.** $f(x) = x^3 - 2x$

90. $f(x) = 7x^2 - 4x + 1$

91. A farmer has 50 meters of fence with which to build a rectangular pen with one side along a brook. Let x be the length in meters of each side perpendicular to the brook and let $A(x)$ be the area of the pen. Write an expression for $A(x)$. Find $A(3)$.

92. The strength of a beam is equal to a constant k times the product of the width and the square of the depth all divided by the length. Let x be the depth in meters and let $S(x)$ be the strength. If the width is $\frac{1}{3}$ meter and the length is 12 meters, write an expression for $S(x)$. In this case, the units for $S(x)$ are called newtons; find $S(2)$.

93. It will cost a bicycle company $180,000 to set up production for a new model. Once the new model is in production, it will cost $50 to make and distribute each bicycle, and each bicycle will sell for $110. Let x be the number of new model bicycles the company expects to sell the first year. Let $P(x)$ be the profit the company expects to make the first year. Write an expression for $P(x)$. For what value of x will $P(x)$ be zero? Find $P(5000)$.

94. A bank account earns 6% interest each year. Let x be the number of dollars in the account at the beginning of a certain year, and let $I(x)$ be the interest earned on x dollars that year. Write an expression for $I(x)$. Find $I(1000)$. Assume that there are no deposits and no withdrawals, and assume that $I(x)$ is added to the account at the end of the year. Write an expression for the amount $A(x)$ in the account at the end of the year.

CHECK LIST FOR THE STUDENT

If you have learned the material in Section 2.1, you should be able to do the following:

1. State the two things that must be given in the definition of any particular function.

2. State the definition of the range of a function.

3. Illustrate several common ways of giving the rule of assignment of a function.

4. Find $f(a)$, if $f(x)$ is given and a is in the domain of f.

5. Simplify expressions such as

$$\frac{f(x + h) - f(x)}{h} \quad \text{or} \quad \frac{f(x_1 + \Delta x) - f(x_1)}{\Delta x}$$

where $f(x)$ is a specific second- or third-degree polynomial.

6. Recognize that there is a set of ordered pairs associated with any function.

7. Identify the domain and range and give the rule of assignment of a function that is given as a set of ordered pairs.

8. Recognize that some authors define a function as a certain set of ordered pairs.

9. Recognize that some authors define a function as a certain correspondence or rule.

10. State and use the domain convention for determining the domain of a function whose domain is not explicitly stated.

11. State the definition of a real function.

12. Plot points on a Cartesian coordinate system.

13. State the names of the coordinates of a point on a Cartesian coordinate system.

14. State the definition of the graph of a real function f.

15. Sketch the graph of any real function f with a given domain and with $f(x) = x^2$.

16. Make a table of values for a given real function.

17. Sketch the graph of any real function f with a given domain and with $f(x) = \sqrt{x}$ or with $f(x) = -\sqrt{x}$.

18. State the provisions under which two functions are equal.

19. State the definition of the graph of an equation in two variables.

20. State the definition of a relation.

21. Tell from looking at the graph of a relation whether or not it represents a function.

SECTION QUIZ 2.1

1. Let f be the function whose domain is $\{-3, 0, 1, 2\}$ and which is such that $f(x) = x^2 + 3$. Find the range of f.

2. If 0, 2, and real number a are in the domain of f, and if $f(x) = x^2 - 3x + 5$, find $f(0)$, $f(2)$, and $f(a)$.

3. State whether or not the following set represents a function. If it does represent a function, give the domain, show how the rule of assignment can be written with arrows, and give the range.

$$\{(\tfrac{1}{3}, -4), (1, -2), (0, -5), (2, -2)\}$$

4. According to the domain convention, what is the domain of the function f in each of the following?
   (a) $f(x) = 1/(x + 5)$ (b) $f(x) = \sqrt{x + 5}$

5. (a) What is the sign of the first coordinate of a point in the third quadrant, and what is the sign of the second coordinate of a point in the fourth quadrant?
   (b) Let $f(x) = |x|$, let $y = f(x)$, and plot the points (x, y) on the graph of f for $x = -3, -1, 0, 1,$ and 3.

6. Let f be the function whose domain is $\{-1, 0, 1, 2, 3\}$ and which is such that $f(x) = x^2$. Sketch the graph of f.

7. Let f be the function whose domain is $\{x \mid x \geq -1\}$ and which is such that $f(x) = x^2$. Sketch the graph of f.

8. Sketch the graph of each of the following relations, and state whether or not the relation is a function.
   (a) $\{(x, y) \mid y = x^2$ and $0 \leq x \leq 4\}$
   (b) $\{(x, y) \mid y^2 = x$ and $0 \leq x \leq 9\}$

9. Tell whether or not each of the following graphs represents a function.

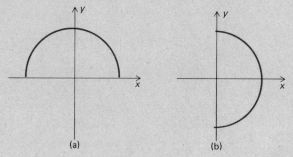

(a)   (b)

10. Let $f(x) = x^2 - 5x + 2$, let h be nonzero, and let $x + h$ be in the domain of f. Express

$$\frac{f(x + h) - f(x)}{h}$$

as a polynomial in x and h.

2.2 Lines; Linear Functions

The purpose of this section is to show that a given linear equation in one or two variables algebraically represents a line, to show how an equation can be found to represent a given line, to define the slope of a line, and to define and graph linear functions.

We have said (Section 2.1) that the graph of an equation in the two variables x and y is the set of all points (x, y) whose coordinates satisfy the equation. The graph of any linear equation in x and y is a line that is not parallel to either axis. Such a line is called an **oblique** line. For example, the graph of

$$3x + 2y - 12 = 0$$

is an oblique line. In general, if A, B, and C are real numbers with A and B not zero, it will be proved later in this section that the graph of

$$Ax + By + C = 0$$

is an oblique line. We say that $Ax + By + C = 0$ is an equation *of* the line that is its graph, and sometimes we refer to the equation itself as the line.

The graph of $3x + 2y - 12 = 0$ is shown in Figure 2.8. Since two points determine a line, to sketch the graph of a linear equation in two variables we need only to plot two points whose coordinates satisfy the equation and then draw the line through them. Any number of points on the line may be found simply by assigning convenient values to one of the variables and then solving for the other. In the table of values in Figure 2.8, first the value 0 was assigned to x, and y was found to be 6.

$$3 \cdot 0 + 2y - 12 = 0$$
$$2y = 12$$
$$y = 6$$

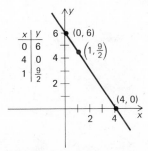

FIGURE 2.8   Graph of $3x + 2y - 12 = 0$

Next the value 0 was assigned to y, and x was found to be 4.

$$3x + 2 \cdot 0 - 12 = 0$$
$$3x = 12$$
$$x = 4$$

Finally, as a "check," the value 1 was assigned to x, and y was found to be $\frac{9}{2}$.

$$3 \cdot 1 + 2y - 12 = 0$$
$$2y = 9$$
$$y = \tfrac{9}{2}$$

If the check point $(1, \frac{9}{2})$ had not seemed to lie on the line drawn through the first two points plotted, we would have gone back and checked our computations.

EXAMPLE 1

Find the point of intersection of the lines $2x - 3y + 10 = 0$ and $3x - y + 1 = 0$. Sketch the graphs of the lines, and label the point of intersection.

Solution

To find the point of intersection, we must find the point whose coordinates satisfy both equations. We can do this by solving the system of simultaneous equations consisting of the two given equations. Using the method of choice of multiplier and addition that was presented in Section 1.10, let us first choose multipliers and eliminate y.

$$
\begin{array}{r|l}
1 & 2x - 3y + 10 = 0 \\
-3 & 3x - \ y + \ 1 = 0
\end{array}
$$

$$2x - 3y + 10 = 0$$
$$\underline{-9x + 3y - \ 3 = 0}$$
$$-7x \qquad + \ 7 = 0$$
$$- 7x = -7$$
$$x = 1$$

Substituting 1 for x in the first equation and solving for y, we find that $y = 4$.

$$2 - 3y + 10 = 0$$
$$- 3y = -12$$
$$y = 4$$

Check

$$2x - \ 3y + 10 = 0 \qquad\qquad 3x - y + 1 = 0$$
$$2(1) - 3(4) + 10 = 0 \ \text{ True} \qquad 3(1) - 4 + 1 = 0 \ \text{ True}$$

The solution is $x = 1$, $y = 4$. In other words, the coordinates of the point (1, 4) satisfy both equations, so the point (1, 4) lies on both lines. The point of intersection is (1, 4). It is shown with the lines in Figure 2.9.

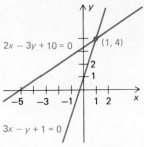

FIGURE 2.9 Graphs of $2x - 3y + 10 = 0$ and $3x - y + 1 = 0$

The points in which a line intersects the coordinate axes are called its **intercepts**. The point in which a line intersects the x-axis is called its **x-intercept**. The point in which a line intersects the y-axis is called its **y-intercept**. To find the x-intercept of a line whose equation is given, let $y = 0$ and solve for x. To find the y-intercept, let $x = 0$ and solve for y. This was done in sketching the graph of the equation $3x + 2y - 12 = 0$ in Figure 2.8. The x-intercept of that line is (4, 0), and the y-intercept is (0, 6). Sometimes just the first coordinate of the x-intercept is called the x-intercept, and sometimes just the second coordinate of the y-intercept is called the y-intercept. We might say that the x-intercept of $3x + 2y - 12 = 0$ is 4 and the y-intercept of $3x + 2y - 12 = 0$ is 6. In any case, it should be clear from context whether the word intercept refers to the point itself, the coordinates of the point, or just the pertinent coordinate of the point.

Exercise 2.2, 1–10

The graph of any equation in the one variable x is the set of all points (x, y) whose coordinates satisfy the equation. Since no y appears in the equation, any point whose x-coordinate satisfies the equation will be on the graph, and the graph consists of all such points. The graph of a *linear* equation in x is a line parallel to the y-axis. For example, the graph of the equation $3x - 6 = 0$ is shown in Figure 2.10. It is the line every point of which has x-coordinate 2.

$$3x - 6 = 0$$
$$3x = 6$$
$$x = 2$$

In other words, the graph of $3x - 6 = 0$ is the graph of the equation $x = 2$, and the graph of $x = 2$ is a line parallel to the y-axis and two units to the right of it.

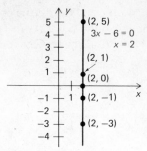

FIGURE 2.10   Graph of $3x - 6 = 0$
(Graph of $x = 2$)

In general, if *a* is a nonzero real number,

$x = a$ represents a line parallel to the *y*-axis.

$x = 0$ represents the *y*-axis.

Note that the graph of $x = a$ is not just a single point. It is the *line* every point of which has *x*-coordinate *a*.

EXAMPLE 2

Sketch the graph of the equation $x = -2$. What are its intercepts?

Solution

The equation $x = -2$ represents a line parallel to the *y*-axis and two units to the left of it. It is shown in Figure 2.11. Its *x*-intercept is $(-2, 0)$. It has no *y*-intercept.

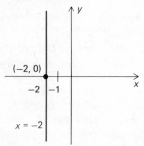

FIGURE 2.11   Graph of $x = -2$

The graph of any equation in the one variable *y* is the set of all points (x, y) whose coordinates satisfy the equation. Since no *x* appears in the equation, any point whose *y*-coordinate satisfies the equation will be on the graph. The graph of a *linear* equation in *y* is a line parallel to the *x*-axis. For example, the graph of $2y + 8 = 0$ is the line every point of which has *y*-coordinate -4.

$$2y + 8 = 0$$
$$2y = -8$$
$$y = -4$$

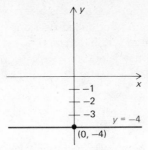

FIGURE 2.12   Graph of $2y + 8 = 0$
(Graph of $y = -4$)

That is, the graph of $2y + 8 = 0$ is the graph of $y = -4$. It is shown in Figure 2.12.

In general, if b is a nonzero real number,

$y = b$ represents a line parallel to the x-axis.

$y = 0$ represents the x-axis.

Note that the graph of $y = b$ is *not* just a point. It is the line every point of which has y-coordinate b.

EXAMPLE 3

Sketch the graph of the equation $y = 4$. What are its intercepts?

Solution

Exercise 2.2, 11–15

The graph of $y = 4$ is the line parallel to the x-axis and four units above it. It is shown in Figure 2.13. Its y-intercept is $(0, 4)$. It has no x-intercept.

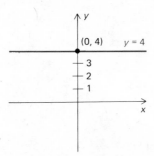

FIGURE 2.13   Graph of $y = 4$

We have noted that any given linear equation in one or two variables represents a line, and we have seen how to sketch the graph of that line. Conversely, if we are given a line, we can find a linear equation in one or two variables to represent that line. Sometimes a line is given by giving two points on it, and we shall see later in this section how to find an equation for a line if we are given two of its points. But first let us define the *slope* of a line because sometimes a line is given by giving just one of its points together with the number called its slope.

The slope of a segment of a nonvertical line is defined in terms of the coordinates of its endpoints. The **slope** of the nonvertical line segment between point P_1 with coordinates (x_1, y_1) and point P_2 with coordinates (x_2, y_2) is

$$\frac{y_2 - y_1}{x_2 - x_1}.$$

To get a feeling for the geometric meaning of slope, consider Figures 2.14 and 2.15. In those figures lines have been drawn through P_1 and P_2 parallel to the *x*- and *y*-axes, respectively. In each figure the point of intersection of these drawn parallels has been labeled *M* and has co-ordinates (x_2, y_1). In each figure, if we move along the line from P_1 to P_2, the numerator of the slope formula indicates how many units we move in the *y*-direction (positive, if up; negative, if down), and the denominator indicates how many units we move in the *x*-direction (positive, if right; negative, if left).

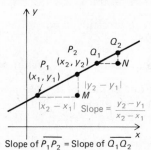

Slope of $\overline{P_1 P_2}$ = Slope of $\overline{Q_1 Q_2}$

FIGURE 2.14 Slope of line segment

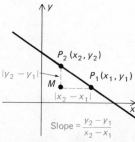

FIGURE 2.15 Slope of line segment

It can be shown by constructing similar right triangles that the slope of any segment of an oblique line is the same as the slope of any other seg-ment of it. Figure 2.14 suggests how this can be shown. In that figure let

Q_1 and Q_2 be the endpoints of some other segment of the line. Note that the triangles P_1MP_2 and Q_1NQ_2 are similar triangles, so the ratios of the lengths of corresponding sides are equal and

$$\frac{MP_2}{P_1M} = \frac{NQ_2}{Q_1N}.$$

The slope of the segment with endpoints Q_1 and Q_2 must be the same as the slope of the segment with endpoints P_1 and P_2. Similar arguments show that

The slope of any segment of a nonvertical line is the same as the slope of any other segment of it.

Consequently, we define the **slope of a nonvertical line** as the slope of any segment of the line. The slope of a line is usually denoted by the lower-case letter m. Thus if (x_1, y_1) and (x_2, y_2) are any two points on a nonvertical line, we may say that

$$m = \frac{y_2 - y_1}{x_2 - x_1}.$$ Slope of nonvertical line containing (x_1, y_1) and (x_2, y_2)

The slope of a vertical line is not defined. Indeed, if we tried to apply the slope formula to two points on a vertical line, we would obtain a zero in the denominator, because all points on any vertical line have the same x-coordinate and hence the $x_2 - x_1$ would be zero.

The slope of any horizontal line is zero. For example, the slope of the horizontal line $y = 3$ is zero. This is consistent with the slope formula because the y-coordinates of any two distinct points on $y = 3$ are both 3 so that the numerator $y_2 - y_1$ is $3 - 3 = 0$. In particular, the points $(1, 3)$ and $(5, 3)$ are both on the line $y = 3$. Using the slope formula, the slope of $y = 3$ is

$$\frac{3 - 3}{5 - 1} = \frac{0}{4} = 0.$$

EXAMPLE 4

Show that $(3, 1)$ and $(8, 4)$ are both points on the line $3x - 5y - 4 = 0$. Find the slope, and sketch the line.

Solution

To show that the given points are on the line represented by the given equation, we must show that their coordinates satisfy the equation.

$3x - 5y - 4 = 0$	$3x - 5y - 4 = 0$
$3(3) - 5(1) - 4 = 0$	$3(8) - 5(4) - 4 = 0$
$9 - 5 - 4 = 0$ True	$24 - 20 - 4 = 0$ True
$(3, 1)$ is on the line.	$(8, 4)$ is on the line.

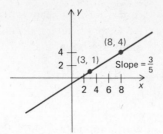

FIGURE 2.16 Graph of $3x - 5y - 4 = 0$

The slope of the line containing (3, 1) and (8, 4) is

$$\frac{4 - 1}{8 - 3} = \frac{3}{5}.$$

The points (3, 1) and (8, 4) have been plotted and the line they determine has been drawn in Figure 2.16.

Note that in using the slope formula to find the slope of a line through two given points, it does not matter which of the points we call P_1 and which we call P_2. In Example 4, if we call (3, 1) the point P_1 and (8, 4) the point P_2, then

$$m = \frac{4 - 1}{8 - 3} = \frac{3}{5}.$$

If we call (8, 4) the point P_1 and (3, 1) the point P_2, we still get the same number for the slope.

$$m = \frac{1 - 4}{3 - 8} = \frac{-3}{-5} = \frac{3}{5}$$

EXAMPLE 5

Show that the points (3, 5) and (8, 2) are both on the line $3x + 5y - 34 = 0$. Find the slope, and sketch the line.

Solution

Let us first show that (3, 5) and (8, 2) satisfy the given equation.

$$3x + 5y - 34 = 0 \qquad\qquad 3x + 5y - 34 = 0$$
$$3(3) + 5(5) - 34 = 0 \qquad\qquad 3(8) + 5(2) - 34 = 0$$
$$9 + 25 - 34 = 0 \quad \text{True} \qquad 24 + 10 - 34 = 0 \quad \text{True}$$

(3, 5) is on the line.   (8, 2) is on the line.

Using the slope formula, we can call (3, 5) the point P_1 and (8, 2) the point P_2.

$$m = \frac{2 - 5}{8 - 3} = \frac{-3}{5} = -\frac{3}{5}$$

If we call (8, 2) the point P_1 and (3, 5) the point P_2, we still get $-\frac{3}{5}$ for the slope.

$$m = \frac{5 - 2}{3 - 8} = \frac{3}{-5} = -\frac{3}{5}$$

The points (3, 5) and (8, 2) have been plotted and the line they determine has been sketched in Figure 2.17.

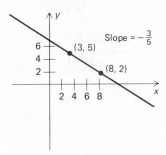

FIGURE 2.17 Graph of $3x + 5y - 34 = 0$

The line in Figure 2.16 slopes up from left to right, and its slope is positive. The line in Figure 2.17 slopes down from left to right, and its slope is negative. Any line that slopes up from left to right has positive slope. Any line that slopes down from left to right has negative slope.

The next example utilizes the functional notation that was introduced in Section 2.1.

EXAMPLE 6

Find an expression for the slope of the line segment with one end at the point with coordinates $(x, f(x))$ and with the other end at the point with coordinates $(x + h, f(x + h))$ where $h \neq 0$.

Solution

Using the slope formula, the slope is

$$\frac{f(x + h) - f(x)}{x + h - x} = \frac{f(x + h) - f(x)}{h}.$$

Exercise 2.2, 16–25

If two nonvertical lines are parallel, they have exactly the same slope. More formally, if nonvertical lines ℓ_1 and ℓ_2 are parallel and have slopes m_1 and m_2 respectively,

$$m_1 = m_2. \qquad \text{Lines } \ell_1 \text{ and } \ell_2 \text{ are parallel.}$$

The slopes of perpendicular oblique lines are also related in a simple way. If two oblique lines are perpendicular, it can be shown that the slope of either one of them is the negative reciprocal of the slope of the other. For example, if two lines are perpendicular and the slope of one of them is $\frac{2}{3}$, then the slope of the other one must be $-\frac{3}{2}$. Another way of describing this relationship is to say that the product of the slopes of perpendicular oblique lines is -1. In more general terms, if oblique lines l_1 and l_2 are perpendicular with slopes m_1 and m_2 respectively, then

$$m_1 m_2 = -1. \qquad \text{Lines } l_1 \text{ and } l_2 \text{ are perpendicular.}$$

In other words,

$$m_1 = \frac{-1}{m_2}$$

and

$$m_2 = \frac{-1}{m_1}.$$

If we know one point on a line and we also know the slope of the line, then we can draw the line. For example, suppose that the point is $(1, 4)$ and the slope is $-\frac{2}{3}$. To draw the line, all we need to do is to plot the given point and locate another point on the line by starting at the given point and moving three units to the right (the positive x-direction) and then moving two units down (the negative y-direction) in accordance with the given slope of $-\frac{2}{3} = \frac{-2}{3}$. Then we just draw the line through the given point and the newly located point. This was done in Figure 2.18. We could also have located a second point on the line by considering the slope $-\frac{2}{3}$ to be $\frac{2}{-3}$ and moving from the given point three units left followed by two units up.

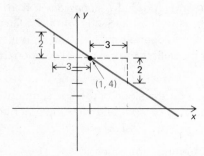

FIGURE 2.18   The line through $(1, 4)$ with slope $-\frac{2}{3}$

We may now show how to find an equation of a line if we are given one of its points together with its slope. Suppose that the given point is $(1, 4)$ and the slope is $\frac{-2}{3}$, as in Figure 2.18. To find an equation satisfied by any point (x, y) on the line, first employ the slope formula and acknowledge that the slope of the segment from $(1, 4)$ to any other point (x, y) must be equal to $\frac{-2}{3}$. In other words, if $x \neq 1$,

$$\frac{y - 4}{x - 1} = \frac{-2}{3}.$$

Multiplying both sides of this equation by $(x - 1)$, we have

$$y - 4 = \frac{-2}{3}(x - 1).$$

This is an equation of the line through $(1, 4)$ with slope $\frac{-2}{3}$. It can be written in the standard form $Ax + By + C = 0$ by first multiplying both sides through by 3.

$$3(y - 4) = -2(x - 1)$$
$$3y - 12 = -2x + 2$$
$$2x + 3y - 14 = 0$$

This is the standard form of an equation of the line through $(1, 4)$ with slope $\frac{-2}{3}$.

To obtain a general formula for an equation of the line through a given point with a given slope, suppose that the given point is the point (x_1, y_1) and that the given slope is m. Then, just as we did in the particular example in the preceding paragraph, to find an equation satisfied by any point (x, y) on the line, first employ the slope formula and acknowledge that the slope of the segment from (x_1, y_1) to any other point (x, y) must be equal to m. In other words, if $x \neq x_1$,

$$\frac{y - y_1}{x - x_1} = m.$$

Multiplying both sides of this equation by $(x - x_1)$, we have

$$y - y_1 = m(x - x_1). \qquad \text{Point-slope form}$$

This is called the **point-slope form** of an equation of a line. It is an equation of the line through (x_1, y_1) with slope m.

Exercise 2.2, 26–40

If the given point is the y-intercept, the point-slope form reduces to a very useful formula. For example, suppose that the given point is $(0, 2)$ and that the slope is $\frac{3}{5}$. Using the point-slope form, an equation of the line through $(0, 2)$ with slope $\frac{3}{5}$ is

$$y - 2 = \tfrac{3}{5}(x - 0),$$
or
$$y = \tfrac{3}{5}x + 2.$$

In this form, note that the coefficient of x is the slope and the nonvariable term is the y-intercept (that is, it is the second coordinate of the y-intercept).

To obtain a general formula for an equation of the line with a given slope and a given y-intercept, suppose that the given slope is m and that the given y-intercept is b—in other words, the y-intercept is $(0, b)$. Using the point-slope form just as we did in the particular example in the preceding paragraph,

$$y - b = m(x - 0),$$

or
$$y = mx + b \qquad \text{Slope—}y\text{-intercept form}$$

Slope y-intercept
m $(0, b)$

The slope—y-intercept form is particularly useful if we are given a linear equation in x and y and are required to find the slope of the line it represents. All we need do is write the given linear equation in the form $y = mx + b$ and read off the coefficient of x. In other words, given a linear equation in the two variables x and y, to find the slope of its line, solve for y in terms of x and read off the coefficient of x.

EXAMPLE 7

Find the slope of $3x - 5y + 7 = 0$.

Solution

Solving for y in terms of x, we have

$$-5y = -3x - 7$$
$$y = \tfrac{3}{5}x + \tfrac{7}{5}.$$

Comparing this with

$$y = mx + b,$$

we see that the slope m is $\tfrac{3}{5}$. (We also see that the y-intercept is $(0, \tfrac{7}{5})$.)

We can also find the slope of $3x - 5y + 7 = 0$ by locating any two points on it and using the definition of slope, but this involves more computation than simply writing the equation in the slope—y-intercept form.

It is now easy to prove that if A, B, and C are real numbers with A and B not zero, then, as was asserted earlier in this section, the graph of

$$Ax + By + C = 0$$

is an oblique line. All we need do is solve for y in terms of x and note that we have the slope—y-intercept form of an equation of a line.

$$By = -Ax - C$$
$$y = \frac{-A}{B}x - \frac{C}{B}$$

Thus $Ax + By + C = 0$ represents a line with slope $-A/B$ and with y-intercept $(0, -C/B)$.

EXAMPLE 8

Find the slope of a line parallel to $2x - 7y + 5 = 0$, and find the slope of a line perpendicular to $2x - 7y + 5 = 0$.

Solution

First let us find the slope of the given line.

$$-7y = -2x - 5$$
$$y = \tfrac{2}{7}x + \tfrac{5}{7}$$

The slope of $2x - 7y + 5 = 0$ is $\tfrac{2}{7}$. The slope of any line parallel to $2x - 7y + 5 = 0$ is also $\tfrac{2}{7}$. The slope of any line perpendicular to $2x - 7y + 5 = 0$ is the negative reciprocal of $\tfrac{2}{7}$ or $-\tfrac{7}{2}$.

Exercise 2.2, 41–50

If we are given two points on a nonvertical line, we can find an equation of the line by first computing its slope and then using the point-slope form with either one of the given points together with the computed slope. For example, suppose that the two given points are $(-1, 2)$ and $(4, 8)$. The slope of the line through $(-1, 2)$ and $(4, 8)$ is

$$\frac{8 - 2}{4 - (-1)} = \frac{6}{5}.$$

Using the given point $(-1, 2)$ and the slope $\tfrac{6}{5}$, the point-slope form yields an equation of the line.

$$y - 2 = \tfrac{6}{5}(x + 1).$$

Writing this in standard form, we have

$$5(y - 2) = 6(x + 1)$$
$$5y - 10 = 6x + 6$$
$$6x - 5y + 16 = 0.$$

If we had used the other given point, $(4, 8)$, we would have obtained the same standard form. (Students should verify this for themselves.)

EXAMPLE 9

Find an equation of the line with intercepts $(0, -2)$ and $(3, 0)$.

Solution

Here we have a choice of ways to proceed. We can first compute the slope and then use the slope–y-intercept form, or we can first compute the slope and then use the point-slope form. The slope is

$$\frac{0 - (-2)}{3 - 0} = \frac{2}{3}.$$

Since the y-intercept is $(0, -2)$, if we now use the slope–y-intercept form, we have

$$y = \tfrac{2}{3}x - 2.$$

Writing this in standard form, we have

$$3y = 2x - 6$$
$$2x - 3y - 6 = 0.$$

If we use the point-slope form and choose the point (3, 0), we have

$$y - 0 = \tfrac{2}{3}(x - 3).$$

Writing this in standard form should produce the same result as the one we obtained by using the slope–y-intercept form.

$$y = \tfrac{2}{3}(x - 3)$$
$$3y = 2x - 6$$
$$2x - 3y - 6 = 0$$

The two results do agree.

EXAMPLE 10

If a and b are nonzero real numbers, find an equation of the line with intercepts (0, b) and (a, 0).

Solution

We can proceed just as we did in Example 9 by first finding the slope and then using either the slope–y-intercept form or the point-slope form. The slope is

$$\frac{0 - b}{a - 0} = \frac{-b}{a}.$$

Since the y-intercept is (0, b), the slope–y-intercept form gives us

$$y = \frac{-b}{a}x + b.$$

If we divide this through by b and isolate the nonvariable term on the right side of the equation, we will have it written in a form that is often seen in classical analytic geometry texts.

$$\frac{y}{b} = \frac{-bx}{ba} + \frac{b}{b}$$

$$\frac{y}{b} = \frac{-x}{a} + 1$$

$$\frac{x}{a} + \frac{y}{b} = 1$$

This last form is sometimes called the *intercept form* of the equation of a line.

Exercise 2.2, 51–65

A **linear function** is a function f such that

1. the domain is the set of all real numbers

2. $f(x) = mx + b$

where m and b are specific real numbers. Since the graph of linear function f is the graph of the equation

$$y = mx + b,$$

we see that the graph of a linear function f is a line with slope m and y-intercept $(0, b)$.

EXAMPLE 11   Sketch the graph of linear function f such that $f(x) = 3x - 2$.

Solution   The graph of f is the graph of the equation $y = 3x - 2$ (see Figure 2.19). The graph is the line with slope 3 and y-intercept $(0, -2)$.

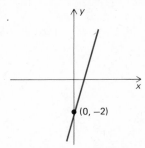

FIGURE 2.19   Graph of f: $f(x) = 3x - 2$
(Graph of $y = 3x - 2$)

We saw in Section 2.1 that any function is associated with a set of ordered pairs and that some authors define a function to be its associated set of ordered pairs. If a function is defined as a certain set of ordered pairs, set-builder notation can be used to define a linear function as

$$\{(x, y)\,|\,y = mx + b\}$$

where m and b are specific real numbers. Using this notation, the linear function f of Example 11 could be written as follows:

$$f = \{(x, y)\,|\,y = 3x - 2\}.$$

We may say that any linear equation in two variables, say x and y, *defines* a linear function because the graph of the equation is the graph of a linear function. We can solve such an equation for y in terms of x to obtain the rule of assignment, and we can use the domain convention to determine that the domain is the set of all real numbers. For example, if f is the function defined by the equation

$$3x - 7y + 2 = 0,$$

then $$f(x) = \tfrac{3}{7}x + \tfrac{2}{7}$$

because if we solve the equation for y in terms of x, we get

$$-7y = -3x - 2$$
$$y = \tfrac{3}{7}x + \tfrac{2}{7}.$$

Defining a function as a certain set of ordered pairs, we could say that

Exercise 2.2, 66–80

$$f = \{(x, y)\,|\,y = \tfrac{3}{7}x + \tfrac{2}{7}\}.$$

In this section we introduced a formula for the slope of a line segment. The next example utilizes this formula and again uses notation and algebraic manipulations that are so frequently encountered in mathematics and its applications.

EXAMPLE 12

If $f(x) = 5x^2 + 3x - 7$ and $h \neq 0$, find an expression for the slope of the line segment with endpoints $(x, f(x))$ and $(x + h, f(x + h))$ and write this expression for the slope as a polynomial in x and h.

Solution

The slope is

$$\frac{f(x + h) - f(x)}{x + h - x} = \frac{f(x + h) - f(x)}{h}.$$

Since

$$f(x) = 5x^2 + 3x - 7,$$
$$f(x + h) = 5(x + h)^2 + 3(x + h) - 7.$$

It follows that the slope of the line segment with endpoints $(x, f(x))$ and $(x + h, f(x + h))$ can be written as follows:

$$\frac{f(x + h) - f(x)}{h} = \frac{5(x + h)^2 + 3(x + h) - 7 - (5x^2 + 3x - 7)}{h}$$

$$= \frac{5(x^2 + 2xh + h^2) + 3x + 3h - 7 - 5x^2 - 3x + 7}{h}$$

$$= \frac{\cancel{5x^2} + 10xh + 5h^2 + \cancel{3x} + 3h - \cancel{7} - \cancel{5x^2} - \cancel{3x} + \cancel{7}}{h}$$

$$= \frac{10xh + 5h^2 + 3h}{h}$$

$$= 10x + 5h + 3$$

That is, where $f(x) = 5x^2 + 3x - 7$, the slope of the line segment from $(x, f(x))$ to $(x + h, f(x + h))$ is

$$\frac{f(x + h) - f(x)}{h} = 10x + 5h + 3.$$

Exercise 2.2, 81–85
Verbal Problems
Exercise 2.2, 86–89

EXERCISE 2.2 LINES; LINEAR FUNCTIONS

▶ Sketch the graph of each of the following and label its intercepts.

EXAMPLE

$-2x - y + 6 = 0$

Solution

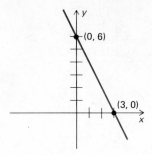

1. $x - 3y + 6 = 0$ **2.** $5x - 2y - 10 = 0$

3. $3x - 4y = 7$ **4.** $2x = 5y + 3$

5. $7x - 8y + 1 = 0$

▶ In each of the following, find the point of intersection of the given lines, sketch the graphs of the lines, and label the point of intersection.

EXAMPLE

$x - 4y + 7 = 0$
$3x + y - 5 = 0$

Solution

$$x - 4y + 7 = 0$$
$$\underline{12x + 4y - 20 = 0}$$
$$13x - 13 = 0$$
$$x = 1$$
$$1 - 4y + 7 = 0$$
$$-4y = -8$$
$$y = 2$$

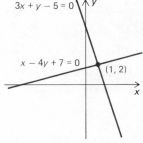

6. $x - 3y - 10 = 0,\ 2x + y + 1 = 0$

7. $x + y + 3 = 0,\ x - y + 1 = 0$

8. $5x - 2y = 4,\ 3x + 4y - 18 = 0$

9. $2x = 7y - 32,\ 4x = 16 - 5y$

10. $3x + 7y = 5,\ 9x - 2y = -31$

▶ Sketch the graph of each of the following and label its intercept.

EXAMPLE $2y + 2 = 0$

Solution $2y + 2 = 0$
                   $y = -1$

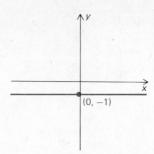

(0, −1)

11. $3y - 6 = 0$ **12.** $3x + 6 = 0$

13. $x = -1$ **14.** $y = 3$

15. $x = 0$

▶ In each of the following, find the slope of the line containing the given points.

EXAMPLE $(2, -4), (-5, 3)$

Solution $m = \dfrac{3 - (-4)}{-5 - 2} = \dfrac{7}{-7} = -1$

16. $(3, -5), (-4, 2)$ **17.** $(3, 0), (-2, 1)$

18. $(1, -2), (7, -2)$ **19.** $(0, 4), (3, 0)$

20. $(-2, 5), (-1, -4)$

▶ In each of the following, show that the given points lie on the given line, find the slope, and sketch the line.

EXAMPLE $(-1, 2), (3, -4), 3x + 2y - 1 = 0$

Solution $3(-1) + 2(2) - 1 = 0$ $3(3) + 2(-4) - 1 = 0$
                   $-3 + 4 \quad - 1 = 0$ True $9 - 8 \quad - 1 = 0$ True
               $(-1, 2)$ is on the line. $(3, -4)$ is on the line.

$$m = \frac{-4 - 2}{3 - (-1)} = \frac{-6}{4}$$

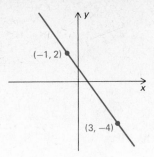

(−1, 2)

(3, −4)

21. (1, 2), (3, 4), $x - y + 1 = 0$

22. (1, −2), (4, 5), $7x - 3y - 13 = 0$

23. (2, 1), (−3, 4), $3x + 5y - 11 = 0$

24. (5, 0), (0, −3), $3x - 5y - 15 = 0$

25. (−3, 2), (1, 1), $x + 4y - 5 = 0$

▶ In each of the following, sketch the graph of the line through the given point with the given slope.

EXAMPLE   (4, 5), $\frac{-3}{2}$

Solution

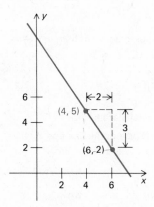

26. (3, 5), $\frac{-4}{3}$ **27.** (2, 7), $\frac{-3}{5}$ **28.** (−1, 3), $\frac{3}{4}$

29. (3, 3), 1   **30.** (−1, 3), −1

▶ In each of the following, find the standard form of an equation of the line through the given point and with the given slope.

EXAMPLES   (a) (2, 3), $\frac{-1}{2}$ (b) (0, 2), $\frac{3}{4}$

Solutions

(a) $y - 3 = \frac{-1}{2}(x - 2)$ (b) $y = \frac{3}{4}x + 2$

$2y - 6 = -x + 2$ $4y = 3x + 8$

$x + 2y - 8 = 0$ $3x - 4y + 8 = 0$

31. $(3, 4)$, $\frac{-2}{5}$ **32.** $(-2, 4)$, $\frac{3}{7}$ **33.** $(-3, 2)$, -1

34. $(-2, -3)$, $\frac{-4}{5}$ **35.** $(-1, 5)$, $\frac{3}{8}$ **36.** $(0, 4)$, $\frac{2}{3}$

37. $(0, -5)$, $\frac{-3}{4}$ **38.** $(0, 6)$, $\frac{3}{7}$ **39.** $(0, 3)$, 1

40. $(0, -1)$, $\frac{-4}{7}$

▶ Write each of the following lines in the slope—y-intercept form and then find the slope and the y-intercept.

EXAMPLE $2x - 3y + 12 = 0$

Solution

$-3y = -2x - 12$

$y = \frac{2}{3}x + 4$

Slope $= \frac{2}{3}$; y-intercept is $(0, 4)$.

41. $2x - 7y + 28 = 0$ **42.** $x - 6y + 12 = 0$

43. $5x + y - 11 = 0$ **44.** $x + y - 3 = 0$

45. $5x - 8y + 12 = 0$

▶ Find the slope of a line parallel to each of the given lines, and then find the slope of a line perpendicular to it.

EXAMPLE $3x - 7y + 2 = 0$

Solution

$-7y = -3x - 2$ Slope of parallel line $= \frac{3}{7}$.

$y = \frac{3}{7}x + \frac{2}{7}$ Slope of perpendicular line $= \frac{-7}{3}$.

46. $5x - 3y + 1 = 0$ **47.** $2x + 4y - 5 = 0$

48. $3x + 7y + 8 = 0$ **49.** $2x + y - 6 = 0$

50. $3x + 4y + 9 = 0$

▶ In each of the following, find the standard form of an equation of the line through the two given points.

EXAMPLE $(2, 3)$, $(4, -5)$

Solution   Slope $= \dfrac{-5 - 3}{4 - 2} = \dfrac{-8}{2} = -4$

$$y - 3 = -4(x - 2)$$
$$y - 3 = -4x + 8$$
$$4x + y - 11 = 0$$

51. $(3, -4), (2, 1)$ **52.** $(2, 5), (3, 7)$

53. $(-1, -3), (4, -2)$ **54.** $(2, -3), (4, -3)$

55. $(-2, 6), (7, 11)$ **56.** $(0, 4), (-5, 0)$

57. $(0, 1), (1, 0)$ **58.** $(0, -2), (-3, 0)$

59. $(0, 3), (2, 0)$ **60.** $(0, \frac{1}{2}), (\frac{-1}{3}, 0)$

61. Find the standard form of an equation of the line through $(0, 0)$ and parallel to $3x - 4y + 1 = 0$.

62. Find the standard form of an equation of the line through $(0, 0)$ and perpendicular to $4x - 3y + 7 = 0$.

63. Find the standard form of an equation of the line through $(0, 0)$ and the point of intersection of $x = 2$ and $3x - 8y + 2 = 0$.

64. Find the standard form of an equation of the line with y-intercept 7 and perpendicular to $y = -3x + 4$.

65. Find the standard form of an equation of the line with x-intercept -2 and with y-intercept 2.

▶ In each of the following, sketch the graph of the linear function f.

EXAMPLES

Solutions

(a) $f(x) = 2x - 3$ (b) $f = \{(x, y)\,|\,y = x + 3\}$

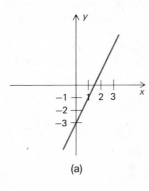

(a)

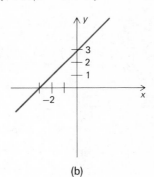

(b)

66. $f(x) = 5x + 2$ **67.** $f(x) = -5x + 1$

68. $f(x) = x + 5$ **69.** $f(x) = -x + 5$

70. $f(x) = 4x + 3$ **71.** $f = \{(x, y)\,|\,y = 3x + 2\}$

72. $f = \{(x, y) | y = \frac{2}{3}x - 1\}$ **73.** $f = \{(x, y) | y = -x + 2\}$

74. $f = \{(x, y) | y = x + 2\}$ **75.** $f = \{(x, y) | y = \frac{-2}{3}x + 1\}$

▶ In each of the following, let f be the function defined by the given equation and find $f(x)$, $f(1)$, and $f(0)$.

EXAMPLE

$2x - 5y + 6 = 0$

Solution

$-5y = -2x - 6$

$y = \frac{2}{5}x + \frac{6}{5}$

$f(x) = \frac{2}{5}x + \frac{6}{5}$

$f(1) = \frac{8}{5}$

$f(0) = \frac{6}{5}$

76. $3x - 7y + 10 = 0$ **77.** $2x - y + 18 = 0$

78. $3x + 7y - 21 = 0$ **79.** $x + y - 6 = 0$

80. $-x - 3y + 8 = 0$

▶ In each of the following assume that h is a nonzero real number. Find an expression for the slope of the line segment with endpoints $(x, f(x))$ and $(x + h, f(x + h))$, and write this expression for the slope as a polynomial in x and h.

81. $f(x) = 2x^2 + 4x - 7$ **82.** $f(x) = 5x^2 - x + 3$

83. $f(x) = 7x^2 + 3x - 5$ **84.** $f(x) = x^2 - 4x + 3$

85. $f(x) = 4x^2 - 3x + 2$

86. A 15% tip on a dinner is computed by multiplying the cost of the dinner by 0.15. Let x be the cost of the dinner and let $C(x)$ be the cost of the dinner plus a 15% tip. Write an expression for $C(x)$. Find $C(0)$. Find $C(20)$. Would it make any sense to have negative numbers in the domain of C? If the domain of C is $\{x | x \geq 0\}$, what is the graph of $y = C(x)$?

87. If $C(x)$ represents the circumference of a circle of radius x meters, then $C(x) = 2\pi x$. Find $C(0)$. Find $C(10)$. Would it make any sense to have negative numbers in the domain of C? If the domain of C is $\{x | x \geq 0\}$, what is the graph of $y = C(x)$?

88. The profit made by selling x mopeds is $P(x) = 75x - 150,000$. For what value of x is $P(x)$ equal to zero? If the domain of P is $\{x | P(x) \geq 0\}$, what is the graph of $y = P(x)$?

89. A couple finds that their annual taxes are 60% of their income. They aso pay $4000 annually for various kinds of insurance. Write an expression for the amount $A(x)$ left from their annual income x after they pay their taxes and insurance. If you drew the graph of $y = A(x)$, what would be the x-intercept? What is the significance of the x-intercept for this couple?

CHECK LIST FOR THE STUDENT

If you have learned the material in Section 2.2, you should be able to do the following:

1. Sketch the graph of a linear equation in two variables.

2. Sketch the graph of a linear equation in one variable.

3. Find the point of intersection of two lines if their equations are given.

4. Find the intercepts of a line if its equation is given.

5. Find the slope of a nonvertical line through two given points.

6. Show whether or not a given point lies on a line whose equation is given.

7. Tell from looking at the graph of a line whether its slope is positive, negative, zero, or not defined.

8. Find the slope of a line parallel to a line with known slope.

9. Find the slope of a line perpendicular to a line with known slope.

10. Draw the line with a given slope through a given point.

11. Find an equation of the line with a given slope through a given point.

12. Write an equation of a line in standard form.

13. Find an equation of the line with a given slope and a given y-intercept.

14. Write an equation of a line in slope–y-intercept form and find its slope and y-intercept.

15. Find an equation of the line through two given points.

16. Find an equation of a line whose intercepts are given.

17. Define a linear function.

18. Sketch the graph of a linear function.

19. Find the rule of assignment of the linear function defined by a given linear equation in two variables.

20. Find a polynomial expression for the slope of a line segment with endpoints at $(x, f(x))$ and $(x + h, f(x + h))$ for a given function f where $f(x) = ax^2 + bx + c$ and where nonzero h, nonzero a, b, and c are real numbers.

SECTION QUIZ 2.2

1. Sketch the graph of $2x + 5y - 10 = 0$, and label its intercepts.

2. Find the point of intersection of the following lines, sketch the graphs of the lines, and label the point of intersection.

$$2x - 3y + 10 = 0$$
$$x - 4y + 14 = 0$$

3. Sketch the graph of each of the following, and label its intercept.
(a) $3x + 3 = 0$ (b) $y = 4$

4. (a) Find the slope of the line containing the points $(3, -2)$ and $(-1, 5)$.

(b) Find an expression for the slope of the line containing the points $(a, f(a))$ and $(a + h, f(a + h))$ where $h \neq 0$.

5. Sketch the graph of the line through $(6, 2)$ with slope $\frac{-1}{3}$.

6. Find the standard form of an equation of the line through the point $(3, -1)$ and with slope $\frac{2}{5}$.

7. Write the line $3x - 2y + 10 = 0$ in the slope–y-intercept form and then find the slope and the y-intercept.

8. (a) Find the slope of a line parallel to $2x - 9y + 4 = 0$.

(b) Find the slope of a line perpendicular to $2x - 9y + 4 = 0$.

9. Find the standard form of an equation of the line through the points $(3, 5)$ and $(1, -3)$.

10. (a) Sketch the graph of the linear function f such that $f(x) = -x + 2$.

(b) Let f be the function defined by the equation $3x - 6y + 5 = 0$. Find $f(x)$, $f(0)$, and $f(1)$.

*2.3 Some Properties of Graphs

The purpose of this section is to demonstrate some properties of graphs that may provide information that can be used in sketching a graph. In particular, intercepts, symmetry, asymptotes, and excluded areas will be discussed and illustrated.

INTERCEPTS

In Section 2.2 we saw that the points in which a line intersects the co-ordinate axes are called its intercepts. Likewise, the point(s) in which any graph intersects the x-axis is (are) called its x-intercept(s), and the point(s) in which any graph intersects the y-axis is (are) called its y-intercept(s). To find the x-intercept(s) of the graph of an equation in x and y, let $y = 0$ and solve for x. To find the y-intercept(s), let $x = 0$ and solve for y.

EXAMPLE 1   Find the intercepts of the graph of $4x^2 + 9y^2 = 36$.

Solution   To find the x-intercepts, let $y = 0$ and solve for x.

$$4x^2 = 36$$
$$x^2 = 9$$
$$x = \pm 3$$

The x-intercepts are (3, 0) and (−3, 0). To find the y-intercepts, let $x = 0$ and solve for y.

$$9y^2 = 36$$
$$y^2 = 4$$
$$y = \pm 2$$

The y-intercepts are (0, 2) and (0, −2).

SYMMETRY

Some graphs are symmetric with respect to one or both of the coordinate axes. Roughly speaking, a graph is symmetric with respect to a line provided that if the paper were folded on the line, the two halves would exactly match. The graph in Figure 2.20 is symmetric with respect to the x-axis. Technically speaking, **a graph is symmetric with respect to the x-axis** provided that whenever the point (x, y) is on the graph, the point $(x, -y)$ is also on the graph. In other words, for any point (x, y) on the graph, there is another point $(x, -y)$ on the graph such that the x-axis is the perpendicular bisector of the segment joining the two points. There is an easy way to test an equation to see whether or not its graph is symmetric with respect to the x-axis.

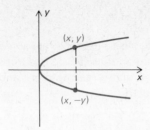

FIGURE 2.20 Graph of $y^2 = x$
(Symmetric with respect to x-axis)

TEST FOR SYMMETRY WITH RESPECT TO x-AXIS

Construct an equation by replacing y by $-y$ in the given equation. If the constructed equation is equivalent to the given equation, then the graph of the given equation is symmetric with respect to the x-axis. Otherwise there is no symmetry with respect to the x-axis.

EXAMPLE 2

Test $y^2 = x$ for symmetry with respect to the x-axis.

Solution

Replacing y by $-y$ in $y^2 = x$, we obtain an equivalent equation.

$$(-y)^2 = x$$

or $$y^2 = x$$

It follows that for any point (x, y) on the graph, the point $(x, -y)$ is also on the graph, so the graph is symmetric with respect to the x-axis. The graph of $y^2 = x$ is shown in Figure 2.20.

A graph is symmetric with respect to the y-axis provided that whenever the point (x, y) is on the graph, the point $(-x, y)$ is also on the graph. The graph in Figure 2.21 is symmetric with respect to the y-axis.

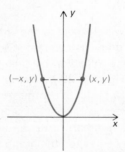

FIGURE 2.21 Graph of $y = x^2$
(Symmetric with respect to y-axis)

The test for symmetry with respect to the *y*-axis is similar to the *x*-axis symmetry test.

TEST FOR SYMMETRY WITH RESPECT TO *y*-AXIS

Construct an equation by replacing *x* by −*x* in the given equation. If the constructed equation is equivalent to the given equation, then the graph of the given equation is symmetric with respect to the *y*-axis. Otherwise there is no symmetry with respect to the *y*-axis.

EXAMPLE 3

Test $y = x^2$ for symmetry with respect to the *y*-axis and with respect to the *x*-axis.

Solution

To y-axis   Replacing *x* by −*x* in $y = x^2$, we obtain an equivalent equation.

$$y = (-x)^2$$
or
$$y = x^2$$

It follows that for any point (x, y) on the graph, the point $(-x, y)$ is also on the graph, so the graph is symmetric with respect to the *y*-axis.

To x-axis   Replacing *y* by −*y* in $y = x^2$ to test for symmetry with respect to the *x*-axis, we obtain

$$(-y) = x^2$$
or
$$y = -x^2.$$

This equation is not equivalent to the given equation, so the graph is not symmetric with respect to the *x*-axis. The graph of $y = x^2$ is shown in Figure 2.21.

EXAMPLE 4

Test $y = x^3$ for symmetry with respect to the axes.

Solution

To x-axis   Replacing *y* by −*y*, we obtain

$$(-y) = x^3$$
or
$$y = -x^3.$$

This equation is not equivalent to the given equation. The graph is not symmetric with respect to the *x*-axis.

To y-axis   Replacing *x* by −*x*, we again obtain an equation that is not equivalent to the given equation.

$$y = (-x)^3$$
or
$$y = -x^3$$

The graph is not symmetric with respect to the *y*-axis. The graph is not symmetric with respect to either axis, and it is shown in Figure 2.22.

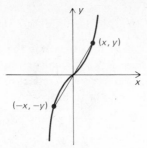

FIGURE 2.22   Graph of $y = x^3$
(Symmetric with respect to the origin)

Some graphs are symmetric with respect to the origin. The graph in Figure 2.22 is symmetric with respect to the origin. **A graph is symmetric with respect to the origin** provided that whenever (*x, y*) is on the graph, the point (−*x*, −*y*) is also on the graph. In other words, for any point (*x, y*) on the graph, there is another point (−*x*, −*y*) on the graph such that the origin is the midpoint of the segment joining them.

TEST FOR SYMMETRY WITH RESPECT TO ORIGIN

Construct an equation by replacing *x* by −*x* and *y* by −*y* in the given equation. If the constructed equation is equivalent to the given equation, then the graph of the given equation is symmetric with respect to the origin. Otherwise there is no symmetry with respect to the origin.

EXAMPLE 5

Test $y = x^3$ for symmetry with respect to the origin.

Solution

Replacing *x* by −*x* and *y* by −*y* in $y = x^3$, we obtain an equivalent equation.

$$(-y) = (-x)^3$$

or $\qquad\qquad -y = -x^3$

or $\qquad\qquad y = x^3$

It follows that for any point (*x, y*) on the graph, the point (−*x*, −*y*) is also on the graph, so the graph is symmetric with respect to the origin.

EXAMPLE 6

Test $4x^2 + 9y^2 = 36$ for symmetry with respect to the origin and the axes.

Solution

To x-axis Replacing y by $-y$, we obtain an equivalent equation.

$$4x^2 + 9(-y)^2 = 36$$

or $\qquad\qquad 4x^2 + 9y^2 = 36$

The graph of $4x^2 + 9y^2 = 36$ is symmetric with respect to the x-axis.

To y-axis Replacing x by $-x$, we obtain an equivalent equation.

$$4(-x)^2 + 9y^2 = 36$$

or $\qquad\qquad 4x^2 + 9y^2 = 36$

The graph of $4x^2 + 9y^2 = 36$ is symmetric with respect to the y-axis.

To origin Replacing x by $-x$ and y by $-y$, we obtain an equivalent equation.

$$4(-x)^2 + 9(-y)^2 = 36$$

or $\qquad\qquad 4x^2 + 9y^2 = 36$

The graph of $4x^2 + 9y^2 = 36$ is symmetric with respect to the origin.

The graph is shown in Figure 2.23. Its intercepts were found in Example 1. It will be discussed more fully in Example 9. It is called an *ellipse*.

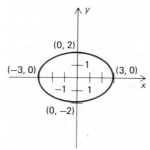

FIGURE 2.23 Graph of $4x^2 + 9y^2 = 36$
(Symmetric with respect to both axes and the origin)

Note that if a graph is symmetric with respect to *both* axes, then it must be symmetric with respect to the origin, but Examples 4 and 5 show that a graph can be symmetric with respect to the origin without being symmetric with respect to *either* axis.

Intercepts
and Symmetry
Exercise 2.3, 1–18, 41–45

ASYMPTOTES

Some graphs have one or more *asymptotes*. An **asymptote** is a line that a graph appears to approach in the sense that as we go farther and farther out on the graph, the graph looks more and more like the line. The graph in Figure 2.24 has the line $x = 1$ as a vertical asymptote, and it has the x-axis as a horizontal asymptote.

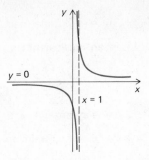

FIGURE 2.24 Graph of $y = \dfrac{1}{x - 1}$

Asymptotes: $x = 1$ and $y = 0$

To get the feeling that the graph in Figure 2.24 really does approach the line $x = 1$, consider a table of values such as Table 2.1, in which x successively takes on values nearer and nearer 1 through values larger than 1. Also consider a table of values such as Table 2.2, in which x successively takes on values nearer and nearer 1 through values smaller than 1. Then plot the points from each of these tables of values, and note that the points appear to be strung together like beads on a graph that more nearly assumes the position of the line $x = 1$ as the values of x get nearer and nearer 1. Plot points, taking x nearer and nearer 1, first to the right of 1 and then to the left of 1, until you see that you can make the points on the graph be as close to the line $x = 1$ as you please simply by taking values of x sufficiently close to 1.

To see intuitively that the graph of $y = 1/(x - 1)$ approaches the x-axis, consider a table of values such as Table 2.3, in which x succes-

TABLE 2.1

x	$y = \dfrac{1}{x - 1}$
2	1
$1\frac{1}{2}$	2
$1\frac{1}{4}$	4
$1\frac{1}{8}$	8
$1\frac{1}{16}$	16
$1\frac{1}{32}$	32

TABLE 2.2

x	$y = \dfrac{1}{x - 1}$
0	-1
$1 - \frac{1}{2}$	-2
$1 - \frac{1}{4}$	-4
$1 - \frac{1}{8}$	-8
$1 - \frac{1}{16}$	-16
$1 - \frac{1}{32}$	-32

TABLE 2.3		TABLE 2.4	
x	$y = \dfrac{1}{x-1}$	x	$y = \dfrac{1}{x-1}$
3	$\frac{1}{2}$	-1	$\frac{-1}{2}$
4	$\frac{1}{3}$	-2	$\frac{-1}{3}$
5	$\frac{1}{4}$	-3	$\frac{-1}{4}$
10	$\frac{1}{9}$	-4	$\frac{-1}{5}$
100	$\frac{1}{99}$	-10	$\frac{-1}{11}$
		-100	$\frac{-1}{101}$

sively takes on larger positive values. Also consider a table such as Table 2.4, in which x successively takes on smaller negative values—that is, x successively becomes larger in absolute value but is negative. Then note in Figure 2.24 that if we plot the points from each of these tables, the points appear to be strung together like beads on a graph that more nearly assumes the position of the x-axis as the values of x get larger and larger positively or as the values of x are negative and get larger and larger in absolute value. The students should plot points, extending Tables 2.3 and 2.4, if necessary, until they see that they can make the points on the graph of $y = 1/(x-1)$ be as close to the x-axis as they please simply by taking sufficiently large positive values of x or by taking negative values of x with sufficiently large absolute values.

Precise definitions of vertical and horizontal asymptotes and ways of finding them can be given early in the study of calculus. Some unsophisticated ways of finding vertical or horizontal asymptotes for the graphs of some equations of the form $y = P(x)/D(x)$ are given and illustrated below.

TO FIND VERTICAL ASYMPTOTES FOR $y = P(x)/D(x)$

Find any real numbers x for which $D(x) = 0$. If a is a real number such that $D(a) = 0$ and $P(a) \neq 0$, then $x = a$ is a vertical asymptote. If $D(x) = 0$ has no real solutions, there are no vertical asymptotes.

EXAMPLE (i)

$$y = \frac{3x + 2}{4x - 1}$$

$$4x - 1 = 0$$

$$x = \tfrac{1}{4}$$

The line $x = \frac{1}{4}$ is the vertical asymptote.

EXAMPLE (ii)

$$y = \frac{8}{\sqrt{x^2 - 4}}$$

Here the denominator $D(x)$ is zero when $x = 2$ or -2. The lines $x = 2$ and $x = -2$ are the vertical asymptotes.

TO FIND HORIZONTAL ASYMPTOTES FOR $y = P(x)/D(x)$ WHERE $P(x)$ AND $D(x)$ ARE POLYNOMIALS IN x

Case 1   *The degree of P(x) is lower than the degree of D(x).* The horizontal asymptote is the x-axis, $y = 0$.

Example: $y = \dfrac{1}{x - 1}$

The degree of $P(x)$ is zero; the degree of $D(x)$ is 1. The horizontal asymptote is the x-axis, $y = 0$.

Case 2   *The degree of P(x) is the same as the degree of D(x).* Suppose that $P(x)$ and $D(x)$ are both of degree n. Then the horizontal asymptote is $y = A/B$ where A is the coefficient of x^n in $P(x)$ and B is the coefficient of x^n in $D(x)$.

Example: $y = \dfrac{3x + 2}{4x - 1}$

$P(x)$ and $D(x)$ are both of degree 1. The coefficient of x in $P(x)$ is 3; the coefficient of x in $D(x)$ is 4. The horizontal asymptote is $y = \frac{3}{4}$.

Case 3   *The degree of P(x) is higher than the degree of D(x).* There are no horizontal asymptotes.

Example: $y = \dfrac{x^2}{x - 1}$

The degree of $P(x)$ is 2; the degree of $D(x)$ is 1. In any table of values in which x successively takes on larger absolute values, y also takes on successively larger absolute values and fails to approach any definite number. There are no horizontal asymptotes.

If the equation $y = P(x)/D(x)$ can be solved for x in terms of y so that $x = Q(y)/E(y)$, then any horizontal asymptotes that the graph of $y = P(x)/D(x)$ may have can be found by considering $E(y)$. In particular, if b is a real number for which $E(y) = 0$ and $Q(y) \neq 0$, then $y = b$ is a horizontal asymptote. If no real values of y make $E(y)$ equal to zero, then there are no horizontal asymptotes.

EXAMPLE 7   Discuss the graph of

$$y = \frac{3x + 2}{4x - 1},$$

taking into consideration intercepts, symmetry with respect to axes and origin, and vertical and horizontal asymptotes. Sketch the graph.

Solution

Intercepts If $x = 0$, $y = -2$. The y-intercept is $(0, -2)$.
If $y = 0$, $3x + 2 = 0$, so $x = \frac{-2}{3}$. The x-intercept is $(\frac{-2}{3}, 0)$.

Symmetry

To x-axis Replacing y by $-y$, we obtain an equation that is not equivalent to the given equation.

$$-y = \frac{3x + 2}{4x - 1} \quad \text{or} \quad y = \frac{-3x - 2}{4x - 1}$$

There is no symmetry with respect to the x-axis.

To y-axis Replacing x by $-x$, we obtain an equation that is not equivalent to the given equation.

$$y = \frac{3(-x) + 2}{4(-x) - 1} \quad \text{or} \quad y = \frac{-3x + 2}{-4x - 1}$$

There is no symmetry with respect to the y-axis.

To origin Replacing x by $-x$ and y by $-y$, we obtain an equation that is not equivalent to the given equation.

$$-y = \frac{3(-x) + 2}{4(-x) - 1} \quad \text{or} \quad -y = \frac{-3x + 2}{-4x - 1} \quad \text{or} \quad y = \frac{3x - 2}{-4x - 1}$$

There is no symmetry with respect to the origin.

Asymptotes

Vertical $4x - 1 = 0$

$x = \frac{1}{4}$ is the vertical asymptote.

Horizontal The numerator and denominator are both polynomials of degree 1 with the coefficients of their first-degree terms being 3 and 4, respectively. As we saw in the illustration of Case 2, the horizontal asymptote is $y = \frac{3}{4}$. In this example the horizontal asymptote can also readily be determined by solving the given equation for x in terms of y and then examining the new denominator.

$$y = \frac{3x + 2}{4x - 1}$$

$$4xy - y = 3x + 2$$

$$4xy - 3x = y + 2$$

$$x(4y - 3) = y + 2$$

$$x = \frac{y + 2}{4y - 3}$$

The new denominator $4y - 3$ equals zero when $y = \frac{3}{4}$. The horizontal asymptote is $y = \frac{3}{4}$, as we have already discovered.

EXCLUDED AREAS

The intercepts and asymptotes for

$$y = \frac{3x + 2}{4x - 1}$$

have been marked in Figure 2.25. It is apparent that more information will be needed in order to sketch the graph. In this example, as is often the case, areas can be excluded by considering what signs the variables must have on each side of each asymptote or intercept. We shall now demonstrate this technique.

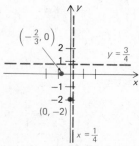

FIGURE 2.25 Intercepts and asymptotes of $y = \dfrac{3x + 2}{4x - 1}$

In this case, the x-intercept is at $(\frac{-2}{3}, 0)$ and the vertical asymptote is at $x = \frac{1}{4}$. We see that if x is greater than $\frac{1}{4}$, the denominator $4x - 1$ is positive and the numerator $3x + 2$ is also positive, so y must be positive.

$$\text{If } x > \tfrac{1}{4}, \text{ then } y > 0.$$

We can therefore cross out the portion of the plane for $x > \frac{1}{4}$ and y negative, because there is no part of the graph there. This has been done in Figure 2.26.

If x is between $\frac{1}{4}$ and $\frac{-2}{3}$ (that is, $x < \frac{1}{4}$ and $x > \frac{-2}{3}$), the denominator $4x - 1$ is negative and the numerator $3x + 2$ is positive, so y must be negative.

$$\text{If } \tfrac{-2}{3} < x < \tfrac{1}{4}, \text{ then } y < 0.$$

This means that y is not positive for any x between $\frac{-2}{3}$ and $\frac{1}{4}$. We can therefore cross out the portion of the plane where y would be positive between $x = \frac{-2}{3}$ and $x = \frac{1}{4}$, because there is no part of the graph there.

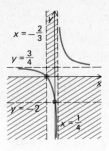

$x = -\frac{2}{3}$

$y = \frac{3}{4}$

$y = -2$

$x = \frac{1}{4}$

FIGURE 2.26 Graph of $y = \dfrac{3x + 2}{4x - 1}$

Finally, if x is less than $\frac{-2}{3}$, the numerator $3x + 2$ is negative and the denominator $4x - 1$ is also negative; since the numerator and denominator are of the same sign, y must be positive.

$$\text{If } x < \tfrac{-2}{3}, \text{ then } y > 0.$$

In other words, y cannot be negative for $x < \frac{-2}{3}$. The portion of the plane to the left of $\frac{-2}{3}$ where y is negative can be crossed out, because there is no part of the graph there.

We could now make a table of values and plot a few points and sketch the graph. In this example, however, as we saw earlier, it happens to be particularly easy to solve for x in terms of y and rule out portions of the plane by looking at the signs that x must have on each side of the y-intercept $(0, -2)$ and each side of the horizontal asymptote $y = \frac{3}{4}$. Let us do just that.

Solving the given equation for x in terms of y, we found that

$$x = \frac{y + 2}{4y - 3}.$$

From this we see that the signs of the numerator $y + 2$ and the denominator $4y - 3$ change at $y = -2$ and $y = \frac{3}{4}$. If $y > \frac{3}{4}$ (for points above the horizontal asymptote), the denominator $4y - 3$ is positive, and the numerator $y + 2$ is also positive so x must be positive.

$$\text{If } y > \tfrac{3}{4}, \text{ then } x > 0.$$

We can rule out the portion of the plane above the horizontal asymptote where x would be negative. Some of this was already ruled out before and was crossed out in Figure 2.26.

For points between the horizontal asymptote and the y-intercept (that is, for $-2 < y < \frac{3}{4}$), the numerator $y + 2$ is positive and the denominator is negative, so x must be negative.

$$\text{If } -2 < y < \tfrac{3}{4}, \text{ then } x < 0.$$

That means that x cannot be positive for $-2 < y < \tfrac{3}{4}$. We can therefore rule out all the portion of the plane to the right of the y-axis between $y = -2$ and $y = \tfrac{3}{4}$ because there is no part of the graph there.

For points below the y-intercept (that is, for $y < -2$), the numerator $y + 2$ is negative and the denominator is also negative, so x must be positive.

$$\text{If } y < -2, \text{ then } x > 0.$$

In other words, x cannot be negative for $y < -2$, so we can rule out all that part of the plane below $y = -2$ where x would be negative.

It is now fairly evident that the graph must look like the one sketched. Constructing a table of values and plotting points would verify this.

EXAMPLE 8   Discuss the graph of

$$y = \frac{3x^2}{x - 1},$$

taking into consideration intercepts, symmetry with respect to axes and origin, vertical and horizontal asymptotes, and excluded areas. Sketch the graph.

Solution

Intercepts   If $x = 0$, $y = 0$. The y-intercept is $(0, 0)$.
If $y = 0$, $3x^2 = 0$, so $x = 0$. The x-intercept is $(0, 0)$.

Symmetry
To x-axis   Replacing y by $-y$, we obtain an equation not equivalent to the given equation. There is no symmetry with respect to the x-axis.

To y-axis   Replacing x by $-x$, we obtain

$$y = \frac{3x^2}{-x - 1},$$

which is not equivalent to the given equation. There is no symmetry with respect to the y-axis.

To origin   Replacing x by $-x$ and y by $-y$, we obtain

$$-y = \frac{3x^2}{-x - 1} \quad \text{or} \quad y = \frac{3x^2}{x + 1},$$

which is not equivalent to the given equation. There is no symmetry with respect to the origin.

Asymptotes

Vertical $x - 1 = 0$

$x = 1$ is the vertical asymptote.

Horizontal   The numerator and denominator are polynomials, and the degree of the numerator is higher than the degree of the denominator. There are no horizontal asymptotes.

Excluded areas   In this example the numerator $3x^2$ is zero when $x = 0$, but otherwise the numerator is always positive. The denominator $x - 1$ is positive if $x > 1$; the denominator is negative if $x < 1$. It follows that

if $x > 1$, y is positive;

if $x < 1$ and $x \neq 0$, y is negative;

if $x = 0$, $y = 0$.

We can therefore *exclude* those areas for which

$x > 1$ and y is negative,

and *exclude* those areas for which

$x < 1$, $x \neq 0$, and y is positive

because there is no part of the graph there. These areas have been excluded in Figure 2.27.

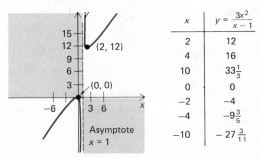

x	$y = \dfrac{3x^2}{x-1}$
2	12
4	16
10	$33\frac{1}{3}$
0	0
−2	−4
−4	$-9\frac{3}{5}$
−10	$-27\frac{3}{11}$

FIGURE 2.27   Graph of $y = \dfrac{3x^2}{x - 1}$

We know from our considerations that $x = 1$ is an asymptote and that the graph of

$$y = \frac{3x^2}{x - 1}$$

goes through (0, 0), although it does not cross the x-axis there. We could now tediously make an extensive table of values and sketch the graph as it is shown in Figure 2.27. We could see from an extensive table that the point (2, 12) appears to be the lowest point of the branch of the graph in the first quadrant. We could also see from an extensive table that if we take larger and larger values of x beyond 2, the corresponding values of y increase without bound and the graph rises to the right as shown.

An extensive table of values would also indicate that (0, 0) is the highest point on the branch of the graph to the left of $x = 1$ and that if we successively take negative values of x with larger and larger absolute values, the corresponding values of y are negative and also successively take on larger and larger absolute values. The graph would fall to the left as shown.

In Example 8, the precise location and identification of the points (2, 12) and (0, 0) as a relative low point and a relative high point can be readily accomplished by using elementary calculus. Here we are merely providing some experience with common properties of graphs that can be examined with algebraic methods.

We have seen that in sketching graphs, we can sometimes exclude areas by considering the signs of the variables involved. At other times, we can exclude intervals and hence areas because values of one of the variables in those intervals would produce imaginary values for the other variable. This is illustrated in the next example.

EXAMPLE 9   Discuss the graph of $4x^2 + 9y^2 = 36$, taking into consideration intercepts, symmetry with respect to axes and origin, vertical and horizontal asymptotes, and excluded areas. Sketch the graph.

Solution

Intercepts   We saw in Example 1 that the x-intercepts are (3, 0) and (-3, 0) and that the y-intercepts are (0, 2) and (0, -2).

Symmetry   We saw in Example 6 that the graph of $4x^2 + 9y^2 = 36$ is symmetric with respect to the x-axis, the y-axis, and the origin.

Asymptotes
Vertical   If we solve $4x^2 + 9y^2 = 36$ for y in terms of x, we have

$$9y^2 = 36 - 4x^2$$

$$y^2 = \frac{36 - 4x^2}{9}$$

$$y = \frac{\pm\sqrt{36 - 4x^2}}{3}$$

$$= \frac{\pm 2\sqrt{9 - x^2}}{3}.$$

There are no real values of x for which the denominator is zero, so there are no vertical asymptotes.

Horizontal   If we solve $4x^2 + 9y^2 = 36$ for x in terms of y, we have

$$4x^2 = 36 - 9y^2$$

$$x^2 = \frac{36 - 9y^2}{4}$$

$$x = \frac{\pm\sqrt{36 - 9y^2}}{2}$$

$$= \frac{\pm 3\sqrt{4 - y^2}}{2}.$$

There are no values of y for which the denominator is zero, so there are no horizontal asymptotes.

Excluded areas   When we solved the given equation for y in terms of x, we found that

$$y = \frac{\pm 2\sqrt{9 - x^2}}{3}.$$

We have agreed to restrict ourselves to relations in which the ordered pairs (x, y) are pairs of real numbers. Consequently, we must *exclude* all values of x for which $9 - x^2$ is negative, because for such values of x the corresponding values of y would be imaginary.

$$\text{Exclude } x \text{ such that } 9 - x^2 < 0 \quad \text{or} \quad 9 < x^2.$$

In other words, we must exclude the portion of the plane for which $x^2 > 9$, that is, the areas where $x > 3$ or $x < -3$.

When we solved the given equation for x in terms of y, we obtained

$$x = \frac{\pm 3\sqrt{4 - y^2}}{2}.$$

Thus to avoid imaginary values for x, we must *exclude* the portion of the plane for which $4 - y^2$ is negative.

$$\text{Exclude } y \text{ such that } 4 - y^2 < 0 \quad \text{or} \quad 4 < y^2.$$

In other words, exclude the portion of the plane for which $y^2 > 4$, that is, the areas where $y > 2$ or $y < -2$.

The intercepts and excluded areas are shown, together with the graph, in Figure 2.28. By virtue of the symmetry with respect to both axes, in order to obtain a sketch of the entire graph it is only necessary to sketch the part in the first quadrant corresponding to values of x between 0 and 3. The rest of the graph can be obtained by reflection in the x-axis

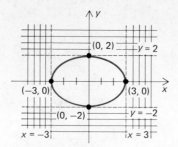

FIGURE 2.28 Graph of $4x^2 + 9y^2 = 36$
Excluded areas: $|x| > 3$ and $|y| > 2$

Excluded Areas
Exercise 2.3, 41–45

Miscellaneous
Exercise 2.3, 1–50

and then in the *y*-axis. As we noted in Example 6, this graph is called an ellipse.

EXERCISE 2.3 SOME PROPERTIES OF GRAPHS

▶ Discuss the graph of each of the following, taking into consideration intercepts, symmetry with respect to axes and origin, vertical and horizontal asymptotes, and excluded areas. Sketch the graph. Label any intercepts or vertical or horizontal asymptotes that the graph has.

1. $y^2 = x - 1$ **2.** $y^2 = x + 1$

3. $y^2 = 4x$ **4.** $y^2 = 6x$

5. $y^2 = 4(x - 1)$ **6.** $y = x^2 + 1$

7. $y = x^2 - 1$ **8.** $y = 3x^2$

9. $4y = x^2$ **10.** $4(y - 1) = x^2$

11. $y = 2x^3$ **12.** $y = -x^3$

13. $y = -2x^3$ **14.** $y = x^3 + 1$

15. $y = x^5$ **16.** $y = x^7$

17. $y = -x^5$ **18.** $y = -x^7$

19. $y = \dfrac{1}{x + 1}$ **20.** $y = \dfrac{1}{x - 3}$

21. $y = \dfrac{1}{x + 3}$ **22.** $y = \dfrac{2}{x - 4}$

23. $y = \dfrac{3}{x - 1}$ **24.** $xy = 1$

25. $xy = -2$ **26.** $y = \dfrac{2x + 3}{4x - 1}$

27. $y = \dfrac{3x - 6}{2x + 1}$

28. $y = \dfrac{x - 5}{x + 1}$

29. $y = \dfrac{5x + 10}{x - 3}$

30. $y = \dfrac{x}{x + 5}$

31. $y = \dfrac{2x - 8}{3x + 6}$

32. $y = \dfrac{x - 1}{2x}$

33. $y = \dfrac{x^2}{3x^2 - 12}$

34. $y = \dfrac{x^2}{x^2 - 9}$

35. $y = \dfrac{3x^2}{x^2 - 4}$

36. $y = \dfrac{4x^2}{x - 3}$

37. $y = \dfrac{x^2}{x - 1}$

38. $y = \dfrac{2x^2}{x - 3}$

39. $y = \dfrac{3x^2}{x + 1}$

40. $y = \dfrac{4x^2}{x + 3}$

41. $9x^2 + 4y^2 = 36$

42. $4x^2 + 4y^2 = 36$

43. $9x^2 + y^2 = 9$

44. $9x^2 + 9y^2 = 9$

45. $x^2 + 9y^2 = 9$

46. $9x^2 - 4y^2 = 36$

47. $4x^2 - 9y^2 = 36$

48. $25x^2 + 16y^2 = 400$

49. $16x^2 + 25y^2 = 400$

50. $25x^2 - 16y^2 = 400$

CHECK LIST FOR THE STUDENT

If you have learned the material in Section 2.3, you should be able to do the following:

.1. Find the intercepts of the graphs of some equations.

2. Test for symmetry with respect to the x-axis.

3. Test for symmetry with respect to the y-axis.

4. Test for symmetry with respect to the origin.

5. Find vertical asymptotes for $y = P(x)/D(x)$ provided that any exist and provided that you can solve $D(x) = 0$.

6. Find horizontal asymptotes for $y = P(x)/D(x)$ where $P(x)$ and $D(x)$ are polynomials in x provided that any horizontal asymptotes exist.

7. Determine excluded areas for some equations by considering signs of the variables.

8. Determine excluded areas for some equations by ruling out imaginary values of the variables.

9. Sketch the graphs of equations similar to those illustrated in the examples of Section 2.3.

SECTION QUIZ 2.3

1. Find the intercepts of the graph of $4x^2 + 25y^2 = 100$.

2. Determine whether or not the graph of each of the following is symmetric with respect to the x-axis.
   (a) $y^2 = 3x$ (b) $y = 3x^2$ (c) $4x^2 + 25y^2 = 100$ (d) $y = 2x^3$

3. Determine whether or not the graph of each of the following is symmetric with respect to the y-axis.
   (a) $y^2 = 3x$ (b) $y = 3x^2$ (c) $4x^2 + 25y^2 = 100$ (d) $y = 2x^3$

4. Determine whether or not the graph of each of the following is symmetric with respect to the origin.
   (a) $y^2 = 3x$ (b) $y = 3x^2$ (c) $4x^2 + 25y^2 = 100$ (d) $y = 2x^3$

5. In each of the following, find any vertical asymptote(s) that the graph has.

   (a) $y = \dfrac{x - 1}{x^2 - 9}$ (b) $y = \dfrac{x - 1}{x^2 + 9}$ (c) $y = \dfrac{x - 1}{x - 9}$

6. In each of the following, find any horizontal asymptote(s) that the graph has.

   (a) $y = \dfrac{3x}{x^2 + 4}$ (b) $y = \dfrac{2x - 3}{5x + 1}$ (c) $y = \dfrac{x^2 + 4}{3x + 1}$

7. Determine areas from which the graph of

$$y = \frac{5x + 5}{x - 4}$$

is excluded by considering signs of the variables.

8. Determine areas from which the graph of $4x^2 + 25y^2 = 100$ is excluded by ruling out imaginary values of the variables.

Discuss the graph of each of the following, taking into consideration intercepts, symmetry with respect to axes and origin, vertical and horizontal asymptotes, and excluded areas. Sketch the graph. Label any intercepts or vertical or horizontal asymptotes that the graph has.

9. $25x^2 + 4y^2 = 100$

10. $y = \frac{3x + 6}{2x - 8}$

*2.4 Distance and Midpoint Formulas; Circles; Translation

The purpose of this section is to present formulas for the distance between two points and for the midpoint of a line segment, to consider equations of circles, and to show how to transform an equation of a graph by translating the coordinate axes.

DISTANCE AND MIDPOINT FORMULAS

The distance between two points on a line parallel to the x-axis is the absolute value of the difference in their x-coordinates. For example, the distance between $(2, 3)$ and $(7, 3)$ on the line $y = 3$ is $|7 - 2| = 5$. In Figure 2.29 we see that $|7 - 2|$ is the number of units between $(2, 3)$ and $(7, 3)$. Note that $|7 - 2| = |2 - 7|$.

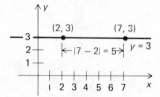

FIGURE 2.29 $|7 - 2| = |2 - 7| = 5$ The distance between $(2, 3)$ and $(7, 3)$

The distance between two points on a line parallel to the y-axis is the absolute value of the difference in their y-coordinates. In Figure 2.30 the distance between $(2, 1)$ and $(2, 5)$ on the line $x = 2$ is $|5 - 1| = 4$. Note that $|5 - 1| = |1 - 5|$.

FIGURE 2.30 $|5 - 1| = |1 - 5| = 4$ The distance between $(2, 1)$ and $(2, 5)$

It is easy to find the distance between two points on an oblique line by constructing a right triangle and making a direct application of the theorem from geometry called the *Pythagorean theorem.* This theorem states that in any right triangle, the square of the length of the hypotenuse is equal to the sum of the squares of the lengths of the other two sides. To demonstrate, let us use the Pythagorean theorem to find the distance between (1., 2) and (4, 6). Let P be the point (1, 2), and let Q be the point (4, 6), as shown in Figure 2.31. Draw a line through P parallel to the x-axis and a line through Q parallel to the y-axis. Let the point of intersection of these two parallels be M. Then M has coordinates (4, 2), and the triangle PMQ is a right triangle with hypotenuse $\overline{PQ}$. The length of the hypotenuse is the distance d between P and Q. The length of $\overline{PM}$ is $|4 - 1| = 3$; the length of $\overline{MQ}$ is $|6 - 2| = 4$. Using the Pythagorean theorem,

$$d^2 = 3^2 + 4^2$$
$$= 9 + 16$$
$$= 25.$$

Since the distance between two points is nonnegative, $d = 5$.

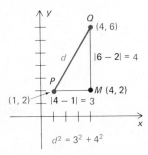

FIGURE 2.31 The distance between (1, 2) and (4, 6)

Precisely the same technique can be used to derive a general formula for the distance between point P_1 with coordinates (x_1, y_1) and point P_2 with coordinates (x_2, y_2) where P_1 and P_2 are on an oblique line. In Figure 2.32, triangle P_1MP_2 is a right triangle with hypotenuse $\overline{P_1P_2}$. The coordinates of M are (x_2, y_1); the length of $\overline{P_1M}$ is $|x_2 - x_1|$; the length of $\overline{MP_2}$ is $|y_2 - y_1|$. By the Pythagorean theorem,

$$d^2 = |x_2 - x_1|^2 + |y_2 - y_1|^2.$$

But $|x_2 - x_1|^2$ is the same as $(x_2 - x_1)^2$, and $|y_2 - y_1|^2$ is the same as $(y_2 - y_1)^2$, so we can write

$$d^2 = (x_2 - x_1)^2 + (y_2 - y_1)^2.$$

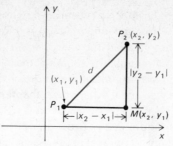

FIGURE 2.32 $d^2 = (x_2 - x_1)^2 + (y_2 - y_1)^2$

And since the distance between two points is nonnegative,

$$d = \sqrt{(x_2 - x_1)^2 + (y_2 - y_1)^2}.$$ Distance between (x_1, y_1) and (x_2, y_2)

EXAMPLE 1   Find the distance between $(-2, 3)$ and $(10, 8)$.

Solution   If we call $(-2, 3)$ the point P_1 and $(10, 8)$ the point P_2, we have

$$\begin{aligned} d &= \sqrt{(10 - (-2))^2 + (8 - 3)^2} \\ &= \sqrt{12^2 + 5^2} \\ &= \sqrt{144 + 25} \\ &= \sqrt{169} \\ &= 13. \end{aligned}$$

Note that we get the same answer if we call $(10, 8)$ the point P_1 and $(-2, 3)$ the point P_2. That is because $(x_2 - x_1)^2$ is the same as $(x_1 - x_2)^2$, and $(y_2 - y_1)^2$ is the same as $(y_1 - y_2)^2$.

$$\begin{aligned} d &= \sqrt{(-2 - 10)^2 + (3 - 8)^2} \\ &= \sqrt{(-12)^2 + (-5)^2} \\ &= \sqrt{144 + 25} \\ &= 13 \end{aligned}$$

Exercise 2.4, 1–10

We can construct appropriate similar triangles and use a geometric argument to prove that the *x*-coordinate of the midpoint of a line segment is the *average* of the *x*-coordinates of the endpoints, and the *y*-coordinate of the midpoint is the *average* of the *y*-coordinates of the endpoints. More precisely, as displayed in Figure 2.33, the midpoint of the line segment with endpoints at (x_1, y_1) and (x_2, y_2) is

$$\left(\frac{x_1 + x_2}{2}, \frac{y_1 + y_2}{2} \right).$$ Midpoint of segment with endpoints (x_1, y_1) and (x_2, y_2)

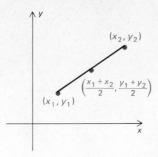

FIGURE 2.33 Midpoint: $\left(\dfrac{x_1 + x_2}{2}, \dfrac{y_1 + y_2}{2}\right)$

EXAMPLE 2

Use the midpoint formula to find the coordinates of the midpoint of the line segment with endpoints $(-2, 3)$ and $(4, -5)$. Use the distance formula to verify that the midpoint is equidistant from each end.

Solution

The midpoint of the line segment with endpoints $(-2, 3)$ and $(4, -5)$ is

$$\left(\frac{-2 + 4}{2}, \frac{3 + (-5)}{2}\right).$$

Simplifying these coordinates, we obtain $(\frac{2}{2}, -\frac{2}{2})$ or

$$(1, -1).$$

Using the distance formula, we can show that the distance between $(1, -1)$ and $(-2, 3)$ is the same as the distance between $(1, -1)$ and $(4, -5)$. The distance between $(1, -1)$ and $(-2, 3)$ is

$$\sqrt{(-2 - 1)^2 + (3 - (-1))^2} = \sqrt{3^2 + 4^2} = \sqrt{25} = 5.$$

The distance between $(1, -1)$ and $(4, -5)$ is

$$\sqrt{(4 - 1)^2 + (-5 - (-1))^2} = \sqrt{3^2 + (-4)^2} = \sqrt{25} = 5.$$

Exercise 2.4, 11–15

The point $(1, -1)$ *is* equidistant from each end.

CIRCLES

A **circle** is the set of all points in a plane equally distant from a fixed point. The fixed point is called the **center**. The distance from the center to any point on the circle is called the **radius**. Twice the radius is called the **diameter**.

We can use the distance formula to find an equation of the circle with a given center and radius. That is, we can find an equation that is satisfied by every point on the circle and by no other point. To demonstrate, let us find an equation of the circle with center at $(-2, 3)$ and with radius 5. In

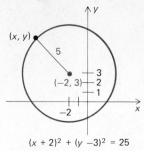

$(x + 2)^2 + (y - 3)^2 = 25$

FIGURE 2.34 Circle with center $(-2, 3)$ and radius 5

Figure 2.34, let (x, y) be any point on the circle. Then the distance from (x, y) to the center $(-2, 3)$ must be 5. Using the distance formula,

$$\sqrt{(x + 2)^2 + (y - 3)^2} = 5.$$

This is an equation of the circle with center at $(-2, 3)$ and radius 5. We can write the equation in an equivalent form by squaring both sides.

$$(x + 2)^2 + (y - 3)^2 = 25$$

Using precisely the same technique we just used for the specific circle in the previous paragraph, we can derive a general form for an equation of a circle with center at (h, k) and radius r where h, k, and nonnegative r represent specific real numbers. To do this, we again let (x, y) be any point on the circle, and we use the distance formula to express the fact that the distance from (x, y) to the center (h, k) must be equal to the radius.

$$\sqrt{(x - h)^2 + (y - k)^2} = r$$

Squaring both sides, we obtain the "standard" form

$$(x - h)^2 + (y - k)^2 = r^2 \qquad \text{Circle: Center } (h, k), \text{ radius } r$$

If the center is at the origin, $h = 0$ and $k = 0$ so that the standard form becomes much simpler.

Exercise 2.4, 16–20

$$x^2 + y^2 = r^2 \qquad \text{Circle: Center } (0, 0), \text{ radius } r$$

EXAMPLE 3   What are the center and radius of the circle $(x - 1)^2 + (y + 4)^2 = 36$?

Solution   Comparing the given equation with the standard form, we see that $h = 1$, $k = -4$, and $r^2 = 36$. The center is therefore at $(1, -4)$ and the radius is 6. (The radius must be nonnegative.)

If we multiply out the standard form $(x - h)^2 + (y - k)^2 = r^2$ and collect terms, we obtain the equation

$$x^2 + y^2 - 2hx - 2ky + (h^2 + k^2 - r^2) = 0.$$

We see that this is a special case of the second-degree equation

$$Ax^2 + By^2 + Cx + Dy + E = 0$$

where A, B, C, D, and E represent specific real numbers. In our special case, $A = B = 1$, $C = 2h$, $D = 2k$, and $E = h^2 + k^2 - r^2$. If we are given such a second-degree equation with $A = B$ and A and B nonzero, we can complete squares and write the equation in standard form as illustrated in Exercise 1.7 and in the next example. This next example shows that such a second-degree equation represents either a circle, a point, or no graph at all, depending on whether the constant term on the right side of the standard form is positive, zero, or negative.

Exercise 2.4, 21–25

EXAMPLE 4

Write each of the following equations in the form $(x - h)^2 + (y - k)^2 = r^2$. If the equation has a graph, sketch or plot the graph; if the equation has no graph, simply state "No graph."

(a) $2x^2 + 2y^2 - 16x - 20y + 64 = 0$

(b) $2x^2 + 2y^2 - 16x - 20y + 82 = 0$

(c) $2x^2 + 2y^2 - 16x - 20y + 100 = 0$

Solution

(a) $2x^2 + 2y^2 - 16x - 20y + 64 = 0$

We will use the completing-the-square process discussed in Sections 1.5 and 1.7 to complete the square in x and to complete the square in y.

1. Divide through by the common coefficients of x and y.

$$x^2 + y^2 - 8x - 10y + 32 = 0$$

2. Isolate the nonvariable term on the right; write the terms in x and the terms in y together.

$$x^2 - 8x + y^2 - 10y = -32$$

3. To both sides of the equation add the square of half the coefficient of x and also the square of half the coefficient of y.

$$x^2 - 8x + (-4)^2 + y^2 - 10y + (-5)^2 = -32 + (-4)^2 + (-5)^2$$
$$x^2 - 8x + 16 + y^2 - 10y + 25 = -32 + 16 + 25$$
$$x^2 - 8x + 16 + y^2 - 10y + 25 = 9$$

4. Factor the first three terms; factor the second three terms.

$$(x - 4)^2 + (y - 5)^2 = 9$$

This is the standard form of an equation of a circle with center at $(4, 5)$ and radius 3. It is sketched in Figure 2.35.

(b) $2x^2 + 2y^2 - 16x - 20y + 82 = 0$

Completing the square in x and also in y we have the following.

$$x^2 - 8x \qquad\quad + y^2 - 10y \qquad\qquad = -41$$
$$x^2 - 8x + 16 + y^2 - 10y + 25 = -41 + 16 + 25$$
$$(x - 4)^2 \qquad\quad + (y - 5)^2 \qquad\qquad = 0$$

This equation represents the single point (4, 5). It is shown in Figure 2.35.

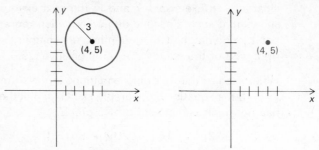

FIGURE 2.35   Circle: $(x - 4)^2 + (y - 5)^2 = 9$
Point: $(x - 4)^2 + (y - 5)^2 = 0$

(c) $2x^2 + 2y^2 - 16x - 20y + 100 = 0$

Again completing squares, we have:

$$x^2 - 8x \qquad\quad + y^2 - 10y \qquad\qquad = -50$$
$$x^2 - 8x + 16 + y^2 - 10y + 25 = -50 + 16 + 25$$
$$(x - 4)^2 \qquad\quad + (y - 5)^2 \qquad\qquad = -9$$

Since the sum of two squares is nonnegative, there are no points (x, y) whose coordinates satisfy this equation. This equation has no graph.

Exercise 2.4, 26–35

TRANSLATION

We have seen that the equation of a circle is much simpler when the origin of the coordinate system is at the center of the circle. An equation of a circle or of any graph depends on the orientation or placement of the graph relative to the coordinate axes. In calculus, as well as in most applications, calculations can frequently be simplified by proper location of coordinate axes. The circle with radius 5 in Figure 2.34 has the equation $(x + 2)^2 + (y - 3)^2 = 25$ because the coordinate axes are labeled the x-axis and the y-axis and are located so that the center of the circle is at $(-2, 3)$. If the coordinate axes were labeled x'-y' and were located so that the center of the circle was at the origin, that same circle would have the simpler equation $x'^2 + y'^2 = 25$. Figure 2.36 demonstrates this. In that figure we see that the center Q has coordinates $(-2, 3)$ relative to the x-y coordinate system, but Q has coordinates $(0, 0)$ relative to the x'-y' coordinate system.

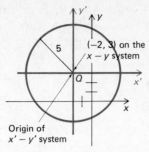

FIGURE 2.36   The circle $(x + 2)^2 + (y - 3)^2 = 25$
(The circle $x'^2 + y'^2 = 25$)

The process of transforming an equation of a graph by referring the graph to a new set of axes parallel to and having the same directions as the original axes is called **translation of axes**. We can transform the equation $(x + 2)^2 + (y - 3)^2 = 25$ of the circle in Figure 2.34 by replacing $(x + 2)$ by x' and replacing $(y - 3)$ by y'. These algebraic replacements effectively refer the circle to a new set of axes parallel to the original axes, having the same directions as the original axes, and having the new origin at the old point $(-2, 3)$. In general, to translate axes so that the new axes are labeled x'-y' and have their origin at the old point (h, k), replace $(x - h)$ by x' and replace $(y - k)$ by y'.

$$x - h = x' \qquad y - k = y'$$

In other words, to translate axes so that the new origin is at (h, k), let

$$x = x' + h \quad \text{and} \quad y = y' + k.$$

These are called the **translation equations**. The coordinates of P in Figure 2.37 are shown relative to both the x-y axes and the x'-y' axes.

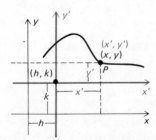

FIGURE 2.37 $x = x' + h$
$y = y' + k$
Translation of axes

EXAMPLE 5

Write the equation

$$3x^2 + 3y^2 - 18x + 30y + 90 = 0$$

in standard form; find the center and radius; transform the equation by translating axes so that the new origin is at the center of the circle; draw and label both sets of axes; sketch the graph.

Solution

$$3x^2 + 3y^2 - 18x + 30y + 90 \qquad\qquad = 0$$
$$x^2 + y^2 - 6x + 10y + 30 \qquad\qquad = 0$$
$$x^2 - 6x \qquad\quad + y^2 + 10y \qquad\qquad = -30$$
$$x^2 - 6x + (\tfrac{-6}{2})^2 + y^2 + 10y + (\tfrac{10}{2})^2 = -30 + (\tfrac{-6}{2})^2 + (\tfrac{10}{2})^2$$
$$x^2 - 6x + 9 \qquad + y^2 + 10y + 25 \qquad = -30 + 9 + 25$$
$$(x - 3)^2 \qquad\qquad + (y + 5)^2 \qquad\qquad = 4$$

This is the standard form.

Center: $(3, -5)$
Radius: 2

To translate axes so that the new origin is at $(3, -5)$, let $x = x' + 3$ and $y = y' - 5$, so that $x - 3 = x'$ and $y + 5 = y'$. The standard form becomes

$$x'^2 + y'^2 = 4.$$

The axes and graph are shown in Figure 2.38.

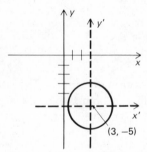

FIGURE 2.38 $3x^2 + 3y^2 - 18x + 30y + 90 = 0$
$(x - 3)^2 + (y + 5)^2 = 4$
$x'^2 + y'^2 = 4$

EXAMPLE 6

Transform and simplify the equation $y - x^2 - 6x - 9 = 0$ by translating the axes so that the new origin is at $(-3, 0)$. Draw and label both sets of axes, and sketch the graph.

Solution

The translation equations are $x = x' - 3$ and $y = y' + 0$. The given equation becomes

$$y' - (x' - 3)^2 - 6(x' - 3) - 9 = 0.$$

This may be simplified as follows:

$$y' - (x'^2 - 6x' + 9) - 6x' + 18 - 9 = 0$$
$$y' - x'^2 + 6x' - 9 - 6x' + 18 - 9 = 0$$
$$y' = x'^2$$

The graph of this equation was sketched in Figure 2.5 (Section 2.1) and again in Figure 2.21 (Section 2.3). It is shown with both sets of axes in Figure 2.39. Here the x and x' axes coincide, but the new origin is three units to the left of the original origin. The graph will be identified as a parabola in Section 2.5.

Exercise 2.4, 36–65
Verbal Problems
Exercise 2.4, 66–67

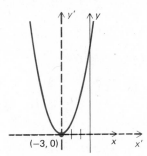

FIGURE 2.39 $y - x^2 - 6x - 9 = 0$
$$y' = x'^2$$

EXERCISE 2.4 DISTANCE AND MIDPOINT FORMULAS; CIRCLES; TRANSLATION

▶ In each of the following, find the distance between the two given points.

EXAMPLE $(-1, 2), (11, 7)$

Solution $\sqrt{(11 - (-1))^2 + (7 - 2)^2} = \sqrt{144 + 25}$
$$= 13$$

1. $(-1, 2), (7, 2)$ **2.** $(4, -3), (4, 5)$

3. $(-2, 5), (4, 12)$ **4.** $(1, -2), (7, 5)$

5. $(2, -4), (6, 0)$ **6.** $(-3, 1), (6, 13)$

7. $(-2, \frac{3}{5}), (10, \frac{7}{5})$ **8.** $(5, -7), (10, 5)$

9. $(-3, \frac{-2}{3}), (\frac{2}{3}, 10)$ **10.** $(-4, 3), (0, 6)$

▶ In each of the following, find the midpoint of the segment joining the two given points.

EXAMPLE $(2, -3), (7, 1)$

Solution $\left(\dfrac{2 + 7}{2}, \dfrac{-3 + 1}{2} \right)$ or $(\frac{9}{2}, -1)$

11. $(3, 7), (9, 1)$ **12.** $(1, -2), (6, 0)$

13. $(-3, 1), (4, 6)$ **14.** $(\frac{-2}{7}, \frac{3}{4}), (\frac{10}{7}, \frac{5}{4})$

15. $(-3, -2), (-4, 3)$

▶ In each of the following, find the standard form of an equation of the circle with the given center and radius.

EXAMPLE   Center $(3, -7)$, radius 6

Solution $(x - 3)^2 + (y + 7)^2 = 36$

16. Center $(-2, 3)$, radius 4   **17.** Center $(-4, -5)$, radius $\frac{3}{4}$

18. Center $(-3, \frac{1}{2})$, radius 8   **19.** Center $(5, 0)$, radius $\frac{7}{4}$

20. Center $(2, -5)$, radius 9

▶ Find the center and radius of each of the following circles.

EXAMPLE $(x + 2)^2 + (y - 7)^2 = 19$

Solution   Center $(-2, 7)$, radius $\sqrt{19}$

21. $(x + 3)^2 + (y - 5)^2 = 36$ **22.** $(x - 7)^2 + (y - 8)^2 = 13$

23. $(x - 6)^2 + (y + 2)^2 = 80$ **24.** $(x + \frac{1}{2})^2 + (y - \frac{5}{2})^2 = 10$

25. $(x - 7)^2 + (y + 8)^2 = 2$

▶ Write each of the following in the form $(x - h)^2 + (y - k)^2 = r^2$. If the equation has a graph, sketch or plot the graph; otherwise state "No graph."

EXAMPLE $x^2 + y^2 - 10x + 6y - 15 = 0$

Solution

$$x^2 - 10x + y^2 + 6y = 15$$
$$x^2 - 10x + 25 + y^2 + 6y + 9 = 15 + 34$$
$$(x - 5)^2 + (y + 3)^2 = 49$$

Center $(5, -3)$
Radius 7

26. $x^2 + y^2 - 8x - 2y - 8 = 0$

27. $x^2 + y^2 + 2x - 6y + 10 = 0$

28. $x^2 + y^2 - 4x + 10y + 65 = 0$

29. $x^2 + y^2 - 12x - 2y = -33$

30. $x^2 + y^2 + 14x + 12y + 76 = 0$

31. $4x^2 + 4y^2 - 4x + 24y + 56 = 0$

32. $16x^2 + 16y^2 + 8x - 160y + 145 = 0$

33. $2x^2 + 2y^2 - 16x - 12y + 50 = 0$

34. $64x^2 + 64y^2 - 16x + 128y = 191$

35. $2x^2 + 2y^2 = 8$

▶ In each of the following, transform the equation by translating axes so that the new origin is at the center of the circle. Draw and label both sets of axes, and sketch the graph.

EXAMPLE

$(x - 7)^2 + (y + 6)^2 = 16$

Solution

$x' = x - 7 \qquad y' = y + 6$

$x'^2 + y'^2 = 16$

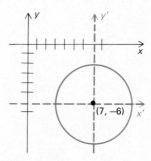

36. $(x - 8)^2 + (y - 7)^2 = 25$ **37.** $(x + 1)^2 + (y + 4)^2 = 9$

38. $(x - 1)^2 + y^2 = 4$ **39.** $(x + 2)^2 + (y - 5)^2 = 1$

40. $x^2 + (y + 4)^2 = 9$ **41.** $(x - 5)^2 + (y + 2)^2 = 49$

42. $x^2 + (y + 4)^2 = 25$ **43.** $(x + 3)^2 + y^2 = 36$

44. $(x + 1)^2 + (y + 2)^2 = 81$ **45.** $(x + 6)^2 + (y - 8)^2 = 64$

▶ In each of the following, transform and simplify the given equation by translating the axes so that the new origin is at the given point. Draw and label both sets of axes, and sketch the graph.

EXAMPLE

$y - 1 = -\sqrt{x - 2}, \ (2, 1)$

Solution

$x = x' + 2$

$y = y' + 1$

$y' + 1 - 1 = -\sqrt{x' + 2 - 2}$

$\qquad y' = -\sqrt{x'}$ \quad (See Example 8, Section 2.1)

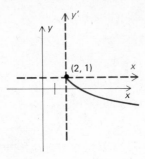

46. $y - x^2 - 2x - 3 = 0$, $(-1, 2)$

47. $y = x^2 + 3$, $(0, 3)$ \qquad **48.** $y - 3 = -\sqrt{x - 4}$, $(4, 3)$

49. $y - x^2 + 6x = 9$, $(3, 0)$ \qquad **50.** $y - 1 = \sqrt{x - 2}$, $(2, 1)$

51. $y = x^2 + 6x + 8$, $(-3, -1)$ \qquad **52.** $y - 2 = |x + 3|$, $(-3, 2)$

53. $y = |x + 3|$, $(-3, 0)$

54. $y = x^3 + 6x^2 + 12x + 8$, $(-2, 0)$

55. $y + 1 = (x - 2)^3$, $(2, -1)$ \qquad **56.** $(y - 1)^2 = x - 2$, $(2, 1)$

57. $y = 3x + 4$, $(0, 4)$ \qquad **58.** $y = 3x - 4$, $(0, -4)$

59. $y = -|x - 2|$, $(2, 0)$ \qquad **60.** $y + 3 = -|x - 4|$, $(4, -3)$

61. $y = \dfrac{1}{x - 1}$, $(1, 0)$ \qquad **62.** $y = \dfrac{1}{x + 1}$, $(-1, 0)$

63. $y + 2 = \dfrac{1}{x - 3}$, $(3, -2)$ \qquad **64.** $y - 2 = \dfrac{1}{x + 3}$, $(-3, 2)$

65. $y + 1 = \dfrac{3(x - 2)^2}{x - 3}$, $(2, -1)$

66. In a sodium chloride crystal, two sodium atoms and two chlorine atoms are located at the vertices of a rectangle with the sodium atoms at opposite vertices. If the dimensions of the rectangle are 6 units by 8 units, how far apart are the sodium atoms?

67. A cross section of a cubic potassium chloride crystal is put on a Cartesian coordinate system so that there is a chlorine atom located at each of the points $(0, 0)$, $(8, 0)$, and $(8, 8)$. A cesium atom is located half-way between the chlorine atoms at $(0, 0)$ and $(8, 8)$. What are the coordinates of the location of the cesium atom?

CHECK LIST FOR THE STUDENT

If you have learned the material in Section 2.4, you should be able to do the following:

1. Find the distance between two given points.

2. Find the midpoint of a line segment whose endpoints are given.

3. Find an equation of a circle whose center and radius are given.

4. Find the center and radius of a circle whose equation is given in standard form.

5. Complete squares and write an equation of the form $Ax^2 + By^2 + Cx + Dy + E = 0$ in the form $(x - h)^2 + (y - k)^2 = r^2$. If the equation has a graph, sketch or plot the graph; otherwise state "No graph."

6. Transform an equation in x and y by translating axes so that the new origin is at a given point.

7. Translate axes, and draw and label both sets of axes.

SECTION QUIZ 2.4

1. (a) Find the distance between the points $(-2, 5)$ and $(2, 8)$.
   (b) Find the midpoint of the segment joining the points $(5, -3)$ and $(3, 9)$.

2. Find the standard form of an equation of the circle with center at $(2, -5)$ and radius 7.

3. Find the center and radius of the circle $(x + 4)^2 + (y - 1)^2 = 36$.

4. Write each of the following in the form $(x - h)^2 + (y - k)^2 = r^2$. If the equation has a graph, sketch or plot the graph; otherwise state "No graph."
   (a) $x^2 + y^2 - 14x + 4y - 53 = 0$
   (b) $x^2 + y^2 - 2x + 4y + 14 = 0$

5. Transform the equation $(x + 6)^2 + (y - 4)^2 = 9$ by translating axes so that the new origin is at $(-6, 4)$. Draw and label both sets of axes. Sketch the graph.

*2.5 Parabolas; Quadratic Functions

The purpose of this section is to define a parabola and derive equations for parabolas, to consider the graphs of certain parabolas, and to define and graph quadratic functions. An example at the end of the section indicates one way in which this material is put to use.

A **parabola** is the set of all points equally distant from a fixed point and a fixed line. The fixed point is called the **focus**. The fixed line is called the **directrix**. The line through the focus and perpendicular to the directrix is called the **axis** of the parabola. The point on the axis midway between the focus and the directrix is called the **vertex**. These points and lines are labeled for the parabola shown in Figure 2.40.

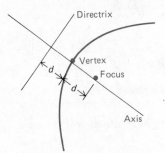

FIGURE 2.40   A parabola

If the coordinate axes are located so that the focus is on one of them and so that the directrix is parallel to the other with the origin halfway between the focus and the directrix, we can use the distance formula and find an equation of the parabola in a relatively simple form. For example, suppose that the focus is the point (3, 0) on the *x*-axis and that the directrix is the line $x = -3$ parallel to the *y*-axis, as shown in Figure 2.41. Let (*x*, *y*) be

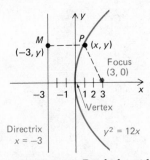

FIGURE 2.41   Parabola with focus (3, 0) and directrix $x = -3$

any point on the parabola, and call (x, y) the point P. Then the distance from P to the directrix must equal the distance from P to the focus. By the distance from a point to a line, we mean the distance along the perpendicular to the line, so the distance from P to the directrix must be the distance between P and M shown in Figure 2.41. This distance is $|x - (-3)| = |x + 3|$.

$$|x + 3| \qquad \text{Distance from } (x, y) \text{ to}$$
$$\text{directrix in Figure 2.41}$$

By the distance formula, the distance from P to the focus is

$$\sqrt{(x - 3)^2 + y^2} \qquad \text{Distance from } (x, y) \text{ to}$$
$$\text{focus in Figure 2.41}$$

Since the distance from P to the directrix must equal the distance from P to the focus, we have

$$|x + 3| = \sqrt{(x - 3)^2 + y^2}.$$

This equation can be simplified by squaring both sides and collecting terms. (Note that $|x + 3|^2$ is the same as $(x + 3)^2$.)

$$|x + 3|^2 = (x - 3)^2 + y^2$$
$$x^2 + 6x + 9 = x^2 - 6x + 9 + y^2$$
$$12x - y^2 = 0$$
$$y^2 = 12x$$

The equation $y^2 = 12x$ is called the *standard form* of an equation of the parabola with focus at $(3, 0)$ and with directrix $x = -3$. The graph is shown in Figure 2.41. We see that the vertex is at the origin and that the parabola opens to the right.

Precisely the same technique can be used to derive an equation for any parabola whose focus is on the x-axis, whose vertex is at the origin, and whose directrix is perpendicular to the x-axis. For example, suppose that the focus is on the x-axis to the *right* of the origin with coordinates $(p, 0)$, as shown in Figure 2.42. Since the focus is to the *right* of the origin, p must be *positive*. Since the origin is midway between the focus and the directrix, the directrix must be the line $x = -p$. Now, again let (x, y) be any point on the parabola and call (x, y) the point P. Then in Figure 2.42, point M has coordinates $(-p, y)$, and the distance from P to the directrix is the distance from P to M, or $|x - (-p)| = |x + p|$.

$$|x + p| \qquad \text{Distance from } (x, y) \text{ to}$$
$$\text{directrix in Figure 2.42}$$

By the distance formula, the distance from (x, y) to the focus is

$$\sqrt{(x - p)^2 + (y - 0)^2} \qquad \text{Distance from } (x, y) \text{ to}$$
$$\text{focus in Figure 2.42}$$

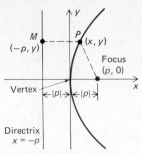

FIGURE 2.42 $y^2 = 4px$; $p > 0$, opens right

Since the distance from (x, y) to the directrix must equal the distance from (x, y) to the focus, we have

$$|x + p| = \sqrt{(x - p)^2 + (y - 0)^2}.$$

Squaring both sides, using the fact that $|x + p|^2 = (x + p)^2$, and collecting terms, we have the following.

$$(x + p)^2 = (x - p)^2 + (y - 0)^2$$
$$x^2 + 2px + p^2 = x^2 - 2px + p^2 + y^2$$
$$4px - y^2 = 0$$
$$y^2 = 4px$$

Any point whose coordinates satisfy this equation is equally distant from the focus $(p, 0)$ and the directrix $x = -p$. Notice that with the focus to the *right* of the origin, p is positive and the parabola opens to the *right*.

If the directrix is perpendicular to the x-axis and the focus is on the x-axis to the *left* of the origin with coordinates $(p, 0)$, p must be negative and the parabola opens to the *left*, as shown in Figure 2.43. In either case, $|p|$ is the distance between the vertex and the directrix.

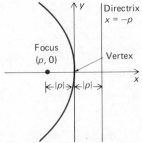

FIGURE 2.43 $y^2 = 4px$; $p < 0$, opens left

The equation

$$y^2 = 4px$$

Parabola: Vertex: (0, 0)
Axis: x-axis
$p > 0$, opens right
$p < 0$, opens left

is called the standard form of an equation of a parabola with vertex at the origin, with focus on the x-axis at $(p, 0)$, with axis along the x-axis, and with directrix $x = -p$ perpendicular to the x-axis.

EXAMPLE 1

Find the vertex, axis, focus, and directrix of $y^2 = 12x$. Sketch the graph, plot the focus, and draw the directrix.

Comparing $y^2 = 12x$ with the standard form $y^2 = 4px$, we see that $y^2 = 12x$ represents a parabola with vertex at the origin and with axis along the x-axis. We also see that

$$4p = 12$$
$$p = 3.$$

Thus the focus must be (3, 0) and the directrix must be $x = -3$. The graph was sketched in Figure 2.41.

If the x-y coordinate axes are located so that a parabola with axis parallel to or along the x-axis has vertex at (h, k) instead of at the origin, we can translate the axes and readily write the standard form of an equation of the parabola by referring to a new set of axes labeled x'-y' with origin at the old point (h, k). Relative to the x'-y' axes, the parabola would have the equation

$$y'^2 = 4px'^2.$$

Using the translation equations

$$x - h = x'$$
$$y - k = y'$$

developed in the last section, on the x-y coordinate system the parabola must have the equation

$$(y - k)^2 = 4p(x - h).$$

Parabola: Vertex: (h, k)
Axis parallel to or along x-axis.
$p > 0$, opens right
$p < 0$, opens left

Figure 2.44 shows such a parabola.

EXAMPLE 2

Find the vertex, axis, focus, and directrix of $(y - 5)^2 = 12(x + 4)$. Transform this equation by translating axes so that the new origin is at

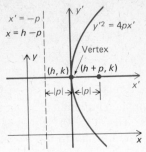

FIGURE 2.44 $(y - k)^2 = 4p(x - h);\quad p > 0$
$$y'^2 = 4px'$$

$(-4, 5)$. Draw and label both sets of axes, plot the focus, draw the directrix, and sketch the graph.

Solution

Comparing

$$(y - 5)^2 = 12(x + 4)$$

with

$$(y - k)^2 = 4p(x - h),$$

we see that the vertex is at $(-4, 5)$ and that $4p = 12$ so $p = 3$. Since the axis is parallel to the x-axis and goes through $(-4, 5)$, the axis must be the line $y = 5$. The focus is located on the axis a distance of $|p| = 3$ units to the right of the vertex and so must be at the point $(-4 + 3, 5)$, or $(-1, 5)$. The directrix is perpendicular to the x-axis and is located $|p| = 3$ units to the *left* of the vertex, so the directrix must be the line $x = -4 - 3$, or $x = -7$. The parabola is shown in Figure 2.45.

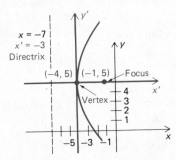

FIGURE 2.45 $(y - 5)^2 = 12(x + 4)$
$$y'^2 = 12x'$$

If we transform the equation $(y - 5)^2 = 12(x + 4)$ by translating axes so that the new origin is at $(-4, 5)$, the translation equations are

$$x + 4 = x'$$
$$y - 5 = y',$$

and the equation becomes

$$y'^2 = 12x'.$$

This is an equation of the parabola with vertex at the origin of the x'-y' coordinate system—that is, at the old point $(-4, 5)$. We could have sketched the graph by first translating the axes and then sketching the graph of $y'^2 = 12x'$ on the x'-y' coordinate system. The equation $y^2 = 12x$ was discussed in Example 1 and sketched in Figure 2.41.

We have seen that if the coordinate axes are located so that the vertex of a parabola is at the origin and the axis is along the x-axis, the parabola has the equation $y^2 = 4px$ where $|p|$ is the distance between the vertex and the directrix. In like manner, if the coordinate axes are located so that the vertex is at the origin and the axis is along the y-axis, the roles of x and y are interchanged, and the parabola has the equation

$$x^2 = 4py \qquad \text{Parabola: Vertex: } (0, 0)$$
$$\text{Axis: } y\text{-axis}$$
$$p > 0, \text{ opens up}$$
$$p < 0, \text{ opens down}$$

where $|p|$ is again the distance between the vertex and the directrix, where the focus is on the y-axis at $(0, p)$, and where the directrix is the line $y = -p$. If p is positive, the parabola $x^2 = 4py$ opens up; if p is negative, it opens down. Figures 2.46 and 2.47 show both cases.

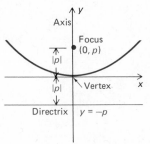

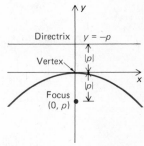

FIGURE 2.46 $x^2 = 4py; p > 0$, opens up

FIGURE 2.47 $x^2 = 4py; p < 0$, opens down

EXAMPLE 3

Find the vertex, axis, focus, and directrix of $x^2 = 12y$. Plot the focus, draw the directrix, and sketch the graph.

Solution

Comparing

$$x^2 = 12y$$

with

$$x^2 = 4py,$$

we see that $4p = 12$ so $p = 3$. The vertex is at the origin, the axis is along the y-axis, and since p is positive the parabola opens up. Since the focus is located on the axis $|p| = 3$ units above the origin, the focus must be at (0, 3). Since the directrix is perpendicular to the y-axis and $|p| = 3$ units below the origin, the directrix must be the line $y = -3$. The graph is shown in Figure 2.48.

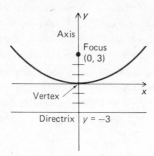

FIGURE 2.48 $x^2 = 12y$

If the x-y coordinate axes are located so that a parabola with axis parallel to or along the y-axis has vertex at (h, k) instead of at the origin, on the x-y coordinate system the parabola has the equation

$$(x - h)^2 = 4p(y - k).$$ Parabola: Vertex: (h, k)

Axis parallel to or along y-axis

$p > 0$, opens up

$p < 0$, opens down

Relative to x'-y' axes with origin at the old point (h, k), the parabola would have the equation

$$x'^2 = 4py'.$$

Such a parabola is shown with both sets of axes in Figure 2.49.

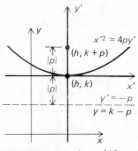

FIGURE 2.49 $(x - h)^2 = 4p(y - k);$ $p > 0$
$$x'^2 = 4py'$$

EXAMPLE 4

Find the vertex, axis, focus, and directrix of $(x + 4)^2 = -12(y - 5)$. Transform this equation by translating axes so that the new origin is at $(-4, 5)$. Draw and label both sets of axes, plot the focus, draw the directrix, and sketch the graph.

Solution

Comparing

$$(x + 4)^2 = -12(y - 5)$$

with

$$(x - h)^2 = 4p(y - k),$$

we see that the vertex is at $(-4, 5)$ and that $4p = -12$ so $p = -3$. Since the axis is parallel to the y-axis and goes through $(-4, 5)$, the axis must be the line $x = -4$. In Figure 2.50 the focus is located on the axis a distance of $|p| = 3$ units below the vertex and so must be at the point $(-4, 5 - 3)$, or $(-4, 2)$. The directrix is perpendicular to the y-axis and is located $|p| = 3$ units above the vertex, so the directrix must be the line $y = 5 + 3$, or $y = 8$.

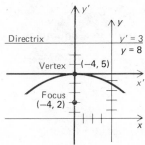

FIGURE 2.50 $(x + 4)^2 = -12(y - 5)$
$$x'^2 = -12y'$$

If we transform the equation $(x + 4)^2 = -12(y - 5)$ by translating axes so that the new origin is at $(-4, 5)$, the translation equations are

$$x + 4 = x'$$
$$y - 5 = y',$$

and the equation becomes

$$x'^2 = -12y'.$$

This is an equation of the parabola with vertex at the origin of the x'-y' coordinate system—that is, at the old point $(-4, 5)$. We could have sketched the graph by first translating the axes and then sketching the graph of $x'^2 = -12y'$ on the x'-y' coordinate system.

Exercise 2.5, 1–20

We have seen that parabolas with axes parallel to or along the co-ordinate axes have equations of the form

$$(y - k)^2 = 4p(x - h)$$

or of the form

$$(x - h)^2 = 4p(y - k).$$

These are called *standard forms*. If we multiply out the standard forms, we obtain the following second-degree equations.

$$y^2 - 4px - 2ky + (k^2 - 4ph) = 0$$
$$x^2 - 2hx - 4py + (h^2 + 4pk) = 0$$

The first of these second-degree equations is of the form

$$By^2 + Cx + Dy + E = 0$$

where B, C, D, and E represent specific real numbers. In this special case,

$$B = 1, \quad C = -4p, \quad D = -2k, \quad \text{and} \quad E = k^2 + 4ph.$$

The second of these second-degree equations is of the form

$$Ax^2 + Cx + Dy + E = 0$$

where in this case

$$A = 1, \quad C = -2h, \quad D = -4p, \quad \text{and} \quad E = h^2 + 4pk.$$

Any second-degree equation of the form

$$By^2 + Cx + Dy + E = 0$$

or of the form

$$Ax^2 + Cx + Dy + E = 0$$

where the letters A, B, C, D, and E represent specific real numbers is an equation of a parabola with its axis parallel to or along a coordinate axis. Such a second-degree equation can be written in a standard form by completing the square. This is illustrated in the next example.

EXAMPLE 5

Write the equation $2x^2 - 4x - 3y - 10 = 0$ in standard form. Find the vertex and axis and sketch the graph.

Solution

We begin to complete the square in x by factoring the coefficient of x^2 out of the terms involving x and by isolating the remaining terms on the other side of the equation.

$$2(x^2 - 2x \quad) = 3y + 10$$

To complete the square in x, we must insert the number 1 in the parentheses. This is equivalent to adding $2 \times 1 = 2$, so we insert the 1 in the parentheses and add 2 to the other side of the equation.

$$2(x^2 - 2x + 1) = 3y + 10 + 2$$
$$2(x - 1)^2 = 3(y + 4)$$
$$(x - 1)^2 = \tfrac{3}{2}(y + 4)$$

This is in standard form. Comparing it with

$$(x - h)^2 = 4p(y - k),$$

we see that the vertex is at $(1, -4)$. The axis is parallel to the y-axis and goes through the vertex, so the axis is the line $x = 1$. The graph is shown in Figure 2.51.

Exercise 2.5, 21–30

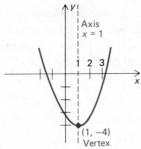

FIGURE 2.51 $2x^2 - 4x - 3y - 10 = 0$
$$(x - 1)^2 = \tfrac{3}{2}(y + 4)$$

Parabolas with axes along or parallel to the y-axis are graphs of *quadratic functions*. A **quadratic function** f is a function such that

1. The domain is the set of all real numbers,

and

2. $f(x) = ax^2 + bx + c,$

where a, b, and c represent specific real numbers and $a \neq 0$.
 The graph of quadratic function f is the graph of the equation

$$y = ax^2 + bx + c.$$

This can be written $ax^2 + bx - y + c = 0$, so it is a second-degree equation of one of the two forms we have just been considering. Since we can complete the square in x and write this equation in the standard form

$$(x - h)^2 = 4p(y - k),$$

we know that the graph of quadratic function f is a parabola with axis parallel to or along the *y*-axis and opening either up or down. Furthermore, when we complete the square to put the equation in standard form, we see that the $4p$ of the standard form is $1/a$ where a is the coefficient of x^2 in the equation $y = ax^2 + bx + c$. Thus the sign of coefficient a determines whether the graph of f opens up or down. In short, we can say that the graph of the quadratic function f such that

$$f(x) = ax^2 + bx + c$$

is a parabola with axis parallel to or along the *y*-axis; the parabola opens up if a is positive, and it opens down if a is negative. This is illustrated in the next example.

EXAMPLE 6

Find the vertex and axis and sketch the graph of the function f such that $f(x) = -2x^2 + 12x - 17$.

Solution

The given function f is a quadratic function with the coefficient of x^2 negative, so we know in advance that the graph is a parabola with axis parallel to the *y*-axis and opening down. To find the vertex and axis, we complete the square in x and put the equation $y = -2x^2 + 12x - 17$ in standard form.

$$-2(x^2 - 6x \quad) = y + 17$$
$$-2(x^2 - 6x + 9) = y + 17 - 18$$
$$-2(x - 3)^2 = y - 1$$
$$(x - 3)^2 = -\tfrac{1}{2}(y - 1)$$

The vertex is at $(3, 1)$, and the axis is the line $x = 3$. The graph is shown in Figure 2.52.

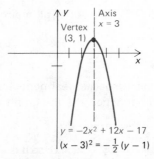

FIGURE 2.52 Graph of f: $f(x) = -2x^2 + 12x - 17$

In the previous example we found the vertex of the graph of the quadratic function f by completing the square in x and writing the equation $y = f(x)$ in standard form. Early in beginning calculus a more efficient method for locating the vertex of the graph of a quadratic function is encountered. That method is used more generally to locate relative high and low points on the graphs of functions other than quadratic functions.

One of the most common applications of parabolas has to do with the path of a projectile. For example, if a projectile such as a bullet is fired from the ground, its distance s (in feet) above the ground t seconds later is given approximately by the equation

$$s = -16t^2 + kt$$

where k is a constant depending on the initial velocity and the angle from the horizontal to the direction of fire. Analyzing such a problem sometimes requires the ability to sketch the graph of the equation relating s and t. The graph is a parabola. This is demonstrated for a particular case of the projectile problem in the next example.

EXAMPLE 7

A projectile is fired from the ground and its distance s above the ground at t seconds later is given by the equation $s = -16t^2 + 64t$. Sketch the graph of this equation on a Cartesian coordinate system having the t-axis for its horizontal axis and the s-axis for its vertical axis. At what time t does the projectile reach its maximum height? At what time t does the projectile return to the ground?

Solution

The function f such that $f(t) = -16t^2 + 64t$ is a quadratic function. The graph of $s = -16t^2 + 64t$ on a coordinate system having t as the horizontal axis is a parabola opening down. To find the vertex of the parabola, we complete the square in t and put the equation in standard form.

$$-16(t^2 - 4t \qquad) = s$$
$$-16(t^2 - 4t + 4) = s - 64$$
$$-16(t - 2)^2 = s - 64$$
$$(t - 2)^2 = -\tfrac{1}{16}(s - 64)$$

The vertex is at $t = 2$, $s = 64$; the axis is $t = 2$. Note that one t-intercept is at $t = 0$, $s = 0$. To find the other, let $s = 0$ and solve for t.

$$-16t^2 + 64t = 0$$
$$-16t(t - 4) = 0$$
$$t = 0 \quad \text{or} \quad t = 4$$

As shown in Figure 2.53, the graph crosses the t-axis at $(0, 0)$ and also at $(4, 0)$.

We see in Figure 2.53 that the maximum height (the largest s) is reached at the vertex. At the vertex, $t = 2$. Thus it takes the projectile 2 seconds to reach its maximum height.

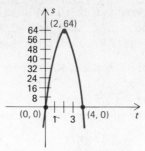

FIGURE 2.53 Graph of $s = -16t^2 + 64t$
(Graph of $(t - 2)^2 = -\frac{1}{16}(s - 64)$)

The projectile is on the ground when $s = 0$; that is, it is on the ground at $t = 0$ and also at $t = 4$. Therefore it takes the projectile 4 seconds to return to the ground.

Exercise 2.5, 36–40
Verbal Problems
Exercise 2.5, 41–44

For convenience of reference, the various forms of equations of parabolas considered in this section are summarized as follows:

$y^2 = 4px$ Parabola: Vertex: (0, 0)
            Axis: x-axis
            $p > 0$, opens right
            $p < 0$, opens left

$x^2 = 4py$ Parabola: Vertex: (0, 0)
            Axis: y-axis
            $p > 0$, opens up
            $p < 0$, opens down

$(y - k)^2 = 4p(x - h)$ Parabola: Vertex: (h, k)
            Axis parallel to or along
               x-axis
            $p > 0$, opens right
            $p < 0$, opens left

$(x - h)^2 = 4p(y - k)$ Parabola: Vertex: (h, k)
            Axis parallel to or along y-axis
            $p > 0$, opens up
            $p < 0$, opens down

$By^2 + Cx + Dy + E = 0$ Parabola: Axis parallel to or along x-axis
$Ax^2 + Cx + Dy + E = 0$ Parabola: Axis parallel to or along y-axis
$y = ax^2 + bx + c$ Parabola: Axis parallel to or along y-axis
                             $a > 0$, opens up
                             $a < 0$, opens down

EXERCISE 2.5 PARABOLAS; QUADRATIC FUNCTIONS

▶ In each of the following, find the vertex, axis, focus, and directrix. Sketch the graph, plot the focus, and draw the directrix.

EXAMPLE $\quad x^2 = -8y$

Solution

$x^2 = 4py$

$p = -2$, opens down
Vertex: (0, 0)
Axis: y-axis
Focus: (0, −2)
Directrix: $y = 2$

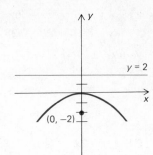

1. $y^2 = -8x$ **2.** $x^2 = 8y$ **3.** $y^2 = 8x$ **4.** $y^2 = 4x$

5. $y^2 = -4x$ **6.** $x^2 = 4y$ **7.** $x^2 = -4y$ **8.** $x^2 = -y$

9. $y^2 = -x$ **10.** $x^2 = -\frac{1}{4}y$

▶ In each of the following, find the vertex, axis, focus, and directrix. Transform the equation by translating axes so that the new origin is at the vertex. Draw and label both sets of axes, plot the focus, draw the directrix, and sketch the graph.

EXAMPLE $\quad y^2 = -16(x - 5)$

Solution

Vertex: (5, 0)
Axis: x-axis
$p = -4$, opens left
Focus: (1, 0)
Directrix: $x = 9$
$y'^2 = -16x'$

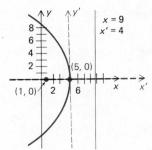

11. $y^2 = -12(x - 5)$ **12.** $y^2 = 16(x - 5)$

13. $x^2 = -16(y - 5)$ **14.** $(x - 1)^2 = 4(y + 3)$

15. $(y - 3)^2 = 8(x + 1)$ **16.** $(x - 4)^2 = -20(y + 7)$

17. $(y - 2)^2 = 2(x + 3)$ **18.** $(x - 3)^2 = -(y + 4)$

19. $(y + 1)^2 = (x + 4)$ **20.** $(x - 3)^2 = 3(y + 2)$

▶ Write each of the following equations in standard form. Find the vertex and axis and sketch the graph.

EXAMPLE

Solution

$y^2 - 2x + 2y + 7 = 0$

$y^2 + 2y = 2x - 7$

$y^2 + 2y + 1 = 2x - 6$

$(y + 1)^2 = 2(x - 3)$

Vertex: $(3, -1)$

Axis: $y = -1$

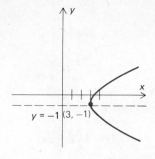

21. $x^2 - 2x - 4y - 7 = 0$ 22. $4x^2 - 8x - 3y - 2 = 0$

23. $4y^2 - 3x + 8y + 10 = 0$ 24. $2x^2 - 8x - 3y - 4 = 0$

25. $4y^2 - 7x - 40y + 142 = 0$ 26. $x^2 - 6x - 7y + 2 = 0$

27. $2y^2 - 5x - 20y + 30 = 0$ 28. $y^2 - 2x - 12y + 28 = 0$

29. $3x^2 - 12x - 5y + 7 = 0$ 30. $2x^2 - 28x - 9y + 53 = 0$

▶ In each of the following, find the vertex and axis and sketch the graph of the function f.

EXAMPLE

Solution

$f(x) = 4x^2 - 40x + 101$

$y = 4x^2 - 40x + 101$

$4x^2 - 40x = y - 101$

$4(x^2 - 10x \qquad) = y - 101$

$4(x^2 - 10x + 25) = y - 1$

$(x - 5)^2 = \frac{1}{4}(y - 1)$

Vertex: $(5, 1)$

Axis: $x = 5$

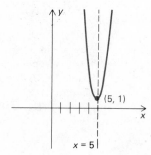

31. $f(x) = \frac{1}{4}x^2 - \frac{5}{2}x + \frac{29}{4}$ 32. $f(x) = 4x^2 + 40x + 102$

33. $f(x) = -2x^2 - 16x - 29$ 34. $f(x) = -x^2 + 2x - 4$

35. $f(x) = \frac{2}{7}x^2 + \frac{8}{7}x + \frac{15}{7}$

▶ In each of the following, assume that a projectile is fired from the ground and that its distance s above the ground at t seconds later is given by the equation $s = -16t^2 + kt$. For each given value of k, sketch the graph of $s = -16t^2 + kt$, determine the time it takes the projectile to reach its maximum height, and determine the time it takes the projectile to return to the ground.

EXAMPLE $k = 96$

Solution

$s = -16t^2 + 96t$

$-16(t^2 - 6t) = s$

$(t - 3)^2 = \frac{-1}{16}(s - 144)$

Vertex at $t = 3$, $s = 144$
If $s = 0$, $t = 0$ or 6.
Reaches maximum height at $t = 3$.
Returns to ground at $t = 6$.

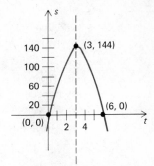

36. $k = 128$ **37.** $k = 160$ **38.** $k = 192$ **39.** $k = 48$ **40.** $k = 80$

41. A farmer has 40 meters of fence with which to build a rectangular pen with one side along a brook. Let x be the length in meters of each side perpendicular to the brook and let $A(x)$ be the area of the pen. Write an expression for $A(x)$. What value of x will give the greatest possible area? What is that area?

42. A farmer has 100 meters of fence with which to build a rectangular pen. Let x be the width of the pen. Write an expression for the area $A(x)$. What value of x will give the greatest possible area? What is that area?

43. The sum of two numbers is 12. Let x be one of the numbers. Write an expression for the product $P(x)$ of the numbers. What value of x will give the greatest possible $P(x)$?

44. The difference of two numbers is 12. Let x be one of the numbers. Write an expression for the product $P(x)$ of the numbers. What value of x will give the smallest possible $P(x)$? What is the smallest possible product?

CHECK LIST FOR THE STUDENT

If you have learned the material in Section 2.5, you should be able to do the following:

1. Define a parabola.

2. Recognize the standard form of an equation of a parabola.

3. Find the vertex, axis, focus, and directrix of a parabola from its equation in standard form.

4. Tell from looking at the standard form of an equation of a parabola whether it opens up, down, right, or left.

5. Sketch the graph of a parabola if its equation is given in standard form.

6. Transform the equations $(y - k)^2 = 4p(x - h)$ and $(x - h)^2 = 4p(y - k)$ by translating axes so that the new origin is at (h, k). Draw and label both sets of axes, and sketch the graphs of the equations.

7. Recognize equations of the form

$$By^2 + Cx + Dy + E = 0 \text{ or } Ax^2 + Cx + Dy + E = 0$$

as equations of parabolas and write such equations in standard form. (A, B, C, D, and E represent specific real numbers.)

8. Define a quadratic function.

9. Sketch the graph of a given quadratic function.

10. Tell from looking at the coefficient of x^2 in $f(x) = ax^2 + bx + c$ whether the graph of f opens up or down.

11. Sketch the graph of the equation $s = -16t^2 + kt$ for a given real number k.

12. Find the time it takes a projectile fired from the ground to reach its maximum height if its distance s above the ground at t seconds after firing is given by $s = -16t^2 + kt$ for some given constant k.

SECTION QUIZ 2.5

In each of the following, find the vertex, axis, focus, and directrix. Sketch the graph, plot the focus, and draw the directrix.
1. $x^2 = 20y$

2. $y^2 = -20x$

In each of the following, label the vertex and sketch the graph.
3. $(x + 3)^2 = -20(y - 2)$

4. $(y + 3)^2 = -20(x - 2)$

Write each of the following equations in standard form. Label the vertex and sketch the graph.
5. $y^2 - 6y + 9 - 20x = 0$

6. $x^2 - 2x - 8y - 7 = 0$

7. In each of the following, name the graph of the function f and tell whether it opens up or down.
   (a) $f(x) = 2x^2 - 12x + 17$
   (b) $f(x) = -2x^2 + 12x - 19$

8. If $f(x) = -x^2 + 1$, sketch the graph of f.

9. Sketch the graph of $y = x^2 + 2$.

10. A projectile is fired from the ground, and its distance s above the ground at t seconds after firing is given by $s = -16t^2 + 224t$. Find the time it takes the projectile to reach its maximum height.

*2.6 Ellipses and Hyperbolas

The purpose of this section is to define ellipses and hyperbolas, to derive equations for ellipses and hyperbolas, and to demonstrate their graphs.

ELLIPSES

An **ellipse** is the set of all points such that the *sum* of the distances from each of the points to two given fixed points is a constant. Each of the two fixed points is called a **focus** of the ellipse. The midpoint of the line segment joining the two foci (plural of focus) is called the **center** of the ellipse. An ellipse with foci F and F' is shown in Figure 2.54.

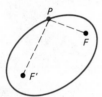

FIGURE 2.54 Ellipse: $PF + PF' = $ constant

If the coordinate axes are located so that they intersect at the center and so that one of the axes goes through the foci, we can use the distance formula and obtain an equation of the ellipse in a relatively simple form. For example, suppose that the axes are located as shown in Figure 2.55 so that the foci are on the *x*-axis at the points F and F' with coordinates

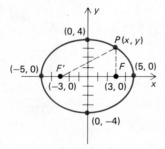

FIGURE 2.55 Ellipse with foci $(3, 0)$ and $(-3, 0)$

$$PF + PF' = 10$$

$$\frac{x^2}{25} + \frac{y^2}{16} = 1$$

$(3, 0)$ and $(-3, 0)$. Further suppose that the sum of the distances from any point P on the ellipse to F and to F' is equal to the constant 10.

$$PF + PF' = 10$$

According to the distance formula,

$$PF = \sqrt{(x - 3)^2 + y^2} \quad \text{and} \quad PF' = \sqrt{(x + 3)^2 + y^2},$$

so $$\sqrt{(x - 3)^2 + y^2} + \sqrt{(x + 3)^2 + y^2} = 10.$$

This equation represents the ellipse in Figure 2.55; it is satisfied by every point P on the ellipse and by no other point. It is customary to simplify such an equation by writing it in a form containing no radicals, as was done in Example 2 of Section 1.8. (Recall that to do this, we isolate a radical on one side, square both sides, collect like terms, and repeat the procedure.) Without radicals, this equation can be simplified to

$$16x^2 + 25y^2 = 400.$$

It can now be written in *standard form*

$$\frac{x^2}{a^2} + \frac{y^2}{b^2} = 1$$

by dividing both sides through by 400.

$$\frac{x^2}{25} + \frac{y^2}{16} = 1$$

Some properties of graphs that were discussed in Section 2.3 and that apply to the ellipse shown in Figure 2.55 can be readily deduced from the equation $x^2/25 + y^2/16 = 1$. (See Example 9 of Section 2.3.) For example, the ellipse is symmetric with respect to both axes and the origin. (Replacing x by $-x$ produces an equivalent equation, and replacing y by $-y$ produces an equivalent equation.) The intercepts are $(5, 0)$, $(-5, 0)$, $(0, 4)$, and $(0, -4)$. All points (x, y) for which $x > 5$ or for which $x < -5$ are excluded from the graph, because for such values of x the corresponding values of y would be imaginary. Likewise, all points (x, y) for which $y > 4$ or for which $y < -4$ are excluded from the graph, because for such values of y the corresponding values of x would be imaginary.

In general, suppose that the axes are located as shown in Figure 2.56 so that the foci are at the points F and F' with coordinates $(c, 0)$ and $(-c, 0)$ rather than at the specific points $(3, 0)$ and $(-3, 0)$. Also suppose that the sum of the distances from any point on the ellipse to F and F' equals the positive constant $2a$ where $a > c$, rather than the specific constant 10. (Specifying that $a > c$ makes the constant $2a$ greater than the distance between F and F'.) Then precisely the same type of algebraic manipulations as were performed in Example 2, Section 1.8, for our

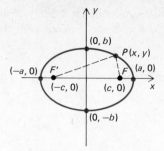

FIGURE 2.56 Ellipse with foci $(c, 0)$ and $(-c, 0)$

$$PF + PF' = 2a$$

$$\frac{x^2}{a^2} + \frac{y^2}{b^2} = 1, \ b^2 = a^2 - c^2$$

specific case will produce an equation of the ellipse in the standard form

$$\frac{x^2}{a^2} + \frac{y^2}{b^2} = 1 \qquad \begin{array}{l}\text{Ellipse} \\ \text{Foci at } (c, 0) \text{ and } (-c, 0)\end{array}$$

where $b^2 = a^2 - c^2$. (Note that since $a^2 = b^2 + c^2$, a^2 is the larger of the two denominators in the standard form.)

If the axes are located as shown in Figure 2.57 so that the foci are at the points F and F' with coordinates $(0, 3)$ and $(0, -3)$, and if $PF + PF' = 10$, an equation of the ellipse can be shown to be

$$\frac{x^2}{16} + \frac{y^2}{25} = 1.$$

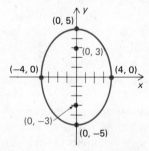

FIGURE 2.57 Ellipse with foci $(0, 3)$ and $(0, -3)$

$$PF + PF' = 10$$

$$\frac{x^2}{16} + \frac{y^2}{25} = 1$$

In general, as in Figure 2.58, if the foci are at $(0, c)$ and $(0, -c)$, and if $PF + PF' = 2a$ for positive $a > c$, an equation of the ellipse is

$$\frac{x^2}{b^2} + \frac{y^2}{a^2} = 1 \qquad \text{Ellipse}$$
Foci at $(0, c)$ and $(0, -c)$

where again $b^2 = a^2 - c^2$ and a^2 is the larger of the two denominators in the standard form.

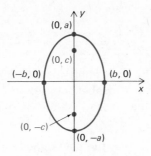

FIGURE 2.58   Ellipse with foci $(0, c)$ and $(0, -c)$
$$PF + PF' = 2a$$
$$\frac{x^2}{b^2} + \frac{y^2}{a^2} = 1, \, b^2 = a^2 - c^2$$

The line segment through the foci of an ellipse and with endpoints on the ellipse is called the **major axis** of the ellipse. The line segment through the center, perpendicular to the major axis, and with endpoints on the ellipse is called the **minor axis**. The major axis is longer than the minor axis. We see in Figure 2.56 that if the center is at the origin and the major axis is on the x-axis, the larger denominator a^2 is under x^2 in the standard form. Likewise, we see in Figure 2.58 that if the center is at the origin and the major axis is on the y-axis, the larger denominator a^2 is under y^2. If the center is at the origin, the ends of the axes are the intercepts. Using the equation in standard form, it is easy to sketch the graph simply by locating the intercepts and hence the ends of the axes.

Exercise 2.6, 1–10

If the center of an ellipse is at some point (h, k) rather than necessarily at the origin, and if the axes of the ellipse are parallel to the coordinate axes, the standard-form equation becomes

$$\frac{(x - h)^2}{a^2} + \frac{(y - k)^2}{b^2} = 1, \, a > b \qquad \text{Ellipse}$$
Center: (h, k)
Major axis parallel to or along x-axis

or $\dfrac{(x - h)^2}{b^2} + \dfrac{(y - k)^2}{a^2} = 1, \; a > b.$ Ellipse
Center: (h, k)
Major axis parallel to or
along y-axis

Note that the larger denominator a^2 is under $(x - h)^2$ if the major axis is parallel to or along the x-axis, and the larger denominator a^2 is under $(y - k)^2$ if the major axis is parallel to or along the y-axis.

EXAMPLE 1   Locate the center of the ellipse

$$\frac{(x - 2)^2}{25} + \frac{(y - 3)^2}{16} = 1.$$

Transform the equation by translating coordinate axes so that the new origin is at the center. Draw and label both sets of coordinate axes. Label the ends of the axes of the ellipse. Sketch the graph.

Solution   The center is at $(2, 3)$.

To translate the axes so that the new origin will be at $(2, 3)$, let

$$x' = x - 2 \quad \text{and} \quad y' = y - 3$$

or $\qquad\qquad x' + 2 = x \quad \text{and} \quad y' + 3 = y$

so that the equation becomes

$$\frac{x'^2}{25} + \frac{y'^2}{16} = 1.$$

This is an equation of the ellipse with center at the new origin and with major axis of length $2\sqrt{25} = 10$ and minor axis of length $2\sqrt{16} = 8$. In other words, the intercepts on the x'-y' axes are at $x' = 5$ and $y' = 0$, at $x' = -5$ and $y' = 0$, at $x' = 0$ and $y' = 4$, and at $x' = 0$ and $y' = -4$. On the x-y axes those are the points $(7, 3)$, $(-3, 3)$, $(2, 7)$, and $(2, -1)$, respectively. The graph and both sets of axes are sketched in Figure 2.59. We can check the location of the ends of the major axis by starting at the center $(2, 3)$ and counting off $\sqrt{25} = 5$ units to the right and then to the left on the x'-axis (the line $y = 3$).

If the standard forms

$$\frac{(x - h)^2}{a^2} + \frac{(y - k)^2}{b^2} = 1 \quad \text{and} \quad \frac{(x - h)^2}{b^2} + \frac{(y - k)^2}{a^2} = 1$$

are multiplied out, the equations respectively become

$$\underset{\substack{\| \\ A}}{b^2 x^2} + \underset{\substack{\| \\ B}}{a^2 y^2} - 2b^2 h x - 2a^2 k y + b^2 h^2 + a^2 k^2 - a^2 b^2 = 0$$

and

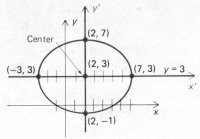

FIGURE 2.59 $\dfrac{(x-2)^2}{25} + \dfrac{(y-3)^2}{16} = 1$

$$\dfrac{x'^2}{25} + \dfrac{y'^2}{16} = 1$$

$$a^2 x^2 + b^2 y^2 - 2a^2 hx - 2b^2 ky + a^2 h^2 + b^2 k^2 - a^2 b^2 = 0.$$
$$\underset{A}{\parallel} \qquad \underset{B}{\parallel}$$

These are both of the form

$$Ax^2 + By^2 + Cx + Dy + E = 0$$

where A and B are different since $a^2 > b^2$ and either $A = b^2$ and $B = a^2$, as they are in the first equation, or $A = a^2$ and $B = b^2$, as they are in the second equation. Note also that A and B are both of the same sign.

Where A, B, C, D, and E are all specific real numbers, any second degree equation of the form

$$Ax^2 + By^2 + Cx + Dy + E = 0, \qquad A \neq B \text{ and } A \text{ and } B \text{ same sign:}$$
$$\text{Ellipse, point, or no graph}$$

where $A \neq B$ and A and B are of the same sign, represents an ellipse, a single point, or no graph. The equation can be written in standard form by completing the squares in x and y, as was done in an illustrative example in Exercise 1.7 and as is illustrated again in the next example. After the squares are completed, if the constant term is isolated on one side of the equation and is positive, the equation represents an ellipse; if the isolated constant term is zero, the equation represents a single point; if the isolated constant term is negative, the equation represents no graph at all. Each of these cases is illustrated in the next example.

EXAMPLE 2

Write each of the following equations in standard form. If the equation has a graph, sketch it. If the equation has no graph, simply state "No graph."

(a) $16x^2 + 9y^2 + 160x - 36y + 292 = 0$

(b) $16x^2 + 9y^2 + 160x - 36y + 436 = 0$

(c) $16x^2 + 9y^2 + 160x - 36y + 580 = 0$

Solution

(a) $16x^2 + 160x \qquad + 9y^2 - 36y \qquad = -292$

$16(x^2 + 10x \quad) + 9(y^2 - 4y \quad) = -292$

Completing the squares in both parentheses by adding the appropriate numbers to both sides of the equation produces the following:

$$16(x^2 + 10x + 25) + 9(y^2 - 4y + 4) = -292 + 400 + 36$$

$$16(x + 5)^2 \qquad + 9(y - 2)^2 \qquad = 144$$

$$\frac{(x + 5)^2}{9} \qquad + \frac{(y - 2)^2}{16} \qquad = 1$$

This is an ellipse with center at $(-5, 2)$ and major axis parallel to the y-axis. The graph is sketched in Figure 2.60.

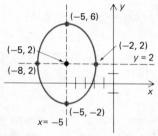

FIGURE 2.60 Ellipse: $16x^2 + 9y^2 + 160x - 36y + 292 = 0$

$$\frac{(x + 5)^2}{9} + \frac{(y - 2)^2}{16} = 1$$

(b) $16x^2 + 9y^2 + 160x - 36y + 436 = 0$

$$16(x^2 + 10x + 25) + 9(y^2 - 4y + 4) = -436 + 400 + 36$$

$$16(x + 5)^2 \qquad + 9(y - 2)^2 \qquad = 0$$

$$\frac{(x + 5)^2}{9} \qquad + \frac{(y - 2)^2}{16} \qquad = 0$$

The only point that satisfies this equation is the point $(-5, 2)$. To sketch the graph we need only to plot this single point. The graph is shown in Figure 2.61.

(c) $16x^2 + 9y^2 + 160x - 36y + 580 = 0$

$$16(x^2 + 10x + 25) + 9(y^2 - 4y + 4) = -580 + 400 + 36$$

$$16(x + 5)^2 \qquad + 9(y - 2)^2 \qquad = -144$$

$$\frac{(x + 5)^2}{9} \qquad + \frac{(y - 2)^2}{16} \qquad = -1$$

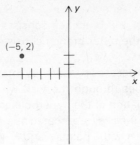

FIGURE 2.61 $16x^2 + 9y^2 + 160x - 36y + 436 = 0$

$$\frac{(x + 5)^2}{9} + \frac{(y - 2)^2}{16} = 0$$

The point $(-5, 2)$

This equation states that the sum of two squares is negative. There are no real numbers *x* and *y* for which this is true. There is no graph.

We have seen that if *A* and *B* are of the same sign but $A \neq B$, the equation $Ax^2 + By^2 + Cx + Dy + E = 0$ represents an ellipse, a single point, or no graph. Note that if *A* and *B* are equal, the equation represents a circle, a single point, or no graph, because then the equation can be divided through by the common value of *A* and *B* and the resulting equation can be put in the form $(x - h)^2 + (y - k)^2 = r^2$, as was illustrated in Example 4 of Section 2.4.

Exercise 2.6, 11–30

HYPERBOLAS

A **hyperbola** is the set of all points such that the absolute value of the *difference* of the distances from each of the points to two given fixed points is a constant. As in an ellipse, each of the two fixed points is called a **focus**. The midpoint of the line segment joining the two foci is called the **center**. A hyperbola with foci *F* and *F'* is shown in Figure 2.62. Notice

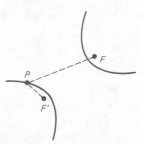

FIGURE 2.62 Hyperbola: $|PF - PF'| = $ constant

that a hyperbola has two separate branches. Since $|PF - PF'|$ is a constant (call it C), one of the branches consists of the points P such that $PF - PF' = C$, and the other branch consists of the points P such that $PF - PF' = -C$.

If the coordinate axes are located so that they intersect at the center and so that one of the axes goes through the foci, we can use the distance formula and obtain an equation of the hyperbola as we did for an ellipse. For example, suppose that the axes are located as shown in Figure 2.63 so that the foci are on the x-axis at the points F and F' with coordinates $(5, 0)$ and $(-5, 0)$. Also suppose that the absolute value of the difference of the distances from any point P on the hyperbola to F and to F' is equal to the constant 8.

$$|PF - PF'| = 8$$

Then, since $PF = \sqrt{(x - 5)^2 + y^2}$ and $PF' = \sqrt{(x + 5)^2 + y^2}$, an equation of the hyperbola is

$$\left|\sqrt{(x - 5)^2 + y^2} - \sqrt{(x + 5)^2 + y^2}\right| = 8.$$

In other words,

$$\sqrt{(x - 5)^2 + y^2} - \sqrt{(x + 5)^2 + y^2} = \pm 8$$

where the $+8$ is associated with points on one branch and the -8 is associated with points on the other branch. By systematically isolating a radical and squaring both sides as we did for an ellipse, we may write this as a single equation in standard form containing no radicals and no double signs. Let us outline these manipulations and show that we do obtain a single equation satisfied by points on both branches of the hyperbola.

Isolating a radical:

$$\sqrt{(x - 5)^2 + y^2} = \pm 8 + \sqrt{(x + 5)^2 + y^2}$$

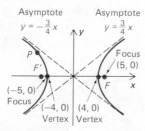

FIGURE 2.63 Hyperbola: $|PF - PF'| = 8$

$$\frac{x^2}{16} - \frac{y^2}{9} = 1$$

Squaring both sides: (Note that $(\pm 8)^2 = 64$.)

$$(x - 5)^2 + y^2 = 64 \pm 16\sqrt{(x + 5)^2 + y^2} + (x + 5)^2 + y^2$$

Collecting like terms and isolating a radical:

$$-64 - 20x = \pm 16\sqrt{(x + 5)^2 + y^2}$$

or $\qquad\qquad 16 + 5x = \mp 4\sqrt{(x + 5)^2 + y^2}$

Squaring both sides: (Note that $(\mp 4)^2 = 16$.)

$$(16 + 5x)^2 = 16[(x + 5)^2 + y^2]$$

Collecting like terms:

$$9x^2 - 16y^2 = 144$$

This equation may be written in the standard form

$$\frac{x^2}{a^2} - \frac{y^2}{b^2} = 1$$

by dividing through by 144.

$$\frac{x^2}{16} - \frac{y^2}{9} = 1$$

Some properties of the graph of the hyperbola considered in the previous paragraph are fairly evident. We see from the equation $x^2/16 - y^2/9 = 1$ that the graph is symmetric with respect to both coordinate axes and with respect to the origin. The x-intercepts are $(4, 0)$ and $(-4, 0)$. There are no y-intercepts, because if we let $x = 0$ we have $y^2 = -9$, and there are no such real numbers y. We can locate an area excluded by points on the graph by solving for y in terms of x and noting that some values of x would produce imaginary values of y. In particular, if we do solve for y in terms of x, we obtain

$$\frac{y^2}{9} = \frac{x^2}{16} - 1$$

$$y^2 = \frac{9}{16}(x^2 - 16)$$

$$y = \pm\frac{3}{4}\sqrt{x^2 - 16}.$$

From this we see that there are no points on the graph for which $-4 < x < 4$, because for such values of x the radicand $x^2 - 16$ is negative and y is imaginary.

The two points on a hyperbola that are nearest its center are called its **vertices** (plural of *vertex*). The vertices are always on the line through the foci. In Figure 2.63 the vertices are also the intercepts $(4, 0)$ and $(-4, 0)$. Note that in the equation $x^2/16 - y^2/9 = 1$, the coefficient of x^2 is positive and the vertices and foci are on the x-axis. It is true in general that if the vertices and foci are on the x-axis, then the coefficient of x^2 in the standard form is positive.

In Figure 2.63 the lines $y = \frac{3}{4}x$ and $y = -\frac{3}{4}x$ appear to be asymptotes of the hyperbola. It can be shown that these lines *are* its asymptotes. It can also be shown that an easy mechanical way to find equations of the asymptotes is to factor the left side of the standard form and set each of the factors equal to zero. That is, factor $x^2/16 - y^2/9$, and set each of its factors equal to zero.

$$\frac{x^2}{16} - \frac{y^2}{9} = \left(\frac{x}{4} - \frac{y}{3}\right)\left(\frac{x}{4} + \frac{y}{3}\right)$$

$$\frac{x}{4} - \frac{y}{3} = 0 \qquad \frac{x}{4} + \frac{y}{3} = 0$$

$$\frac{y}{3} = \frac{x}{4} \qquad \frac{y}{3} = -\frac{x}{4}$$

$$y = \frac{3}{4}x \qquad y = -\frac{3}{4}x$$

Knowing these asymptotes makes it much easier to sketch the graph of the hyperbola.

We have just considered a particular hyperbola with foci at F and F', with coordinates $(5, 0)$ and $(-5, 0)$, and such that $|PF - PF'| = 8$. Its center was at the origin, and in standard form its equation became

$$\frac{x^2}{16} - \frac{y^2}{9} = 1;$$

its vertices were located four units from the center on the x-axis at its only intercepts. In general, if the foci are at $(c, 0)$ and $(-c, 0)$, and if $|PF - PF'| = 2a$ with $a < c$, the center is at the origin and its equation in standard form is

$$\frac{x^2}{a^2} - \frac{y^2}{b^2} = 1 \qquad \text{Hyperbola}$$
$$\text{Vertices on } x\text{-axis}$$

where the number $c^2 - a^2$ is denoted by b^2. Specifying that $a < c$ makes the constant $2a$ less than the distance between F and F'.

$$c^2 - a^2 = b^2$$

(Note that a^2 is not prohibited from being equal to b^2. This is contrary to the case in the standard form of an equation of an ellipse, where $a^2 > b^2$.) As in the specific example, the graph is symmetric with respect to both axes, the only intercepts are on the x-axis, and the vertices are located a units from the center at the intercepts. Again as in the specific example, the asymptotes may be found mechanically by factoring $x^2/a^2 - y^2/b^2$ and setting each factor equal to zero. The asymptotes are

$$y = \frac{b}{a}x \quad \text{and} \quad y = -\frac{b}{a}x.$$

The hyperbola is sketched in Figure 2.64. Note that, as in the specific example, the coefficient of x^2 in the standard form is positive and the vertices and foci are on the x-axis.

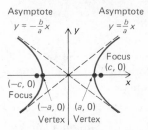

FIGURE 2.64 $\quad \dfrac{x^2}{a^2} - \dfrac{y^2}{b^2} = 1$

If the foci of a hyperbola are on the y-axis at $(0, c)$ and $(0, -c)$, the standard form becomes

$$\frac{y^2}{a^2} - \frac{x^2}{b^2} = 1 \qquad \begin{array}{l}\text{Hyperbola} \\ \text{Vertices on } y\text{-axis}\end{array}$$

Again the graph is symmetric with respect to both axes, but this time the only intercepts are on the y-axis. The intercepts are at $(0, a)$ and $(0, -a)$. The vertices are located on the line through the foci and are a units from the center, so the vertices are also the points $(0, a)$ and $(0, -a)$. Here again we may use our mechanical technique to determine the asymptotes. The asymptotes are

$$\frac{y}{a} - \frac{x}{b} = 0 \quad \text{and} \quad \frac{y}{a} + \frac{x}{b} = 0.$$

That is, the asymptotes are

$$y = \frac{a}{b}x \quad \text{and} \quad y = -\frac{a}{b}x.$$

The graph is sketched in Figure 2.65. Note that in the equation $y^2/a^2 - x^2/b^2 = 1$, the coefficient of y^2 is positive and the vertices and foci are on the y-axis.

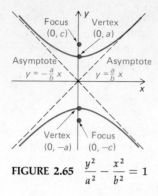

FIGURE 2.65 $\dfrac{y^2}{a^2} - \dfrac{x^2}{b^2} = 1$

Exercise 2.6, 31–40

If the center of a hyperbola is at the origin, it is easy to sketch the graph from the standard form by first locating the vertices (the intercepts) and drawing the asymptotes.

If the center of a hyperbola is at some point (h, k) rather than necessarily at the origin, and if the foci are on a line parallel to the x-axis or to the y-axis, then the standard-form equation becomes

$$\frac{(x - h)^2}{a^2} - \frac{(y - k)^2}{b^2} = 1 \qquad \text{Hyperbola}$$

Center: (h, k)
Vertices on line parallel to or along x-axis

$$\text{or} \quad \frac{(y - k)^2}{a^2} - \frac{(x - h)^2}{b^2} = 1. \qquad \text{Hyperbola}$$

Center: (h, k)
Vertices on line parallel to or along y-axis

(Note that when the coefficient of $(x - h)^2$ is positive, the vertices are on a line parallel to the x-axis; when the coefficient of $(y - k)^2$ is positive, the vertices are on a line parallel to the y-axis.) The same mechanical technique that we used to determine the asymptotes of hyperbolas with centers at the origin and vertices on a coordinate axis can be used here. That is, to find equations for the asymptotes, factor the left side of the standard form and set each factor equal to zero.

EXAMPLE 3

Locate the center and vertices of the hyperbola

$$\frac{(y - 5)^2}{4} - \frac{(x + 2)^2}{9} = 1.$$

Transform the equation by translating coordinate axes so that the new origin is at the center. Draw and label both sets of coordinate axes. Label the vertices. Find equations of the asymptotes. Draw the asymptotes and sketch the graph.

Solution

The center is at $x = -2$, $y = 5$. The vertices are on a line through the center parallel to the y-axis and $\sqrt{4} = 2$ units from the center. The vertices must be at $x = -2$, $y = 7$ and at $x = -2$, $y = 3$.

To transform the equation by translating axes so that the new origin is at the center, let

$$x' = x + 2 \quad \text{and} \quad y' = y - 5.$$

The equation becomes

$$\frac{y'^2}{4} - \frac{x'^2}{9} = 1.$$

One way to find equations for the asymptotes is to factor $y'^2/4 - x'^2/9$ and set each factor equal to zero. Doing this, we have

$$\frac{y'}{2} - \frac{x'}{3} = 0 \quad \text{and} \quad \frac{y'}{2} + \frac{x'}{3} = 0,$$

or

$$y' = \frac{2}{3}x' \quad \text{and} \quad y' = -\frac{2}{3}x'.$$

Since $x' = x + 2$ and $y' = y - 5$, equations of the asymptotes referred to the x-y-axes are

$$y - 5 = \frac{2}{3}(x + 2) \quad \text{and} \quad y - 5 = -\frac{2}{3}(x + 2).$$

Another way to find equations of the asymptotes referred to the x-y axes is to factor $(y - 5)^2/4 - (x + 2)^2/9$ and set each factor equal to zero. If we do this, we have

$$\frac{y - 5}{2} - \frac{x + 2}{3} = 0 \quad \text{and} \quad \frac{y - 5}{2} + \frac{x + 2}{3} = 0,$$

or $y - 5 = \frac{2}{3}(x + 2)$ and $y - 5 = -\frac{2}{3}(x + 2)$, as we found before. The graph is sketched in Figure 2.66.

If the standard forms

$$\frac{(x - h)^2}{a^2} - \frac{(y - k)^2}{b^2} = 1 \quad \text{and} \quad \frac{(y - k)^2}{a^2} - \frac{(x - h)^2}{b^2} = 1$$

are multiplied out, the equations respectively become

$$\underset{\substack{\| \\ A}}{b^2 x^2} - \underset{\substack{\| \\ B}}{a^2 y^2} - 2b^2 hx + 2a^2 ky + b^2 h^2 - a^2 k^2 - a^2 b^2 = 0$$

and

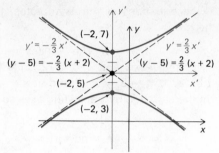

FIGURE 2.66 $\dfrac{(y-5)^2}{4} - \dfrac{(x+2)^2}{9} = 1$

$$\dfrac{y'^2}{4} - \dfrac{x'^2}{9} = 1$$

$$\underset{\substack{\| \\ A}}{-a^2 x^2} + \underset{\substack{\| \\ B}}{b^2 y^2} + 2a^2 hx - 2b^2 ky + b^2 k^2 - a^2 h^2 - a^2 b^2 = 0.$$

These are both of the form

$$Ax^2 + By^2 + Cx + Dy + E = 0$$

where A and B are of opposite signs because either $A = b^2$ and $B = -a^2$ or $A = -a^2$ and $B = b^2$. (Recall that in an ellipse, A and B are different but of the same sign. Here A and B might possibly have the same absolute values, but A and B are of opposite sign.) Where A, B, C, D, and E are specific real numbers and A and B have opposite signs, any equation of the form

$$Ax^2 + By^2 + Cx + Dy + E = 0 \qquad \begin{array}{l}A \text{ and } B \text{ opposite sign} \\ \text{Hyperbola or two intersecting} \\ \text{lines}\end{array}$$

represents a hyperbola or a pair of intersecting lines. The equation can be put in standard form by completing squares in x and y. When the squares are completed, if the constant term is isolated on one side of the equation and is not zero, the equation represents a hyperbola; if the isolated constant term is zero, the equation represents two intersecting lines. Both of these cases are illustrated in the next example.

EXAMPLE 4

Write each of the following equations in standard form and sketch the graph.

(a) $9x^2 - 4y^2 - 90x - 16y + 173 = 0$

(b) $9x^2 - 4y^2 - 90x - 16y + 209 = 0$

Solution

(a)
$$9x^2 - 90x \quad\quad - (4y^2 + 16y \quad) = -173$$
$$9(x^2 - 10x \quad\quad) - 4(y^2 + 4y \quad\quad) = -173$$
$$9(x^2 - 10x + 25) - 4(y^2 + 4y + 4) = -173 + 225 - 16$$
$$9(x - 5)^2 - 4(y + 2)^2 \quad\quad = 36$$
$$\frac{(x - 5)^2}{4} - \frac{(y + 2)^2}{9} \quad = 1$$

This is a hyperbola with center at $(5, -2)$ and with vertices located two units from the center on a line parallel to the x-axis. The graph is shown in Figure 2.67. Contrast this with the graph in Figure 2.66.

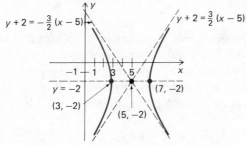

$y + 2 = -\frac{3}{2}(x - 5)$

$y + 2 = \frac{3}{2}(x - 5)$

$y = -2$

$(3, -2)$

$(7, -2)$

$(5, -2)$

FIGURE 2.67 $\quad 9x^2 - 4y^2 - 90x - 16y + 173 = 0$
$$\frac{(x - 5)^2}{4} - \frac{(y + 2)^2}{9} = 1$$

(b)
$$9x^2 - 90x \quad\quad - (4y^2 + 16y \quad) = -209$$
$$9(x^2 - 10x \quad\quad) - 4(y^2 + 4y \quad\quad) = -209$$
$$9(x^2 - 10x + 25) - 4(y^2 + 4y + 4) = -209 + 225 - 16$$
$$9(x - 5)^2 - 4(y + 2)^2 \quad\quad = 0$$
$$\frac{(x - 5)^2}{4} - \frac{(y + 2)^2}{9} \quad = 0$$

The left side may be factored as the difference of two squares.
$$\left(\frac{x - 5}{2} - \frac{y + 2}{3}\right)\left(\frac{x - 5}{2} + \frac{y + 2}{3}\right) = 0$$

This equation is satisfied by all points (x, y) for which
$$\frac{x - 5}{2} - \frac{y + 2}{3} = 0 \quad \text{or} \quad \frac{x - 5}{2} + \frac{y + 2}{3} = 0.$$

These equations represent two lines intersecting at $(5, -2)$. These two lines constitute the graph of the given equation. The graph is shown in Figure 2.68.

Exercise 2.6, 41–60
Miscellaneous exercises,
Exercise 2.6, 61–70

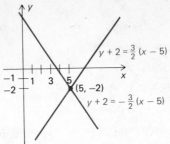

$$y + 2 = \frac{3}{2}(x - 5)$$

(5, −2)

$$y + 2 = -\frac{3}{2}(x - 5)$$

FIGURE 2.68 $9x^2 - 4y^2 - 90x - 16y + 209 = 0$

$$\frac{(x - 5)^2}{4} - \frac{(y + 2)^2}{9} = 0$$

The various forms of equations of ellipses and hyperbolas considered in this section and similar equations encountered in Sections 2.4 and 2.5 are summarized in the following list. The graphs of all of these equations are called **conic sections,** because each is the cross section of a right circular cone cut by a plane. (See Figure 2.69.)

$$\frac{x^2}{a^2} + \frac{y^2}{b^2} = 1, \; a > b$$

Ellipse
Center: (0, 0)
Major axis on x-axis

$$\frac{x^2}{b^2} + \frac{y^2}{a^2} = 1, \; a > b$$

Ellipse
Center: (0, 0)
Major axis on y-axis

$$\frac{(x - h)^2}{a^2} + \frac{(y - k)^2}{b^2} = 1, \; a > b$$

Ellipse
Center: (h, k)
Major axis parallel
to or along x-axis

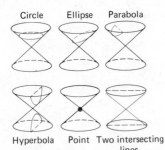

Circle   Ellipse   Parabola

Hyperbola   Point   Two intersecting
                              lines

FIGURE 2.69 The conic sections

$$\frac{(x - h)^2}{b^2} + \frac{(y - k)^2}{a^2} = 1, \ a > b$$

Ellipse
Center: (h, k)
Major axis parallel
to or along y-axis

$$\frac{x^2}{a^2} - \frac{y^2}{b^2} = 1$$

Hyperbola
Center: $(0, 0)$
Vertices on x-axis

$$\frac{y^2}{a^2} - \frac{x^2}{b^2} = 1$$

Hyperbola
Center: $(0, 0)$
Vertices on y-axis

$$\frac{(x - h)^2}{a^2} - \frac{(y - k)^2}{b^2} = 1$$

Hyperbola
Center: (h, k)
Vertices on line
parallel to or along
x-axis

$$\frac{(y - k)^2}{a^2} - \frac{(x - h)^2}{b^2} = 1$$

Hyperbola
Center: (h, k)
Vertices on line
parallel to or along
y-axis

$$y^2 = 4px$$

Parabola
Vertex: $(0, 0)$
Axis: x-axis

$$x^2 = 4py$$

Parabola
Vertex: $(0, 0)$
Axis: y-axis

$$(y - k)^2 = 4p(x - h)$$

Parabola
Vertex: (h, k)
Axis parallel to or
along x-axis

$$(x - h)^2 = 4p(y - k)$$

Parabola
Vertex: (h, k)
Axis parallel to or
along y-axis

$$y = ax^2 + bx + c$$

Parabola
Axis parallel to or
along y-axis

$Ax^2 + By^2 + Cx + Dy + E = 0$, $A \neq B$ and   Ellipse or Point or
A and B of same sign   No graph

$Ax^2 + By^2 + Cx + Dy + E = 0$, A and B of   Hyperbola or Two
opposite sign   intersecting lines

$Ax^2 + By^2 + Cx + Dy + E = 0$, $A = B \neq 0$ Circle or Point or
                                                    No graph

$By^2 + Cx + Dy + E = 0$, $C \neq 0$, $B \neq 0$ Parabola

$Ax^2 + Cx + Dy + E = 0$, $D \neq 0$, $A \neq 0$ Parabola

EXERCISE 2.6 ELLIPSES AND HYPERBOLAS

▶ Find the center, label the ends of the axes, and sketch the graph of each of the following.

EXAMPLE

$$\frac{4}{9}x^2 + \frac{16}{81}y^2 = 1$$

Solution

$$\frac{x^2}{\frac{9}{4}} + \frac{y^2}{\frac{81}{16}} = 1$$

Center: $(0, 0)$

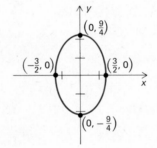

1. $4x^2 + 9y^2 = 36$ 2. $9x^2 + 4y^2 = 36$

3. $9x^2 + 4y^2 = 1$ 4. $4x^2 + 9y^2 = 1$

5. $\frac{16}{81}x^2 + \frac{4}{9}y^2 = 1$ 6. $9x^2 + 16y^2 = 144$

7. $16x^2 + 9y^2 = 144$ 8. $9x^2 + y^2 = 9$

9. $x^2 + 9y^2 = 9$ 10. $4x^2 + 100y^2 = 400$

▶ In each of the following, locate the center and then transform the equation by translating coordinate axes so that the new origin is at the center. Draw and label both sets of coordinate axes. Label the ends of the axes of the ellipse. Sketch the graph.

EXAMPLE

$$\frac{(x - 2)^2}{16} + \frac{(y - 3)^2}{25} = 1$$

Solution

Center $(2, 3)$

$$\frac{x'^2}{16} + \frac{y'^2}{25} = 1$$

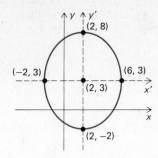

11. $\dfrac{(x + 3)^2}{9} + \dfrac{(y - 4)^2}{16} = 1$ 12. $\dfrac{(x - 3)^2}{1} + \dfrac{(y - 2)^2}{9} = 1$

13. $\dfrac{(x + 5)^2}{4} + \dfrac{(y + 2)^2}{1} = 4$ 14. $9(x - 6)^2 + (y + 3)^2 = 9$

15. $(x - 1)^2 + 9(y + 2)^2 = 9$ 16. $\frac{1}{9}(x - 1)^2 + \frac{1}{4}(y + 4)^2 = 1$

17. $\frac{16}{25}(x + 3)^2 + \frac{4}{9}(y - 1)^2 = 1$

18. $100(x - 5)^2 + 4(y - 4)^2 = 400$

19. $16(x + 3)^2 + 4(y - 5)^2 = 64$ 20. $4(x + 3)^2 + 16(y - 5)^2 = 64$

▶ Write each of the following equations in standard form. If the equation has a graph, sketch it. If the equations has no graph, simply state "No graph."

EXAMPLE

$4x^2 + 9y^2 - 16x + 90y + 242 = 0$

Solution

$4(x^2 - 4x\ \ ) + 9(y^2 + 10y\ \ ) = -242$

$\qquad 4(x - 2)^2 + 9(y + 5)^2 = -242 + 16 + 225$

$\qquad 4(x - 2)^2 + 9(y + 5)^2 = -1 \qquad$ No graph

21. $4x^2 + 9y^2 - 16x + 90y + 240 = 0$

22. $4x^2 + 9y^2 - 16x + 90y + 241 = 0$

23. $4x^2 + 9y^2 - 24x + 18y + 9 = 0$

24. $4x^2 + 9y^2 - 24x + 18y + 81 = 0$

25. $4x^2 + 9y^2 - 24x + 18y + 45 = 0$

26. $4x^2 + 25y^2 + 24x - 100y + 36 = 0$

27. $x^2 + 9y^2 - 10x - 56 = 0$

28. $4x^2 + 25y^2 + 24x - 100y + 236 = 0$

29. $x^2 + 9y^2 - 72y + 63 = 0$

30. $4x^2 + 25y^2 + 24x - 100y + 136 = 0$

Find the center, vertices, and equations of the asymptotes of each of the following. Draw the asymptotes and sketch the graph.

EXAMPLE

$$\frac{y^2}{16} - \frac{x^2}{9} = 1$$

Solution

Center: (0, 0)
Vertices: (0, 4) and (0, −4)

$$\frac{y^2}{16} - \frac{x^2}{9} = 0$$

$$\left(\frac{y}{4} - \frac{x}{3}\right)\left(\frac{y}{4} + \frac{x}{3}\right) = 0$$

Asymptotes: $y = \frac{4}{3}x$ and $y = -\frac{4}{3}x$

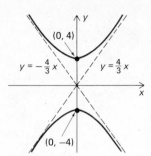

31. $4x^2 - 9y^2 = 36$ **32.** $9x^2 - 4y^2 = 36$

33. $9y^2 - 4x^2 = 1$ **34.** $4y^2 - 9x^2 = 1$

35. $\frac{16}{81}x^2 - \frac{4}{9}y^2 = 1$ **36.** $\frac{16}{81}y^2 - \frac{4}{9}x^2 = 1$

37. $4y^2 - 100x^2 = 400$ **38.** $100x^2 - 4y^2 = 400$

39. $x^2 - 9y^2 = 9$ **40.** $y^2 - 9x^2 = 9$

▶ Find the center and vertices of each of the following. Transform each equation by translating coordinate axes so that the new origin is at the center. Draw and label both sets of coordinate axes. Find equations of the asymptotes referred to the new coordinate axes, and also find equations of the asymptotes referred to the original coordinate axes. Draw the asymptotes and sketch the graph.

EXAMPLE

$$\frac{(x + 2)^2}{4} - \frac{(y - 5)^2}{9} = 1$$

Solution

Center: (−2, 5)

$$\frac{x'^2}{4} - \frac{y'^2}{9} = 1$$

Vertices: $x' = 2, y' = 0$ and $x' = -2, y' = 0$
$\qquad\qquad x = 0, y = 5$ and $x = -4, y = 5$

Asymptotes: $y' = \pm\frac{3}{2}x'$
$\qquad\qquad y - 5 = \pm\frac{3}{2}(x + 2)$

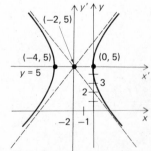

41. $\frac{(x + 3)^2}{9} - \frac{(y - 4)^2}{16} = 1$ **42.** $\frac{(y - 4)^2}{16} - \frac{(x + 3)^2}{9} = 1$

43. $\dfrac{(x + 5)^2}{4} - \dfrac{(y + 2)^2}{1} = 4$ **44.** $\dfrac{(y + 2)^2}{1} - \dfrac{(x + 5)^2}{4} = 4$

45. $9(x - 6)^2 - (y + 3)^2 = 9$ **46.** $(y + 3)^2 - 9(x - 6)^2 = 9$

47. $\frac{1}{9}(x - 1)^2 - \frac{1}{4}(y + 4)^2 = 1$

48. $100(x - 5)^2 - 4(y - 4)^2 = 400$

49. $4(y - 4)^2 - 100(x - 5)^2 = 400$

50. $4(x + 3)^2 - 16(y - 5)^2 = 64$

▶ Write each of the following equations in standard form and sketch the graph.

EXAMPLE

$4x^2 - 9y^2 - 16x - 90y - 209 = 0$

Solution

$$4(x^2 - 4x\ \ ) - 9(y^2 + 10y\ \ ) = 209$$
$$4(x - 2)^2 - 9(y + 5)^2 = 209 + 16 - 225$$
$$4(x - 2)^2 - 9(y + 5)^2 = 0$$
$$(y + 5) = \pm\tfrac{2}{3}(x - 2)$$

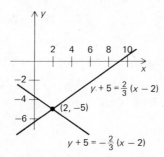

51. $4x^2 - 9y^2 - 16x - 90y - 210 = 0$

52. $4x^2 - 9y^2 - 24x - 18y + 27 = 0$

53. $-4x^2 + 24x + 9y^2 + 18y - 28 = 0$

54. $4x^2 - 9y^2 - 24x - 18y + 26 = 0$

55. $x^2 - y^2 + 2x + 1 = 0$

56. $y^2 - x^2 - 10x - 25 = 0$

57. $x^2 - y^2 + 2x = 0$

58. $4x^2 - 9y^2 - 24x - 18y - 9 = 0$

59. $-4x^2 + 9y^2 + 24x + 18y - 63 = 0$

60. $x^2 - y^2 - 2x + 2y = 0$

▶ In each of the following, if the equation has a graph, identify it. If the equation has no graph, state "No graph."

EXAMPLES	Solutions
(a) $x^2 + 9y^2 = 36$	(a) Ellipse
(b) $x^2 - 9y^2 = 36$	(b) Hyperbola
(c) $x^2 + 9y^2 = -36$	(c) No graph
(d) $x^2 - 9y^2 = -36$	(d) Hyperbola
(e) $9y^2 - x^2 = 36$	(e) Hyperbola
(f) $9x^2 + 9y^2 = 36$	(f) Circle
(g) $9x^2 - 9y^2 = 36$	(g) Hyperbola
(h) $x^2 + 9y^2 = 0$	(h) Point
(i) $x^2 - 9y^2 = 0$	(i) Two intersecting lines
(j) $x^2 - 9y = 0$	(j) Parabola
(k) $9y^2 - x = 0$	(k) Parabola

61. $4x^2 + 25y^2 = 100$ **62.** $4x^2 - 25y^2 = 100$

63. $4x^2 + 4y^2 = 100$ **64.** $4x^2 + 25y^2 = -100$

65. $4x^2 - 25y^2 = -100$ **66.** $4x^2 + 25y^2 = 0$

67. $4x^2 - 25y^2 = 0$ **68.** $25y^2 - 100x = 0$

69. $4x^2 + 4y^2 - 40x + 100 = 0$ **70.** $25x^2 + 100x + 4y = 0$

CHECK LIST FOR THE STUDENT

If you have learned the material in Section 2.6, you should be able to do the following:

1. Define an ellipse.

2. Recognize the standard form of an equation of an ellipse.

3. Find the center and the ends of the axes of an ellipse from its equation in standard form.

4. Sketch the graph of an ellipse if its equation is given in standard form.

5. Transform the standard form of an equation of an ellipse with center at (h, k) by translating axes so that the new origin is at (h, k). Draw and label both sets of axes and sketch the graph.

6. Write in standard form an equation of the form $Ax^2 + By^2 + Cx + Dy + E = 0$ with A and B of the same sign and $A \neq B$.

7. Define a hyperbola.

8. Recognize the standard form of an equation of a hyperbola.

9. Find the center, vertices, and asymptotes of a hyperbola from its equation in standard form.

10. Sketch the graph of a hyperbola if its equation is given in standard form.

11. Transform the standard form of an equation of a hyperbola with center at (h, k) by translating axes so that the new origin is at (h, k). Draw and label both sets of axes and sketch the graph.

12. Write in standard form an equation of the form $Ax^2 + By^2 + Cx + Dy + E = 0$ with A and B of opposite sign.

13. Identify an equation of the form $Ax^2 + By^2 + Cx + Dy + E = 0$ with A and B not both zero as a circle, parabola, ellipse, hyperbola, point, two lines, or no graph. Sketch the graph if it exists.

1. Find the center, label the ends of the axes, and sketch the graph of

$$\frac{x^2}{16} + \frac{y^2}{9} = 1.$$

2. Transform the equation

$$\frac{(x-3)^2}{9} + \frac{(y-6)^2}{25} = 1$$

by translating coordinate axes so that the new origin is at the center. Draw and label both sets of coordinate axes. Label the ends of the axes of the ellipse. Sketch the graph.

Write each of the following equations in standard form and identify it.

3. $4x^2 + 9y^2 - 24x + 72y + 144 = 0$

4. $4x^2 - 9y^2 - 24x - 72y - 144 = 0$

5. Find the center, vertices, and equations of the asymptotes of

$$\frac{x^2}{25} - \frac{y^2}{9} = 1.$$

Draw the asymptotes and sketch the graph.

6. Transform the equation

$$\frac{(y-6)^2}{25} - \frac{(x-3)^2}{9} = 1$$

by translating coordinate axes so that the new origin is at the center. Draw and label both sets of coordinate axes. Find equations of the asymptotes referred to the new coordinate axes, and also find equations of the asymptotes referred to the original coordinate axes. Draw the asymptotes and sketch the graph.

In each of the following, if the equation has a graph, identify it. If the equation has no graph, state "No graph."

7. (a) $4x^2 + 9y^2 = 36$ (b) $4x^2 + 4y^2 = 36$

8. (a) $4x^2 - 9y^2 = 36$ (b) $4x^2 - 9y^2 = 0$

9. (a) $4x^2 + 9y^2 = -36$ (b) $4x^2 = 9y$

10. (a) $4x^2 - 9y^2 = -36$ (b) $9y^2 = 4x$

*2.7 Inverse Relations and One-to-One Functions

The purpose of this section is to reinforce the definitions of relation and function, to define inverse relations and one-to-one functions, and to demonstrate that the inverse of any one-to-one function is itself a function.

We noted in Section 2.1 that a *relation* is defined provided that a set called the domain is given and provided that there is given a rule of assignment that assigns one or more things to each element in the domain. We saw that a relation may be defined as any set of ordered pairs. We saw that if the ordered pairs are pairs of real numbers, the graph of the relation is the set of all points associated with its ordered pairs.

The lines, circles, parabolas, ellipses, and hyperbolas that we considered in the last few sections are all graphs of relations. For example, the ellipse in Figure 2.55 is the graph of the relation

$$\left\{ (x,\ y) \,\middle|\, \frac{x^2}{25} + \frac{y^2}{16} = 1 \right\}.$$

We say that the equation $x^2/25 + y^2/16 = 1$ *defines* this relation.

We also saw in Section 2.1 that a *function* is defined provided that a set called the domain is given and provided that there is given a rule of assignment that assigns a *single* thing to each element in the domain. In other words, a function is a special relation in which the rule of assignment assigns exactly one thing to each element in the domain. Thus a function may be defined as a set of ordered pairs with the restriction that no two different ordered pairs have the same first member.

The circles, ellipses, hyperbolas, and many of the parabolas that we considered in the last few sections are graphs of relations but are *not* graphs of functions. The ''sweeping vertical line'' test demonstrated in Figure 2.7 verifies this. Parabolas that open up or down, however, *are* graphs of functions. For example, the parabola opening up in Figure 2.7 is the graph of the function

$$\{ (x,\ y) \mid y = x^2 \}.$$

We saw in Section 2.5 that such a function is called a quadratic function.

Parts of circles, ellipses, hyperbolas, or parabolas opening left or right might represent functions. For example, the lower half of the parabola $y^2 = x$ in Figure 2.7 is the graph of the function

$$\{ (x,\ y) \mid y = -\sqrt{x} \text{ and } x \geq 0 \}.$$

Exercise 2.7, 1–10 (This is the function h sketched in Figure 2.4.)

The **inverse** of any given relation is the relation obtained by interchanging the members in each of its ordered pairs. If a relation is defined by a given equation in x and y, an equation defining its inverse can be found simply by interchanging *x* and *y* in the given equation. The inverse of the relation

$$\left\{(x,\ y)\ \middle|\ \frac{x^2}{25} + \frac{y^2}{16} = 1\right\}$$

is the relation

$$\left\{(x,\ y)\ \middle|\ \frac{y^2}{25} + \frac{x^2}{16} = 1\right\}.$$

The inverse of a relation *r* is called *r*-inverse and is denoted symbolically by r^{-1}. Here the symbol -1 stands for "inverse"; it does *not* mean "negative one."

r^{-1} represents *r*-inverse

EXAMPLE 1

Sketch the graph of the relation $r = \{(x,\ y)\,|\,2x + 3y - 18 = 0\}$. State whether or not this relation is a function. Find the inverse of *r*. State whether or not the inverse of *r* is a function. Sketch the graph of r^{-1} on the same coordinate system as the graph of *r*.

Solution

The graph of *r* is the oblique line shown in Figure 2.70 with intercepts (0, 6) and (9, 0). We saw in Section 2.2 that any linear equation in *x* and *y* defines a linear function, so the relation *r* is a linear function.

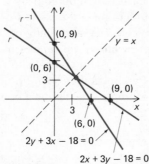

FIGURE 2.70 $r = \{(x,\ y)\,|\,2x + 3y - 18 = 0\}$
$r^{-1} = \{(x,\ y)\,|\,2y + 3x - 18 = 0\}$

The inverse of *r* is the relation

$$r^{-1} = \{(x,\ y)\,|\,2y + 3x - 18 = 0\}.$$

The relation r^{-1} is also a linear function. Its graph is the line in Figure 2.70 with intercepts (6, 0) and (0, 9). The sweeping vertical line test verifies that both *r* and r^{-1} are functions.

EXAMPLE 2

We have seen that the relation $f = \{(x, y)|y = x^2\}$ is a function. Find the inverse of f. State whether or not the inverse of f is a function.

Solution

$$f = \{(x, y)|y = x^2\}$$
$$f^{-1} = \{(x, y)|x = y^2\}$$

Exercise 2.7, 11–30

Here f^{-1} is not a function. Its graph is shown in Figure 2.7, where the sweeping vertical line test verifies that it is not a function.

Examples 1 and 2 demonstrate that the inverse of a relation that is a function may or may not be a function. The reason for this is that although the rule of assignment of a function must assign only one single thing to each domain element, that rule may assign the same single thing to more than one domain element. For example, the function f of Example 2 assigns the single number 9 to both 3 and -3. That is, the points $(3, 9)$ and $(-3, 9)$ are both on the graph of f. This forces the points $(9, 3)$ and $(9, -3)$ both to be on the graph of f^{-1}, so the graph of f^{-1} does not represent a function. The relation f^{-1} has different ordered pairs with the same first member, so f^{-1} is not a function.

If the rule of assignment of a function does not assign the same single number to more than one domain element, the function is said to be a *one-to-one function,* and its inverse has to be a one-to-one function. The function r of Example 1 is a one-to-one function. A **one-to-one function** is a relation whose rule of assignment assigns exactly one thing to each domain element *and* each of whose range elements is the assignment made to exactly one domain element. There is what is known as a one-to-one correspondence between the domain and the range of a one-to-one function. No two different ordered pairs have the same first member, *and* no two different ordered pairs have the same second member.

It is easy to tell from the graph of a function whether or not it is one-to-one by applying a "sweeping horizontal line" test similar to the sweeping vertical line test that we use to determine whether or not a relation is a function. If no horizontal line intersects the graph of a function in more than one point, then the function is one-to-one.

EXAMPLE 3

Use the sweeping horizontal line test to determine whether or not the function $g = \{(x, y)|y = \sqrt{x}$ and $x \geq 0\}$ is one-to-one.

Solution

The graph of g is shown in Figure 2.71. It is the upper half of the parabola $y^2 = x$. The sweeping vertical line test verifies that g is a function. Furthermore, no horizontal line intersects the graph of g in more than one point, so g is a one-to-one function.

Exercise 2.7, 31–50

We have said that the inverse of a one-to-one function is itself a one-to-one function. In Example 1 we found an equation defining the inverse of a given one-to-one function simply by interchanging x and y in the

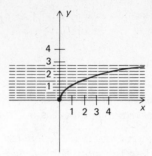

FIGURE 2.71   A one-to-one function, $g = \{(x, y)|y = \sqrt{x} \text{ and } x \geq 0\}$

equation defining the given function. Such interchanging is equivalent to interchanging the domain and range and "reversing" the rule of assignment. For example, let *f* be the function such that

$$f(x) = 2x - 4,$$
$$\text{Domain of } f = \{x|x \geq 0\}.$$

The graph of *f* is that part of the line $y = 2x - 4$ for which $x \geq 0$. In Figure 2.72 it is clear that the range of *f* is $\{y|y \geq -4\}$.

$$\text{Range of } f = \{y|y \geq -4\}$$

To find the domain and range of f^{-1}, we simply interchange the domain and range of *f*. As usual, we use the letter *x* to represent any domain element, and we use the letter *y* to represent the corresponding range element.

$$\text{Domain of } f^{-1} = \{x|x \geq -4\}$$
$$\text{Range of } f^{-1} = \{y|y \geq 0\}$$

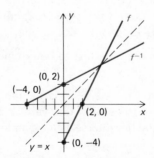

FIGURE 2.72 $f(x) = 2x - 4,\ x \geq 0$

$$f^{-1}(x) = \frac{x + 4}{2},\ x \geq -4$$

To find the rule of assignment of f^{-1}, we "reverse" the rule of f. Since f takes any element x from its domain, multiplies it by 2, and then subtracts 4, the rule of f^{-1} must reverse this procedure by first adding 4 and then dividing by 2. In other words,

$$f^{-1}(x) = \frac{x + 4}{2}.$$

The graph of f^{-1} is shown, together with the graph of f, in Figure 2.72.

If f is a one-to-one function and the equation $y = f(x)$ can easily be solved for x in terms of y, the rule of assignment of function f^{-1} can easily be found mechanically by interchanging x and y and then solving for y in terms of x. For example, if $f(x) = 2x - 4$, we can find $f^{-1}(x)$ mechanically as follows:

$$f(x) = 2x - 4$$

1. Let $y = f(x)$. $y = 2x - 4$ Defines f where $y = f(x)$

2. Interchange x and y. $x = 2y - 4$ Defines f^{-1} where $y = f^{-1}(x)$

3. Solve for y. $2y = x + 4$

$$y = \frac{x + 4}{2}$$

$$f^{-1}(x) = \frac{x + 4}{2}$$

The next example demonstrates that when this mechanical procedure is utilized to find $f^{-1}(x)$, the algebraic manipulations involved must take the range of f^{-1} into consideration.

EXAMPLE 4

Let f be the function such that

$$\text{Domain of } f = \{x \mid x \leq -2\},$$
$$f(x) = x^2.$$

Sketch the graph of f. Find the domain and range of f^{-1}. Find $f^{-1}(x)$. Sketch the graph of f^{-1} on the same coordinate system as the graph of f.

Solution

The graph of f is shown in Figure 2.73. It is that part of the parabola $y = x^2$ to the left of and including $(-2, 4)$.

The range of f is $\{y \mid y \geq 4\}$, so the domain of f^{-1} is $\{x \mid x \geq 4\}$. The range of f^{-1} is the domain of f, so the range of f^{-1} is $\{y \mid y \leq -2\}$.

To find $f^{-1}(x)$, interchange x and y in $y = x^2$ and then solve for y, taking into account the range of f^{-1}.

$$x = y^2$$
$$y = \pm\sqrt{x}$$

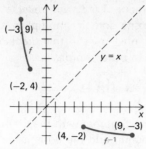

FIGURE 2.73 $f(x) = x^2, \; x \le -2$
$f^{-1}(x) = -\sqrt{x}, \; x \ge 4$

Either $y = +\sqrt{x}$ or $y = -\sqrt{x}$ will satisfy the equation $x = y^2$. But the range of f^{-1} is a set of negative numbers, in fact it is the set of all numbers $y \le -2$, so for the rule of f^{-1} we need the minus sign.

$$f^{-1}(x) = -\sqrt{x}$$

The graph of f^{-1} is shown in Figure 2.73. It is that part of the lower half of the parabola $y^2 = x$ to the right of and including $(4, -2)$.

In each of Figures 2.71, 2.72, and 2.73 we have sketched the graphs of a one-to-one function and its inverse on the same coordinate system. In each of these figures the line $y = x$ has been drawn as a dotted line to show that in each case the graph of f^{-1} is the reflection of the graph of f in the line $y = x$. In other words, if the paper were folded on the line $y = x$, the graphs of f and f^{-1} would exactly match.

EXAMPLE 5

Let $f(x) = x^2$, and let the domain of f be $\{x \,|\, x \ge 0\}$. Sketch the graphs of f and f^{-1} on the same coordinate system.

Solution

The graph of f is the right half of the parabola $y = x^2$. It is shown in Figure 2.74. The line $y = x$ has been drawn as a dotted line in Figure 2.74. The reflection of f in $y = x$ is the graph of f^{-1}. The graph of f^{-1} is the upper half of the parabola $x = y^2$.

Exercise 2.7, 51–60

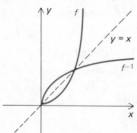

FIGURE 2.74 $f(x) = x^2, \; x \ge 0$
$f^{-1}(x) = \sqrt{x}, \; x \ge 0$

EXERCISE 2.7 INVERSE RELATIONS
AND ONE-TO-ONE FUNCTIONS

▶ Sketch the graph of each of the following, and state whether or not it is the graph of a function.

EXAMPLE

$\{(x, y) | x^2 + y^2 = 9, x \geq 0, \text{ and } y \geq 0\}$

Solution

Graph of a function.

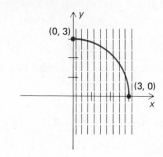

1. $\{(x, y) | x^2 + y^2 = 9 \text{ and } x \geq 0\}$

2. $\{(x, y) | x^2 + y^2 = 9, x \leq 0, \text{ and } y \geq 0\}$

3. $\{(x, y) | x^2 + y^2 = 9\}$

4. $\{(x, y) | x^2 + y^2 = 9 \text{ and } y \geq 0\}$

5. $\{(x, y) | x^2 + y^2 = 9, x \geq 0, \text{ and } y \leq 0\}$

6. $\{(x, y) | x^2 + y^2 = 9, x \leq 0, \text{ and } y \leq 0\}$

7. $\left\{(x, y) \left| \dfrac{y^2}{9} - \dfrac{x^2}{16} = 1 \text{ and } y \geq 0 \right.\right\}$

8. $\left\{(x, y) \left| \dfrac{y^2}{9} + \dfrac{x^2}{16} = 1 \text{ and } y \geq 0 \right.\right\}$

9. $\left\{(x, y) \left| \dfrac{x^2}{9} - \dfrac{y^2}{16} = 1 \text{ and } x \geq 0 \right.\right\}$

10. $\left\{(x, y) \left| \dfrac{x^2}{9} - \dfrac{y^2}{16} = 1 \text{ and } y \geq 0 \right.\right\}$

▶ Find the inverse of each of the following relations:

EXAMPLE

$r = \{(x, y) | 9x^2 - 4y^2 = 36\}$

Solution

$r^{-1} = \{(x, y) | 9y^2 - 4x^2 = 36\}$

11. $r = \{(x, y) | 7x - 8y + 9 = 0\}$

12. $r = \{(x, y) | 9x^2 + 4y^2 = 36\}$

13. $r = \{(x, y) \mid x^2 + y^2 = 9\}$

14. $r = \{(x, y) \mid y = 2x\}$

15. $r = \{(x, y) \mid y = x^2 - 6x + 9\}$

16. $r = \{(x, y) \mid y^2 - 8y + 16 = x\}$

17. $r = \{(x, y) \mid y = 10^x\}$

18. $r = \{(x, y) \mid y = 2^x\}$

19. $r = \{(x, y) \mid y = e^x\}$

20. $r = \{(x, y) \mid y = \sqrt{1 - x^2}\}$

▶ Find the inverse of each of the following functions, and state whether or not the inverse is a function.

EXAMPLES   (a) $f = \{(x, y) \mid y = 3x\}$ (b) $f = \{(x, y) \mid y = 3x^2\}$

Solutions   (a) $f^{-1} = \{(x, y) \mid x = 3y\}$ A function
              (b) $f^{-1} = \{(x, y) \mid x = 3y^2\}$ Not a function

21. $f = \{(x, y) \mid y = 3x - 7\}$ **22.** $f = \{(x, y) \mid y = 3x^2 - 6x + 3\}$

23. $f = \{(x, y) \mid 3x - 7y + 10 = 0\}$

24. $f = \{(x, y) \mid y = x^3\}$ **25.** $f = \{(x, y) \mid y = x^4\}$

26. $f = \{(x, y) \mid y = x^5\}$ **27.** $f = \{(x, y) \mid y = \sqrt{x}\}$

28. $f = \{(x, y) \mid y = -\sqrt{x}\}$ **29.** $\{(x, y) \mid y = -x\}$

30. $\{(x, y) \mid y = 5x^2 + 10x + 5\}$

▶ State whether or not the function in each of the following is one-to-one.

EXAMPLES   (a) $\{(x, y) \mid y = 3x^2\}$ (b) $\{(x, y) \mid y = 3x^2 \text{ and } x \geq 0\}$

Solutions   (a) Not one-to-one   (b) One-to-one

31. $\{(x, y) \mid y = 3x\}$ **32.** $\{(x, y) \mid y = x^2 - 6x + 9\}$

33. $\{(x, y) \mid y = x^2 - 6x + 9 \text{ and } x \geq 0\}$

34. $\{(x, y) \mid y = x^2 - 6x + 9 \text{ and } x \geq 3\}$

35. $\{(x, y) \mid x^2 + y^2 = 9, x \geq 0 \text{ and } y \geq 0\}$

36. Domain of $f = \{x \mid x \geq 0\}$

$f(x) = 3x$ for $x \leq 1$

$f(x) = 3$ for $x > 1$

37. Domain of $f = \{x \mid x \geq -1\}$

$f(x) = x^2$

38. Domain of $f = \{x \mid -1 \le x \le 1\}$
$f(x) = x^2$

39. Domain of $f = \{x \mid x \ge 1\}$
$f(x) = x^2$

40. Domain of $f = \{x \mid -4 \le x \le 4\}$
$f(x) = x^2$ for $x \le 1$
$f(x) = 3$ for $x > 1$

► Use the sweeping horizontal line test to determine whether or not each of the following graphs represents a one-to-one function.

41.

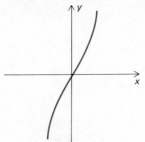

42.

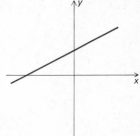

43.

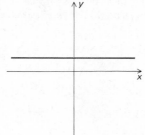

44.

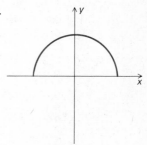

45.

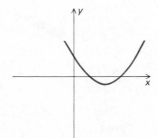

46.

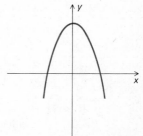

47.

48.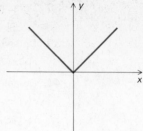

49.

50.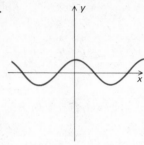

▶   In each of the following, find the domain and range of *f*-inverse, and find $f^{-1}(x)$. Sketch the graphs of *f* and f^{-1} on the same coordinate system. (If the domain is not explicitly stated, use the domain convention introduced in Section 2.1.)

EXAMPLE

$f(x) = 3x^2$

Domain of $f = \{x \mid x \geq 1\}$

Solution

Range of $f = \{y \mid y \geq 3\}$
Domain of $f^{-1} = \{x \mid x \geq 3\}$
Range of $f^{-1} = \{y \mid y \geq 1\}$

$$y = 3x^2$$
$$x = 3y^2$$
$$y = +\sqrt{\frac{x}{3}}, \ f^{-1}(x) = \sqrt{\frac{x}{3}}$$

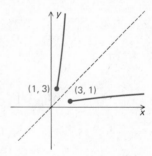

51. $f(x) = 3x + 5$
Domain of $f = \{x \mid x \geq -1\}$

52. $f(x) = 3x + 5$

53. $f(x) = x^2$
Domain of $f = \{x \mid x \geq 2\}$

54. $f(x) = \sqrt{x}$
Domain of $f = \{x \mid x \geq 9\}$

55. $f(x) = 2x - 5$
Domain of $f = \{x \mid x \geq 0\}$

56. $f(x) = 2x - 5$

57. $f(x) = x^2$
Domain of $f = \{x \mid x \leq 0\}$

58. $f(x) = -x$
Domain of $f = \{x \mid x \leq 0\}$

59. $f(x) = 3x - 4$

60. $f(x) = -\sqrt{x}$

CHECK LIST FOR THE STUDENT

If you have learned the material in Section 2.7, you should be able to do the following:

1. Tell from looking at the graph of a relation whether or not the relation is a function.

2. Define the inverse of a relation.

3. Find an equation defining the inverse of a relation that is defined by a given equation.

4. Use the notation r^{-1} for the inverse of a relation r.

5. Recognize that the inverse of a function may or may not be a function.

6. Define a one-to-one function.

7. Recognize that the inverse of a one-to-one function is itself a function.

8. Tell from looking at the graph of a function whether or not the function is one-to-one.

9. Find the domain, range, and rule of assignment of the inverse of a given one-to-one function.

10. Sketch the graph of the inverse of a one-to-one function whose graph is given by reflecting the given graph in the line $y = x$.

SECTION QUIZ 2.7

1. Sketch the graph of each of the following, and state whether or not it is the graph of a function.
   (a) $\{(x, y) \mid x^2 + y^2 = 25 \text{ and } x \geq 0\}$
   (b) $\{(x, y) \mid x^2 + y^2 = 25, x \geq 0, \text{ and } y \geq 0\}$

2. Find the inverse of each of the following, and state whether or not the inverse is a function.
   (a) $\{(x, y) \mid y = 2x + 7\}$ (b) $\{(x, y) \mid y = 8x^2\}$

3. State whether or not each of the following functions is one-to-one.
   (a) $\{(x, y) \mid y = 5x\}$ (b) $\{(x, y) \mid y = 5x^2\}$

4. In each of the following, tell whether or not the graph represents a one-to-one function.

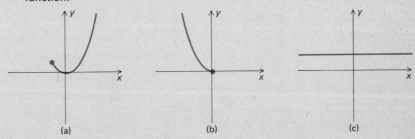

(a)   (b)   (c)

5. Let the domain of f be $\{x \mid x \geq 0\}$, and let $f(x) = 2x + 3$. Find the domain and range of f-inverse, and find $f^{-1}(x)$. Sketch the graphs of f and f^{-1} on the same coordinate system.

*2.8 Exponential and Logarithmic Functions

The purpose of this section is to define and graph exponential functions and to introduce logarithmic functions as the inverses of exponential functions.

If b is any positive real number other than the number 1, then the function f such that

$$f(x) = b^x$$

is called the **exponential function** with base b. In Sections 1.2 and 1.3 we defined b^x for integral x and more generally for rational x. We saw in those sections that integral and rational exponents satisfy five basic exponent laws. Using a concept usually taken up in calculus, the expression b^x can also be defined for irrational numbers x so that b^x represents a real number and so that the five exponent laws hold for real numbers x whether they are rational or irrational. The expression b^x is called a **power** of b. Let us agree that the exponent laws hold for real-number exponents and that b^x represents a real number for any real number x. Then according to our domain convention, the domain of an exponential function is the set of all real numbers.

$$f(x) = b^x$$

Domain of f is the set of all real numbers

For example, the function f such that

$$f(x) = 2^x$$

is the exponential function with base 2. Its domain is the set of all real numbers.

The graph of function f such that $f(x) = 2^x$ is the graph of the equation $y = 2^x$. If we make a table of values and plot points on the graph of $y = 2^x$, we see that the points all appear to be strung together like beads as shown in Figure 2.75. Note that the graph lies entirely above the

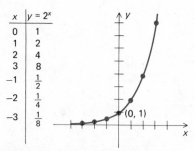

x	$y = 2^x$
0	1
1	2
2	4
3	8
−1	$\frac{1}{2}$
−2	$\frac{1}{4}$
−3	$\frac{1}{8}$

FIGURE 2.75 Graph of $f: f(x) = 2^x$
(Graph of $y = 2^x$)

x-axis. This is true of the graph of any exponential function, because for any real number *x* the expression $y = b^x$ is positive. In other words, the range of any exponential function is the set of all positive real numbers.

$$f(x) = b^x$$

$$\text{Range of } f = \{y \mid y > 0\}$$

The *y*-intercept of the graph of any exponential function is easy to find. For example, the *y*-intercept of the graph of $y = 2^x$ is (0, 1) because when $x = 0$, $y = 2^0 = 1$. The point (0, 1) is also the *y*-intercept of the graph of any exponential function because if $b > 0$, then $b^0 = 1$. In other words, the graph of any exponential function intersects the *y*-axis at (0, 1).

Note in Figure 2.75 that as *x* takes on larger and larger positive values, $y = 2^x$ also takes on larger and larger positive values. But as *x* takes on negative values that become larger and larger in absolute value, $y = 2^x$ takes on positive values that become nearer and nearer to zero. As we continue the sketch farther and farther to the left, the graph approaches the position of the left half of the *x*-axis. The *x*-axis is an asymptote. This is all true of the graph of any exponential function with base greater than the number 1. Tracing the graph of such a function from *left to right,* we see that as *x* increases, $y = b^x$ with $b > 1$ also increases so that the graph leaves its asymptote, the *x*-axis, goes through its intercept (0, 1), and rises off to the right.

If the base *b* of an exponential function is less than the number 1, if $0 < b < 1$, then the graph falls from left to right and approaches the position of the right half of the *x*-axis. This is demonstrated by the graph of $y = (\frac{1}{2})^x$ shown in Figure 2.76. Note that $(\frac{1}{2})^x = (2^{-1})^x = 2^{-x}$ so that the graph of $y = (\frac{1}{2})^x$ is precisely the graph of $y = 2^{-x}$.

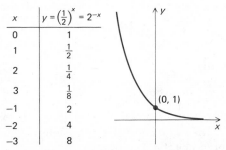

x	$y = \left(\dfrac{1}{2}\right)^x = 2^{-x}$
0	1
1	$\frac{1}{2}$
2	$\frac{1}{4}$
3	$\frac{1}{8}$
−1	2
−2	4
−3	8

FIGURE 2.76 Graph of $f: f(x) = (\frac{1}{2})^x$
(Graph of $y = (\frac{1}{2})^x$)

In calculus, as well as in the natural sciences and in many physical applications, it is frequently convenient to use the number *e* as the base of an exponential function. The number *e* was introduced in Section 1.1. Recall that *e* is an irrational number approximately equal to 2.718281828.

$$e \approx 2.718281828$$

Since *e* is greater than 1, the graph of $y = e^x$ will rise from left to right as did the graph of $y = 2^x$ in Figure 2.75. The graph of $y = e^x$ is shown in Figure 2.77

Exercise 2.8, 1–10

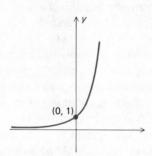

(0, 1)

FIGURE 2.77 Graph of $f: f(x) = e^x$
(Graph of $y = e^x$)

Any exponential function is a one-to-one function. The sweeping horizontal line test verifies this, because the graph of any exponential function either looks similar to the graph of $y = 2^x$ in Figure 2.75 or similar to the graph of $y = (\frac{1}{2})^x$ in Figure 2.76. Thus any exponential function has an inverse that is itself a one-to-one function.

The inverse of any exponential function is called a *logarithmic function*. Let us find the inverse of the function *f* such that $f(x) = b^x$ where *b* is a positive real number other than the number 1. We know the domain, rule of assignment, and range of *f*.

$$\text{Domain of } f = \{x|x \text{ is a real number}\}$$
$$f(x) = b^x \text{ where } b > 0 \text{ and } b \neq 1$$
$$\text{Range of } f = \{y|y > 0\}$$

In other words,

$$f = \{(x, y)|y = b^x \text{ where } b > 0 \text{ and } b \neq 1\}. \quad \text{Exponential function}$$

To find *f*-inverse, we interchange the domain and range of *f* and "reverse" the rule of *f*. That is, to find *f*-inverse we interchange the members in each of the ordered pairs of *f* so that

$$f^{-1} = \{(x, y)|x = b^y \text{ where } b > 0 \text{ and } b \neq 1\}. \quad \text{Logarithmic function}$$

The domain of f^{-1} is the range of *f*, and the range of f^{-1} is the domain of *f*.

$$\text{Domain of } f^{-1} = \{x|x > 0\}$$
$$\text{Range of } f^{-1} = \{y|y \text{ is a real number}\}$$

If $x = b^y$, the number y is called the *logarithm of x to the base b* and is written $y = \log_b x$.

$$y = \log_b x \text{ provided that } x = b^y$$

Using this notation, we may write

$$f^{-1} = \{(x, y) \mid y = \log_b x \text{ where } b > 0 \text{ and } b \neq 1\}.$$

This function f^{-1} is called the *logarithm function to the base b*. For convenience, let us denote it by the letter g. The logarithm function to the base b is the function g such that

$$\text{Domain of } g = \{x \mid x > 0\},$$
$$\text{Range of } g = \{y \mid y \text{ is a real number}\},$$

and
$$g(x) = \log_b x.$$

The graph of g is the graph of $y = \log_b x$.

EXAMPLE 1

Give the domain, range, and rule of assignment of the inverse of function f such that $f(x) = 10^x$. Sketch the graphs of f and its inverse on the same coordinate axes.

Solution

f

Domain of $f = \{x \mid x \text{ is a real number}\}$

Range of $f = \{y \mid y > 0\}$

Rule of f: $y = 10^x$

f^{-1}

Domain of $f^{-1} = \{x \mid x > 0\}$

Range of $f^{-1} = \{y \mid y$ is a real number$\}$

Rule of f^{-1}: $x = 10^y$ or $y = \log_{10} x$

The graph of f goes through $(0, 1)$. The graph of f^{-1} goes through $(1, 0)$. The graph of f^{-1} is the reflection of the graph of f in the line $y = x$. Both graphs are shown in Figure 2.78.

Exercise 2.8, 11–15

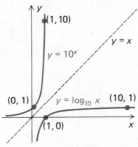

FIGURE 2.78   Graphs of $y = 10^x$ and $y = \log_{10} x$.

The graph of any logarithm function *with base b greater than 1* looks similar to the graph of $y = \log_{10} x$ shown in Figure 2.78. The graph lies entirely to the right of the x-axis and has the y-axis as asymptote. The graph rises from left to right, intersecting the x-axis at (1, 0). It is the reflection of the graph of $y = b^x$ in the line $y = x$.

The graph of $y = \log_b x$ *for 0 < b < 1* is again the reflection of the graph of $y = b^x$ in the line $y = x$. For example, the graph of $y = \log_{1/2} x$ is the reflection in the line $y = x$ of the graph of $y = (\frac{1}{2})^x$ shown in Figure 2.76. Logarithms with bases less than 1 are not often encountered.

EXAMPLE 2

Find the values of $\log_2 8$, $\log_2 \frac{1}{8}$, $\log_{10} 100$, $\log_{10} 1000$, $\log_{10} 1$, and $\log_e e$.

Solution

By definition,

$\log_b x = y$	provided that	$b^y = x$.
$\log_2 8 = 3$	because	$2^3 = 8$
$\log_2 \dfrac{1}{8} = -3$	because	$2^{-3} = \dfrac{1}{2^3} = \dfrac{1}{8}$
$\log_{10} 100 = 2$	because	$10^2 = 100$
$\log_{10} 1000 = 3$	because	$10^3 = 1000$
$\log_{10} 1 = 0$	because	$10^0 = 1$
$\log_e e = 1$	because	$e^1 = e$

Most of the logarithms encountered in calculus, in the natural sciences, and in many applications are to the base e, because using e as a base considerably simplifies many formulas. The graph of the logarithm function to the base e is sketched in the next example.

Exercise 2.8, 16–30

EXAMPLE 3

Sketch the graph of $y = \log_e x$.

Solution

The graph of $y = \log_e x$ is the reflection of the graph of $y = e^x$ in the line $y = x$. Since $\log_e x$ is only defined for positive real numbers x, the graph lies entirely to the right of the y-axis. The x-intercept is (1, 0).

Exercise 2.8, 31–40
Verbal Problems
Exercise 2.8, 41–44

(Log$_e$ 1 = 0 because $e^0 = 1$.) The graph is shown in Figure 2.79.

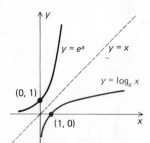

FIGURE 2.79   Graphs of $y = e^x$ and $y = \log_e x$.

▶ Sketch the graph of each of the following.

EXAMPLE

$y = 3^x$

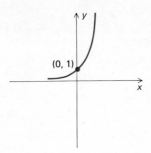

(0, 1)

1. $y = (\frac{1}{3})^x$ **2.** $y = 5^x$ **3.** $y = (\frac{1}{5})^x$

4. Function f: domain of $f = \{x \mid x \geq -1\}$, $f(x) = 2^x$

5. Function f: domain of $f = \{x \mid x \geq -1\}$, $f(x) = 2^{-x}$

6. Function f: domain of $f = \{x \mid x \geq -1\}$, $f(x) = e^x$

7. $y = e^{-x}$ **8.** $y = (\frac{1}{10})^x$

9. Function f: domain of $f = \{x \mid x \geq 0\}$, $f(x) = 10^x$

10. Function f: domain of $f = \{x \mid x \geq 0\}$, $f(x) = e^{-x}$

▶ In each of the following, give the domain, range, and rule of assignment of the inverse of f. Sketch the graphs of f and its inverse on the same coordinate axes.

EXAMPLE

$f(x) = (\frac{1}{2})^x$
Domain of $f = \{x \mid x$ is a real number$\}$
Range of $f = \{y \mid y > 0\}$
Rule of f: $y = (\frac{1}{2})^x$

Solution

Domain of $f^{-1} = \{x \mid x > 0\}$
Range of $f^{-1} = \{y \mid y$ is a real number$\}$
Rule of f^{-1}: $y = \log_{1/2} x$

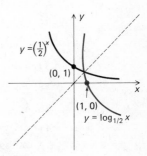

$y = (\frac{1}{2})^x$

(0, 1)

(1, 0)

$y = \log_{1/2} x$

11. $f(x) = 2^x$ **12.** $f(x) = 3^x$ **13.** $f(x) = (\frac{1}{3})^x$ **14.** $f(x) = e^{-x}$

15. $f(x) = 7^x$

▶ Find the value of each of the following.

EXAMPLES

EXAMPLES

Solutions

(a) $\log_3 9$ (b) $\log_e 1$

(a) $\log_3 9 = 2$ (because $3^2 = 9$)   (b) $\log_e 1 = 0$ (because $e^0 = 1$)

16. $\log_5 25$ **17.** $\log_5 1$ **18.** $\log_{25} 25$ **19.** $\log_e e^4$

20. $\log_{10} \frac{1}{10}$ **21.** $\log_{10} 0.01$ **22.** $\log_e \dfrac{1}{e}$ **23.** $\log_7 \frac{1}{7}$

24. $\log_4 64$ **25.** $\log_4 4$ **26.** $\log_{10} 10$ **27.** $\log_7 49$

28. $\log_{10} 1000$ **29.** $\log_3 81$ **30.** $\log_9 81$

▶ Sketch the graph of each of the following and label its intercept.

EXAMPLE

Solution

$y = \log_3 x$

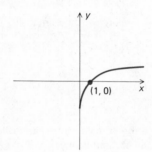

(1, 0)

31. $y = \log_4 x$ **32.** $y = \log_5 x$ **33.** $y = \log_{1/4} x$

34. Function f: $f(x) = \log_{10} x$, domain of $f = \{x \mid x \geq 1\}$

35. Function f: $f(x) = \log_e x$, domain of $f = \{x \mid x \geq 1\}$

36. Function f: $f(x) = \log_{10} x$, domain of $f = \{x \mid 0 < x \leq 1\}$

37. Function f: $f(x) = \log_2 x$, domain of $f = \{x \mid x \geq 1\}$

38. Function f: $f(x) = \log_e x$, domain of $f = \{x \mid 0 < x \leq 1\}$

39. $y = \log_7 x$ **40.** $y = \log_{1/7} x$

41. A culture grows in such a way that if k is the number of bacteria present at the beginning of an experiment, the number present x minutes later is $k(3)^x$. If $k = 20{,}000$, how many bacteria are present at the end of 2 minutes?

42. A chemical evaporates in such a way that the amount left of one unit volume at the end of x days is 2^{-x}. How much is left at the end of the third day?

43. A bank account containing $100 grows at 6% interest in such a way that the amount in the account at the end of n interest periods is $100(1.06)^n$. By how much does the account grow during the second interest period?

44. A radioactive material decays in such a way that if k is the amount initially present, then the amount present t seconds later is $ke^{-0.6t}$. If k is 2 grams, how much is present at the end of 10 seconds?

CHECK LIST FOR THE STUDENT

If you have learned the material in Section 2.8, you should be able to do the following:

1. Define an exponential function and give its domain.

2. Tell what numbers may be used as bases for exponential functions.

3. Give the range of an exponential function.

4. Sketch the graph of an exponential function with base $b > 1$ and label the intercept of the graph.

5. Sketch the graph of an exponential function with base $0 < b < 1$ and label the intercept of the graph.

6. Tell whether or not all exponential functions are one-to-one functions.

7. Sketch the graph of $y = e^x$.

8. Find the inverse of a given exponential function.

9. Sketch the graph of an exponential function and its inverse on the same coordinate system.

10. Complete the statement: $y = \log_b x$ provided that _____.

11. Find the values of $\log_b x$ for some given values of x and b.

12. Define the logarithm function to the base b and give its domain.

13. Give the range of the logarithm function to the base b.

14. Sketch the graph of $y = \log_b x$ with $b > 1$.

15. Sketch the graph of $y = \log_b x$ with $0 < b < 1$.

16. Sketch the graph of $y = \log_e x$.

SECTION QUIZ 2.8

Sketch the graph of each of the following.

1. $y = 6^x$

2. $y = (\frac{1}{6})^x$

3. Let $f(x) = 3^x$. Give the domain, range, and rule of assignment of the inverse of f. Sketch the graphs of f and its inverse on the same coordinate axes.

4. Find the value of each of the following.
   (a) $\log_2 64$ (b) $\log_3 1$ (c) $\log_7 7$ (d) $\log_2 \frac{1}{2}$ (e) $\log_{10} 0.001$

5. Sketch the graph of $y = \log_7 x$ and label its intercept.

*2.9 Logarithms; Common and Natural Logarithms

The purpose of this section is to reinforce the definition of a logarithm, to provide manipulative practice with logarithmic notation, to present essential properties of logarithms, to show how to find values of common and natural logarithms, and to show how common logarithms may be used in computations. Computations made with tables are compared with those made on a hand calculator.

We have seen that if b is a positive real number other than 1, then for any positive real number x,

$$\log_b x = y \quad \text{provided that} \quad b^y = x.$$

—Positive real number

—Positive real number other than 1

In other words, the logarithm of positive number x to the base b is the exponent to which the base must be raised in order to produce the number x.

In Examples 2 and 3 of Section 2.8, we saw that $\log_{10} 1 = 0$ and that $\log_e 1 = 0$. In fact, if b is any acceptable base (b is any positive number other than 1), then

$$\log_b 1 = 0.$$

This is true because $b^0 = 1$.

In Example 2 of Section 2.8 we saw that $\log_e e = 1$. It is likewise true that if b is *any* acceptable base, then

$$\log_b b = 1.$$

This is true because $b^1 = b$.

In Example 2 of Section 2.8 we also saw that $\log_2 2^3 = 3$ and $\log_{10} 10^2 = 2$. In general, if b is any acceptable base and k is any real number,

$$\log_b b^k = k.$$

This is true because k is the exponent to which the base b must be raised in order to produce b^k.

If the base of a logarithm is the number 10, then the logarithm is called a **common logarithm**. It is customary to omit the subscript 10 from the notation for a common logarithm. For example, $\log_{10} x$ is customarily written $\log x$.

$$\log x \quad \text{means} \quad \log_{10} x$$

If the base of a logarithm is the number e, then the logarithm is called a **natural logarithm**. It is customary to write the symbols ln in place of $\log_e$ in the notation for a natural logarithm.

$$\ln x \quad \text{means} \quad \log_e x$$

EXAMPLE 1

Write $\log 10^{-1}$ and $\ln e^{x^2}$ in equivalent forms without using the symbols log or ln.

Solution

$$\log 10^{-1} = -1$$

(This means that -1 is the exponent to which the base 10 must be raised in order to produce the number 10^{-1}.)

$$\ln e^{x^2} = x^2$$

(This means that x^2 is the exponent to which the base e must be raised in order to produce e^{x^2}.)

EXAMPLE 2

Write $y = e^x$ in logarithmic notation.

Solution

$y = e^x$ is equivalent to $x = \ln y$.

Exercise 2.9, 1–40

We have said that for positive number x, $\log_b x$ is the exponent to which the base must be raised to produce x. That is,

$$x = b^{\log_b x}.$$

This form of the definition is utilized in the next three examples.

EXAMPLE 3

Write 10 as a power of e. Write 10^x as a power of e.

Solution

$$10 = e^{\ln 10}$$
$$10^x = (e^{\ln 10})^x$$
$$\quad = e^{x \ln 10} \qquad \text{Exponent Law 3: } (a^r)^s = a^{rs}$$
$$10^x = e^{x \ln 10}$$

EXAMPLE 4

If a is any positive real number and x is any real number, write a^x as a power of e.

Solution

$$a = e^{\ln a}$$
$$a^x = (e^{\ln a})^x$$
$$\quad = e^{x \ln a} \qquad \text{Law 3: } (a^r)^s = a^{rs}$$
$$a^x = e^{x \ln a}$$

EXAMPLE 5	Let x be any positive real number, and let n be any real number. Write x as a power of e. Write x^n as a power of e.
Solution	$x = e^{\ln x}$
	$x^n = (e^{\ln x})^n$
Exercise 2.9, 41–50	$= e^{n \ln x}$ $\quad$ Law 3: $(a^r)^s = a^{rs}$

Familiarity with the meaning of $\ln |x|$ is sometimes needed in applications. From the definition of absolute value, we know that

$$\ln |x| = \ln x, \text{ if } x \text{ is positive}$$
$$\ln |x| = \ln(-x), \text{ if } x \text{ is negative.}$$

Use of this notation is demonstrated in the next two examples.

EXAMPLE 6	Find the value of $\ln |x|$ for $x = e^3$ and also for $x = -e^3$.
Solution	$\ln |e^3| = \ln e^3 \qquad (e^3 \text{ is positive})$
	$= 3$
	$\ln |-e^3| = \ln (-(-e^3)) \qquad (-e^3 \text{ is negative})$
	$= \ln e^3$
	$= 3$

EXAMPLE 7	Where x is any real number, why is it true that $\ln |1 + x^2| = \ln(1 + x^2)$? Is $\ln |x^3|$ equal to $\ln x^3$ for all real numbers x? Why or why not?
Solution	If x is any real number, then $1 + x^2$ is positive. If $1 + x^2$ is positive, then $|1 + x^2| = 1 + x^2$. It follows that

$$\ln |1 + x^2| = \ln(1 + x^2).$$

It is not true that $\ln |x^3|$ is equal to $\ln x^3$ for all real numbers x, because for some values of x the expression x^3 is nonpositive so that $\ln x^3$ is not even defined. If x^3 is *positive,* however, $|x^3| = x^3$, and then

Exercise 2.9, 51–55 $\ln |x^3| = \ln x^3$.

Three basic properties of logarithms to any acceptable base b may be derived directly from the exponent laws and the definition of a logarithm. These three properties are:

Property 1 $\quad \log_b MN = \log_b M + \log_b N$

Property 2 $\quad \log_b \dfrac{M}{N} = \log_b M - \log_b N$

Property 3 $\quad \log_b M^k = k \log_b M.$

Here M and N are understood to be positive real numbers, and k is any real number. In Property 3, $\log_b M^k$ means $\log_b (M^k)$. Note that this is not the same as $(\log_b M)^k$.

To illustrate how these three properties follow from the exponent laws and the definition of a logarithm, let us prove Property 1 for the case in which the base is the number e.

Property 1 $\ln MN = \ln M + \ln N$

Proof Let $x = \ln M$.   Then $M = e^x$.   Definition of $\ln M$

Let $y = \ln N$.   Then $N = e^y$.   Definition of $\ln N$

$$MN = e^x e^y$$

$$= e^{x+y}$$ Law 1: $a^r a^s = a^{r+s}$

But $MN = e^{x+y}$ implies that

$$\ln MN = x + y.$$ Definition of $\ln MN$

Since $x = \ln M$ and $y = \ln N$, we have

$$\ln MN = \ln M + \ln N.$$

EXAMPLE 8 Use Property 3 ($\log_b M^k = k \log_b M$) and prove that $\ln e^u = u$ for any real number u.

Solution $\ln e^u = u \ln e$ Property 3: $\log_b M^k = k \log_b M$

Exercise 2.9, 56–60 $= u$ $\ln e = 1$

Thus far we have only found the numerical values of the logarithms of numbers that are certain powers of the base. For example, we found $\log_2 8 = \log_2 2^3 = 3$, we found $\log 100 = \log 10^2 = 2$, and we found $\ln e^3 = 3$. Let us now see how to find the common logarithm of any positive number.

The common logarithm of a number that is any integral power of 10 may be found by direct application of the definition of a logarithm as an exponent. Some examples follow.

$$\log 10 = 1 \quad \text{because} \quad 10^1 = 10$$
$$\log 100 = 2 \qquad\qquad 10^2 = 100$$
$$\log 1000 = 3 \qquad\qquad 10^3 = 1000$$
$$\log 10000 = 4 \qquad\qquad 10^4 = 10000$$
$$\log 10^0 = \log 1 = 0 \qquad\qquad 10^0 = 1$$
$$\log 0.1 = -1 \qquad\qquad 10^{-1} = 0.1$$
$$\log 0.01 = -2 \qquad\qquad 10^{-2} = 0.01$$
$$\log 0.001 = -3 \qquad\qquad 10^{-3} = 0.001$$
$$\log 0.0001 = -4 \qquad\qquad 10^{-4} = 0.0001$$

The common logarithm of a number between 1 and 10 can be found approximately with a hand calculator or in a table of common logarithms. For example, Table I in Appendix B gives four-decimal-place approximations to the common logarithms of numbers between 1 and 10. In particular, in Table I we read that

$$\log 5 = 0.6990 \quad \text{and} \quad \log 5.23 = 0.7185.$$

With a hand calculator we would find that to six decimal places

$$\log 5 = 0.698970 \quad \text{and} \quad \log 5.23 = 0.718502.$$

Even though these are approximate values of log 5 and log 5.23, we adopt common convention and use "equals" signs instead of "approximately equals" signs.

Values of logarithms are found much more easily and to more decimal places with a hand calculator than with a table, but not all students have calculators, so we shall explain how to use logarithm tables.

The common logarithm of any positive number x that is not an integral power of 10 and that is not between 1 and 10 can be found approximately by first expressing x as the product of a number between 1 and 10 and an integral power of 10 and then by using Property 1 ($\log_b MN = \log_b M + \log_b N$), the definition of a logarithm as an exponent, and a table of common logarithms. This is demonstrated in the next example.

EXAMPLE 9   Use Table I to find log 523 to four decimal places.

Solution   We shall first express 523 as the product of a number between 1 and 10 and an integral power of 10. Then we will use Property 1 ($\log_b MN = \log_b M + \log_b N$).

$$523 = 5.23 \times 10^2$$
$$\log 523 = \log (5.23 \times 10^2)$$

$$\quad = \log 5.23 + \log 10^2 \qquad \begin{array}{l} \text{Property 1:} \\ \log_b MN = \log_b M + \log_b N \end{array}$$

$$\quad = 0.7185 + \log 10^2 \qquad \text{From Table I}$$
$$\quad = 0.7185 + 2 \qquad \text{Definition of logarithm}$$
$$\log 523 = 2.7185$$

On a calculator, to six decimal places, log 523 = 2.718502.

EXAMPLE 10   Use Table I to find log 0.00523 to four decimal places.

Solution
$$0.00523 = 5.23 \times 10^{-3}$$
$$\log 0.00523 = \log (5.23 \times 10^{-3})$$
$$\quad = \log 5.23 + \log 10^{-3}$$
$$\quad = 0.7185 + \log 10^{-3}$$
$$\quad = 0.7185 + (-3)$$
$$\log 0.00523 = 0.7185 - 3$$
$$\quad = -2.2815$$

On a calculator, to six decimal places, log 0.00523 = −2.281498.

We have seen in Examples 9 and 10 that log 523 = 0.7185 + 2 and log 0.00523 = 0.7185 + (−3). Any common logarithm can be expressed as the sum of a nonnegative decimal less than 1 and an integer. The nonnegative decimal is called the **mantissa.** The integer is called the **characteristic.** For a given positive number x, the mantissa of log x can be determined from a table of common logarithms. The characteristic of log x can be found by first expressing x as the product of a number between 1 and 10 and an integral power of 10 and then reading off the power of 10.

Our Table I only permits us to read off directly the common logarithms of numbers having three so-called significant digits. For example, we can read log 5.23, but we cannot read log 5.234. If the logarithm of a number with more than three significant digits is required, a technique called *interpolation* could be used, but it would be better to use a more extensive table or a hand calculator.

Exercise 2.9, 61–75

Table I may also be used to approximate **antilogarithms,** numbers whose logarithms are known. For example, if we know that log x = 2.6749, we can find an approximation to x by looking in the table for mantissa 0.6749, reading off the associated three digits, and then placing the decimal point so that the characteristic of log x will be 2. In Table I we read

$$\log 4.73 = 0.6749.$$

It follows that

$$\log 473 = 2.6749,$$

so x is approximately 473. On a calculator, $x = 10^{2.6749}$, or approximately 473.0423.

If log x is known but the mantissa of log x is not given in Table I, we can approximate x by finding the two tabular values nearest the mantissa of log x. This will be demonstrated in Examples 13, 14, and 15. Better approximations can be obtained, however, by using a more extensive table or a calculator.

Exercise 2.9, 81–85

To find the value of the natural logarithm of some given positive number x, we may use a table of natural logarithms such as Table II in Appendix B, we may use a calculator, or we may use the fact that the natural logarithm of x is the common logarithm of x divided by log e.

$$\ln x = \frac{\log x}{\log e}$$

We can prove this by first writing x as a power of e,

$$x = e^{\ln x},$$

and then taking the common logarithm of x:

$$\log x = \log (e^{\ln x})$$
$$= (\ln x)(\log e). \quad \text{Property 3:}$$
$$\log_b M^k = k \log_b M$$

Solving for $\ln x$:

$$\log x = (\ln x)(\log e)$$

$$\ln x = \frac{\log x}{\log e}.$$

This formula enables us to express a natural logarithm in terms of common logarithms. In other words, it enables us to change from base 10 to base e.

The formula $\ln x = \log x / \log e$ also enables us to change from base e to base 10, because it may be written $\log x = (\ln x)(\log e)$. In this form it simply says that the common logarithm of x is the product of the number $\log e$ and the natural logarithm of x.

Using general bases b and a instead of 10 and e, we may similarly derive a formula that enables us to change from base b to any other base a.

First writing x as a power of a:

$$x = a^{\log_a x}.$$

Then taking the logarithm to the base b of x:

$$\log_b x = \log_b (a^{\log_a x})$$
$$= (\log_a x)(\log_b a) \quad \text{Property 3:}$$
$$\log_b M^k = k \log_b M$$

Solving for $\log_a x$:

$$\log_b x = (\log_a x)(\log_b a)$$

$$\log_a x = \frac{\log_b x}{\log_b a} \qquad \text{Change of base formula}$$

This is called the *change of base formula*. As an example, take a to be 2 and b to be 10. The change of base formula then says:

$$\log_2 x = \frac{\log x}{\log 2}.$$

If x is 5,

$$\log_2 5 = \frac{\log 5}{\log 2}.$$

The number $\log e$ is approximately equal to 0.4343, so the formula for changing from base e to base 10 or from base 10 to base e may be written approximately as

$$\ln x = \frac{\log x}{0.4343}$$

or

$$\log x = (0.4343)(\ln x)$$

EXAMPLE 11   Use Table I to find $\ln 7$ to four decimal places.

Solution $\ln 7 = \dfrac{\log 7}{\log e}$

In Table I we read $\log 7 = 0.8451$. Using $\log e = 0.4343$, we have

$$\ln 7 = \frac{0.8451}{0.4343}.$$

To four decimal places,

$$\ln 7 = 1.9459.$$

On a hand calculator having natural logarithms, we could read directly Exercise 2.9, 76–80 that $\ln 7 = 1.945910$ to six decimal places. In Table II, $\ln 7 = 1.9459$.

Common logarithms may be used to facilitate computations involving products, quotients, powers, and roots in the decimal system. But due to the advent of inexpensive hand calculators and high-speed computers, logarithms are no longer the essential computational tool that they once were. The next three examples show how logarithms may be used for computations.

EXAMPLE 12   Use logarithms and Table I to find an approximation to $\sqrt{76.8}$.

Solution
$$\sqrt{76.8} = (76.8)^{1/2}$$
$$\log \sqrt{76.8} = \tfrac{1}{2}\log (76.8) \qquad \text{Property 3: } \log_b M^k = k \log_b M$$
$$= \tfrac{1}{2}(1.8854) \qquad \text{Using Table I}$$
$$= 0.9427$$

Now to find an approximation to the number $\sqrt{76.8}$, whose logarithm is 0.9427, we look in Table I for mantissa 0.9427. In Table I we can find 0.9425 and 0.9430, but we cannot find 0.9427. We have

$$\log 8.76 = 0.9425$$
$$\log \sqrt{76.8} = 0.9427$$
$$\log 8.77 = 0.9430.$$

We see that $\sqrt{76.8}$ must be between 8.76 and 8.77. Since 0.9427 is closer to 0.9425 than it is to 0.9430, we say that

$$\sqrt{76.8} \text{ is approximately } 8.76.$$

A better approximation can be obtained from Table I by using a process called *linear interpolation.* A still better approximation can be obtained by using a more extensive table or a hand calculator. It would be much more efficient to perform such computations with a calculator. On a calculator, $\sqrt{76.8}$ is approximately 8.763561.

EXAMPLE 13   Use logarithms and Table I to find an approximation to $(53.2)^{10}$.

Solution

$$\begin{aligned}
\log (53.2)^{10} &= 10 \log (53.2) \\
&= 10(1.7259) \\
&= 17.2590
\end{aligned}$$

Using Table I, we have

$$\begin{aligned}
\log (1.81 \times 10^{17}) &= 17.2577 \\
\log (53.2)^{10} &= 17.2590 \\
\log (1.82 \times 10^{17}) &= 17.2601.
\end{aligned}$$

We see that $(53.2)^{10}$ is between 1.81×10^{17} and 1.82×10^{17}. Since 0.2590 is closer to 0.2601 than it is to 0.2577, we say that

$$(53.2)^{10} \text{ is approximately } 1.82 \times 10^{17}.$$

On a calculator, $(53.2)^{10}$ is approximately 1.8160×10^{17}.

EXAMPLE 14   Use logarithms and Table I to find an approximation to

$$\frac{762 \times 0.0342}{83.9}.$$

Solution

$$\log \frac{762 \times 0.0342}{83.9} = \log (762 \times 0.0342) - \log 83.9$$

$$\text{Property 2: } \log_b \frac{M}{N} = \log_b M - \log_b N$$

$$= \log 762 + \log 0.0342 - \log 83.9$$

$$\text{Property 1: } \log_b MN = \log_b M + \log_b N$$

$$= (2.8820) + (0.5340 - 2) - (1.9238)$$

$$= 1.4160 - (1.9238)$$

$$= -0.5078$$

$$= -0.5078 + 1 - 1$$

$$= 0.4922 - 1$$

We added and subtracted 1 from -0.5078 in order to express the logarithm of the quotient we are computing as the sum of an integer and a *nonnegative* four-place decimal so that we can use Table I. Using Table I, we have

$$\log 0.310 = 0.4914 - 1$$

$$\log \frac{762 \times 0.0342}{83.9} = 0.4922 - 1$$

$$\log 0.311 = 0.4928 - 1.$$

Since 0.4922 is closer to 0.4928 than to 0.4914, we say that

$$\frac{762 \times 0.0342}{83.9} \text{ is approximately } 0.311.$$

Exercise 2.9, 86–95   On a calculator, the given quotient is approximately 0.31061.

EXERCISE 2.9 LOGARITHMS; COMMON AND NATURAL LOGARITHMS

▶ Express each of the following in the exponential form.

EXAMPLE $\log_{10} x = y$

Solution $10^y = x$

1. $\log_3 9 = 2$ 2. $\log_4 64 = 3$ 3. $\log_5 25 = 2$

4. $\log x = \frac{1}{2}$ 5. $\log x = 5$ 6. $\ln x = y$

7. $\ln x = 5$ 8. $\ln e^7 = 7$ 9. $\ln x^2 = 3$

10. $\log_2 32 = 5$

▶ Express each of the following in logarithmic notation.

EXAMPLE $y = 3^4$

Solution $\log_3 y = 4$

11. $y = 2^6$ 12. $y = 10^{-7}$ 13. $y = e^{x^2}$ 14. $y = e^u$

15. $2^3 = 8$ 16. $y = e^{5x}$ 17. $y = e^{x+h}$ 18. $49 = 7^2$

19. $y = e^{m+n}$ 20. $y = e^{2x}$

► Find the value of each of the following.

EXAMPLE $\ln e^{x^3}$

Solution $\ln e^{x^3} = x^3$

21. $\ln e^u$ **22.** $\ln e$ **23.** $\log 10$

24. $\log_2 2$ **25.** $\log_2 1$ **26.** $\log 10^x$

27. $\log_2 2^x$ **28.** $\log_3 3^{x^2}$ **29.** $\ln e^x$

30. $\ln e^{-x}$ **31.** $e^{\ln x}$ **32.** $10^{\log x}$

33. $2^{\log_2 x}$ **34.** $e^{\ln x^2}$ **35.** $e^{\ln 2x}$

36. $\log 100000$ **37.** $\log 0.00001$ **38.** $\ln e^5$

39. $\ln e^{-5}$ **40.** $\log 10^{-8}$

► Write each of the following as a power of e.

EXAMPLES   (a) 2^x (b) x^3

Solutions   (a) $2 = e^{\ln 2}$ (b) $x = e^{\ln x}$
 $2^x = e^{x \ln 2}$ $x^3 = e^{3 \ln x}$

41. 5^x **42.** 2^{3x} **43.** 2^{x^2} **44.** $(\tfrac{1}{2})^x$ **45.** $(\sqrt{2})^x$

46. x^2 **47.** x^5 **48.** $x^{1/2}$ **49.** $x^{1/3}$ **50.** $x^{\sqrt{2}}$

► In each of the following, tell whether or not $\ln |P(x)|$ is equal to $\ln (P(x))$ for all real numbers x.

EXAMPLE $P(x) = x - 1$

Solution   Here $\ln |x - 1| \neq \ln (x - 1)$ for all real numbers x because $(x - 1)$ is nonpositive when $x \leq 1$, and for such values of x, $\ln (x - 1)$ is not defined.

51. $P(x) = x^4$ **52.** $P(x) = x^5$ **53.** $P(x) = x^4 + 1$

54. $P(x) = x - 3$ **55.** $P(x) = x + 2$

► Where M and N are positive real numbers, prove each of the following.
56. $\log MN = \log M + \log N$

57. $\ln \dfrac{M}{N} = \ln M - \ln N$

58. $\ln M^k = k \ln M$ for any real number k

59. $\log M^k = k \log M$ for any real number k

60. $\log_b \dfrac{M}{N} = \log_b M - \log_b N$ for positive number $b \neq 1$

▶ Use Table I in Appendix B to find an approximation to each of the following.

EXAMPLES

(a) $\log 92.3$ (b) $\ln |-8|$

Solutions

(a) $\log 92.3 = 1.9652$ (b) $\ln |-8| = \ln 8$

$$= \frac{\log 8}{\log e}$$

$$= \frac{0.9031}{0.4343}$$

$$= 2.0794$$

61. $\log 923$ **62.** $\log 0.0923$ **63.** $\log 0.000578$

64. $\log 5780$ **65.** $\log 62500$ **66.** $\log 0.00625$

67. $\ln 347$ **68.** $\ln |-73.4|$ **69.** $\ln 27$

70. $\ln |-10|$ **71.** $\ln 0.134$ **72.** $\ln |-260|$

73. $\log_2 10$ **74.** $\log_5 21$ **75.** $\log_3 21$

▶ Express each of the following in terms of a natural logarithm.

EXAMPLES

(a) $\log 273$ (b) $\log (x^2 + 7)$

Solutions

(a) $\log 273 = (\log e)(\ln 273)$ (b) $\log (x^2 + 7) = (\log e)(\ln (x^2 + 7))$

76. $\log 35.9$ **77.** $\log (3x + 1)$ **78.** $\log 2$

79. $\log 38$ **80.** $\log (x^2 - 9)$

▶ Use Table I to find an approximation to x in each of the following.

EXAMPLE

$\log x = 1.4032$

Solution

$\log 25.3 = 1.4031$

$\quad \log x = 1.4032$

$\log 25.4 = 1.4048$

x is approximately 25.3

81. $\log x = 3.9167$ **82.** $\log x = 0.8175 - 2$

83. $\log x = 4.5530$ **84.** $\log x = 0.3931 - 4$

85. $\log x = 5.2718$

▶ Use logarithms and Table I to find an approximation to each of the following.

EXAMPLE $\sqrt[3]{768}$

Solution

$$\log \sqrt[3]{768} = \tfrac{1}{3} \log 768$$
$$= \tfrac{1}{3}(2.8854)$$
$$= 0.9618$$

$\log 9.15 = 0.9614$

$\log \sqrt[3]{768} = 0.9618$

$\log 9.16 = 0.9619$

$\sqrt[3]{768}$ is approximately 9.16.

86. $\sqrt{827}$ **87.** $(25.2)^6$ **88.** $\dfrac{359 \times \sqrt{2}}{21.7}$

89. $\sqrt[5]{4060}$ **90.** $\dfrac{32.8}{2.1 \times 7.6}$ **91.** $(0.0054)^{-1}$

92. $\dfrac{2370 \times 350}{5\sqrt{3}}$ **93.** $\sqrt[3]{4670}$ **94.** $\dfrac{3\sqrt{7}}{52}$

95. $(0.0121)^5$

CHECK LIST FOR THE STUDENT

If you have learned the material in Section 2.9, you should be able to do the following:

1. Define the logarithm of positive number x to the base b (b is positive and not equal to 1) as an exponent.

2. Find $\log_b 1$ for any acceptable base b.

3. Find $\log_b b$ for any acceptable base b.

4. Find $\log_b b^k$ for any acceptable base b and any real number k.

5. Define a common logarithm.

6. Write the customary notation for the common logarithm of x.

7. Define a natural logarithm.

8. Write the customary notation for the natural logarithm of x.

9. Write $y = e^x$ in logarithmic notation.

10. Write any positive real number a as a power of e.

11. Write x^n as a power of e.

12. Find the value of $\ln |x|$ for certain given values of x.

13. Tell whether or not $\ln |P(x)|$ is equal to $\ln (P(x))$ for certain given polynomials $P(x)$.

14. State, use, and derive the following basic properties of logarithms:

   1. $\log_b MN = \log_b M + \log_b N$

   2. $\log_b \dfrac{M}{N} = \log_b M - \log_b N$

   3. $\log_b M^k = k \log_b M$

15. Find the common logarithm of a number that is an integral power of 10.

16. Use a table of logarithms to find an approximation to the common logarithm of a number between 1 and 10.

17. Use a table of logarithms and find an approximation to the common logarithm of any positive number.

18. Express a given common logarithm as the sum of a nonnegative decimal less than 1 and an integer.

19. Give the meanings of the words *mantissa* and *characteristic*.

20. Use a table of logarithms to find an approximation to a number whose common logarithm is given.

21. Derive the formula $\ln x = \log x / \log e$.

22. Express a given natural logarithm in terms of common logarithms.

23. Express a given common logarithm in terms of natural logarithms.

24. State and derive the formula for expressing a logarithm to the base a in terms of logarithms to the base b.

25. Use common logarithms in computations involving products, quotients, powers, and roots in the decimal system.

SECTION QUIZ 2.9

1. (a) Express $\log_2 a = b$ in exponential notation.
   (b) Express $y = 3^5$ in logarithmic notation.
   (c) Write 3 as a power of e.
   (d) Express $\log_{10} 37$ in terms of natural logarithms.

2. Find the value of each of the following.
   (a) $\ln 1$ (b) $\log_{10} 10$ (c) $\ln e^5$
   (d) $e^{\ln 5}$ (e) $10^{\log_{10} 2}$

Suppose that $\log_{10} 7 = 0.8451$ and $\log_{10} 5 = 0.6990$. Find the value of each of the following.

3. (a) $\log_{10} 700$ (b) $\log_{10} 0.7$

4. (a) $\log_{10} (5 \times 7)$ (b) $\log_{10} \frac{7}{5}$

5. (a) $\log_{10} 7^3$ (b) $\log_{10} \sqrt[3]{5}$

*2.10 Operations on Functions

The purpose of this section is to define addition, subtraction, multiplication, division, and composition of functions, and to provide practice with the notation.

Let f and g be functions. Then the **sum** of f and g is denoted by $f + g$ and is defined to be the function whose rule of assignment assigns the number $f(x) + g(x)$ to x and whose domain is the set of all numbers common to the domains of f and g. That is, $f + g$ is the function such that

$$(f + g)(x) = f(x) + g(x),$$

$$\text{Domain of } f + g = \left\{ x \,\middle|\, \begin{array}{l} x \text{ is in the domain of } f \\ \text{and also in the domain of } g \end{array} \right\}.$$

To illustrate, let f be the function such that

$$f(x) = \ln x,$$

Domain of f is the set of all positive real numbers.

Let g be the function such that

$$g(x) = \sqrt{-x + 2},$$

$$\text{Domain of } g = \{x \mid x \leq 2\}.$$

Compare the domains of f and g graphically in Figure 2.80. The only numbers that are in the domain of f and also in the domain of g are those numbers x such that $0 < x \leq 2$. (Note that 2 is included in this set, but 0 is not.) For this f and g, $f + g$ is the function such that

$$(f + g)(x) = \ln x + \sqrt{-x + 2},$$

$$\text{Domain of } f + g = \{x \mid 0 < x \leq 2\}.$$

The **difference** of functions f and g is denoted by $f - g$ and is defined to be the function whose rule of assignment assigns the number $f(x) - g(x)$ to x and whose domain is the set of all numbers common to the domains of f and g. That is, $f - g$ is the function such that

$$(f - g)(x) = f(x) - g(x),$$

$$\text{Domain of } f - g = \left\{ x \,\middle|\, \begin{array}{l} x \text{ is in the domain of } f \\ \text{and also in the domain of } g \end{array} \right\}.$$

Domain of $f = \{x \mid x > 0\}$

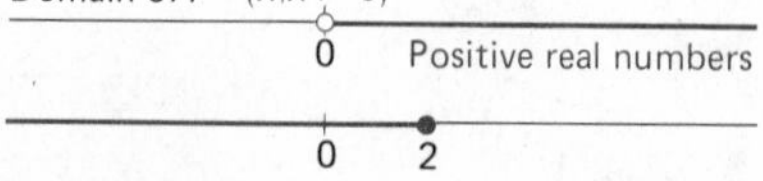

$$\text{0} \qquad \text{Positive real numbers}$$

$$\text{0} \quad \text{2}$$

Domain of $g = \{x \mid x \leq 2\}$

FIGURE 2.80

As an illustration, again consider f and g such that $f(x) = \ln x$ and $g(x) = \sqrt{-x + 2}$. For this f and g, $f - g$ is the function such that

$$(f - g)(x) = \ln x - \sqrt{-x + 2},$$
$$\text{Domain of } f - g = \{x \mid 0 < x \leq 2\}.$$

The **product** of functions f and g is denoted by fg and is defined to be the function whose rule of assignment assigns the number $f(x) \cdot g(x)$ to x and whose domain is the set of all numbers common to the domains of f and g. That is, fg is the function such that

$$(fg)(x) = f(x) \cdot g(x),$$
$$\text{Domain of } fg = \left\{x \;\middle|\; \begin{array}{l} x \text{ is in the domain of } f \\ \text{and also in the domain of } g \end{array}\right\}.$$

For example, if we again let f and g be such that $f(x) = \ln x$ and $g(x) = \sqrt{-x + 2}$, fg is the function such that

$$(fg)(x) = \ln x \cdot \sqrt{-x + 2}$$
$$\text{Domain of } fg = \{x \mid 0 < x \leq 2\}.$$

The **quotient** of functions f and g is denoted by f/g and is defined to be the function whose rule of assignment assigns the number $f(x)/g(x)$ to x and whose domain is the set of all numbers common to the domains of f and g and for which $g(x)$ is not zero. That is, f/g is the function such that

$$\left(\frac{f}{g}\right)(x) = \frac{f(x)}{g(x)},$$
$$\text{Domain of } \frac{f}{g} = \left\{x \;\middle|\; \begin{array}{l} x \text{ is in the domain of } f \\ \text{and also in the domain of } g, \\ \text{and } g(x) \neq 0 \end{array}\right\}.$$

If we once more let f and g be such that $f(x) = \ln x$ and $g(x) = \sqrt{-x + 2}$, the function f/g is such that

$$\left(\frac{f}{g}\right)(x) = \frac{\ln x}{\sqrt{-x + 2}},$$
$$\text{Domain of } \frac{f}{g} = \{x \mid 0 < x < 2\}.$$

The number 2 is not in the domain of f/g because $g(2) = 0$.

The **composite** of functions f and g is denoted by $f \circ g$ (read "f circle g") and is defined to be the function whose rule of assignment assigns the number $f(g(x))$ to x and whose domain is the set of all numbers in the domain of g to which the rule of g assigns a number that is in the domain of f. That is, $f \circ g$ is the function such that

$$(f \circ g)(x) = f(g(x)),$$

$$\text{Domain of } f \circ g = \left\{ x \left| \begin{array}{l} x \text{ is in the domain of } g \\ \text{and } g(x) \text{ is in the} \\ \text{domain of } f \end{array} \right. \right\}$$

The process of finding the composite of two functions is sometimes called *composition*. To find $(f \circ g)(x) = f(g(x))$, we may think of working from right to left. First we take x from the domain of g. The rule of g assigns the number $g(x)$ to the x. Finally, if x is such that the number $g(x)$ is in the domain of f, the rule of f assigns the number $(f \circ g)(x)$ to the $g(x)$. If we let $y = f(u)$ and $u = g(x)$, this procedure may be described schematically as follows:

$$x \xrightarrow{\ g\ } u \xrightarrow{\ f\ } y$$
$$\| \qquad \|$$
$$g(x) \qquad f(u)$$
$$\|$$
$$f(g(x))$$

$$x \xrightarrow{\ \ \ f \circ g\ \ \ } y$$
$$\|$$
$$f(g(x))$$

To illustrate the composition of functions, let functions f and g be such that

$$f(x) = \sqrt{x},$$

Domain of f is the set of all nonnegative real numbers;

$$g(x) = x - 3,$$

Domain of g is the set of all real numbers.

The only numbers in the domain of g for which $g(x) = x - 3$ is nonnegative (is in the domain of f) are those numbers x for which $x \geq 3$. For this f and g, $f \circ g$ is the function such that

$$(f \circ g)(x) = f(g(x))$$
$$= f(x - 3)$$
$$= \sqrt{x - 3},$$

Domain of $f \circ g = \{x | x \geq 3\}$.

We may describe the composition of this f and g schematically as follows:

$$x \xrightarrow{\ g\ } (x - 3) \xrightarrow{\ f\ } \sqrt{x - 3}$$
$$x \xrightarrow{\ \ \ f \circ g\ \ \ } \sqrt{x - 3}$$

As an additional illustration, let f and g be such that $f(x) = \ln x$ and $g(x) = \sqrt{-x + 2}$, as they were earlier in this section. In this case the domain of f is the set of all positive real numbers and the domain of g is the set of all numbers less than or equal to 2. The only numbers in the domain of g for which $g(x) = \sqrt{-x + 2}$ is positive (is in the domain of f) are the numbers x for which $x < 2$.

$$(f \circ g)(x) = f(g(x))$$
$$= f(\sqrt{-x + 2})$$
$$= \ln \sqrt{-x + 2},$$
$$\text{Domain of } f \circ g = \{x \mid x < 2\}.$$

Note that $x = 2$ is not in the domain of $f \circ g$ because $g(2) = 0$ is not in the domain of f; $f(x) = \ln x$ is not defined for $x = 0$.

EXAMPLE 1   Let functions f and g be such that $f(x) = \sqrt{x}$ and $g(x) = x - 3$. Find $(g \circ f)(x)$.

Solution

$$(g \circ f)(x) = g(f(x))$$
$$= g(\sqrt{x})$$
$$= \sqrt{x} - 3$$

In Example 1 we found that $(g \circ f)(x) = \sqrt{x} - 3$. For the same f and g, we found earlier that $(f \circ g)(x) = \sqrt{x - 3}$. This proves that $f \circ g$ is not necessarily the same as $g \circ f$.

Exercise 2.10, 1–20

EXERCISE 2.10 OPERATIONS ON FUNCTIONS

▶   In each of the following, find the rule of assignment and state the domain of $f + g$, $f - g$, fg, f/g, $f \circ g$, and $g \circ f$. If the domain of f or g is not explicitly stated, use the domain convention.

EXAMPLE

$$f(x) = e^x$$
$$g(x) = \sqrt{x}$$

Solution

Domain of f is set of all real numbers.

Domain of g is set of all nonnegative real numbers.

$(f + g)(x) = e^x + \sqrt{x}$	Domain of $f + g = \{x \mid x \geq 0\}$
$(f - g)(x) = e^x - \sqrt{x}$	Domain of $f - g = \{x \mid x \geq 0\}$
$(fg)(x) = e^x \sqrt{x}$	Domain of $fg = \{x \mid x \geq 0\}$
$\left(\dfrac{f}{g}\right)(x) = \dfrac{e^x}{\sqrt{x}}$	Domain of $\dfrac{f}{g} = \{x \mid x > 0\}$
$(f \circ g)(x) = f(g(x))$ $= e^{\sqrt{x}}$	Domain of $f \circ g = \{x \mid x \geq 0\}$

$$(g \circ f)(x) = g(f(x))$$
$$= \sqrt{e^x} \qquad$$ Domain of $g \circ f$ is the set of all real numbers
(e^x is positive for all real numbers x)

1. $f(x) = e^{3x}$, $g(x) = \sqrt{-x}$

2. $f(x) = x^3$, $g(x) = 5x^2 + 2$

3. $f(x) = \ln |x|$, $g(x) = x^3$

4. $f(x) = \ln x$, $g(x) = \sqrt[3]{x}$

5. $f(x) = \ln x$; $g(x) = 2x$, domain of $g = \{x | -1 \le x \le 1\}$

6. $f(x) = 3x^2 + 2x$, $g(x) = e^{-x}$

7. $f(x) = \sqrt{x + 3}$, $g(x) = x^2$

8. $f(x) = -\sqrt{x}$, domain of $f = \{x | 0 \le x \le 9\}$; $g(x) = 2x$

9. $f(x) = x^2$, domain of $f = \{x | x \le 0\}$; $g(x) = x + 5$

10. $f(x) = \log x$, domain of $f = \{x | x \ge 1\}$; $g(x) = x + 2$, domain of $g = \{x | x \ge 0\}$.

EXAMPLE

Let $f(x) = \sqrt[3]{x}$ and $g(x) = 2x + 3$. Find

(a) $(f + g)(8)$ (b) $(f - g)(8)$

(c) $(fg)(8)$ (d) $(f/g)(8)$

(e) $(f \circ g)(8)$ (f) $(g \circ f)(8)$

Solution

(a) $(f + g)(8) = \sqrt[3]{8} + 2(8) + 3$ (b) $(f - g)(8) = \sqrt[3]{8} - (16 + 3)$
$\qquad\qquad\quad = 2 + 16 + 3 \qquad\qquad\qquad\qquad\qquad = 2 - 19$
$\qquad\qquad\quad = 21 \qquad\qquad\qquad\qquad\qquad\qquad\quad = -17$

(c) $(fg)(8) = \sqrt[3]{8}(16 + 3)$ (d) $\left(\dfrac{f}{g}\right)(8) = \dfrac{\sqrt[3]{8}}{16 + 3}$
$\qquad\qquad = 2(19) \qquad\qquad\qquad\qquad\qquad\quad = \frac{2}{19}$
$\qquad\qquad = 38$

(e) $(f \circ g)(8) = f(g(8))$ (f) $(g \circ f)(8) = g(f(8))$
$\qquad\qquad\quad = \sqrt[3]{16 + 3} \qquad\qquad\qquad\qquad\quad = 2(\sqrt[3]{8}) + 3$
$\qquad\qquad\quad = \sqrt[3]{19} \qquad\qquad\qquad\qquad\qquad\quad = 4 + 3$
$\qquad\qquad\qquad\qquad\qquad\qquad\qquad\qquad\qquad\quad = 7$

Let $f(x) = \sqrt{x + 1}$ and $g(x) = x^2$. Find the value of each of the following.

11. $(f + g)(3)$ 12. $(f - g)(3)$ 13. $(fg)(3)$ 14. $\left(\dfrac{f}{g}\right)(3)$

15. $(f \circ g)(3)$ 16. $(g \circ f)(3)$ 17. $(f \circ g)(8)$ 18. $(g \circ f)(8)$

19. $(f \circ g)(0)$ 20. $(g \circ f)(0)$

CHECK LIST FOR THE STUDENT

If you have learned the material in Section 2.10, you should be able to do the following:

1. Define the sum of two functions.

2. Find the rule of assignment and the domain of $f + g$ for given functions f and g.

3. Define the difference of two functions.

4. Find the rule of assignment and the domain of $f - g$ for given functions f and g.

5. Define the product of two functions.

6. Find the rule of assignment and the domain of fg for given functions f and g.

7. Define the quotient of two functions.

8. Find the rule of assignment and the domain of f/g for given functions f and g.

9. Define the composite of two functions.

10. Find the rule of assignment and the domain of $f \circ g$ for given functions f and g.

11. Prove that $f \circ g$ is not necessarily the same as $g \circ f$.

SECTION QUIZ 2.10

1. Let $f(x) = \sqrt{x}$, and let the domain of f be $\{x \mid x \geq 0\}$. Let $g(x) = \sqrt{-x + 3}$, and let the domain of g be $\{x \mid x \leq 3\}$.
   (a) What is the domain of $f + g$?
   (b) What is the domain of f/g?

Let $f(x) = \sqrt[3]{x + 1}$, and let $g(x) = 5x - 4$. Find the value of each of the following.

2. (a) $(f + g)(7)$ (b) $(f - g)(7)$ (c) $(fg)(7)$

3. (a) $\left(\dfrac{f}{g}\right)(7)$ (b) $f(g(7))$ (c) $g(f(7))$

Let $f(x) = \sqrt{x}$, and let the domain of f be $\{x \mid x \geq 0\}$. Let $g(x) = x^2 + 9$, and let the domain of g be the set of all real numbers. Find the rule of assignment and state the domain of each of the following.

4. (a) $f + g$ (b) $f - g$ (c) fg

5. (a) $\dfrac{f}{g}$ (b) $f \circ g$ (c) $g \circ f$

Part 2 Review Exercises

1. Let f be the function whose domain is $\{-1, 0, 1, 2\}$ and which is such that $f(x) = x^3 - x + 1$. Find the range of f and sketch the graph of f.

2. If 0, 1, 2, a, and $a + h$ are in the domain of f, and if $f(x) = x^2 - x + 1$, find $f(0)$, $f(1)$, $f(2)$, $f(a)$, and $f(a + h)$. Express $f(a + h)$ as a polynomial in a and h.

3. State whether or not each of the following sets represents a function. If it does represent a function, give the domain, show how the rule of assignment can be written with arrows, and give the range.
   (a) $\{(1, 1), (2, 32), (-2, -32), (3, 243)\}$
   (b) $\{(1, 1), (1, -1), (81, 3), (81, -3), (0, 0)\}$

4. According to the domain convention, what is the domain of the function f in each of the following?
   (a) $f(x) = \dfrac{1}{x^2 - 4}$
   (b) $f(x) = \sqrt{x - 4}$
   (c) $f(x) = \sqrt{x^2 - 4}$
   (d) $f(x) = \dfrac{1}{\sqrt{x - 7}}$
   (e) $f(x) = x^5$
   (f) $f(x) = x^4$

5. Sketch the graph of $y = |x|$ for $-4 \le x \le 5$.

6. Let f be the function whose domain is $\{x \mid -1 \le x \le 3\}$ and which is such that $f(x) = -x^2$. Sketch the graph of f.

7. Let f be the function whose domain is $\{-1, 0, 1, 2, 3\}$ and which is such that $f(x) = -x^2$. Sketch the graph of f.

8. Tell whether or not the graph of each of the following represents a function.
   (a) $x^2 + y^2 = 36$
   (b) $y = x^2$
   (c) $y^2 = x$
   (d) $y = x$
   (e) $y = 3$
   (f) $x = 3$

9. Let $f(x) = 3x^2 - x - 4$, let h be nonzero, and let $x + h$ be in the domain of f. Express $\dfrac{f(x + h) - f(x)}{h}$ as a polynomial in x and h.

10. Let $f(x) = x^2 - 3x - 2$. Let Δx represent a nonzero real number. Express $\dfrac{f(x_1 + \Delta x) - f(x_1)}{\Delta x}$ as a polynomial in x_1 and Δx.

11. Sketch the graph of each of the following and label any intercept(s).
    (a) $3x - 2y + 8 = 0$
    (b) $3x + 6 = 0$
    (c) $2y - 8 = 0$
    (d) $x - y = 0$
    (e) $y = 3x - 1$

12. Find the point of intersection of the following lines.
$$5x - 7y + 8 = 0$$
$$10x + 3y - 1 = 0$$

13. Find an expression for the slope of the line through the point with coordinates $(x, f(x))$ and the point with coordinates $(x + h, f(x + h))$.

14. Find the slope of the line through the points $(-5, 7)$ and $(8, -1)$.

In each of the following, find the standard form of an equation of the line with the given specifications.

15. Through the point $(2, -7)$ and with slope $\frac{3}{8}$.

16. Through the points $(3, -5)$ and $(-1, 6)$.

17. With intercepts $(3, 0)$ and $(0, 5)$.

18. With slope $\frac{2}{3}$ and y-intercept -5.

19. Through the origin and perpendicular to the line with slope $\frac{3}{4}$.

20. Through the point $(0, -1)$ and parallel to the x-axis.

21. Through the point $(-1, 0)$ and parallel to the y-axis.

22. Through the point $(-2, 0)$ and parallel to $y = x$.

23. Through the point $(0, -2)$ and parallel to $y = x$.

24. Find the slope and y-intercept of $7x - 9y + 11 = 0$.

25. Sketch the graph of the linear function f such that $f(x) = 3x - 5$.

26. Find the intercepts of the graph of each of the following:
  (a) $49x^2 + y^2 = 49$ (b) $49x^2 - y^2 = 49$

27. Determine whether or not the graph of each of the following is symmetric with respect to the x-axis, the y-axis, or the origin.
  (a) $y = x^2 - 3$ (b) $xy = 3$ (c) $x^4 - x^2 + y = 0$
  (d) $3x^2 - 3y^2 = 1$ (e) $y = x^5$

28. In each of the following, find any vertical or horizontal asymptote(s) that the graph has.

  (a) $y = \dfrac{3}{x - 7}$ (b) $y = \dfrac{3x}{x - 7}$ (c) $y = \dfrac{3x^2}{x - 7}$

  (d) $y = \dfrac{3}{x^2 + 7}$

29. Let $y = \dfrac{x - 1}{3x + 6}$. Classify y as positive or negative for (a) $x > 1$, (b) $-2 < x < 1$, and (c) $x < -2$.

30. Let $y = \dfrac{1}{\sqrt{x - 1}}$. Determine all values of x excluded by ruling out imaginary values of the variables.

31. Sketch the graph of each of the following. Label any intercept(s) or vertical or horizontal asymptote(s) that the graph has.

  (a) $y = \dfrac{5x + 10}{x + 1}$ (b) $y = \dfrac{4}{x^2 + 4}$

32. Let A and B be the points: $A(-6, 4)$, $B(6, 9)$.
  (a) Find the distance between A and B.
  (b) Find the midpoint of the segment joining A and B.

(c) Find an equation of the perpendicular bisector of the segment joining A and B.

(d) Find an equation of the circle having the segment AB as a diameter.

33. Find an equation of the circle having center at $(2, 3)$ and passing through the origin.

34. Find the center and radius of $4x^2 + 4y^2 + 12x + 32y + 29 = 0$.

35. Transform the equation

$$\frac{(x + 7)^2}{25} + \frac{(y - 5)^2}{36} = 1$$

by translating axes so that the new origin is at $(-7, 5)$. Draw and label both sets of axes. Sketch the graph, and label the ends of its axes relative to the x-y coordinate system.

36. In each of the following, find the vertex, axis, focus, and directrix. Sketch the graph, plot the focus, and draw the directrix.
    (a) $x^2 = 28y$ (b) $x^2 = -28y$ (c) $y^2 = 28x$ (d) $y^2 = 28x$
    (e) $(x - 1)^2 = 32(y + 4)$ (f) $(y + 4)^2 = -32(x - 1)$
    (g) $x^2 - 4x + 28y + 32 = 0$ (h) $y^2 + 2y - 28x + 57 = 0$

37. In each of the following, sketch the graph of the function f.
    (a) $f(x) = x^2 - 6x + 9$ (b) $f(x) = -x^2 + 6$

38. Sketch the graph of each of the following.
    (a) $y = x^2 - 4x + 8$ (b) $y = -x^2 + 4x + 8$

39. A projectile is fired from the ground, and its distance s above the ground at t seconds after firing is given by $s = -16t^2 + 288t$ feet. Find the time it takes the projectile to reach its maximum height. What is its maximum height?

40. A publisher estimates that the number of copies of a text sold per year t years after publication is given by formula $s = -1000t^2 + 6000t$. According to this formula, how many years after publication does it take for the number of copies sold per year to reach its maximum? What is the maximum number of copies sold per year?

Sketch the graph of each of the following. If it is a circle, find its center and radius. If it is an ellipse, find its center and label the ends of its axes. If it is a hyperbola, find equations of its asymptotes and draw its asymptotes. If it is a parabola, label its vertex.

41. $9x^2 + y^2 - 36x + 10y + 25 = 0$

42. $9x^2 - y^2 - 36x - 10y - 25 = 0$

43. $x^2 + y^2 - 4x + 10y + 28 = 0$

44. $x^2 - 4x + 4y = 0$

45. $x + \dfrac{(y + 5)^2}{4} = 1$

In each of the following, if the equation has a graph, identify it. If the equation has no graph, state "No graph."

46. $7x^2 + 7y^2 = 1$ $\qquad$ **47.** $7x^2 - 7y^2 = 1$

48. $7x^2 + 7y^2 = -1$ $\qquad$ **49.** $7x^2 - 7y^2 = 0$

50. $7x^2 - 7y^2 = -1$ $\qquad$ **51.** $7x^2 + 5y^2 = 1$

52. $7x^2 = y$ $\qquad$ **53.** $7y^2 = x$

Sketch the graph of each of the following, and state whether or not it is the graph of a function.

54. $\{(x, y) \mid x^2 + y^2 = 49,\ \text{and}\ x \leq 0\}$

55. $\{(x, y) \mid x^2 + y^2 = 49,\ x \leq 0,\ \text{and}\ y \geq 0\}$

State whether or not each of the following functions is one-to-one.

56. $f(x) = x^2$ and the domain of f is $\{x \mid x \geq 1\}$.

57. $f(x) = 3x + 8$ and the domain of f is $\{x \mid x \geq 0\}$.

58. $f(x) = x^3$

59. $f(x) = x^4$

In each of the following, find the domain of f^{-1}, and find $f^{-1}(x)$.

60. The domain of f is $\{x \mid -1 \leq x \leq 3\}$, and $f(x) = -2x + 3$.

61. The domain of f is $\{x \mid x \geq 0\}$, and $f(x) = x^2 + 1$.

Sketch the graph of each of the following.

62. $y = 11^x$ $\qquad$ **63.** $y = \left(\dfrac{1}{11}\right)^x$ $\qquad$ **64.** $y = \log_{11} x$

Find the value of each of the following.

65. $\log_6 36$ $\qquad\qquad$ **66.** $\log_6 \frac{1}{36}$

67. $\log_6 6^8$ $\qquad\qquad$ **68.** $\ln e^{25}$

Suppose that $\log_{10} 8.7 = 0.9395$ and $\log_{10} 5.6 = 0.7482$. Find the value of each of the following.

69. $\log_{10} (560 \times 8.7)$ $\qquad$ **70.** $\log_{10} \dfrac{0.0087}{5.6}$

71. $\log_{10} \sqrt{5.6}$ $\qquad\qquad$ **72.** $\log_{10} \sqrt[5]{870}$

73. $\log_{10} (8.7)^{10}$

In each of the following, let $f(x) = \sqrt[3]{x}$, let the domain of f be $\{x \mid x \geq 8\}$, let $g(x) = \sqrt{x - 2}$, and let the domain of g be $\{x \mid x \geq 2\}$. Find each of the following.

74. (a) The domain of $f + g$ $\qquad$ (b) $(f + g)(x)$

$\qquad$ (c) The domain of $\dfrac{f}{g}$ $\qquad$ (d) $\left(\dfrac{f}{g}\right)(x)$

75. (a) The domain of $f \circ g$ $\qquad$ (b) $f(g(x))$

$\qquad$ (c) The domain of $g \circ f$ $\qquad$ (d) $g(f(x))$

Part 2 Final Test

1. Let f be the function whose domain is $\{-2, 0, 1, 3\}$ and which is such that $f(x) = x + 3$.
   (a) Find the range of f.   (b) Sketch the graph of f.

2. According to the domain convention, what is the domain of function f in each of the following?
   (a) $f(x) = \sqrt{x - 4}$ (b) $f(x) = 1/x^2$

3. Let f be the function whose domain is $\{x \mid x \geq -1\}$ and which is such that $f(x) = -x^2$. Sketch the graph of f.

4. Find the point of intersection of the following lines, sketch the graphs of the lines, and label the point of intersection.

$$x - 2y - 5 = 0 \quad \text{and} \quad 3x + 2y - 7 = 0$$

5. Let A and B be the points: $A(-3, 7)$, $B(2, 4)$.
   (a) Find the slope of the line through A and B.
   (b) Find the slope of a line perpendicular to the line through A and B.
   (c) Find the standard form of an equation of the line through A and B.

6. (a) Find the standard form of an equation of the line through the point $(2, 5)$ and with slope 3.
   (b) Find the slope and y-intercept of the line $5x - 3y - 7 = 0$.
   (c) Sketch the graph of the linear function f such that $f(x) = x - 3$, and label both its x- and y-intercepts.

7. In each of the following, find the intercepts of the graph, determine whether or not the graph is symmetric with respect to the x-axis, with respect to the y-axis, or with respect to the origin, find any vertical asymptote(s) that the graph has, and find any horizontal asymptote(s) that the graph has.
   (a) $4x^2 - y^2 = 4$ (b) $y = \dfrac{5x}{x^2 - 4}$

8. Sketch the graph of

$$y = \frac{3}{x - 4}.$$

Label any intercept(s) or vertical or horizontal asymptote(s) that the graph has.

9. Let A and B be the points: $A(-2, 3)$, $B(1, 7)$.
   (a) Find the distance between A and B.
   (b) Find the midpoint of the segment joining A and B.

10. (a) Find the standard form of an equation of the circle with center at $(3, -7)$ and radius 6.
    (b) Find the center and radius of the circle $x^2 + y^2 - 10x + 4y - 52 = 0$.

11. Transform the equation $(x + 4)^2 + (y + 2)^2 = 4$ by translating axes so that the new origin is at $(-4, -2)$. Draw and label both sets of axes. Sketch the graph.

12. Sketch the graph of each of the following. Plot and label the focus, and draw and label the directrix.
    (a) $x^2 = 24y$ (b) $y^2 = -24x$

13. Write the following equation in standard form. Label the vertex and sketch the graph.

$$y^2 - 6y - 24x - 39 = 0$$

14. If $f(x) = -3x^2 + 6x - 8$, sketch the graph of the function f.

Sketch the graph of each of the following. If it is an ellipse, label its center and give the lengths of its axes. If it is a hyperbola, give the equations of its asymptotes and draw its asymptotes.

15. $4(x + 5)^2 + 9(y - 4)^2 = 36$

16. $\dfrac{(x - 5)^2}{4} - \dfrac{y^2}{9} = 1$

17. In each of the following, if the equation has a graph, identify it. If the equation has no graph, state "No graph."
    (a) $y = 2x$ (b) $y = 2x^2$
    (c) $2x^2 + 2y^2 = -1$ (d) $2x^2 + 2y^2 = 1$
    (e) $2x^2 + 2y^2 = 0$

18. In each of the following, state whether or not function f is one-to-one.
    (a) $f(x) = 3x^2$ (b) $f(x) = 3x$

19. Let f be the function with domain $\{x \mid 0 \le x \le 2\}$, and let $f(x) = 3x - 1$. Where f^{-1} denotes the inverse of f, find the domain of f^{-1}, and find $f^{-1}(x)$.

20. Sketch the graph of $y = 8^x$.

21. Sketch the graph of $y = \log_8 x$.

22. Find the value of each of the following.
    (a) $\log_7 1$ (b) $\log_7 7$ (c) $\ln e^9$ (d) $\log_{10} 0.0001$ (e) $\log_3 \frac{1}{9}$

23. Suppose that $\log_{10} 3 = 0.4771$ and $\log_{10} 4 = 0.6021$. Find the value of each of the following.
    (a) $\log_{10} 300$ (b) $\log_{10} 0.004$ (c) $\log_{10} 12$
    (d) $\log_{10} \frac{4}{3}$ (e) $\log_{10} \sqrt[3]{4}$ (f) $\log_{10} 3^4$

In each of the following, let $f(x) = \sqrt{x}$, let the domain of f be $\{x \mid x \ge 0\}$, let $g(x) = x - 3$, and let the domain of g be the set of all real numbers.
24. (a) Find $(f \circ g)(x) = f(g(x))$.
    (b) Find $(g \circ f)(x) = g(f(x))$.

25. (a) What is the domain of $f + g$?
    (b) What is the domain of f/g?
    (c) What is the domain of $f \circ g$?
    (d) What is the domain of $g \circ f$?

Part 3 Higher Algebra

Students sometimes have to do more with algebra than just perform straightforward manipulations. For example, they might have to find or approximate the roots of a polynomial equation, they might have to express a rational expression as a sum of partial fractions, they might have to find the probability that an event will occur, or they might be expected to apply the principle of mathematical induction so that they can derive a necessary formula or prove a basic theorem. Part 3 of this text presents topics from college algebra that are especially useful in mathematics and its applications.

3.1 Polynomial Equations

The purpose of this section is to present some useful theorems about polynomial equations, including theorems that will enable us to determine the number of roots of a polynomial equation with real coefficients, to find the rational roots of a polynomial equation with integral coefficients, to locate real roots of polynomial equations with real coefficients, and to approximate irrational roots of polynomial equations with real coefficients.

A **polynomial equation in the one variable x** is an equation that can be written in the form

$$P(x) = 0$$

where $P(x)$ is a polynomial. If $P(x)$ is of degree n, then the equation $P(x) = 0$ is an nth-degree equation. Linear equations such as

$$ax + b = 0$$

and quadratic equations such as

$$ax^2 + bx + c = 0$$

are polynomial equations of degree 1 and 2, respectively. The equation

$$x^4 + 4x^3 - 7x^2 - 22x + 24 = 0$$

is a polynomial equation of fourth degree. As we agreed in Section 1.4, unless otherwise specified, by a *polynomial* we mean a polynomial whose coefficients are all real numbers.

Exercise 3.1, 1–5

The number r is a *solution,* or *root,* of a polynomial equation $P(x) = 0$ provided that $P(r) = 0$. The terms *solution* and *root* are synonyms. If $P(r) = 0$, the number r is also called a *zero* of the polynomial $P(x)$.

We have already seen how to find the solutions or roots of linear and quadratic equations. In fact, in Section 1.7 we even produced formulas for the solutions of $ax + b = 0$ and $ax^2 + bx + c = 0$ where the coefficients a, b, and c are all real numbers and $a \neq 0$. The formulas for these solutions are:

$$ax + b = 0 \qquad ax^2 + bx + c = 0$$

$$x = \frac{-b}{a} \qquad x = \frac{-b \pm \sqrt{b^2 - 4ac}}{2a}.$$

It is also possible to find the solutions of any third- or fourth-degree polynomial equation by using only multiplication, division, addition, subtraction, and extraction of roots, but the methods are extremely long and complicated. Moreover, it has been shown that there is no such straightforward method for finding the solutions of polynomial equations of degree greater than 4. So we shall not be developing additional formulas for the solutions of polynomial equations. Instead, we shall be considering theorems that will enable us to locate or approximate the solutions.

The *remainder theorem,* which was stated and proved in Section 1.4, can be used to determine whether or not a particular number is actually a root of a given polynomial equation. According to the remainder theorem,

if the polynomial $P(x)$ is divided by $x - r$, the remainder equals $P(r)$.

It follows that one way to determine whether or not r is a root of $P(x) = 0$ is to divide $P(x)$ by $(x - r)$ and examine the remainder.

If the remainder $P(r)$ is zero, r must be a root of $P(x) = 0$. If the remainder $P(r)$ is not zero, r is not a root.

We saw in Section 1.4 that synthetic division may be used to advantage in computing the remainder $P(r)$. The remainder theorem and synthetic division are reviewed in the next example.

EXAMPLE 1

Use the remainder theorem and synthetic division to show that -3 is a root and 4 is not a root of $x^4 + 4x^3 - 7x^2 - 22x + 24 = 0$.

Solution

Let $P(x) = x^4 + 4x^3 - 7x^2 - 22x + 24$. To show that -3 is a root, we must show that $P(-3)$ is zero. By the remainder theorem we know that $P(-3)$ is the remainder after $P(x)$ is divided by $x - (-3)$, or $x + 3$. Using synthetic division to divide $P(x)$ by $x + 3$, we have

$$
\begin{array}{r|rrrrr}
-3) & 1 & 4 & -7 & -22 & 24 \\
    &   & -3 & -3 & 30 & -24 \\
\hline
    & 1 & 1 & -10 & 8 & 0
\end{array}
$$

The remainder $P(-3)$ is zero. The number -3 is a root.

To show that 4 is not a root, we compute $P(4)$ by dividing $P(x)$ by $x - 4$.

$$
\begin{array}{r|rrrrr}
4) & 1 & 4 & -7 & -22 & 24 \\
   &   & 4 & 32 & 100 & 312 \\
\hline
   & 1 & 8 & 25 & 78 & 336
\end{array}
$$

Exercise 3.1, 6–10

The remainder $P(4)$ is 336, not zero. The number 4 is not a root.

A very useful theorem called the *factor theorem* follows directly from the remainder theorem.

Factor Theorem   If r is a root of the polynomial equation $P(x) = 0$, then $(x - r)$ is a factor of $P(x)$.

Proof   If $P(x)$ is divided by $x - r$ and $Q(x)$ is the quotient, then

$$P(x) = Q(x)(x - r) + P(r)$$

by the remainder theorem. But if r is a root of $P(x) = 0$, $P(r)$ must be zero, so

$$P(x) = Q(x)(x - r).$$

Exercise 3.1, 11–15   In other words, if r is a root of $P(x) = 0$, then $(x - r)$ is a factor of $P(x)$.

The *converse of the factor theorem* follows directly from the zero factor law, and we have already used it in solving quadratic equations by factoring.

Converse of the   If $(x - r)$ is a factor of polynomial $P(x)$, then r is a root of the equation
Factor Theorem $P(x) = 0$.

Proof   If $(x - r)$ is a factor of $P(x)$, then

$$P(x) = Q(x)(x - r)$$

for some polynomial $Q(x)$. Therefore

$$P(r) = Q(r)(r - r)$$
$$= 0.$$

That is, r is a root of $P(x) = 0$.

EXAMPLE 2   Use the converse of the factor theorem to find the roots of the polynomial equation $(x - 3)(x + 5)x = 0$.

Solution   Let $P(x) = 0$ be the given equation.

Since $x - 3$ is a factor of $P(x)$, 3 is a root of $P(x) = 0$.

Since $x + 5$ is a factor of $P(x)$, -5 is a root of $P(x) = 0$.

Since x is a factor of $P(x)$, 0 is a root of $P(x) = 0$.

Exercise 3.1, 16–20   The roots are 3, -5, and 0.

It may be that a polynomial $P(x)$ with real coefficients has $(x - r)$ as a factor more than once. For example, the polynomial $x^2 - 6x + 9$ has $(x - 3)$ as a factor *twice,* in the sense that

$$x^2 - 6x + 9 = (x - 3)^2.$$

In this case we say that the factor $(x - 3)$ has multiplicity 2, or that the root 3 of the equation $x^2 - 6x + 9 = 0$ has multiplicity 2. In general, if

k is a positive integer, and if polynomial $P(x)$ has $(x - r)^k$ as a factor but has no higher power of $(x - r)$ as a factor, we say that the factor $(x - r)$ has multiplicity k and the solution or root r of the equation $P(x) = 0$ has multiplicity k.

EXAMPLE 3

Make up an equation whose roots are 2, -3, and 0 with multiplicity 1, 2, and 4, respectively.

Solution

Exercise 3.1, 21–30

$$(x - 2)(x + 3)^2 x^4 = 0$$

Now let us consider the question of how many solutions any particular polynomial equation may have. We have seen that any quadratic equation $ax^2 + bx + c = 0$ with real coefficients has either

1. Exactly one real-number solution,

2. Exactly two distinct real-number solutions, or

3. Exactly two distinct imaginary-number solutions.

In the case of exactly one real-number solution, say r, the quadratic is a perfect square, and the equation can be written in the form

$$(x - r)^2 = 0$$

so that the solution r is of multiplicity 2. Thus if we count each solution as many times as its multiplicity, and if we recall that real numbers and imaginary numbers are all complex numbers, we can say that any quadratic equation with real coefficients has exactly 2 complex-number solutions. This is a special case of the *number-of-roots theorem*.

Number-of-Roots Theorem

If each root is counted as many times as its multiplicity, any nth-degree polynomial equation $P(x) = 0$ with real coefficients has exactly n complex-number roots. It follows that any nth-degree polynomial equation $P(x) = 0$ with real coefficients has *at most n distinct* complex-number roots.

EXAMPLE 4

How many distinct roots has the equation $(x - 2)^5 (x + 4)^3 (x - 7) = 0$? What are these roots? If each root is counted as many times as its multiplicity, how many roots does the given equation have?

Solution

The given equation has three distinct roots. They are 2, -4, and 7. If each root is counted as many times as its multiplicity, there are $5 + 3 + 1$ or 9 roots. (Note that the equation is a ninth-degree polynomial equation.)

Exercise 3.1, 31–40

It can be shown that if one of the roots of polynomial equation $P(x) = 0$ is an imaginary number, say $a + bi$, then its *conjugate*, $a - bi$, is also a root of the equation. Imaginary roots occur in pairs.

Conjugate Root Theorem

If imaginary number $a + bi$ is a root of $P(x) = 0$, then $a - bi$ is also a root. Conversely, if $a - bi$ is a root of $P(x) = 0$, then $a + bi$ is also a root.

EXAMPLE 5

The imaginary number $2 - 3i$ is a root of the equation $x^2 - 4x + 13 = 0$. Show that its conjugate $2 + 3i$ is also a root.

Solution

Substituting $2 + 3i$ for x in the given equation, we see that $2 + 3i$ is a root.

$$(2 + 3i)^2 - 4(2 + 3i) + 13 = 0$$
$$4 + 12i + 9i^2 - 8 - 12i + 13 = 0$$
$$4 - 9 - 8 + 13 = 0 \qquad \text{True}$$

EXAMPLE 6

If $3 + 7i$ is a root of a polynomial equation $P(x) = 0$, what are two factors of $P(x)$?

Solution

If $3 + 7i$ is a root, then $3 - 7i$ is also a root. Two factors are $(x - (3 + 7i))$ and $(x - (3 - 7i))$.

In Example 6, we noted that $3 + 7i$ and $3 - 7i$ are roots of $P(x) = 0$ so that two factors of $P(x)$ are $(x - (3 + 7i))$ and $(x - (3 - 7i))$. If we multiply these two factors together, we obtain the quadratic

$$x^2 - 6x + 58.$$

(Students should check this multiplication themselves.) This quadratic $x^2 - 6x + 58$ is a factor of $P(x)$ which has real coefficients, but it cannot be factored into linear factors with *real* coefficients. Its factors $(x - 3 - 7i)$ and $(x - 3 + 7i)$ have *imaginary* coefficients. Any quadratic with real coefficients that cannot be factored into linear factors with real coefficients is called an **irreducible** quadratic. Each pair of conjugate imaginary roots of the polynomial equation $P(x) = 0$ is associated with an irreducible quadratic factor of $P(x)$, just as each real root is associated with a linear factor with real coefficients.

Exercise 3.1, 41–50

Factorization Theorem

Any polynomial $P(x)$ with real coefficients can be factored into a product of linear factors with real coefficients and irreducible quadratic factors.

EXAMPLE 7

Solve $x^3 - 3x^2 + 4x - 12 = 0$ by factoring. (Hint: 3 is a root.)

Solution

If 3 is a root, $(x - 3)$ is a factor. We can factor out $(x - 3)$ by dividing $x^3 - 3x^2 + 4x - 12$ by $x - 3$. Using synthetic division, we have

$$\begin{array}{r|rrrr} 3 & 1 & -3 & 4 & -12 \\ & & 3 & 0 & 12 \\ \hline & 1 & 0 & 4 & 0 \end{array}$$

$$x^3 - 3x^2 + 4x - 12 = (x - 3)(x^2 + 0x + 4)$$

The given equation may be written

$$(x - 3)(x^2 + 4) = 0.$$

(Note that the quadratic $x^2 + 4$ is irreducible.) By the zero factor law, we have

$$x - 3 = 0 \quad \text{or} \quad x^2 + 4 = 0$$
$$x = 3 \qquad\qquad x^2 = -4$$
$$x = \pm 2i.$$

Exercise 3.1, 51–55 The roots are 3, $2i$, and $-2i$. (Note that $2i$ and $-2i$ are conjugates.)

If a given polynomial equation has integral coefficients, we can find any rational roots that it may have by employing the next theorem. (Recall that a rational number is a number that can be written in the form p/q where p and nonzero q are integers.)

Rational Root Theorem If the nth-degree polynomial equation $P(x) = 0$ has integral coefficients, and if $P(x) = 0$ has a rational root p/q where p and q have no common factors other than $+1$ or -1, then the numerator p must be a factor of the constant term of $P(x)$, and the denominator q must be a factor of the coefficient of x^n in $P(x)$.

EXAMPLE 8 List all the possible rational roots of $2x^3 - x^2 - 15x + 18 = 0$.

Solution Possible numerators are factors of 18.

Possible numerators are ± 1, ± 2, ± 3, ± 6, ± 9, ± 18.

Possible denominators are factors of 2.

Possible denominators are ± 1, ± 2.

Possible rational roots are

$$\pm\tfrac{1}{1},\ \pm\tfrac{1}{2},\ \pm\tfrac{2}{1},\ \pm\tfrac{2}{2},\ \pm\tfrac{3}{1},\ \pm\tfrac{3}{2},\ \pm\tfrac{6}{1},\ \pm\tfrac{6}{2},\ \pm\tfrac{9}{1},\ \pm\tfrac{9}{2},\ \pm\tfrac{18}{1},\ \pm\tfrac{18}{2}.$$

If duplicates are omitted, the list of possible rational roots may be written more compactly as

$$\pm 1,\ \pm\tfrac{1}{2},\ \pm 2,\ \pm 3,\ \pm\tfrac{3}{2},\ \pm 6,\ \pm 9,\ \pm\tfrac{9}{2},\ \pm 18.$$

To find any rational roots that a polynomial equation with integral coefficients may have, we first list all the candidates provided by the rational root theorem, and then we may use the remainder theorem and synthetic division to check out each of these candidates. Once one rational root has been found, the task of checking for additional roots is somewhat easier, because we can work with a polynomial equation of lower degree than the given equation. Indeed, suppose that r is found to be a root of

$$P(x) = 0.$$

Then $(x - r)$ is a factor of $P(x)$ so that

$$P(x) = Q(x)(x - r)$$

and the given equation can be written

$$Q(x)(x - r) = 0.$$

Any additional roots of $P(x) = 0$ will be roots of

$$Q(x) = 0,$$

and the root r should be checked in $Q(x) = 0$ to see if it is a multiple root of $P(x) = 0$. The equation $Q(x) = 0$ is called the *depressed equation*. Since $Q(x)$ is the quotient when $P(x)$ is divided by $(x - r)$, the coefficients of $Q(x)$ are the numbers in the third line of the synthetic division of $P(x)$ by $(x - r)$. This is demonstrated in the next example.

EXAMPLE 9

Find the rational roots of $2x^3 - x^2 - 15x + 18 = 0$.

Solution

By the rational root theorem, if the given equation has any rational roots, they must be among the numbers

$$\pm 1, \ \pm\tfrac{1}{2}, \ \pm 2, \ \pm 3, \ \pm\tfrac{3}{2}, \ \pm 6, \ \pm 9, \ \pm\tfrac{9}{2}, \ \pm 18.$$

We see that we have 18 possibilities to check. Using synthetic division to check ± 1, we find that neither of these are roots because the remainder is not zero for either of them. (It is also easy to show by direct substitution in the given equation that $+1$ and -1 are not roots.)

$$
\begin{array}{r|rrrr}
1 & 2 & -1 & -15 & 18 \\
  &   & 2 & 1 & -14 \\
\hline
  & 2 & 1 & -14 & 4 \\
\end{array}
\qquad
\begin{array}{r|rrrr}
-1 & 2 & -1 & -15 & 18 \\
   &   & -2 & 3 & 12 \\
\hline
   & 2 & -3 & -12 & 30 \\
\end{array}
$$

Since it is usually easier to check integral roots, we next use synthetic division to check the candidate 2, and we find that 2 is a root.

$$
\begin{array}{r|rrrr}
2 & 2 & -1 & -15 & 18 \\
  &   & 4 & 6 & -18 \\
\hline
  & 2 & 3 & -9 & 0 \\
\end{array}
$$

The coefficients in the quotient are 2, 3, -9 and the quotient is $2x^2 + 3x - 9$, so the given equation can be written

$$(2x^2 + 3x - 9)(x - 2) = 0.$$

Any additional roots are roots of the depressed equation

$$2x^2 + 3x - 9 = 0.$$

This happens to be a quadratic that we can solve by factoring.

$$(2x - 3)(x + 3) = 0$$

$$2x - 3 = 0 \quad \text{or} \quad x + 3 = 0$$

$$x = \tfrac{3}{2} \qquad\qquad\qquad x = -3.$$

Exercise 3.1, 56–75

The rational roots of the given equation are 2, $\frac{3}{2}$, and -3. Since the given equation is of third degree, we know by the number-of-roots theorem that it can have no more than 3 different roots. Consequently, we have found *all* the roots of the given equation.

Using the rational root theorem in Example 9, we had eighteen candidates to check. We were fortunate to find a rational root on only the third try. The process could be even longer and more tedious for some other given equation. There are several theorems that may sometimes shorten such calculations by eliminating some of the candidates provided by the rational root theorem. One of these is called the *location-of-roots theorem.*

Before formally stating the location-of-roots theorem, let us consider the matter geometrically. Note that the real roots of polynomial equation $P(x) = 0$ are the x-intercepts of the graph of the function defined by the equation $y = P(x)$. That is, they are the x-coordinates of the points of intersection of the x-axis and the graph of $y = P(x)$. For example, to find the x-intercepts of the graph of $y = 4x^3 + 3x^2 - 25x + 6$, we let $y = 0$ and then solve for x. The roots of $4x^3 + 3x^2 - 25x + 6 = 0$ are $-3, \frac{1}{4}$, and 2, so the x-intercepts of the graph of $y = 4x^3 + 3x^2 - 25x + 6$ are $-3, \frac{1}{4}$, and 2. The graph and its intercepts are shown in Figure 3.1.

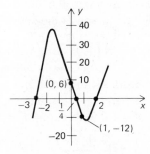

FIGURE 3.1 Graph of $y = 4x^3 + 3x^2 - 25x + 6$

We see in Figure 3.1 that the graph of $y = 4x^3 + 3x^2 - 25x + 6$ is *continuous;* the graph can be drawn without lifting the pencil off the paper. It can be shown that the graph of any *polynomial function*—that is, any function defined by an equation $y = P(x)$ where $P(x)$ is a polynomial with real coefficients—is continuous. Thus if a and b are real numbers such that $P(a)$ and $P(b)$ have opposite signs, the points $(a, P(a))$ and $(b, P(b))$ on the graph must be on opposite sides of the x-axis, and the graph must cross the x-axis at least once, or have at least one x-intercept, between a and b. That is, the polynomial equation $P(x) = 0$ must have at least one real root between a and b. This is the conclusion of the *location-of-roots theorem.*

If $P(x)$ is a polynomial with real coefficients, if a and b are real numbers, and if $P(a)$ and $P(b)$ have opposite signs, then at least one real root of $P(x) = 0$ is located between a and b.

EXAMPLE 10

Use the location-of-roots theorem to show that there is at least one root of $4x^3 + 3x^2 - 25x + 6 = 0$ between 0 and 1.

Solution

Let $P(x) = 4x^3 + 3x^2 - 25x + 6$.

$P(0) = 6$

$P(1) = -12$

$P(0)$ and $P(1)$ are of opposite signs.

There is at least one root of $4x^3 + 3x^2 - 25x + 6 = 0$ between 0 and 1. (Note in Figure 3.1 that the graph goes through $(0, 6)$ and $(1, -12)$, and since the graph is continuous it must cross the x-axis at least once between these two points. Actually, this graph crosses the x-axis *exactly* once between $(0, 6)$ and $(1, -12)$.)

Exercise 3.1, 76–85

If a root located by the location-of-roots theorem happens to be a rational root, it can be found by using the rational root theorem. If a root of $P(x) = 0$ is located between real numbers a and b and is *not* rational, it can be aproximated by using a technique called *linear interpolation*. The idea is to approximate the graph of $y = P(x)$ between $(a, P(a))$ and $(b, P(b))$ with the graph of the line segment between these two points. A first approximation to an irrational root located between a and b is then the x-intercept of the line through $(a, P(a))$ and $(b, P(b))$. (Figure 3.2 demonstrates this for a case in which $P(a)$ is positive.) To find this first approximation, all we need to do is to write an equation of the line through $(a, P(a))$ and $(b, P(b))$ and then find the x-intercept of this line. An equation of the line through $(a, P(a))$ and $(b, P(b))$ is

$$y - P(a) = \left(\frac{P(b) - P(a)}{b - a}\right)(x - a).$$

To find the x-intercept, we let $y = 0$ and solve for x.

$$-P(a) = \left(\frac{P(b) - P(a)}{b - a}\right)(x - a)$$

$$x - a = -P(a)\left(\frac{b - a}{P(b) - P(a)}\right)$$

$$x = a - P(a)\left(\frac{b - a}{P(b) - P(a)}\right) \quad \begin{array}{l}x\text{-intercept of} \\ \text{line through} \\ (a, P(a)) \text{ and} \\ (b, P(b))\end{array}$$

This is illustrated for particular $P(x)$, a, and b in the next example.

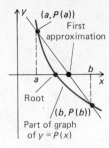

FIGURE 3.2   First approximation by linear interpolation

EXAMPLE 11

Show that there is a root of $x^4 - 6x^3 + 9x^2 - 12x + 14 = 0$ between 1 and 2. Use linear interpolation to find a first approximation to such a root.

Solution

Let $P(x)$ be $x^4 - 6x^3 + 9x^2 - 12x + 14$. We will first compute $P(1)$ and $P(2)$ to show that they are of opposite signs.

$$
\begin{array}{r|rrrrr}
1) & 1 & -6 & 9 & -12 & 14 \\
   &   & 1 & -5 & 4 & -8 \\
\hline
   & 1 & -5 & 4 & -8 & 6 \\
\end{array}
\qquad
\begin{array}{r|rrrrr}
2) & 1 & -6 & 9 & -12 & 14 \\
   &   & 2 & -8 & 2 & -20 \\
\hline
   & 1 & -4 & 1 & -10 & -6 \\
\end{array}
$$

$$P(1) = 6 \qquad\qquad\qquad P(2) = -6$$

There is at least one root between 1 and 2.

To find a first approximation to a root between 1 and 2, we write an equation of the line through $(1, P(1))$ and $(2, P(2))$, that is the line through $(1, 6)$ and $(2, -6)$, and we find its x-intercept. The line is shown in Figure 3.3. An equation of this line is

$$y - 6 = \left(\frac{-6 - 6}{2 - 1}\right)(x - 1)$$

$$y - 6 = -12(x - 1).$$

To find the x-intercept, we let $y = 0$ and solve for x.

$$-6 = -12(x - 1)$$

$$\tfrac{1}{2} = x - 1$$

$$x = 1 + \tfrac{1}{2} \qquad x\text{-intercept of line through} \\ \qquad\qquad (1, 6) \text{ and } (2, -6).$$

Exercise 3.1, 86–95

A first approximation to a root of $P(x) = 0$ between 1 and 2 is 1.5.

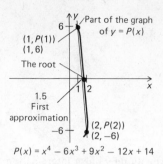

$P(x) = x^4 - 6x^3 + 9x^2 - 12x + 14$

FIGURE 3.3 First approximation to root of $x^4 - 6x^3 + 9x^2 - 12x + 14 = 0$ between 1 and 2

If we let x_1 denote the first approximation to a root of $P(x) = 0$ found by linear interpolation, we can find a second approximation by repeating the linear interpolation process using the line through $(a, P(a))$ and $(x_1, P(x_1))$ or between $(x_1, P(x_1))$ and $(b, P(b))$. This is demonstrated in the next example.

EXAMPLE 12

Use linear interpolation to find a second approximation to the root of $x^4 - 6x^3 + 9x^2 - 12x + 14 = 0$ approximated in Example 11.

Solution

Again let $P(x) = x^4 - 6x^3 + 9x^2 - 12x + 14$, and let $x_1 = 1.5$ be a first approximation to a root between 1 and 2. First we will check to see if the root we are approximating is between 1 and x_1 or between x_1 and 2. We know that $P(1)$ is positive and $P(2)$ is negative. Let us compute $P(x_1)$ or $P(1.5)$.

1.5) 1	-6	9	-12	14
	1.5	-6.75	3.375	-12.9375
1	-4.5	2.25	-8.625	1.0625

$$P(1.5) = 1.0625$$

Since $P(1.5)$ is positive and $P(2)$ is negative, there is a root between 1.5 and 2. So to find our second approximation we write an equation for the line through $(1.5, 1.0625)$ and $(2, -6)$ and we find its x-intercept. An equation of this line is

$$y - 1.0625 = \left(\frac{-6 - 1.0625}{2 - 1.5}\right)(x - 1.5).$$

Letting $y = 0$ and solving for x we have

$$-1.0625 = \frac{-7.0625}{.5}(x - 1.5)$$

$$-1.0625 = -14.1250(x - 1.5)$$

$$x - 1.5 = \frac{-1.0625}{-14.1250}$$

$$x = 1.5 + \frac{1.0625}{14.1250}$$

$$x = 1.575221239.$$

A second approximation to a root of $P(x) = 0$ between 1 and 2 is

1.575.

The linear interpolation process may be repeated again and again to obtain a better and better approximation. (It can be shown that the root between 1 and 2 is actually $3 - \sqrt{2}$, which is 1.585786438 correct to nine decimal places.)

There are methods other than linear interpolation for approximating irrational roots of polynomial equations with real coefficients. One of these, called Newton's method, requires the use of a bit of calculus and is often taken up in beginning calculus courses.

There are also several other theorems that may help locate the roots of a polynomial equation. One theorem says in part that if B is a positive real number, and if the numbers in the third line of the synthetic division of $P(x)$ by $(x - B)$ are all nonnegative, then all the real roots of $P(x) = 0$ are less than or equal to B. Another theorem, called *Descartes' rule of signs*, specifies how to determine the number of positive roots that $P(x) = 0$ may have by considering the signs of the coefficients of $P(x)$. Students who have to do a lot of problems requiring them to locate the real roots of a polynomial equation should look up these theorems in a text on the theory of equations.

EXERCISE 3.1 POLYNOMIAL EQUATIONS

► Tell the degree of each of the following equations.

EXAMPLE

$3x^4 - 5x^2 + 6x^9 - 2x + 5 = 0$

Solution

Degree 9

Exercise 3.1, 96–105

1. $2x^2 - 5x^7 - 3x^4 + 2x^4 - 7 = 0$

2. $x + 7 = 0$ **3.** $2x^2 - 5x + 7 = 0$

4. $x^3 - 2x^2 + 7x - 5 = 0$ **5.** $x - 7x^2 + 6x^4 - 2x^3 = 0$

▶ In each of the following, use the remainder theorem and synthetic division to show that the first number given is a root and the second number is not a root.

EXAMPLE

$2x^3 + 11x^2 + 2x - 15 = 0; \quad -5, 2$

Solution

$$
\begin{array}{r|rrrr}
-5) & 2 & 11 & 2 & -15 \\
 & & -10 & -5 & 15 \\
\hline
 & 2 & 1 & -3 & 0 \\
\end{array}
\qquad
\begin{array}{r|rrrr}
2) & 2 & 11 & 2 & -15 \\
 & & 4 & 30 & 64 \\
\hline
 & 2 & 15 & 32 & 49 \\
\end{array}
$$

Remainder $= 0$; -5 is a root. Remainder $\neq 0$; 2 is not a root.

6. $3x^4 - 22x^3 + 19x^2 - 88x + 28 = 0; \quad 7, -3$

7. $x^4 + 6x^3 - 25x^2 - 54x + 144 = 0; \quad -8, -2$

8. $x^3 + 2x^2 + 15x + 14 = 0; \quad -1, 4$

9. $2x^3 + 3x^2 - 17x + 12 = 0; \quad \frac{3}{2}, 3$

10. $x^4 - 72x^2 - 729 = 0; \quad -9, 2$

▶ In each of the following, use the factor theorem to construct an equation with the lowest possible degree having the given roots.

EXAMPLE

$2, -7, 0$

Solution

$(x - 2)(x + 7)x = 0$

11. $3, -5, 4$ **12.** $0, -2, 1$

13. $3, -3, 4, -4$ **14.** $2 + 3i, 2 - 3i, 5$

15. $2i, -2i, 3$

▶ Use the converse of the factor theorem to find the roots of each of the following equations.

EXAMPLE

$x(x + 7)(x - 2)(x + 3) = 0$

Solution

$0, -7, 2, -3$

16. $(x - 3)(x + 2)(x + 5) = 0$ **17.** $x(x - 2)(x + 4) = 0$

18. $x(x^2 - 4) = 0$ **19.** $x(x^2 + x - 6) = 0$

20. $(x^2 - 9)(x^2 - 4x + 4) = 0$

► Tell the multiplicity of each root in each of the following.

EXAMPLE

$x^3(x - 5)^2(x + 3) = 0$

Solution

0 is of multiplicity 3

5 is of multiplicity 2

−3 is of multiplicity 1

21. $x^2(x + 1)^3(x - 4)^5 = 0$ **22.** $x(x - 1)^2(x + 5)^7 = 0$

23. $(x - 3)^2(x^2 + 1)^3 = 0$ **24.** $x^3(x^2 - 6x + 9) = 0$

25. $x^4(x - 2)^3(x + 9) = 0$

► In each of the following, make up an equation having the given roots with the given multiplicities and having no other roots.

EXAMPLE

0, 3, −2 with multiplicity 4, 1, and 5, respectively

Solution

$x^4(x - 3)(x + 2)^5 = 0$

26. 0, −5, 6 with multiplicity 1, 2, and 3, respectively

27. −5, −4, −1 with multiplicity 2, 3, and 4, respectively

28. 1, 2, −3 with multiplicity 1, 3, and 2, respectively

29. 4, 3*i*, −3*i* with multiplicity 3, 5, and 5, respectively

30. $\frac{2}{3}$, $\frac{4}{5}$, 6 with multiplicity 1, 2, and 3, respectively

► What is the greatest number of distinct complex-number roots that each of the following equations may have? If each root is counted as many times as its multiplicity, how many roots does the given equation have?

EXAMPLE

$x^5 - 3x^2 + 2x - 7 = 0$

Solution

5, 5

31. $2x^3 - 5x^4 - 3x + 7 = 0$ **32.** $x^2 - 5x + 8 = 0$

33. $3x^7 - 2x^5 - 2x + 9x^2 + 8 = 0$

34. $x^3 - 3x^2 + 5x - 8 = 0$ **35.** $2x - 3x^2 + 5x^5 - 3x^4 = 0$

► How many distinct roots does each of the following equations have? What are these roots? If each root is counted as many times as its multiplicity, how many roots does the given equation have?

EXAMPLE

$x(x - 2)^3(x + 1)^5(x - 3) = 0$

Solution   Four distinct roots: 0, 2, −1, 3; 1 + 3 + 5 + 1 = 10 roots

36. $x^2(x - 1)^3(x + 2) = 0$ **37.** $(x + 1)^4(x - 3)^5(x + 4) = 0$

38. $x^3(x - 1)(x + 3)^2 = 0$ **39.** $(x + 5)^2(x - 2)^3(x^2 + 1) = 0$

40. $x^5(x - 1)^4(x + 2)^3 = 0$

▶ In each of the following, show that the conjugate of the given root is also a root.

EXAMPLE $x^2 + 9 = 0,\ \ 3i$

Solution   The conjugate $-3i$ is also a root because $(-3i)^2 + 9 = 0$.

41. $x^2 + 25 = 0,\ 5i$ **42.** $x^2 - 6x + 34 = 0,\ 3 - 5i$

43. $x^2 - 4x + 20 = 0,\ 2 + 4i$ **44.** $x^3 + 36x = 0,\ -6i$

45. $x^2 - 2x + 5 = 0,\ 1 - 2i$

▶ In each of the following, the given number is a root of a polynomial equation $P(x) = 0$. Find an irreducible quadratic factor of $P(x)$.

EXAMPLE $2 - 3i$

Solution $2 + 3i$ is also a root, $(x - 2 + 3i)$ and $(x - 2 - 3i)$ are factors.

$$(x - 2 + 3i)(x - 2 - 3i) = (x - 2)^2 - 9i^2$$
$$= x^2 - 4x + 4 + 9$$

$x^2 - 4x + 13$ is an irreducible quadratic factor.

46. $3 + 4i$ **47.** $2 - 5i$ **48.** $3 + 7i$ **49.** $2 - 6i$ **50.** $9i$

▶ Knowing that the given number is a root of the given equation, solve each of the following by factoring.

EXAMPLE $2,\ x^3 - x - 6 = 0$

Solution

$$
\begin{array}{r|rrrr}
2 & 1 & 0 & -1 & -6 \\
  &   & 2 & 4 & 6 \\
\hline
  & 1 & 2 & 3 & 0
\end{array}
$$

$x^3 - x - 6 = (x - 2)(x^2 + 2x + 3)$

$x^2 + 2x + 3 = 0$

$$x = \frac{-2 \pm \sqrt{4 - 12}}{2}$$

$$= \frac{-2 \pm \sqrt{8}i}{2}$$

$$= \frac{-2 \pm 2\sqrt{2}i}{2}$$

The roots are 2 and $-1 \pm \sqrt{2}i$.

51. 1, $x^3 + 2x^2 + 2x - 5 = 0$ **52.** 3, $x^3 - 2x^2 - 2x - 3 = 0$

53. -2, $x^3 + 3x^2 + 4x + 4 = 0$

54. -1, $x^3 + x^2 + 16x + 16 = 0$

55. 4, $x^3 - 2x^2 - 3x - 20 = 0$

▶ List all the possible rational roots of each of the following.

EXAMPLE

$2x^3 - 3x^2 + 5x - 6 = 0$

Solution

Possible numerators: ± 1, ± 2, ± 3, ± 6

Possible denominators: ± 1, ± 2

Possible rational roots: ± 1, $\pm \frac{1}{2}$, ± 2, ± 3, $\pm \frac{3}{2}$, ± 6

56. $3x^3 - 2x^2 + 5x - 10 = 0$ **57.** $x^4 - 3x^3 + 2x^2 - 8 = 0$

58. $5x^2 - 2x + 7 = 0$ **59.** $7x^3 - 3x^2 + 2x + 15 = 0$

60. $3x^3 - 2x^2 + 7x - 1 = 0$

▶ Find the rational roots of each of the following.

EXAMPLE

$x^3 - 3x - 2 = 0$

Solution

Possible rational roots are ± 2 and ± 1.

$$
\begin{array}{r|rrr}
2 & 1 & 0 & -3 & -2 \\
  &   & 2 & 4 & 2 \\
\hline
  & 1 & 2 & 1 & 0
\end{array}
$$

$(x - 2)(x^2 + 2x + 1) = 0$

$(x - 2)(x + 1)^2 = 0$

2 is a root.   The roots are 2 and -1.

61. $2x^3 - x^2 - 7x + 6 = 0$

62. $12x^3 + 29x^2 - 31x - 30 = 0$

63. $10x^3 + 63x^2 + 87x - 20 = 0$

64. $2x^4 - 15x^3 - 11x^2 + 135x - 63 = 0$

65. $x^4 - 10x^3 + 37x^2 - 60x + 36 = 0$

66. $3x^5 - 142x^3 + 7x^4 - 342x^2 - 245x - 49 = 0$

67. $4x^4 - 5x^3 - 83x^2 + 201x - 45 = 0$

68. $x^5 + 7x^4 + 19x^3 + 25x^2 + 16x + 4 = 0$

69. $x^7 + 2x^6 - 39x^5 - 40x^4 + 400x^3 = 0$

70. $3x^3 - 34x^2 + 115x - 100 = 0$

▶ Find all the roots of each of the following.

EXAMPLE

$5x^3 + 7x^2 + 19x - 15 = 0$

Solution

Possible rational roots: ± 1, $\pm\frac{1}{5}$, ± 3, $\pm\frac{3}{5}$, ± 5, ± 15

$$\frac{3}{5})\overline{\begin{array}{rrrr} 5 & 7 & 19 & -15 \\ & 3 & 6 & 15 \\ \hline 5 & 10 & 25 & 0 \end{array}}$$

$(x - \frac{3}{5})(5x^2 + 10x + 25) = 0$

$x^2 + 2x + 5 = 0$

$x = \dfrac{-2 \pm \sqrt{4 - 20}}{2}$

$= -1 \pm 2i$

The roots are $\frac{3}{5}$ and $-1 \pm 2i$.

71. $11x^3 + 9x^2 + 31x - 6 = 0$ **72.** $9x^3 + 22x^2 + 35x + 12 = 0$

73. $x^3 + 11x^2 + 31x + 6 = 0$ **74.** $14x^3 - 19x^2 + 11x + 2 = 0$

75. $24x^3 - 25x^2 + 14x - 3 = 0$

▶ In each of the following, show that there is at least one real root between the given numbers.

EXAMPLE

$4x^3 - 5x^2 - 47x + 12 = 0$; 0 and 1

Solution

Let $P(x) = 4x^3 - 5x^2 - 47x + 12$

$P(0) = 12$ positive

$P(1) = -36$ negative

By the location-of-roots theorem, there is at least one real root between 0 and 1.

76. $4x^3 - 5x^2 - 47x + 12 = 0$; 3 and 5

77. $4x^3 - 5x^2 - 47x + 12 = 0$; 0 and 1

78. $4x^3 - 9x^2 - 4x + 9 = 0$; 2 and 3

79. $4x^3 - 9x^2 - 4x + 9 = 0$; 0 and 2

80. $4x^4 - 47x^3 + 95x^2 + 200x = 0$; -1 and -2

81. $2x^3 - 6x^2 - 14x + 15 = 0$; 0 and 2

82. $2x^4 - 4x^3 - 20x^2 + x + 15 = 0$; -2 and -3

83. $3x^3 + 7x^2 - 16x - 20 = 0$; -3 and -4

84. $3x^4 + 7x^3 - 16x^2 - 20x = 0$; 1 and 3

85. $3x^4 - 2x^3 - 37x^2 + 28x + 60 = 0$; 0 and -2

▶ In each of the following use linear interpolation to find a one decimal place first approximation to the root of the given equation lying between the given numbers.

EXAMPLE

$x^4 - 2x^3 + 7x^2 - 18x - 18 = 0$; 2 and 3

Solution

Let $P(x) = x^4 - 2x^3 + 7x^2 - 18x - 18$.

$P(2) = -26; P(3) = 18$

Line through $(2, -26)$ and $(3, 18)$ is

$$y + 26 = 44(x - 2),$$

and its x-intercept is $\frac{57}{22} \simeq 2.59$.

First approximation is 2.6.

86. $x^4 - 2x^3 - 12x^2 - 32x - 64 = 0$; 3 and 4

87. $x^5 + x^4 - 12x^3 - 12x^2 + 27x + 27 = 0$; 1 and 2

88. $x^6 - 17x^4 + 82x^2 - 120 = 0$; 3 and 4

89. $x^5 - 2x^4 - 12x^3 - 32x^2 - 64x = 0$; -1 and -2

90. $x^4 - 4x^3 - 6x^2 + 28x - 7 = 0$; 2 and 3

91. $x^4 - 4x^3 - 6x^2 + 28x - 7 = 0$; 3 and 4

92. $x^4 - 4x^3 - 6x^2 + 28x - 7 = 0$; -2 and -3

93. $x^4 - 4x^3 - 6x^2 + 28x - 7 = 0$; 0 and 1

94. $2x^5 - 17x^4 + 18x^3 + 52x^2 + 16x + 69 = 0$; 3 and 4

95. $2x^5 - 17x^4 + 18x^3 + 52x^2 + 16x + 69 = 0$; 6 and 7

In each of the following, first find all the rational roots, then either find the irrational roots or locate the irrational roots between successive integers. Use linear interpolation to find a one-decimal-place approximation to any irrational root that lies between the given numbers.

EXAMPLE

(a) $x^5 + x^3 - 5x^2 - 5 = 0$; 1 and 2

Solution

Possible rational roots are ± 1, ± 5. None of these are roots. There are no rational roots.

   To locate irrational roots, let $P(x) = x^5 + x^3 - 5x^2 - 5$ and compute $P(-1)$, $P(0)$, $P(1)$, $P(2)$, etc.

x	-1	0	1	2
$P(x)$	-12	-5	-8	15

Note: $P(x)$ is positive for all $x > 2$.

   $P(x)$ is negative for all $x < 1$.

   $P(1)$ and $P(2)$ are of opposite signs.

   There is at least one real root between 1 and 2. There are no real roots other than between 1 and 2.

Using linear interpolation to approximate a real root between 1 and 2 we have:

$$y + 8 = 23(x - 1), \text{ x-intercept is } \tfrac{31}{23} \simeq 1.348.$$

Real root between 1 and 2 is about 1.3.

(b) $x^5 - x^4 - 6x^3 + 6x^2 - 7x + 7 = 0$; 3 and 4

Possible rational roots are ± 1, ± 7. Of these, only 1 is a root.

EXAMPLE

Solution

$$
\begin{array}{r|rrrrrr}
1) & 1 & -1 & -6 & 6 & -7 & 7 \\
 & & 1 & 0 & -6 & 0 & -7 \\
\hline
 & 1 & 0 & -6 & 0 & -7 & 0 \\
\end{array}
$$

Depressed equation is $x^4 - 6x^2 - 7 = 0$.

This may be treated as a quadratic in x^2 and may be solved for x^2 by the quadratic formula. Let $y = x^2$.

$$y^2 - 6y - 7 = 0$$

$$y = \frac{6 \pm \sqrt{36 + 28}}{2}$$

$y = 7, -1$; $x^2 = 7$ or $x^2 = -1$; $x = \pm \sqrt{7}$ or $x = \pm i$.

The roots of the given equation are 1, $\pm \sqrt{7}$, $\pm i$. There are no roots between the given numbers.

96. $x^5 + x^3 - 7x^2 - 7 = 0$; 1 and 2

97. $x^3 + 3x^2 - 5x - 15 = 0$; 7 and 8

98. $2x^3 - 3x^2 - 10x + 15 = 0$; -3 and -2

99. $x^6 + 4x^4 - 15x^2 - 60 = 0$; 1 and 2

100. $x^5 - 3x^3 - 20x^2 + 60 = 0$; -2 and -1

CHECK LIST FOR THE STUDENT

If you have learned the material in Section 3.1, you should be able to do the following:

1. Give an example of a polynomial equation of a specified degree.
2. Tell the degree of a given polynomial equation.
3. State and use the remainder theorem.
4. State and use the factor theorem.
5. State and use the converse of the factor theorem.
6. Tell what is meant by a root of multiplicity k where k is a positive integer.
7. Make up an equation with given roots having given multiplicities.
8. State and use the number-of-roots theorem.
9. State and use the conjugate root theorem.
10. Define an irreducible quadratic.
11. State and use the factorization theorem.
12. State the rational root theorem.
13. Tell what is meant by the *depressed equation*, if $P(x)$ is divided by $(x - r)$.
14. Find the rational roots of a polynomial equation with integral coefficients.
15. State and use the location-of-roots theorem.
16. Give a geometric interpretation of the location-of-roots theorem.
17. Use linear interpolation to approximate irrational roots of a polynomial equation with real coefficients.

SECTION QUIZ 3.1

1. What is the degree of the equation $3x^5 - 2x^6 + x^4 - 15x^3 - 8x + 9 = 0$?

2. Give the formula for the solutions of the equation $ax^2 + bx + c = 0$ where a, b, and c are real numbers and $a \neq 0$.

3. Use the remainder theorem to determine whether or not 2 is a root of the equation $4x^3 - 3x^2 + 5x - 30 = 0$.

4. Use the factor theorem to construct an equation with the lowest possible degree having the roots 3, -4, 0, and 1.

5. What are the roots and what is the multiplicity of each root of the equation $x^2(x - 3)^4(x + 1) = 0$?

6. Make up an equation having the roots 0, -4, and 6 with multiplicity 3, 2, and 1, respectively, and having no other roots.

7. What is the greatest number of distinct complex-number roots that the equation $3x^4 - 2x^3 + 5x^2 - 7x + 8 = 0$ may have?

8. If each root is counted as many times as its multiplicity, how many roots does the equation $2x^5 - 3x^4 + 5x^3 - 7 = 0$ have?

9. Show that $2 - i$ and its conjugate are both roots of $x^2 - 4x + 5 = 0$.

10. If $3 - 4i$ is a root of a polynomial equation $P(x) = 0$, what is an irreducible quadratic factor of $P(x)$?

11. Verify that 1 is a root of the equation $x^3 - 5x^2 + 11x - 7 = 0$, and find the other roots.

12. List all the possible rational roots of the equation $7x^5 - 2x^3 + 6x - 15 = 0$.

13. Find the rational roots of the equation $2x^3 - 3x^2 - 11x + 6 = 0$.

14. Find all the roots of the equation $x^3 - 2x^2 + 9x - 18 = 0$.

15. Use the location-of-roots theorem to show that there is a root of $5x^3 + 2x^2 - 16x + 4 = 0$ between 0 and 1, and use linear interpolation to find a first approximation to it.

3.2 Matrices and Simultaneous Linear Equations

The purpose of this section is to introduce matrices and to show how matrices can be used to solve simultaneous linear equations.

A **matrix** is a rectangular array of symbols, usually symbols for numbers. It is conventional to enclose the array in large parentheses or brackets. For example, the coefficients in the system of equations

$$3x - 2y + 6z = -6$$
$$-3x + 10y + 11z = 13$$
$$x - 2y - z = -3$$

form the matrix

$$\begin{pmatrix} 3 & -2 & 6 \\ -3 & 10 & 11 \\ 1 & -2 & -1 \end{pmatrix}.$$

This matrix is called the *coefficient matrix of the system*. The matrix obtained by including the column of constant terms is called the *augmented matrix* or simply *the* matrix of the system. The matrix

$$\begin{pmatrix} 3 & -2 & 6 & -6 \\ -3 & 10 & 11 & 13 \\ 1 & -2 & -1 & -3 \end{pmatrix}$$

is the augmented matrix of the given system.

The numbers making up a matrix are called the **elements** of the matrix. The elements are said to be arranged in *rows* and *columns.* In the matrix

$$\begin{pmatrix} 3 & -2 & 6 & -6 \\ -3 & 10 & 11 & 13 \\ 1 & -2 & -1 & -3 \end{pmatrix}$$

there are three rows and four columns. The first row is $3 \quad -2 \quad 6 \quad -6$, the second row is $-3 \quad 10 \quad 11 \quad 13$, etc. The first column is

$$3$$
$$-3$$
$$1,$$

etc. A matrix with three rows and four columns is known as a three-by-four matrix. A matrix with *m rows* and *n columns* is an *m-by-n* matrix. The expression *m* by *n* is sometimes written $m \times n$. If $m = n$ so that the matrix has the same number of rows as columns, it is said to be a *square* matrix.

The rows and columns of a matrix are numbered from top to bottom and from left to right. The element 8 in the matrix

$$\begin{pmatrix} 2 & 1 & 3 & -5 \\ 6 & 4 & 8 & 2 \\ 5 & -3 & 9 & 7 \end{pmatrix}$$

is in the second row and third column; the element 7 is in the third row and fourth column. The elements of a square matrix located in the first row and first column, the second row and second column, the third row and third column, etc. are said to be in the *main diagonal.* The elements 2 4 9 are in the main diagonal of the matrix

$$\begin{pmatrix} 2 & 1 & 3 \\ 6 & 4 & 8 \\ 5 & -3 & 9 \end{pmatrix}.$$

For convenience, we sometimes refer to the first row of a matrix as R_1, the second row as R_2, the third row as R_3, etc. Likewise, the symbols C_1, C_2, C_3, etc. are used for Column 1, Column 2, Column 3, etc.

Matrices may be used to solve systems of n linear equations in n unknowns. One way to solve such a system is to work with the augmented matrix so as to replace it with the matrix of an equivalent system whose solution is obvious. (Equivalent systems of equations have the same solution.) We work with the rows of the matrix just as we would work with the equations of the system. The following *row operations* replace an augmented matrix with the matrix of an equivalent system.

Row Operations Producing the Matrix of an Equivalent System

1. Interchanging any two rows

2. Multiplying each element of any row by a nonzero real number k

3. Adding to each element of a row the product of a real number k and the corresponding element of some other row

For simplicity, we describe the second of these operations as multiplying the row by k, and we describe the third as replacing the row by the sum of it and k times some other row. We say that these three row operations produce *row-equivalent* matrices.

If we use row operations to replace the augmented matrix of a system by the matrix of an equivalent system whose coefficient matrix has all ones in the main diagonal and zeros elsewhere, we will be able to read off the solution of the system. For example, suppose we used row operations to replace the augmented matrix of a given system of three equations in three unknowns by the matrix

$$\begin{pmatrix} 1 & 0 & 0 & 5 \\ 0 & 1 & 0 & -7 \\ 0 & 0 & 1 & 4 \end{pmatrix}.$$

The given system would then be equivalent to the system

$$1 \cdot x + 0 \cdot y + 0 \cdot z = 5$$
$$0 \cdot x + 1 \cdot y + 0 \cdot z = -7$$
$$0 \cdot x + 0 \cdot y + 1 \cdot z = 4.$$

In other words, the given system would be equivalent to the system

$$x \qquad\qquad = 5$$
$$\qquad y \qquad = -7$$
$$\qquad\qquad z = 4,$$

whose solution is obviously $x = 5$, $y = -7$, $z = 4$. This method is demonstrated in the next example. It is called the **Gaussian elimination method.** In some cases this method takes longer to do by hand than the elimination method presented in Section 1.10, but it is more easily programmed for a high-speed computer.

EXAMPLE 1   Use the Gaussian elimination method to solve the following system of equations.

$$3x - 2y + 6z = -6$$
$$-3x + 10y + 11z = 13$$
$$x - 2y - z = -3$$

(This system of equations was solved in Example 1 of Section 1.10 by using the elimination method presented in that section.)

Solution   The augmented matrix is

$$\begin{pmatrix} 3 & -2 & 6 & -6 \\ -3 & 10 & 11 & 13 \\ 1 & -2 & -1 & -3 \end{pmatrix}.$$

We shall use row operations to produce row-equivalent matrices until we produce the matrix of a system whose coefficient matrix has all ones in the main diagonal and zeros elsewhere. The row operations used will be indicated to the right. For convenience, we shall use the symbols R_1, R_2, and R_3 for Row 1, Row 2, and Row 3, respectively.

$$\begin{pmatrix} 3 & -2 & 6 & -6 \\ -3 & 10 & 11 & 13 \\ 1 & -2 & -1 & -3 \end{pmatrix} \rightarrow \begin{pmatrix} 1 & -2 & -1 & -3 \\ -3 & 10 & 11 & 13 \\ 3 & -2 & 6 & -6 \end{pmatrix}$$

Interchanging R_1 and R_3

$$\rightarrow \begin{pmatrix} 1 & -2 & -1 & -3 \\ 0 & 4 & 8 & 4 \\ 0 & 4 & 9 & 3 \end{pmatrix}$$

$3R_1 + R_2$
$-3R_1 + R_3$

$$\rightarrow \begin{pmatrix} 1 & -2 & -1 & -3 \\ 0 & 1 & 2 & 1 \\ 0 & 4 & 9 & 3 \end{pmatrix}$$

$\left(\dfrac{1}{4}\right) R_2$

$$\rightarrow \begin{pmatrix} 1 & 0 & 3 & -1 \\ 0 & 1 & 2 & 1 \\ 0 & 0 & 1 & -1 \end{pmatrix}$$

$2R_2 + R_1$

$-4R_2 + R_3$

$$\rightarrow \begin{pmatrix} 1 & 0 & 0 & 2 \\ 0 & 1 & 0 & 3 \\ 0 & 0 & 1 & -1 \end{pmatrix}$$

$-3R_3 + R_1$
$-2R_3 + R_2$

The given system is equivalent to the system

$$
\begin{aligned}
x & & & = 2 \\
& y & & = 3 \\
& & z & = -1,
\end{aligned}
$$

whose solution is obviously $x = 2$, $y = 3$, $z = -1$.

Notice that in performing the various row operations to find the solution, we first got a 1 in the first row and first column, and then we successively used row operations to get zeros in the rest of the first column. Next we got a 1 in the second row and second column, and then we successively used row operations to get zeros in the rest of the second column. Finally we got a 1 in the third row and third column and used row operations to get zeros in the rest of the third column.

The system of equations in Example 1 has exactly one solution, namely $x = 2$, $y = 3$, $z = -1$. As we said in Section 1.10, a system of n simultaneous linear equations in n unknowns has either

1. Exactly one solution,

2. No solution, or

3. An unlimited number of solutions.

If we attempt to use the Gaussian elimination method on a system of n linear equations in n unknowns having no solution or having an unlimited number of solutions, we will not be able to produce the matrix of

Exercise 3.2, 1–20

an equivalent system having all ones in the main diagonal and zeros elsewhere in its coefficient matrix. Instead, we can produce the matrix of an equivalent system having a row of all zeros in its coefficient matrix. If we attempt to use the Gaussian elimination method and manage to produce such a row-equivalent matrix, we will know that there is not exactly one solution.

EXAMPLE 2   Try to solve the following system with the Gaussian elimination method.

$$x - 5y = 4$$
$$2x - 10y = 8$$

Solution   The augmented matrix is

$$\begin{pmatrix} 1 & -5 & 4 \\ 2 & -10 & 8 \end{pmatrix}.$$

Multiplying Row 1 by -2 and adding the product to Row 2 produces

$$\begin{pmatrix} 1 & -5 & 4 \\ 0 & 0 & 0 \end{pmatrix}.$$

The row of all zeros in the coefficient matrix indicates that there is not exactly one solution. In fact, there is an unlimited number of solutions. The two given equations are equivalent. (The second may be obtained from the first by multiplying through by 2.) An unlimited number of solutions may be found by arbitrarily assigning values to one of the variables and then solving for the other.

EXAMPLE 3   Try to solve the following system with the Gaussian elimination method.

$$x - 5y = 4$$
$$2x - 10y = 7$$

Solution   The augmented matrix is

$$\begin{pmatrix} 1 & -5 & 4 \\ 2 & -10 & 7 \end{pmatrix}.$$

Multiplying Row 1 by -2 and adding the product to Row 2 produces

$$\begin{pmatrix} 1 & -5 & 4 \\ 0 & 0 & -1 \end{pmatrix}.$$

The row of all zeros in the coefficient matrix indicates that there is not exactly one solution. In this example there is no solution. The given equations are *inconsistent*. The first one tells us that

$$x - 5y = 4$$

or that $2x - 10y$ is 8,

whereas according to the second one,

$$2x - 10y \text{ is } 7.$$

Exercise 3.2, 21–30
Verbal Problems
Exercise 3.2, 31–35

EXERCISE 3.2 MATRICES AND SIMULTANEOUS LINEAR EQUATIONS

▶ Use the Gaussian elimination method to solve each of the following systems of equations.

EXAMPLE

$$x - 5y = 4$$
$$3x - 10y = 2$$

Solution

$$\begin{pmatrix} 1 & -5 & 4 \\ 3 & -10 & 2 \end{pmatrix} \rightarrow \begin{pmatrix} 1 & -5 & 4 \\ 0 & 5 & -10 \end{pmatrix} \qquad -3R_1 + R_2$$

$$\rightarrow \begin{pmatrix} 1 & -5 & 4 \\ 0 & 1 & -2 \end{pmatrix} \qquad \left(\frac{1}{5}\right) R_2$$

$$\rightarrow \begin{pmatrix} 1 & 0 & -6 \\ 0 & 1 & -2 \end{pmatrix} \qquad 5R_2 + R_1$$

$$x = -6$$
$$y = -2$$

1. $2x - 3y = 9$
   $x + 7y = -4$

2. $5x + 2y = 13$
   $3x - 6y = -21$

3. $4x - 3y = -17$
   $x + y = 1$

4. $6x - 5y = 10$
   $8x - 7y = 14$

5. $5x - 4y = 11$
   $3x + 6y = -27$

6. $x - y = -7$
   $3x + 5y = 11$

7. $11x - 9y = 20$
   $6x + 5y = 1$

8. $8x - 6y = 8$
   $3x + 7y = 3$

9. $-x + 5y + 5 = 0$
   $2x + 3y + 16 = 0$

10. $4x - 8y - 4 = 0$
   $4x - 6y - 8 = 0$

11. $x - 6y - 4z = -36$
   $x - 3y - z = -30$
   $-x + 9y + 8z = 41$

12. $x + 2y + 3z = -1$
   $x + 4y + 5z = -3$
   $x + 4y + 6z = 0$

13. $4x + 5y + 23z = -83$
   $x + y + 5z = -18$
   $-3x - 2y - 11z = 40$

14. $5x + 12y - 2z = 6$
   $6x + 18y - 5z = -11$
   $x + 2y = 4$

15.
$$A + C = 4$$
$$3A + B + 2C + D = 6$$
$$3A + 3B + C + D + E = 3$$
$$A + 3B = -5$$
$$B = -2$$

(This system was solved in Example 2 of Section 1.10 by another method.)

16.
$$-A + 4B + 9C = 4$$
$$A - B - 2C = 6$$
$$-2A + 4B + 11C = 1$$

17.
$$-x - 2y - 5z = 3$$
$$x + 3y + 10z = -8$$
$$-2x - 4y - 5z = 6$$

18.
$$x - 2y - 8z = -32$$
$$x - y - z = -5$$
$$2x - 3y - 8z = -33$$

19.
$$A - 3B + C = -3$$
$$A + C = 6$$
$$3A + B = 9$$

20.
$$A + D = -1$$
$$3A + B - 2C + D = -4$$
$$3B + C = -1$$
$$2A - B + D = 1$$

▶ If possible, solve each of the following systems with the Gaussian elimination method. If there is not exactly one solution, say so.

21.
$$2x + 4y = 8$$
$$6x + 12y = 24$$

22.
$$2x + 4y = 8$$
$$6x + 12y = 20$$

23.
$$2x + 4y = 8$$
$$6x + 10y = 22$$

24.
$$3x - 5y = 1$$
$$6x - 10y = 2$$

25.
$$3x - 5y = 1$$
$$6x - 10y = 5$$

26.
$$2x + 3y + z = 1$$
$$-3x + 5y + 3z = -2$$
$$4x + 6y + 2z = 2$$

27.
$$2x + 3y + z = 1$$
$$-3x + 5y + 3z = -2$$
$$4x + 6y + 2z = 4$$

28.
$$2x + 3y + z = 1$$
$$-3x + 5y + 3z = -2$$
$$4x + 6y + 4z = 6$$

29.
$$-A - D = 1$$
$$B + 2C = -1$$
$$-2A - 2D = 3$$
$$2C + 3D = 0$$

30.
$$A + D = 2$$
$$3B + C = -1$$
$$2A + 2D = 4$$
$$4C + 3D = 0$$

31. Diane's Holstein gave 1455 kilograms more milk one year than her Jersey. Together these two cows gave 12,431 kilograms of milk. How many kilograms did each of them give?

32. It took a plane $1\frac{1}{4}$ hours to fly 800 kilometers with the wind and 2 hours to fly 800 kilometers against the same wind. Assume that the plane flew at a constant air speed and that the wind blew at a constant speed. Find the speed of the wind and the air speed of the plane.

33. When using a certain strobe light with an automatic camera, the recommended f-stop for film speed 200 is twice the recommended f-stop for film speed 50, and these two recommended f-stops differ by 2.8. What are these two recommended f-stops?

34. Jeannette deposited $10,000 in a bank by putting part of it in an account paying 5% per year and part of it in an account paying 7% per year. She made no withdrawals and no other deposits. At the end of one year the two accounts had earned a total of $630 in interest. How much did Jeannette deposit in each of the accounts?

35. At a chicken farm you can buy $2\frac{1}{2}$ dozen pullet eggs and 1 dozen large eggs for $2.15. You can also buy 1 dozen pullet eggs and $2\frac{1}{2}$ dozen large eggs for $2.75. How much are the eggs per dozen?

CHECK LIST FOR THE STUDENT

If you have learned the material in Section 3.2, you should be able to do the following:

1. Tell how many rows and how many columns there are in a given matrix.

2. Tell which element is in a specified row and a specified column of a given matrix.

3. List the elements in the main diagonal of a given square matrix.

4. Describe three row operations that produce the matrix of an equivalent system.

5. Tell what is meant by row-equivalent matrices.

6. Use the Gaussian elimination method to solve a system of n linear equations in n unknowns.

7. Tell what is indicated about the solutions when row operations on the matrix of a system produce the matrix of an equivalent system having a row of all zeros in its coefficient matrix.

SECTION QUIZ 3.2

1. How many rows and how many columns are there in the following matrix?

$$\begin{pmatrix} 2 & 7 & 8 & -3 \\ 1 & 0 & 4 & 5 \\ 0 & 1 & -2 & 6 \end{pmatrix}$$

2. What is the element in the second row and third column of the following matrix?

$$\begin{pmatrix} 3 & -4 & 5 \\ 2 & 6 & 9 \\ 1 & 0 & 8 \end{pmatrix}$$

3. What are the elements in the main diagonal of the following matrix?

$$\begin{pmatrix} 2 & -1 & 6 \\ 4 & 7 & 3 \\ 0 & 1 & 5 \end{pmatrix}$$

4. Replace the following matrix with a row-equivalent matrix by interchanging the second and third rows.

$$\begin{pmatrix} 1 & 0 & 6 \\ 0 & 0 & 5 \\ 0 & 1 & 4 \end{pmatrix}$$

5. Replace the following matrix with a row-equivalent matrix by multiplying the second row by -2.

$$\begin{pmatrix} 4 & -3 & 7 \\ 2 & -2 & 3 \\ 5 & 9 & -8 \end{pmatrix}$$

6. Replace the following matrix with a row-equivalent matrix by replacing the third row by the sum of it and two times the first row.

$$\begin{pmatrix} 1 & -1 & 2 \\ 0 & 7 & 9 \\ -2 & 3 & 4 \end{pmatrix}$$

Use the Gaussian elimination method to solve each of the following systems of equations.

7. $3x - 2y = 12$

    $x + 4y = -10$

8. $2x - 3y + 4z = -4$

    $x + y - 2z = 5$

    $4x - y + z = 3$

If possible, solve each of the following systems with the Gaussian elimination method. If there is not exactly one solution, say so.

9. $3x - 4y = 2$

    $6x - 8y = 1$

10. $x - 2y + z = 4$

    $2x - 4y + 2z = 8$

    $x - 3y + 5z = 7$

3.3 Determinants and Simultaneous Linear Equations

The purpose of this section is to introduce determinants, to show how to find the value of a determinant, and to show how determinants can be used to solve systems of simultaneous linear equations.

Associated with each square matrix whose elements are numbers there is a number called the *determinant* of the matrix. The determinant of a matrix is designated by using vertical bars to enclose the matrix rather than large parentheses or brackets. For example, the determinant of the matrix

$$\begin{pmatrix} 3 & -1 & 2 \\ 4 & 5 & 7 \\ 1 & 6 & -8 \end{pmatrix}$$

is written

$$\begin{vmatrix} 3 & -1 & 2 \\ 4 & 5 & 7 \\ 1 & 6 & -8 \end{vmatrix}.$$

If a square matrix is represented by a capital letter, some authors represent its determinant by writing "det" in front of the letter; det A represents the determinant of matrix A. The determinant of an $n \times n$ matrix is said to be of **order** n. A determinant is really a number; sometimes that number is referred to as the *value* of the collection of symbols representing the determinant.

The value of any 2×2 determinant

$$\begin{vmatrix} a & b \\ c & d \end{vmatrix}$$

where a, b, c, and d are numbers is defined as follows.

$$\begin{vmatrix} a & b \\ c & d \end{vmatrix} = ad - bc$$

For example,

$$\begin{vmatrix} 3 & -2 \\ 4 & 5 \end{vmatrix} = (3)(5) - (-2)(4)$$
$$= 15 + 8$$
$$= 23.$$

The value of any $n \times n$ determinant for integral n greater than 2 can be defined concisely after first introducing a subscript notation for the elements in any determinant. According to this notation, the symbol a_{ij}

represents the element in the *ith row* and *jth column* of a matrix. For example, a_{23} represents the element in the second row and third column. Using this notation, any 2 × 2 determinant is written

$$\begin{vmatrix} a_{11} & a_{12} \\ a_{21} & a_{22} \end{vmatrix},$$

and any 3 × 3 determinant is written

$$\begin{vmatrix} a_{11} & a_{12} & a_{13} \\ a_{21} & a_{22} & a_{23} \\ a_{31} & a_{32} & a_{33} \end{vmatrix}.$$

Using this notation, the value of any 2 × 2 determinant is

$$\begin{vmatrix} a_{11} & a_{12} \\ a_{21} & a_{22} \end{vmatrix} = a_{11}a_{22} - a_{12}a_{21}.$$

Any element a_{ij} is said to be in *plus* position if the sum $i + j$ is even, and a_{ij} is said to be in *minus* position if the sum $i + j$ is odd. We see that the positions alternate as shown in the following charts for 2 × 2, 3 × 3, and 4 × 4 determinants.

$$\begin{vmatrix} + & - \\ - & + \end{vmatrix} \quad \begin{vmatrix} + & - & + \\ - & + & - \\ + & - & + \end{vmatrix} \quad \begin{vmatrix} + & - & + & - \\ - & + & - & + \\ + & - & + & - \\ - & + & - & + \end{vmatrix}$$

The **minor** of any element a_{ij} in a given determinant is the determinant that is left after the *i*th row and *j*th column are deleted. For example, to find the minor of a_{12} in a 3 × 3 determinant, simply delete the first row and second column as follows:

$$\begin{vmatrix} a_{11} & a_{12} & a_{13} \\ a_{21} & a_{22} & a_{23} \\ a_{31} & a_{32} & a_{33} \end{vmatrix}$$

The minor of a_{12} in a 3 × 3 determinant is

Exercise 3.3, 1–15

$$\begin{vmatrix} a_{21} & a_{23} \\ a_{31} & a_{33} \end{vmatrix}.$$

The value of any $n \times n$ determinant can be defined in terms of minors. To find the value of any $n \times n$ determinant, we can do the following:

1. Choose any row or column.

2. Multiply each element in the chosen row or column by +1 (if it is in plus position) or by −1 (if it is in minus position), and then multiply it by its minor; add all these products.

This method of finding the value of a determinant is called *expansion by minors* of the elements in the chosen row or column.

EXAMPLE 1

Write down an expression for the value of the following determinant using expansion by minors of the elements in the first row.

$$\begin{vmatrix} a_{11} & a_{12} & a_{13} \\ a_{21} & a_{22} & a_{23} \\ a_{31} & a_{32} & a_{33} \end{vmatrix}$$

Solution

$$\begin{vmatrix} a_{11} & a_{12} & a_{13} \\ a_{21} & a_{22} & a_{23} \\ a_{31} & a_{32} & a_{33} \end{vmatrix} = a_{11} \begin{vmatrix} a_{22} & a_{23} \\ a_{32} & a_{33} \end{vmatrix} - a_{12} \begin{vmatrix} a_{21} & a_{23} \\ a_{31} & a_{33} \end{vmatrix} + a_{13} \begin{vmatrix} a_{21} & a_{22} \\ a_{31} & a_{32} \end{vmatrix}$$

$$= a_{11}(a_{22}a_{33} - a_{23}a_{32}) - a_{12}(a_{21}a_{33} - a_{23}a_{31})$$
$$+ a_{13}(a_{21}a_{32} - a_{22}a_{31})$$

EXAMPLE 2

Find the value of the following determinant using expansion by minors of the elements in the first column.

$$\begin{vmatrix} 3 & 4 & 5 \\ -2 & 1 & -3 \\ -7 & 2 & 6 \end{vmatrix}$$

Solution

$$\begin{vmatrix} 3 & 4 & 5 \\ -2 & 1 & -3 \\ -7 & 2 & 6 \end{vmatrix} = 3 \begin{vmatrix} 1 & -3 \\ 2 & 6 \end{vmatrix} - (-2) \begin{vmatrix} 4 & 5 \\ 2 & 6 \end{vmatrix} + (-7) \begin{vmatrix} 4 & 5 \\ 1 & -3 \end{vmatrix}$$

$$= 3(6 + 6) \quad + 2(24 - 10) \quad - 7(-12 - 5)$$
$$= 3(12) \quad\quad + 2(14) \quad\quad - 7(-17)$$
$$= 36 \quad\quad\quad + 28 \quad\quad\quad + 119$$
$$= 183$$

EXAMPLE 3

Find the value of the following determinant.

$$\begin{vmatrix} 1 & 2 & -1 & 0 \\ 4 & 0 & 5 & 1 \\ 0 & -3 & 1 & 2 \\ 3 & 0 & 6 & 7 \end{vmatrix}$$

Solution

Since we are free to choose any row or column for expanding by minors, here we should choose the second column. It has the most zeros in it, so the computations will be easiest with this choice. Expanding by minors of the elements in the second column, we have

$$\begin{vmatrix} 1 & 2 & -1 & 0 \\ 4 & 0 & 5 & 1 \\ 0 & -3 & 1 & 2 \\ 3 & 0 & 6 & 7 \end{vmatrix} = -(2)\begin{vmatrix} 4 & 5 & 1 \\ 0 & 1 & 2 \\ 3 & 6 & 7 \end{vmatrix} + 0 - (-3)\begin{vmatrix} 1 & -1 & 0 \\ 4 & 5 & 1 \\ 3 & 6 & 7 \end{vmatrix} + 0$$

This reduces our problem to that of computing the values of two 3×3 determinants. Expanding the first of these by minors of the elements in the first column and the second by minors of the elements in the third column, we have

$$\begin{vmatrix} 1 & 2 & -1 & 0 \\ 4 & 0 & 5 & 1 \\ 0 & -3 & 1 & 2 \\ 3 & 0 & 6 & 7 \end{vmatrix} = (-2)\left[4\begin{vmatrix} 1 & 2 \\ 6 & 7 \end{vmatrix} + 3\begin{vmatrix} 5 & 1 \\ 1 & 2 \end{vmatrix} \right] + 3\left[-(1)\begin{vmatrix} 1 & -1 \\ 3 & 6 \end{vmatrix} + 7\begin{vmatrix} 1 & -1 \\ 4 & 5 \end{vmatrix} \right]$$

$$= (-2)[4(7 - 12) + 3(10 - 1)] + 3[-(6 + 3) + 7(5 + 4)]$$
$$= (-2)[4(-5) + 3(9)] + 3[-9 + 7(9)]$$
$$= (-2)[-20 + 27] + 3[-9 + 63]$$
$$= (-2)[7] + 3[54]$$
$$= -14 + 162$$

Exercise 3.3, 16–30

$$= 148.$$

The following properties of determinants will not be proved here, but they are listed because they can sometimes be used to simplify computations.

PROPERTIES OF DETERMINANTS

1. If two rows or two columns are identical, the value of the determinant is zero.

Example:
$$\begin{vmatrix} 1 & 3 & 7 \\ -2 & 4 & 9 \\ 1 & 3 & 7 \end{vmatrix} = 0 \qquad \begin{vmatrix} 3 & 3 & 2 \\ 8 & 8 & 7 \\ 4 & 4 & 5 \end{vmatrix} = 0$$

2. If every element in some row or column is zero, the value of the determinant is zero.

Example:
$$\begin{vmatrix} 1 & 3 & 7 \\ 0 & 0 & 0 \\ 5 & 8 & -4 \end{vmatrix} = 0 \qquad \begin{vmatrix} 3 & 2 & 0 \\ 8 & 6 & 0 \\ 4 & -5 & 0 \end{vmatrix} = 0$$

3. If each element of any row or column is multiplied by a nonzero real number k, the resulting determinant is k times the original one.

Example:

$$\begin{vmatrix} 6 & 1 & 5 \\ -3 & 7 & 8 \\ 9 & 4 & -2 \end{vmatrix} = 3 \begin{vmatrix} 2 & 1 & 5 \\ -1 & 7 & 8 \\ 3 & 4 & -2 \end{vmatrix}$$

$$\begin{vmatrix} 6 & 3 & 15 \\ -1 & 7 & 8 \\ 3 & 4 & -2 \end{vmatrix} = 3 \begin{vmatrix} 2 & 1 & 5 \\ -1 & 7 & 8 \\ 3 & 4 & -2 \end{vmatrix}$$

4. If all the corresponding rows and columns are interchanged, the resulting determinant is equal to the original one.

Example:

$$\begin{vmatrix} 2 & 1 & 5 \\ -1 & 7 & 8 \\ 3 & 4 & -2 \end{vmatrix} = \begin{vmatrix} 2 & -1 & 3 \\ 1 & 7 & 4 \\ 5 & 8 & -2 \end{vmatrix}$$

5. If exactly two rows or exactly two columns are interchanged, the resulting determinant is the negative of the original one.

Example:

$$\begin{vmatrix} 2 & 1 & 5 \\ -1 & 7 & 8 \\ 3 & 4 & -2 \end{vmatrix} = - \begin{vmatrix} 3 & 4 & -2 \\ -1 & 7 & 8 \\ 2 & 1 & 5 \end{vmatrix}$$

$$\begin{vmatrix} 2 & 1 & 5 \\ -1 & 7 & 8 \\ 3 & 4 & -2 \end{vmatrix} = - \begin{vmatrix} 2 & 5 & 1 \\ -1 & 8 & 7 \\ 3 & -2 & 4 \end{vmatrix}$$

6. If the product of a nonzero real number k and each element of some row or column is added to the corresponding element in some other row or column, the resulting determinant is equal to the original one.

Example:

$$\begin{vmatrix} 2 & 1 & 5 \\ -1 & 7 & 8 \\ 3 & 4 & -2 \end{vmatrix} = \begin{vmatrix} 2 & 1 & 5 \\ -1 & 7 & 8 \\ -5 & 0 & -22 \end{vmatrix}$$

$(-4) R_1 + R_3$ (The product of (-4) and each element of R_1 was added to the corresponding element of R_3.)

$$\begin{vmatrix} 2 & 1 & 5 \\ -1 & 7 & 8 \\ 3 & 4 & -2 \end{vmatrix} = \begin{vmatrix} 2 & 15 & 5 \\ -1 & 0 & 8 \\ 3 & 25 & -2 \end{vmatrix}$$

$7C_1 + C_2$ (The product of 7 and each element of C_1 was added to the corresponding element of C_2.)

EXAMPLE 4

Find the value of the following determinant. Use properties of determinants to simplify computation by first reducing the problem to that of finding the value of a 2 × 2 determinant.

$$\begin{vmatrix} 3 & 1 & 7 \\ 2 & -4 & 5 \\ -6 & 9 & 8 \end{vmatrix}$$

Solution

There are an unlimited number of ways to use properties of determinants in finding the value of the given determinant. Two ways are demonstrated here. The various operations on rows and columns are indicated to the right. Here is one way to find the value.

$$\begin{vmatrix} 3 & 1 & 7 \\ 2 & -4 & 5 \\ -6 & 9 & 8 \end{vmatrix} = \begin{vmatrix} 1 & 5 & 2 \\ 2 & -4 & 5 \\ -6 & 9 & 8 \end{vmatrix} \qquad (-1)R_2 + R_1$$

$$= \begin{vmatrix} 1 & 5 & 2 \\ 0 & -14 & 1 \\ 0 & 39 & 20 \end{vmatrix} \qquad \begin{array}{l} (-2)R_1 + R_2 \\ 6R_1 + R_3 \end{array}$$

$$= \begin{vmatrix} -14 & 1 \\ 39 & 20 \end{vmatrix} \qquad \begin{array}{l} \text{Expanding by} \\ \text{minors of } C_1 \end{array}$$

$$= (-14)(20) - (39)$$

$$= -280 - 39$$

$$= -319$$

Here is another way to find the value required.

$$\begin{vmatrix} 3 & 1 & 7 \\ 2 & -4 & 5 \\ -6 & 9 & 8 \end{vmatrix} = - \begin{vmatrix} 1 & 3 & 7 \\ -4 & 2 & 5 \\ 9 & -6 & 8 \end{vmatrix} \qquad \text{Interchanging } C_1 \text{ and } C_2$$

$$= - \begin{vmatrix} 1 & 0 & 0 \\ -4 & 14 & 33 \\ 9 & -33 & -55 \end{vmatrix} \qquad \begin{array}{l} (-3)C_1 + C_2 \\ (-7)C_1 + C_3 \end{array}$$

$$= - \begin{vmatrix} 14 & 33 \\ -33 & -55 \end{vmatrix} \qquad \begin{array}{l} \text{Expanding by} \\ \text{minors of } R_1 \end{array}$$

$$= -[(14)(-55) - (33)(-33)]$$

$$= -[-770 + 1089]$$

$$= -319$$

Exercise 3.3, 31–35

Determinants may be used to solve systems of n simultaneous linear equations in n unknowns. *Cramer's rule* tells how to do this when the determinant of the coefficient matrix is not zero.

Cramer's Rule

If the determinant of the coefficient matrix of a system of n simultaneous linear equations in n unknowns is not zero, then the system has exactly one solution, and the value of each unknown is equal to a quotient of two determinants. The determinant in the denominator is the determinant of the coefficient matrix. The determinant in the numerator is derived from the determinant of the coefficient matrix by replacing the column of coefficients of the unknown sought by the column of constant terms as they appear isolated on the right sides of the equals signs in the equations of the system.

Let us express Cramer's rule symbolically for the case of *two* linear equations in *two* unknowns.

Cramer's Rule for 2 Equations in 2 Unknowns

If $\begin{vmatrix} a_{11} & a_{12} \\ a_{21} & a_{22} \end{vmatrix} \neq 0$, the system of equations

$$a_{11}x + a_{12}y = k_1$$
$$a_{21}x + a_{22}y = k_2$$

has exactly one solution. That solution is

$$x = \frac{\begin{vmatrix} k_1 & a_{12} \\ k_2 & a_{22} \end{vmatrix}}{\begin{vmatrix} a_{11} & a_{12} \\ a_{21} & a_{22} \end{vmatrix}}, \qquad y = \frac{\begin{vmatrix} a_{11} & k_1 \\ a_{21} & k_2 \end{vmatrix}}{\begin{vmatrix} a_{11} & a_{12} \\ a_{21} & a_{22} \end{vmatrix}}.$$

EXAMPLE 5

Use Cramer's rule to find the solution of the following system.

$$x - 5y - 4 = 0$$
$$3x - 10y - 2 = 0$$

Solution

In order to apply Cramer's rule, we first rewrite the equations with the constant terms isolated on the right sides of the equations.

$$x - 5y = 4$$
$$3x - 10y = 2$$

This same system was solved with the Gaussian elimination method in the first example presented in Exercise 3.2. Solving it with Cramer's rule, we have

$$x = \frac{\begin{vmatrix} 4 & -5 \\ 2 & -10 \end{vmatrix}}{\begin{vmatrix} 1 & -5 \\ 3 & -10 \end{vmatrix}} = \frac{-30}{5} = -6,$$

$$y = \frac{\begin{vmatrix} 1 & 4 \\ 3 & 2 \end{vmatrix}}{\begin{vmatrix} 1 & -5 \\ 3 & -10 \end{vmatrix}} = \frac{-10}{5} = -2.$$

This system could also be solved readily with the methods presented in Section 1.10.

In view of the fact that we can solve systems of n simultaneous equations in n unknowns with the methods of Section 1.10, it may seem that it was unnecessary to introduce additional methods for solving systems of linear equations. Cramer's rule and matrix methods are extremely important, however, because they arise in other contexts in mathematics and they are readily programmable for computation on a high-speed computer.

Note that Cramer's rule applies only to the case in which the determinant of the coefficient matrix is not zero. If the determinant of the coefficient matrix is zero, the system fails to have exactly one solution.

Cramer's rule is used to solve *three* linear equations in *three* unknowns in the next example.

EXAMPLE 6

Use Cramer's rule to find the solution of the following system.

$$
\begin{aligned}
3x - 2y + 6z &= -6 \\
-3x + 10y + 11z &= 13 \\
x - 2y - z &= -3
\end{aligned}
$$

Solution

(Note that this system of equations was solved by other methods in Example 1 of Section 1.10 and Example 1 of Section 3.2.) The determinant of the coefficient matrix is

$$
\begin{vmatrix}
3 & -2 & 6 \\
-3 & 10 & 11 \\
1 & -2 & -1
\end{vmatrix}.
$$

Evaluating this determinant, we have

$$
\begin{vmatrix}
3 & -2 & 6 \\
-3 & 10 & 11 \\
1 & -2 & -1
\end{vmatrix} =
\begin{vmatrix}
0 & 4 & 9 \\
0 & 4 & 8 \\
1 & -2 & -1
\end{vmatrix}
\quad
\begin{aligned}
(-3)R_3 + R_1 \\
3R_3 + R_2
\end{aligned}
$$

$$
= \begin{vmatrix}
4 & 9 \\
4 & 8
\end{vmatrix}
$$

$$
= -4.
$$

Since the value of the coefficient matrix is not zero, the given system has exactly one solution. That solution is

$$
x = \frac{\begin{vmatrix} -6 & -2 & 6 \\ 13 & 10 & 11 \\ -3 & -2 & -1 \end{vmatrix}}{-4} = \frac{-8}{-4} = 2,
$$

$$y = \frac{\begin{vmatrix} 3 & -6 & 6 \\ -3 & 13 & 11 \\ 1 & -3 & -1 \end{vmatrix}}{-4} = \frac{-12}{-4} = 3,$$

$$z = \frac{\begin{vmatrix} 3 & -2 & -6 \\ -3 & 10 & 13 \\ 1 & -2 & -3 \end{vmatrix}}{-4} = \frac{4}{-4} = -1.$$

Students should verify that the values of the determinants in the numerators are actually as we have just stated.

Exercise 3.3, 36–55

EXERCISE 3.3 DETERMINANTS AND SIMULTANEOUS LINEAR EQUATIONS

▶ In each of the following, name the element in the indicated row and column of the determinant.

$$\begin{vmatrix} 2 & -1 & 5 \\ 3 & 0 & 4 \\ 6 & -7 & 8 \end{vmatrix}.$$

EXAMPLE   First row, third column

Solution   5

1. First row, second column   2. Second row, first column

3. Second row, third column   4. Third row, second column

5. Third row, third column

▶ Using the subscript notation presented in the text, state the row and column in which each of the following elements lies, and state whether the element is in plus position or minus position.

EXAMPLE a_{34}

Solution   Third row, fourth column

$3 + 4 = 7$, minus position

6. a_{43} 7. a_{23} 8. a_{14} 9. a_{41} 10. a_{13}

► Find the minor of each of the following elements in the determinant

$$\begin{vmatrix} 2 & -1 & 5 \\ 3 & 0 & 4 \\ 6 & -7 & 8 \end{vmatrix}.$$

EXAMPLE

4

Solution

$$\begin{vmatrix} 2 & -1 \\ 6 & -7 \end{vmatrix}$$

11. 5   **12.** −7   **13.** 6   **14.** −1   **15.** 3

► Find the value of each of the following.

EXAMPLES

(a) $\begin{vmatrix} -3 & 8 \\ -2 & 5 \end{vmatrix}$

(b) $\begin{vmatrix} 2 & 4 & 1 \\ 1 & -3 & -2 \\ 3 & 5 & -1 \end{vmatrix}$

Solutions

(a) $(-15) - (-16) = 1$

(b) $\begin{vmatrix} 2 & 4 & 1 \\ 1 & -3 & -2 \\ 3 & 5 & -1 \end{vmatrix} = 2\begin{vmatrix} -3 & -2 \\ 5 & -1 \end{vmatrix} - (1)\begin{vmatrix} 4 & 1 \\ 5 & -1 \end{vmatrix} + 3\begin{vmatrix} 4 & 1 \\ -3 & -2 \end{vmatrix}$

$$= 2(13) - (-9) + 3(-5)$$
$$= 26 + 9 - 15$$
$$= 20$$

16. $\begin{vmatrix} 3 & -5 \\ -4 & 6 \end{vmatrix}$ **17.** $\begin{vmatrix} -1 & 7 \\ 6 & 0 \end{vmatrix}$

18. $\begin{vmatrix} -8 & 9 \\ 6 & 3 \end{vmatrix}$ **19.** $\begin{vmatrix} 1 & -8 \\ 0 & 9 \end{vmatrix}$

20. $\begin{vmatrix} 3 & -6 \\ 5 & -2 \end{vmatrix}$ **21.** $\begin{vmatrix} 4 & 1 & 2 \\ 1 & -5 & 6 \\ -4 & 3 & -5 \end{vmatrix}$

22. $\begin{vmatrix} 1 & 0 & 3 \\ 2 & -5 & 4 \\ 6 & 0 & 8 \end{vmatrix}$ **23.** $\begin{vmatrix} 5 & 7 & 1 \\ -5 & 6 & 2 \\ 1 & 7 & -3 \end{vmatrix}$

24. $\begin{vmatrix} 3 & 7 & -1 \\ 2 & -6 & 5 \\ 0 & 4 & -3 \end{vmatrix}$ **25.** $\begin{vmatrix} 2 & -2 & 1 \\ 4 & -3 & 7 \\ -1 & 3 & -4 \end{vmatrix}$

26.
$$\begin{vmatrix} 3 & 1 & 0 & 2 \\ 4 & -5 & 6 & 0 \\ 2 & -3 & 0 & 1 \\ -3 & -1 & 0 & 7 \end{vmatrix}$$

27.
$$\begin{vmatrix} -3 & 2 & -1 & 5 \\ 0 & 3 & 5 & 6 \\ 1 & 4 & -2 & -4 \\ 0 & 6 & 0 & 8 \end{vmatrix}$$

28.
$$\begin{vmatrix} 0 & 1 & 0 & 1 \\ 2 & 0 & 3 & 0 \\ -1 & 0 & 0 & 1 \\ 4 & -2 & 6 & 0 \end{vmatrix}$$

29.
$$\begin{vmatrix} 1 & -3 & 4 & 5 \\ -2 & 3 & 0 & 7 \\ 6 & -7 & 8 & 2 \\ -1 & -4 & 6 & 1 \end{vmatrix}$$

30.
$$\begin{vmatrix} 1 & 7 & 6 & -3 \\ 2 & 1 & 4 & 3 \\ -2 & 3 & 1 & -4 \\ 6 & 5 & 7 & 1 \end{vmatrix}$$

▶ Using properties of determinants, fill in the blank parentheses in each of the following.

EXAMPLES

(a) $\begin{vmatrix} 3 & 5 & 8 \\ 1 & -3 & 2 \\ 3 & 5 & 8 \end{vmatrix} = (\quad)$

(b) $\begin{vmatrix} 3 & 5 & 8 \\ 1 & -3 & 2 \\ 3 & 5 & 7 \end{vmatrix} = (\quad) \begin{vmatrix} 3 & 1 & 3 \\ 5 & -3 & 5 \\ 8 & 2 & 7 \end{vmatrix}$

Solutions

(a) 0 (Two rows are identical.)

(b) +

31. $\begin{vmatrix} 2 & 5 & 6 \\ 3 & 4 & 1 \\ 4 & -1 & 7 \end{vmatrix} = (\quad) \begin{vmatrix} 5 & 2 & 6 \\ 4 & 3 & 1 \\ -1 & 4 & 7 \end{vmatrix}$

32. $\begin{vmatrix} 2 & 5 & 6 \\ 3 & 4 & 1 \\ 4 & -1 & 7 \end{vmatrix} = (\quad) \begin{vmatrix} 3 & 4 & 1 \\ 2 & 5 & 6 \\ 4 & -1 & 7 \end{vmatrix}$

33. $\begin{vmatrix} 2 & 5 & 6 \\ 3 & 4 & 2 \\ 4 & -1 & 4 \end{vmatrix} = (\quad) \begin{vmatrix} 2 & 5 & 3 \\ 3 & 4 & 1 \\ 4 & -1 & 2 \end{vmatrix}$

34. $\begin{vmatrix} 2 & 5 & 6 \\ 3 & 4 & 1 \\ 0 & 0 & 0 \end{vmatrix} = (\quad)$

35. $\begin{vmatrix} 3 & 7 & 3 \\ 5 & 6 & 5 \\ 1 & 4 & 1 \end{vmatrix} = (\quad)$

▶ **36—55.** Use Cramer's rule to solve each of the systems in items 1—20 of Exercise 3.2.

CHECK LIST FOR THE STUDENT

If you have learned the material in Section 3.3, you should be able to do the following:

1. Show how to denote the determinant of a given square matrix.

2. Give the order of the determinant of an $n \times n$ matrix.

3. Find the value of any 2×2 determinant.

4. State the row and column in which a specified element lies in a given determinant.

5. Use the subscript notation presented in the text, and specify the row and column in which the element a_{ij} lies.

6. Find the minor of a specified element in a given determinant.

7. State whether a specified element in a given determinant is in plus position or in minus position.

8. Use expansion by minors to find the value of a given determinant.

9. Use the properties of determinants given in the text.

10. Use Cramer's rule to solve a given system of n simultaneous linear equations in n unknowns.

SECTION QUIZ 3.3

1. In what row and column does 7 lie in the following determinant?

$$\begin{vmatrix} 2 & -1 & 3 \\ 4 & 8 & 5 \\ -2 & 7 & 9 \end{vmatrix}$$

2. Using the subscript notation presented in the text, in what row and column does the element a_{23} lie, and is a_{23} in plus position or minus position?

3. What is the minor of 9 in the following determinant?

$$\begin{vmatrix} 3 & -1 & 4 \\ 0 & 2 & 8 \\ 4 & 9 & 1 \end{vmatrix}$$

4. Find the value of the following determinant.

$$\begin{vmatrix} 3 & 7 \\ -1 & 8 \end{vmatrix}$$

5. Using expansion by minors of the elements in the first column, find the value of the following determinant.

$$\begin{vmatrix} 3 & 4 & 2 \\ 1 & 7 & 6 \\ -2 & 5 & -1 \end{vmatrix}$$

Using properties of determinants, fill in the blank parentheses in each of the following.

6. (a) $\begin{vmatrix} 2 & 0 & 4 \\ 3 & 0 & 5 \\ 1 & 0 & 7 \end{vmatrix} = (\quad)$ (b) $\begin{vmatrix} 2 & 4 & 4 \\ 3 & 5 & 5 \\ 1 & 7 & 7 \end{vmatrix} = (\quad)$

7. (a) $\begin{vmatrix} 3 & 5 & 7 \\ -2 & 1 & 4 \\ 6 & 8 & -3 \end{vmatrix} = (\quad) \begin{vmatrix} 3 & -2 & 6 \\ 5 & 1 & 8 \\ 7 & 4 & -3 \end{vmatrix}$

   (b) $\begin{vmatrix} 3 & 5 & 7 \\ -4 & 2 & 8 \\ 6 & 8 & -3 \end{vmatrix} = (\quad) \begin{vmatrix} 3 & 5 & 7 \\ -2 & 1 & 4 \\ 6 & 8 & -3 \end{vmatrix}$

8. (a) $\begin{vmatrix} 3 & 5 & 9 \\ -2 & 8 & 7 \\ 5 & 1 & 4 \end{vmatrix} = (\quad) \begin{vmatrix} 5 & 1 & 4 \\ -2 & 8 & 7 \\ 3 & 5 & 9 \end{vmatrix}$

   (b) $\begin{vmatrix} 3 & 5 & 9 \\ -2 & 8 & 7 \\ 5 & 1 & 4 \end{vmatrix} = (\quad) \begin{vmatrix} 5 & -3 & 2 \\ -2 & 8 & 7 \\ 5 & 1 & 4 \end{vmatrix}$

Use Cramer's rule to find the solution of each of the following systems.

9. $3x - 2y = 2$
   $5x - y = -6$

10. $2x - 3y + 5z = -3$
    $x - 2y + 4z = -2$
    $3x + y - 2z = 5$

3.4 Multiplication and Inversion of Matrices; Simultaneous Linear Equations

The purpose of this section is to introduce multiplication and inversion of matrices and to show how they can be used to solve systems of simultaneous linear equations.

If matrix A has the same number of *columns* as the number of *rows* in matrix B, then the product AB is defined. If A is an $m \times n$ matrix and B is an $n \times p$ matrix, then the product AB is an $m \times p$ matrix, and any element c_{ij} of the product can be found by multiplying each pair of corresponding elements in the ith row of A and the jth column of B and then adding all these products. To illustrate, suppose that A is a 2×3 matrix.

$$A = \begin{pmatrix} 3 & 1 & -1 \\ 2 & 4 & -3 \end{pmatrix},$$

and suppose that B is a 3×4 matrix, say

$$B = \begin{pmatrix} 1 & -2 & 4 & 2 \\ 7 & 8 & -5 & -1 \\ 5 & -6 & -7 & 3 \end{pmatrix}.$$

Then the product AB is a 2×4 matrix, and the element in any row and column, say the element c_{23} in the second row and third column, can be found by forming the products of the corresponding elements in the second row of A and the third column of B and then adding these products. The elements in the second row of A are

$$2 \qquad 4 \qquad -3,$$

and the elements in the third column of B are

$$4 \qquad -5 \qquad -7,$$

so
$$\begin{aligned} c_{23} &= (2)(4) + (4)(-5) + (-3)(-7) \\ &= 8 \quad - \quad 20 \quad + \quad 21 \\ &= 9. \end{aligned}$$

This computation is indicated as follows:

$$\begin{pmatrix} 3 & 1 & -1 \\ 2 & 4 & -3 \end{pmatrix} \begin{pmatrix} 1 & -2 & 4 & 2 \\ 7 & 8 & -5 & -1 \\ 5 & -6 & -7 & 3 \end{pmatrix} = \begin{pmatrix} c_{11} & c_{12} & c_{13} & c_{14} \\ c_{21} & c_{22} & 9 & c_{24} \end{pmatrix}$$

$$\begin{array}{ccc} A & B & = & C \\ (2 \times 3) & (3 \times 4) & & (2 \times 4) \end{array}$$

Similarly, to find c_{11} we use the first row of A and the first column of B, to find c_{12} we use the first row of A and the second column of B, to find c_{13} we use the first row of A and the third column of B, and so on.

$$\begin{array}{lccc}\text{Row 1 of } A & 3 & 1 & -1\\ \text{Column 1 of } B & 1 & 7 & 5\end{array}$$

$$\begin{aligned} c_{11} &= (3)(1) + (1)(7) + (-1)(5)\\ &= \quad 3 \ + \ 7 \ - \quad 5\\ &= \quad 5 \end{aligned}$$

$$\begin{array}{lccc}\text{Row 1 of } A & 3 & 1 & -1\\ \text{Column 2 of } B & -2 & 8 & -6\end{array}$$

$$\begin{aligned} c_{12} &= (3)(-2) + (1)(8) + (-1)(-6)\\ &= \quad -6 \ + \ 8 \ + \quad 6\\ &= \quad 8 \end{aligned}$$

$$\begin{array}{lccc}\text{Row 1 of } A & 3 & 1 & -1\\ \text{Column 3 of } B & 4 & -5 & -7\end{array}$$

$$\begin{aligned} c_{13} &= (3)(4) + (1)(-5) + (-1)(-7)\\ &= \quad 12 \ - \quad 5 \ + \quad 7\\ &= \quad 14 \end{aligned}$$

In like manner,

$$\begin{aligned} c_{14} &= (3)(2) + (1)(-1) + (-1)(3)\\ &= \quad 2, \end{aligned}$$

$$\begin{aligned} c_{21} &= (2)(1) + (4)(7) + (-3)(5)\\ &= \quad 15 \end{aligned}$$

$$\begin{aligned} c_{22} &= (2)(-2) + (4)(8) + (-3)(-6)\\ &= \quad 46, \end{aligned}$$

$$\begin{aligned} c_{24} &= (2)(2) + (4)(-1) + (-3)(3)\\ &= \quad -9. \end{aligned}$$

Summarizing, we have

$$\begin{pmatrix} 3 & 1 & -1\\ 2 & 4 & -3 \end{pmatrix} \begin{pmatrix} 1 & -2 & 4 & 2\\ 7 & 8 & -5 & -1\\ 5 & -6 & -7 & 3 \end{pmatrix} = \begin{pmatrix} 5 & 8 & 14 & 2\\ 15 & 46 & 9 & -9 \end{pmatrix}.$$

$$\begin{array}{ccc} A & B & = \qquad C\\ (2 \times 3) & (3 \times 4) & (2 \times 4) \end{array}$$

Note that the product AB of matrices A and B is defined only for the case in which A has the same number of columns as B has rows. In this case, A and B are said to be **conformable** for the multiplication AB. The product AB has the number of rows of A and the number of columns of B.

$$
\begin{array}{ccccc}
A & & B & = & C \\
(m \times n) & & (n \times p) & & (m \times p)
\end{array}
$$

The same

EXAMPLE 1

Find the product AB where

$$
A = \begin{pmatrix} a_{11} & a_{12} \\ a_{21} & a_{22} \end{pmatrix} \quad \text{and} \quad B = \begin{pmatrix} b_{11} & b_{12} \\ b_{21} & b_{22} \end{pmatrix}.
$$

Solution

$$
\begin{pmatrix} a_{11} & a_{12} \\ a_{21} & a_{22} \end{pmatrix} \begin{pmatrix} b_{11} & b_{12} \\ b_{21} & b_{22} \end{pmatrix} = \begin{pmatrix} a_{11}b_{11} + a_{12}b_{21} & a_{11}b_{12} + a_{12}b_{22} \\ a_{21}b_{11} + a_{22}b_{21} & a_{21}b_{12} + a_{22}b_{22} \end{pmatrix}
$$

EXAMPLE 2

Find the product

$$
\begin{pmatrix} 3 & 2 \\ 4 & -5 \end{pmatrix} \begin{pmatrix} x \\ y \end{pmatrix}.
$$

Solution

$$
\begin{pmatrix} 3 & 2 \\ 4 & -5 \end{pmatrix} \begin{pmatrix} x \\ y \end{pmatrix} = \begin{pmatrix} 3x + 2y \\ 4x - 5y \end{pmatrix}
$$

EXAMPLE 3

Find the product

$$
\begin{pmatrix} 2 & 3 & -4 \\ 5 & -3 & 6 \\ 4 & 7 & 9 \end{pmatrix} \begin{pmatrix} x \\ y \\ z \end{pmatrix}.
$$

Solution

$$
\begin{pmatrix} 2 & 3 & -4 \\ 5 & -3 & 6 \\ 4 & 7 & 9 \end{pmatrix} \begin{pmatrix} x \\ y \\ z \end{pmatrix} = \begin{pmatrix} 2x + 3y - 4z \\ 5x - 3y + 6z \\ 4x + 7y + 9z \end{pmatrix}
$$

Exercise 3.4, 1–20

EXAMPLE 4

If

$$
A = \begin{pmatrix} -1 & -3 \\ 2 & 0 \end{pmatrix} \quad \text{and} \quad B = \begin{pmatrix} -2 & 1 \\ 3 & 4 \end{pmatrix},
$$

find the product AB, and then find the product BA.

Solution

$$
AB = \begin{pmatrix} -1 & -3 \\ 2 & 0 \end{pmatrix} \begin{pmatrix} -2 & 1 \\ 3 & 4 \end{pmatrix}
$$

$$= \begin{pmatrix} -7 & -13 \\ -4 & 2 \end{pmatrix}$$

$$BA = \begin{pmatrix} -2 & 1 \\ 3 & 4 \end{pmatrix} \begin{pmatrix} -1 & -3 \\ 2 & 0 \end{pmatrix}$$

$$= \begin{pmatrix} 4 & 6 \\ 5 & -9 \end{pmatrix}$$

Exercise 3.4, 26–30

Example 4 shows that multiplication of matrices is not commutative, that is, the product AB is not necessarily the same as the product BA. In multiplying matrices, we must be careful to observe the order of the multiplication.

The product of a real number c and a matrix A whose elements are real numbers is defined to be the matrix derived from A by multiplying each element of A by c. For example, if c is a real number,

$$c \begin{pmatrix} a_{11} & a_{12} \\ a_{21} & a_{22} \end{pmatrix} = \begin{pmatrix} c\,a_{11} & c\,a_{12} \\ c\,a_{21} & c\,a_{22} \end{pmatrix},$$

More specifically, if

$$A = \begin{pmatrix} 1 & -2 & 3 \\ -1 & 0 & 1 \\ -3 & 5 & 2 \end{pmatrix},$$

then

$$3A = \begin{pmatrix} 3 & -6 & 9 \\ -3 & 0 & 3 \\ -9 & 15 & 6 \end{pmatrix}.$$

As another example,

$$\frac{1}{4} \begin{pmatrix} 1 & 0 & 0 \\ 0 & 1 & 0 \\ 0 & 0 & 1 \end{pmatrix} = \begin{pmatrix} \frac{1}{4} & 0 & 0 \\ 0 & \frac{1}{4} & 0 \\ 0 & 0 & \frac{1}{4} \end{pmatrix}.$$

Exercise 3.4, 21–25

We shall need two properties of multiplication of real numbers and matrices. In particular, if c is a real number, and if A and B are matrices, comformable for the multiplication AB, then

$$1. \; [cA]\,B = c\,[AB],$$

and

$$2. \; A\,[cB] = c\,[AB].$$

These properties are illustrated in the next example.

EXAMPLE 5

Let

$$A = \begin{pmatrix} 2 & 3 \\ 4 & 7 \end{pmatrix}, \quad B = \begin{pmatrix} 1 & -3 \\ 2 & 6 \end{pmatrix}, \quad \text{and} \quad c = 5.$$

Show that $[cA]B = c[AB]$ and that $A[cB] = c[AB]$.

Solution

$$[cA]B = \left[5 \begin{pmatrix} 2 & 3 \\ 4 & 7 \end{pmatrix} \right] \begin{pmatrix} 1 & -3 \\ 2 & 6 \end{pmatrix}$$

$$= \begin{pmatrix} 10 & 15 \\ 20 & 35 \end{pmatrix} \begin{pmatrix} 1 & -3 \\ 2 & 6 \end{pmatrix}$$

$$= \begin{pmatrix} 40 & 60 \\ 90 & 150 \end{pmatrix}$$

$$c[AB] = 5 \left[\begin{pmatrix} 2 & 3 \\ 4 & 7 \end{pmatrix} \begin{pmatrix} 1 & -3 \\ 2 & 6 \end{pmatrix} \right]$$

$$= 5 \begin{pmatrix} 8 & 12 \\ 18 & 30 \end{pmatrix}$$

$$= \begin{pmatrix} 40 & 60 \\ 90 & 150 \end{pmatrix}$$

We see that $[cA]B = c[AB]$.

$$A[cB] = \begin{pmatrix} 2 & 3 \\ 4 & 7 \end{pmatrix} \left[5 \begin{pmatrix} 1 & -3 \\ 2 & 6 \end{pmatrix} \right]$$

$$= \begin{pmatrix} 2 & 3 \\ 4 & 7 \end{pmatrix} \begin{pmatrix} 5 & -15 \\ 10 & 30 \end{pmatrix}$$

$$= \begin{pmatrix} 40 & 60 \\ 90 & 150 \end{pmatrix}$$

Exercise 3.4, 31–35 We see that we also have $A[cB] = c[AB]$.

We may use matrix multiplication to write a system of n simultaneous linear equations in n unknowns as a single matrix equation. In order to do this, we need first to define *equality of $m \times n$ matrices*. Two $m \times n$ matrices are **equal** provided that their corresponding elements are equal. Conversely, if two $m \times n$ matrices are equal, then their corresponding elements are equal. For example, if

$$\begin{pmatrix} 3 & 2 \\ -5 & 7 \end{pmatrix} = \begin{pmatrix} a & b \\ c & d \end{pmatrix},$$

then

$$3 = a, \qquad 2 = b,$$
$$-5 = c, \qquad 7 = d.$$

Likewise if,

$$\begin{pmatrix} a \\ b \end{pmatrix} = \begin{pmatrix} 4 \\ 2 \end{pmatrix},$$

then
$$a = 4,$$

and
$$b = 2.$$

Also, if

$$\begin{pmatrix} x - 5y \\ 3x - 10y \end{pmatrix} = \begin{pmatrix} 4 \\ 2 \end{pmatrix},$$

then
$$x - 5y = 4$$

and
$$3x - 10y = 2.$$

We may now write any system of n simultaneous linear equations in n unknowns as a single matrix equation stating that the product of two matrices is equal to a third matrix. For example, the system

$$x - 5y = 4$$
$$3x - 10y = 2$$

may be written as the matrix equation

$$\begin{pmatrix} 1 & -5 \\ 3 & -10 \end{pmatrix} \begin{pmatrix} x \\ y \end{pmatrix} = \begin{pmatrix} 4 \\ 2 \end{pmatrix}.$$

Indeed, if we perform the multiplication on the left, we have

$$\begin{pmatrix} x - 5y \\ 3x - 10y \end{pmatrix} = \begin{pmatrix} 4 \\ 2 \end{pmatrix}.$$

If we now employ the definition of equality of 2×1 matrices, we have as before

$$x - 5y = 4$$
$$3x - 10y = 2.$$

Note that the matrix

$$\begin{pmatrix} 1 & -5 \\ 3 & -10 \end{pmatrix}$$

is the coefficient matrix of this system, the matrix

$$\begin{pmatrix} x \\ y \end{pmatrix}$$

is the column of unknowns, and the matrix

$$\begin{pmatrix} 4 \\ 2 \end{pmatrix}$$

is the column of constant terms as they appear isolated on the right of the equals signs of the system. To write any system of *n* simultaneous linear equations in *n* unknowns as a matrix equation, set the product of the coefficient matrix and the matrix of the unknowns equal to the matrix of the isolated constant terms.

EXAMPLE 6

Write the following system of equations as a matrix equation. Use matrix multiplication and verify that your matrix equation is correct.

$$3x - 2y + 6z = -6$$
$$-3x + 10y + 11z = 13$$
$$x - 2y - z = -3$$

Solution

$$\begin{pmatrix} 3 & -2 & 6 \\ -3 & 10 & 11 \\ 1 & -2 & -1 \end{pmatrix} \begin{pmatrix} x \\ y \\ z \end{pmatrix} = \begin{pmatrix} -6 \\ 13 \\ -3 \end{pmatrix}$$

To verify that this is correct, we perform the multiplication on the left and obtain

$$\begin{pmatrix} 3x - 2y + 6z \\ -3x + 10y + 11z \\ x - 2y - z \end{pmatrix} = \begin{pmatrix} -6 \\ 13 \\ -3 \end{pmatrix}$$

Exercise 3.4, 36–45

If we now employ the definition of equality of 3×1 matrices, we obtain the given system.

The product of any square matrix *A* with the square matrix of the same order having all 1's in the main diagonal and 0's elsewhere is equal to *A*. For example, suppose that *A* is any 2×2 matrix

$$\begin{pmatrix} a_{11} & a_{12} \\ a_{21} & a_{22} \end{pmatrix}.$$

The product

$$\begin{pmatrix} a_{11} & a_{12} \\ a_{21} & a_{22} \end{pmatrix} \begin{pmatrix} 1 & 0 \\ 0 & 1 \end{pmatrix} = \begin{pmatrix} a_{11} & a_{12} \\ a_{21} & a_{22} \end{pmatrix}$$

is equal to A. As a specific example,

$$\begin{pmatrix} 3 & 2 \\ 5 & 9 \end{pmatrix} \begin{pmatrix} 1 & 0 \\ 0 & 1 \end{pmatrix} = \begin{pmatrix} 3 & 2 \\ 5 & 9 \end{pmatrix}.$$

Similarly, if A is any 3×3 matrix, the product of A and

$$\begin{pmatrix} 1 & 0 & 0 \\ 0 & 1 & 0 \\ 0 & 0 & 1 \end{pmatrix}$$

is equal to A. For example,

$$\begin{pmatrix} 3 & 2 & 7 \\ 1 & -5 & 4 \\ -3 & 6 & 8 \end{pmatrix} \begin{pmatrix} 1 & 0 & 0 \\ 0 & 1 & 0 \\ 0 & 0 & 1 \end{pmatrix} = \begin{pmatrix} 3 & 2 & 7 \\ 1 & -5 & 4 \\ -3 & 6 & 8 \end{pmatrix}.$$

For any given order, the matrix with all 1's in the main diagonal and 0's elsewhere is called the **identity** matrix and is denoted by the letter I. The identity matrices for orders 2, 3, and 4 are

$$\begin{pmatrix} 1 & 0 \\ 0 & 1 \end{pmatrix}, \begin{pmatrix} 1 & 0 & 0 \\ 0 & 1 & 0 \\ 0 & 0 & 1 \end{pmatrix}, \text{ and } \begin{pmatrix} 1 & 0 & 0 & 0 \\ 0 & 1 & 0 & 0 \\ 0 & 0 & 1 & 0 \\ 0 & 0 & 0 & 1 \end{pmatrix}$$

It can be shown that if A is any square matrix and I is the identity matrix of the same order as A, then

$$AI = A,$$

Exercise 3.4, 46–50 *and*
$$IA = A.$$

For any given square matrix A, there may or may not exist a square matrix called A-inverse such that the product of A and A-inverse is the identity matrix. For convenience, we denote A-inverse by the symbol A^{-1}. We say that a given square matrix A *has* an inverse A^{-1} provided that there exists a matrix A^{-1} such that

$$AA^{-1} = I \quad \text{and} \quad A^{-1}A = I.$$

It can be shown that if a matrix has an inverse, then it has only one inverse. For example, we shall see that the matrix

$$A = \begin{pmatrix} 2 & 3 \\ 5 & 4 \end{pmatrix}$$

has

$$A^{-1} = \begin{pmatrix} \frac{-4}{7} & \frac{3}{7} \\ \frac{5}{7} & \frac{-2}{7} \end{pmatrix}$$

for its inverse because

$$\begin{pmatrix} 2 & 3 \\ 5 & 4 \end{pmatrix} \begin{pmatrix} \frac{-4}{7} & \frac{3}{7} \\ \frac{5}{7} & \frac{-2}{7} \end{pmatrix} = I \quad \text{and} \quad \begin{pmatrix} \frac{-4}{7} & \frac{3}{7} \\ \frac{5}{7} & \frac{-2}{7} \end{pmatrix} \begin{pmatrix} 2 & 3 \\ 5 & 4 \end{pmatrix} = I.$$

Furthermore this A^{-1} is the only matrix that shares this property with A.

If a given square matrix A has an inverse, then A is said to be **invertible** or **nonsingular**. It can be shown that a square matrix A is nonsingular provided that det A, the determinant of A, is not zero. For example, the matrix

$$A = \begin{pmatrix} 2 & 3 \\ 5 & 4 \end{pmatrix}$$

is nonsingular and hence has an inverse because the determinant of A,

$$\begin{vmatrix} 2 & 3 \\ 5 & 4 \end{vmatrix} = -7,$$

is not zero. We have just observed that the inverse of this A is

$$\begin{pmatrix} \frac{-4}{7} & \frac{3}{7} \\ \frac{5}{7} & \frac{-2}{7} \end{pmatrix}.$$

Exercise 3.4, 51–60

There are several ways of finding the inverse of a nonsinguler matrix A. The method given below is stated in terms of the *transpose* of A. The **transpose** of square matrix A is the matrix derived from A by interchanging all corresponding rows and columns. The transpose of

$$\begin{pmatrix} 3 & 7 & 8 \\ 2 & 1 & 4 \\ 6 & 9 & 5 \end{pmatrix} \quad \text{is} \quad \begin{pmatrix} 3 & 2 & 6 \\ 7 & 1 & 9 \\ 8 & 4 & 5 \end{pmatrix}.$$

We can find the inverse of a nonsingular matrix A as follows.

To Find the Inverse of Nonsingular A

1. Find det A, the value of the determinant of A.

2. Form the *transpose* of A.

3. Replace each element of the transpose by its minor (if the element is in plus position) or by the negative of its minor (if the element is in minus position).

4. Multiply the resulting matrix by the reciprocal of det A.

This method is not the most easily programmable method, but it does have theoretical significance, and it can be presented here without introducing additional vocabulary. It is illustrated in the next example.

PART THREE | HIGHER ALGEBRA

EXAMPLE 7   Find the inverse of

$$A = \begin{pmatrix} 1 & 0 & -1 \\ 3 & 2 & 1 \\ 0 & 1 & 4 \end{pmatrix}.$$

Solution   *Step 1*   The value of the determinant of A is

$$\det A = \begin{vmatrix} 1 & 0 & -1 \\ 3 & 2 & 1 \\ 0 & 1 & 4 \end{vmatrix} = 7 - 3 = 4.$$

Step 2   The transpose of A is

$$\begin{pmatrix} 1 & 3 & 0 \\ 0 & 2 & 1 \\ -1 & 1 & 4 \end{pmatrix}.$$

Step 3   Replacing each element of the transpose by its minor or the negative of its minor, depending on the position of the element, we obtain

$$\begin{pmatrix} 7 & -(1) & 2 \\ -(12) & 4 & -(4) \\ 3 & -(1) & 2 \end{pmatrix} = \begin{pmatrix} 7 & -1 & 2 \\ -12 & 4 & -4 \\ 3 & -1 & 2 \end{pmatrix}.$$

Step 4   Multiplying by $1/\det A$,

$$A^{-1} = \frac{1}{4} \begin{pmatrix} 7 & -1 & 2 \\ -12 & 4 & -4 \\ 3 & -1 & 2 \end{pmatrix}.$$

Check

$$AA^{-1} = \begin{pmatrix} 1 & 0 & -1 \\ 3 & 2 & 1 \\ 0 & 1 & 4 \end{pmatrix} \times \frac{1}{4} \begin{pmatrix} 7 & -1 & 2 \\ -12 & 4 & -4 \\ 3 & -1 & 2 \end{pmatrix}$$

$$= \frac{1}{4} \times \begin{pmatrix} 1 & 0 & -1 \\ 3 & 2 & 1 \\ 0 & 1 & 4 \end{pmatrix} \begin{pmatrix} 7 & -1 & 2 \\ -12 & 4 & -4 \\ 3 & -1 & 2 \end{pmatrix}$$

$$= \frac{1}{4} \begin{pmatrix} 4 & 0 & 0 \\ 0 & 4 & 0 \\ 0 & 0 & 4 \end{pmatrix}$$

$$= \begin{pmatrix} 1 & 0 & 0 \\ 0 & 1 & 0 \\ 0 & 0 & 1 \end{pmatrix}$$

$$= I$$

$$A^{-1}A = \frac{1}{4} \times \begin{pmatrix} 7 & -1 & 2 \\ -12 & 4 & -4 \\ 3 & -1 & 2 \end{pmatrix} \begin{pmatrix} 1 & 0 & -1 \\ 3 & 2 & 1 \\ 0 & 1 & 4 \end{pmatrix}$$

$$= \frac{1}{4} \begin{pmatrix} 4 & 0 & 0 \\ 0 & 4 & 0 \\ 0 & 0 & 4 \end{pmatrix}$$

$$= \begin{pmatrix} 1 & 0 & 0 \\ 0 & 1 & 0 \\ 0 & 0 & 1 \end{pmatrix}$$

$$= I$$

The next example demonstrates how to use the method just outlined to find the inverse of a 2 × 2 matrix. Note that in a 2 × 2 matrix, the "minor" of any element e is simply the single element that is left when the row and column of e are deleted.

EXAMPLE 8   Find the inverse of $A = \begin{pmatrix} 2 & 3 \\ 5 & 4 \end{pmatrix}$.

Solution   *Step 1* Find det A.

$$\begin{vmatrix} 2 & 3 \\ 5 & 4 \end{vmatrix} = -7$$

Step 2 Find the transpose of A.

$$\text{Transpose of } A = \begin{pmatrix} 2 & 5 \\ 3 & 4 \end{pmatrix}$$

Step 3 Replace each element of the transpose by its minor or the negative of its minor, depending on the position.

$$\begin{pmatrix} 4 & -3 \\ -5 & 2 \end{pmatrix}$$

Step 4 Multiply by 1/det A.

$$A^{-1} = \frac{1}{-7} \begin{pmatrix} 4 & -3 \\ -5 & 2 \end{pmatrix}$$

Check $AA^{-1} = \begin{pmatrix} 2 & 3 \\ 5 & 4 \end{pmatrix} \times \left(\frac{-1}{7}\right) \begin{pmatrix} 4 & -3 \\ -5 & 2 \end{pmatrix}$

$$= \frac{-1}{7} \times \begin{pmatrix} 2 & 3 \\ 5 & 4 \end{pmatrix} \begin{pmatrix} 4 & -3 \\ -5 & 2 \end{pmatrix}$$

$$= \frac{-1}{7} \begin{pmatrix} -7 & 0 \\ 0 & -7 \end{pmatrix}$$

$$= \begin{pmatrix} 1 & 0 \\ 0 & 1 \end{pmatrix}$$

$$= I$$

$$A^{-1}A = \frac{-1}{7} \begin{pmatrix} 4 & -3 \\ -5 & 2 \end{pmatrix} \times \begin{pmatrix} 2 & 3 \\ 5 & 4 \end{pmatrix}$$

$$= \frac{-1}{7} \times \begin{pmatrix} 4 & -3 \\ -5 & 2 \end{pmatrix} \begin{pmatrix} 2 & 3 \\ 5 & 4 \end{pmatrix}$$

$$= \frac{-1}{7} \begin{pmatrix} -7 & 0 \\ 0 & -7 \end{pmatrix}$$

$$= \begin{pmatrix} 1 & 0 \\ 0 & 1 \end{pmatrix}$$

Exercise 3.4, 61–70

$$= I$$

We can now use multiplication and inversion of matrices to solve systems of n simultaneous linear equations in n unknowns. To solve such a system, we can simply write the system as a matrix equation and then multiply both sides of the matrix equation on the left by the inverse of the coefficient matrix. For example, consider the system

$$x - 5y = 4$$
$$3x - 10y = 2.$$

(This system has already been solved by other methods in Exercise 3.2 and in Example 5 of Section 3.3.) We have seen that this system can be written as the matrix equation

$$\begin{pmatrix} 1 & -5 \\ 3 & -10 \end{pmatrix} \begin{pmatrix} x \\ y \end{pmatrix} = \begin{pmatrix} 4 \\ 2 \end{pmatrix}.$$

If we let A be the coefficient matrix, the system is

$$A \begin{pmatrix} x \\ y \end{pmatrix} = \begin{pmatrix} 4 \\ 2 \end{pmatrix}.$$

Now if we multiply both sides on the left by the inverse of A we have

$$A^{-1}A \begin{pmatrix} x \\ y \end{pmatrix} = A^{-1} \begin{pmatrix} 4 \\ 2 \end{pmatrix}.$$

But $$A^{-1}A = \begin{pmatrix} 1 & 0 \\ 0 & 1 \end{pmatrix},$$ and

$$\begin{pmatrix} 1 & 0 \\ 0 & 1 \end{pmatrix} \begin{pmatrix} x \\ y \end{pmatrix} = \begin{pmatrix} x \\ y \end{pmatrix},$$

so the equation becomes

$$\begin{pmatrix} x \\ y \end{pmatrix} = A^{-1} \begin{pmatrix} 4 \\ 2 \end{pmatrix}.$$

After we compute A^{-1} and perform the multiplication on the right, we will be able to employ the definition of equality of 2×1 matrices and read off the solution. Since

$$A = \begin{pmatrix} 1 & -5 \\ 3 & -10 \end{pmatrix},$$

$$A^{-1} = \frac{1}{5} \begin{pmatrix} -10 & 5 \\ -3 & 1 \end{pmatrix}.$$

(The student should check this.) It follows that

$$\begin{pmatrix} x \\ y \end{pmatrix} = \frac{1}{5} \begin{pmatrix} -10 & 5 \\ -3 & 1 \end{pmatrix} \times \begin{pmatrix} 4 \\ 2 \end{pmatrix}$$

$$= \frac{1}{5} \begin{pmatrix} -30 \\ -10 \end{pmatrix}$$

$$= \begin{pmatrix} -6 \\ -2 \end{pmatrix}.$$

That is, $\qquad x = -6, \qquad y = -2.$

EXAMPLE 9

Write the following system as a matrix equation and use the inverse of the coefficient matrix to find the solution.

$$\begin{aligned} 3x - 2y + 6z &= -6 \\ -3x + 10y + 11z &= 13 \\ x - 2y - z &= -3 \end{aligned}$$

Solution

(Note that this same system has been solved by other methods in Example 1 of Section 1.10, Example 1 of Section 3.2, and Example 6 of Section 3.3.) In Example 6 of this section, we saw that the given system can be written

$$\begin{pmatrix} 3 & -2 & 6 \\ -3 & 10 & 11 \\ 1 & -2 & -1 \end{pmatrix} \begin{pmatrix} x \\ y \\ z \end{pmatrix} = \begin{pmatrix} -6 \\ 13 \\ -3 \end{pmatrix}.$$

If we let A be the coefficient matrix, then

$$A \begin{pmatrix} x \\ y \\ z \end{pmatrix} = \begin{pmatrix} -6 \\ 13 \\ -3 \end{pmatrix}.$$

If we multiply both sides on the left by A^{-1}, we have

$$\begin{pmatrix} x \\ y \\ z \end{pmatrix} = A^{-1} \begin{pmatrix} -6 \\ 13 \\ -3 \end{pmatrix}.$$

It can be shown that

$$A^{-1} = \frac{1}{-4} \begin{pmatrix} 12 & -14 & -82 \\ 8 & -9 & -51 \\ -4 & 4 & 24 \end{pmatrix}.$$

(The student should check this.) It follows that

$$\begin{pmatrix} x \\ y \\ z \end{pmatrix} = \frac{-1}{4} \begin{pmatrix} 12 & -14 & -82 \\ 8 & -9 & -51 \\ -4 & 4 & 24 \end{pmatrix} \begin{pmatrix} -6 \\ 13 \\ -3 \end{pmatrix}$$

$$= \frac{-1}{4} \begin{pmatrix} -8 \\ -12 \\ 4 \end{pmatrix}$$

$$= \begin{pmatrix} 2 \\ 3 \\ -1 \end{pmatrix}.$$

Therefore the solution is $x = 2$, $y = 3$, $z = -1$.

The method just illustrated for solving systems of equations works equally well for systems of n linear equations in n unknowns where $n > 3$. If A is the coefficient matrix, if X is the matrix consisting of the column of unknowns, and if C is the matrix consisting of the column of constant terms isolated on the right of the equals signs, then the system can be written

$$AX = C.$$

If A is nonsingular, then A^{-1} exists. If we multiply both sides on the left by A^{-1} and use the facts that $A^{-1}A = I$ and $IX = X$, we have

$$X = A^{-1}C.$$

From this we can employ the definition of equality of $n \times 1$ matrices and read off the solution. If A is not nonsingular, A^{-1} does not exist, and the system fails to have exactly one solution.

Exercise 3.4, 71–90

EXERCISE 3.4 MULTIPLICATION AND INVERSION OF MATRICES; SIMULTANEOUS LINEAR EQUATIONS

▶ Find each of the following products.

EXAMPLES

(a) $\begin{pmatrix} 3 & 7 \\ -2 & 6 \end{pmatrix}\begin{pmatrix} 1 & -4 \\ 5 & 8 \end{pmatrix} = \begin{pmatrix} 38 & 44 \\ 28 & 56 \end{pmatrix}$

(b) $\begin{pmatrix} 3 & -1 & 5 \\ 4 & 2 & 7 \end{pmatrix}\begin{pmatrix} -1 & 0 \\ 1 & -2 \\ -3 & 6 \end{pmatrix} = \begin{pmatrix} -19 & 32 \\ -23 & 38 \end{pmatrix}$

(c) $\begin{pmatrix} 2 & 3 \\ 5 & -1 \end{pmatrix}\begin{pmatrix} x \\ y \end{pmatrix} = \begin{pmatrix} 2x + 3y \\ 5x - y \end{pmatrix}$

(d) $2\begin{pmatrix} 3 & 1 \\ 4 & -1 \end{pmatrix} = \begin{pmatrix} 6 & 2 \\ 8 & -2 \end{pmatrix}$

1. $\begin{pmatrix} 2 & 5 \\ -3 & 7 \end{pmatrix}\begin{pmatrix} -1 & -3 \\ 6 & 0 \end{pmatrix}$ 2. $\begin{pmatrix} -3 & 4 \\ -2 & 1 \end{pmatrix}\begin{pmatrix} 0 & 7 \\ 5 & 9 \end{pmatrix}$

3. $\begin{pmatrix} -1 & 2 \\ 3 & 7 \end{pmatrix}\begin{pmatrix} 6 & 1 \\ 0 & 4 \end{pmatrix}$ 4. $\begin{pmatrix} 6 & -4 \\ 3 & 8 \end{pmatrix}\begin{pmatrix} 2 & 4 \\ 7 & 9 \end{pmatrix}$

5. $\begin{pmatrix} 1 & 0 \\ 0 & 1 \end{pmatrix}\begin{pmatrix} 9 & -8 \\ 7 & 6 \end{pmatrix}$ 6. $\begin{pmatrix} -1 & 2 & 4 \\ 3 & -5 & 6 \end{pmatrix}\begin{pmatrix} 1 & -1 & 2 \\ 4 & 0 & 3 \\ 0 & -5 & 6 \end{pmatrix}$

7. $\begin{pmatrix} -2 & 1 & 3 \\ 4 & -6 & 5 \end{pmatrix}\begin{pmatrix} 1 & 0 \\ -2 & 3 \\ 0 & 5 \end{pmatrix}$

8. $\begin{pmatrix} -1 & 0 & 2 \\ 0 & -2 & 1 \end{pmatrix}\begin{pmatrix} 1 & 0 & -1 & 2 \\ 0 & -2 & 1 & 0 \\ 3 & -1 & 4 & 1 \end{pmatrix}$

9. $\begin{pmatrix} -1 & 1 & 0 \\ 2 & 0 & -3 \\ 5 & -1 & -2 \end{pmatrix}\begin{pmatrix} 1 & -2 & 3 \\ 0 & 1 & 5 \\ -4 & 0 & 2 \end{pmatrix}$

10. $(3 \quad -2 \quad 4)\begin{pmatrix} -1 & 2 \\ 3 & 5 \\ -4 & 1 \end{pmatrix}$ 11. $\begin{pmatrix} 3 & -5 \\ -2 & 4 \end{pmatrix}\begin{pmatrix} x \\ y \end{pmatrix}$

12. $\begin{pmatrix} 4 & -1 \\ 2 & -3 \end{pmatrix} \begin{pmatrix} x \\ y \end{pmatrix}$

13. $\begin{pmatrix} 2 & 3 \\ 4 & 0 \end{pmatrix} \begin{pmatrix} x \\ y \end{pmatrix}$

14. $\begin{pmatrix} 1 & 5 \\ 1 & -3 \end{pmatrix} \begin{pmatrix} x \\ y \end{pmatrix}$

15. $\begin{pmatrix} 1 & 0 \\ 0 & 1 \end{pmatrix} \begin{pmatrix} x \\ y \end{pmatrix}$

16. $\begin{pmatrix} 3 & 5 & 7 \\ 2 & -6 & 4 \\ 1 & -3 & 1 \end{pmatrix} \begin{pmatrix} x \\ y \\ z \end{pmatrix}$

17. $\begin{pmatrix} 2 & -4 & 5 \\ 3 & -7 & 1 \\ 0 & -2 & 6 \end{pmatrix} \begin{pmatrix} x \\ y \\ z \end{pmatrix}$

18. $\begin{pmatrix} 7 & -3 & 4 \\ 5 & -6 & 1 \\ 1 & -4 & 2 \end{pmatrix} \begin{pmatrix} x \\ y \\ z \end{pmatrix}$

19. $\begin{pmatrix} 3 & 5 & 7 \\ 1 & -3 & 4 \\ 1 & -2 & 8 \end{pmatrix} \begin{pmatrix} x \\ y \\ z \end{pmatrix}$

20. $\begin{pmatrix} 1 & 0 & 0 \\ 0 & 1 & 0 \\ 0 & 0 & 1 \end{pmatrix} \begin{pmatrix} x \\ y \\ z \end{pmatrix}$

21. $3 \begin{pmatrix} 2 & 4 \\ -1 & 3 \end{pmatrix}$

22. $5 \begin{pmatrix} 1 & 0 & -1 \\ 2 & 3 & 0 \\ 0 & 1 & -2 \end{pmatrix}$

23. $\dfrac{1}{3} \begin{pmatrix} 3 & 0 & 0 \\ 0 & 3 & 0 \\ 0 & 0 & 3 \end{pmatrix}$

24. $-\dfrac{1}{2} \begin{pmatrix} -2 & 0 \\ 0 & -2 \end{pmatrix}$

25. $(-1) \begin{pmatrix} 3 & 6 & 8 \\ -2 & 4 & -1 \\ -3 & 0 & 5 \end{pmatrix}$

▶ In each of the following, find the product AB and also find the product BA.

EXAMPLE

$A = \begin{pmatrix} 3 & 2 \\ 0 & 1 \end{pmatrix}, \qquad B = \begin{pmatrix} 1 & -4 \\ 2 & 0 \end{pmatrix}$

Solution

$AB = \begin{pmatrix} 7 & -12 \\ 2 & 0 \end{pmatrix}, \qquad BA = \begin{pmatrix} 3 & -2 \\ 6 & 4 \end{pmatrix}$

26. $A = \begin{pmatrix} 2 & -3 \\ 1 & 0 \end{pmatrix}, \qquad B = \begin{pmatrix} 1 & 4 \\ 2 & -3 \end{pmatrix}$

27. $A = \begin{pmatrix} 1 & 3 \\ -2 & 4 \end{pmatrix}, \qquad B = \begin{pmatrix} 0 & 1 \\ 7 & 5 \end{pmatrix}$

28. $A = \begin{pmatrix} 2 & 3 \\ 1 & 2 \end{pmatrix}, \qquad B = \begin{pmatrix} 2 & -3 \\ -1 & 2 \end{pmatrix}$

29. $A = \begin{pmatrix} 1 & 0 \\ 0 & 1 \end{pmatrix}, \qquad B = \begin{pmatrix} 7 & 6 \\ 5 & 4 \end{pmatrix}$

30. $A = \begin{pmatrix} 3 & 8 \\ -1 & 0 \end{pmatrix}, \qquad B = \begin{pmatrix} 2 & 1 \\ -4 & 5 \end{pmatrix}$

▶ Let $A = \begin{pmatrix} 3 & 2 \\ -1 & 0 \end{pmatrix}$, and $B = \begin{pmatrix} 1 & -2 \\ 4 & -1 \end{pmatrix}$. Find each of the following products.

EXAMPLE

$$[3A]B = 3 \times AB = 3\begin{pmatrix} 11 & -8 \\ -1 & 2 \end{pmatrix} = \begin{pmatrix} 33 & -24 \\ -3 & 6 \end{pmatrix}$$

31. $[2A]B$ **32.** $2[AB]$ **33.** $A[2B]$ **34.** $A[3B]$ **35.** $[5A]B$

▶ Write each of the following systems as a matrix equation.

EXAMPLE

$$3x - 5y - 4 = 0$$
$$x + 3y + 1 = 0$$

Solution

First isolate the constant terms.

$$3x - 5y = 4$$
$$x + 3y = -1$$

$$\begin{pmatrix} 3 & -5 \\ 1 & 3 \end{pmatrix}\begin{pmatrix} x \\ y \end{pmatrix} = \begin{pmatrix} 4 \\ -1 \end{pmatrix}$$

36. $2x - 7y = 3$
$\quad\ x - \ y = 4$

37. $3x + 6y - 5 = 0$
$\quad\ x - \ y - 1 = 0$

38. $3x + 2y - 5z - 1 = 0$
$\quad\ x - 3y + \ z + 2 = 0$
$\ 5x - 2y + 3z - 4 = 0$

39. $\ x + y - 5z = 6$
$\ 3x - y + 2z = -3$
$\ x + y + \ z = 2$

40. $\ x + y + 2z - 3w = 1$
$\ x \quad\quad + 5z \quad\quad = 3$
$\ 2x - y + 3z \quad\quad = 4$
$\quad\quad y + 4z - 2w = -5$

▶ Use the definition of equality of $m \times n$ matrices, and write each of the following as a system of equations.

EXAMPLES

(a) $\begin{pmatrix} 3x - 2y \\ x + 5y \end{pmatrix} = \begin{pmatrix} 7 \\ 2 \end{pmatrix}$

(b) $\begin{pmatrix} x \\ y \\ z \end{pmatrix} = \begin{pmatrix} 7 \\ 6 \\ 4 \end{pmatrix}$

Solutions

(a) $3x - 2y = 7$ (b) $x = 7$
      $x + 5y = 2$ $y = 6$
                           $z = 4$

41. $\begin{pmatrix} 2x - 7y \\ 3x + y \end{pmatrix} = \begin{pmatrix} -1 \\ 5 \end{pmatrix}$

42. $\begin{pmatrix} x + y \\ 3x - 2y \end{pmatrix} = \begin{pmatrix} 3 \\ 0 \end{pmatrix}$

43. $\begin{pmatrix} 3x - 2y + 6z \\ x + y - 3z \\ 5x - 4y + 7z \end{pmatrix} = \begin{pmatrix} 2 \\ -1 \\ 3 \end{pmatrix}$

44. $\begin{pmatrix} x \\ y \end{pmatrix} = \begin{pmatrix} -4 \\ 6 \end{pmatrix}$

45. $\begin{pmatrix} x \\ y \\ z \end{pmatrix} = \begin{pmatrix} 3 \\ -4 \\ 7 \end{pmatrix}$

▶ Find each of the following products.

EXAMPLE

$\begin{pmatrix} 3 & 5 & 8 \\ 2 & 7 & 6 \\ 4 & -1 & 9 \end{pmatrix} \begin{pmatrix} 1 & 0 & 0 \\ 0 & 1 & 0 \\ 0 & 0 & 1 \end{pmatrix} = \begin{pmatrix} 3 & 5 & 8 \\ 2 & 7 & 6 \\ 4 & -1 & 9 \end{pmatrix}$

46. $\begin{pmatrix} 1 & 0 & 0 \\ 0 & 1 & 0 \\ 0 & 0 & 1 \end{pmatrix} \begin{pmatrix} 3 & 5 & 8 \\ 2 & 7 & 6 \\ 4 & -1 & 9 \end{pmatrix}$

47. $\begin{pmatrix} 3 & -7 & 4 \\ 2 & 1 & 0 \\ 6 & -3 & 1 \end{pmatrix} \begin{pmatrix} 1 & 0 & 0 \\ 0 & 1 & 0 \\ 0 & 0 & 1 \end{pmatrix}$

48. $\begin{pmatrix} 3 & -5 \\ 2 & -1 \end{pmatrix} \begin{pmatrix} 1 & 0 \\ 0 & 1 \end{pmatrix}$

49. $\begin{pmatrix} 1 & 0 \\ 0 & 1 \end{pmatrix} \begin{pmatrix} 1 & 0 \\ 0 & 1 \end{pmatrix}$

50. $\begin{pmatrix} 3 & 2 & -1 & 7 \\ 4 & -2 & 3 & 5 \\ -4 & 0 & 6 & 1 \\ 2 & 1 & 0 & 3 \end{pmatrix} \begin{pmatrix} 1 & 0 & 0 & 0 \\ 0 & 1 & 0 & 0 \\ 0 & 0 & 1 & 0 \\ 0 & 0 & 0 & 1 \end{pmatrix}$

In each of the following, verify that B is the inverse of A.

$$A = \begin{pmatrix} 3 & 5 \\ -2 & 4 \end{pmatrix}, \qquad B = \frac{1}{22}\begin{pmatrix} 4 & -5 \\ 2 & 3 \end{pmatrix}$$

Solution

$$AB = \begin{pmatrix} 1 & 0 \\ 0 & 1 \end{pmatrix}, \qquad BA = \begin{pmatrix} 1 & 0 \\ 0 & 1 \end{pmatrix}$$

51. $A = \begin{pmatrix} 2 & 7 \\ 1 & 5 \end{pmatrix}, \qquad B = \frac{1}{3}\begin{pmatrix} 5 & -7 \\ -1 & 2 \end{pmatrix}$

52. $A = \begin{pmatrix} 4 & 2 \\ 1 & 2 \end{pmatrix}, \qquad B = \frac{1}{6}\begin{pmatrix} 2 & -2 \\ -1 & 4 \end{pmatrix}$

53. $A = \begin{pmatrix} 0 & 1 \\ 1 & 0 \end{pmatrix}, \qquad B = \begin{pmatrix} 0 & 1 \\ 1 & 0 \end{pmatrix}$

54. $A = \begin{pmatrix} 1 & 0 & -2 \\ 0 & 0 & -1 \\ 1 & 2 & 0 \end{pmatrix}, \qquad B = \frac{1}{2}\begin{pmatrix} 2 & -4 & 0 \\ -1 & 2 & 1 \\ 0 & -2 & 0 \end{pmatrix}$

55. $A = \begin{pmatrix} 0 & 0 & 1 \\ 0 & 1 & 0 \\ 1 & 0 & 0 \end{pmatrix}, \qquad B = \begin{pmatrix} 0 & 0 & 1 \\ 0 & 1 & 0 \\ 1 & 0 & 0 \end{pmatrix}$

Determine whether or not each of the following is nonsingular.

EXAMPLES

(a) $\begin{pmatrix} 3 & 6 \\ 1 & 2 \end{pmatrix}$

(b) $\begin{pmatrix} 1 & 0 & 2 \\ 0 & 2 & 0 \\ -3 & 1 & 0 \end{pmatrix}$

Solutions

(a) $\begin{vmatrix} 3 & 6 \\ 1 & 2 \end{vmatrix} = 0$, not nonsingular

(b) $\begin{vmatrix} 1 & 0 & 2 \\ 0 & 2 & 0 \\ -3 & 1 & 0 \end{vmatrix} = 2(6) = 12 \neq 0$, nonsingular

56. $\begin{pmatrix} 3 & 6 \\ 1 & 4 \end{pmatrix}$

57. $\begin{pmatrix} 2 & 4 \\ 1 & 2 \end{pmatrix}$

58. $\begin{pmatrix} 5 & 0 & 4 \\ 3 & 0 & 1 \\ -2 & 0 & 7 \end{pmatrix}$

59. $\begin{pmatrix} 3 & 2 & 1 \\ -1 & 3 & 0 \\ 6 & 4 & 2 \end{pmatrix}$

60. $\begin{pmatrix} 3 & 0 & 0 \\ 0 & 3 & 0 \\ 0 & 0 & 3 \end{pmatrix}$

▶ If possible, find the inverse of each of the following. If the inverse does not exist, make the statement: "No inverse."

EXAMPLES

(a) $\begin{pmatrix} 4 & -6 \\ -2 & 3 \end{pmatrix}$

(b) $\begin{pmatrix} 1 & 0 & -3 \\ 0 & 2 & 3 \\ 0 & 0 & -1 \end{pmatrix}$

Solutions

(a) $\begin{vmatrix} 4 & -6 \\ -2 & 3 \end{vmatrix} = 0$, no inverse

Step 1 (b) Evaluate the determinant.

$$\begin{vmatrix} 1 & 0 & -3 \\ 0 & 2 & 3 \\ 0 & 0 & -1 \end{vmatrix} = -2$$

Step 2 Find the transpose.

$$\begin{pmatrix} 1 & 0 & 0 \\ 0 & 2 & 0 \\ -3 & 3 & -1 \end{pmatrix}$$

Step 3 Replace each element of transpose by its minor or the negative of its minor, depending on its position.

$$\begin{pmatrix} -2 & 0 & 6 \\ 0 & -1 & -3 \\ 0 & 0 & 2 \end{pmatrix}$$

Step 4 Multiply by reciprocal of determinant. The inverse is $\dfrac{-1}{2}\begin{pmatrix} -2 & 0 & 6 \\ 0 & -1 & -3 \\ 0 & 0 & 2 \end{pmatrix}$.

61. $\begin{pmatrix} 4 & 3 \\ -2 & 1 \end{pmatrix}$

62. $\begin{pmatrix} 1 & 3 \\ 3 & 9 \end{pmatrix}$

63. $\begin{pmatrix} 1 & 0 & 1 \\ 2 & 1 & 2 \\ 3 & 2 & 3 \end{pmatrix}$

64. $\begin{pmatrix} 1 & -1 & 2 \\ 0 & 3 & 4 \\ 0 & 1 & -2 \end{pmatrix}$

65. $\begin{pmatrix} 0 & 0 & 2 \\ 1 & -1 & 3 \\ 2 & 1 & 0 \end{pmatrix}$

66. $\begin{pmatrix} 2 & 0 & 0 \\ 0 & 2 & 0 \\ 0 & 0 & 2 \end{pmatrix}$

67. $\begin{pmatrix} 1 & -1 & 1 \\ -1 & 2 & 1 \\ -3 & 1 & 0 \end{pmatrix}$ **68.** $\begin{pmatrix} 3 & 9 \\ 2 & 6 \end{pmatrix}$

69. $\begin{pmatrix} 1 & 0 & 0 \\ 0 & 3 & 0 \\ -1 & 2 & 4 \end{pmatrix}$ **70.** $\begin{pmatrix} -1 & 0 & 1 & 2 \\ 0 & 0 & 1 & 3 \\ 0 & 1 & 0 & 4 \\ 0 & 2 & 0 & 0 \end{pmatrix}$

▶ **71–90** Write each of the systems in problems 1–20 of Exercise 3.2 as a matrix equation. If possible, use the inverse of the coefficient matrix to find the solution.

EXAMPLE

$2x - y - 5 = 0$

$x + 3y + 8 = 0$

Solution

First isolate the constant terms.

$$2x - y = 5$$
$$x + 3y = -8$$

$$\begin{pmatrix} 2 & -1 \\ 1 & 3 \end{pmatrix}\begin{pmatrix} x \\ y \end{pmatrix} = \begin{pmatrix} 5 \\ -8 \end{pmatrix}$$

The inverse of

$$\begin{pmatrix} 2 & -1 \\ 1 & 3 \end{pmatrix} \text{ is } \frac{1}{7}\begin{pmatrix} 3 & 1 \\ -1 & 2 \end{pmatrix}.$$

$$\begin{pmatrix} x \\ y \end{pmatrix} = \frac{1}{7}\begin{pmatrix} 3 & 1 \\ -1 & 2 \end{pmatrix} \times \begin{pmatrix} 5 \\ -8 \end{pmatrix}$$

$$= \frac{1}{7}\begin{pmatrix} 7 \\ -21 \end{pmatrix}$$

$$= \begin{pmatrix} 1 \\ -3 \end{pmatrix}$$

The solution is $x = 1, y = -3$.

CHECK LIST FOR THE STUDENT

If you have learned the material in Section 3.4. you should be able to do the following.

1. Find the matrix product AB where A has the same number of columns as the number of rows in B.

2. Tell how many rows and columns there are in the matrix product AB, if A is an $m \times n$ matrix, and B is an $n \times p$ matrix.

3. Find the product cA where c is a real number and A is a matrix whose elements are real numbers.

4. Show that the matrix product AB is not necessarily the same as the matrix product BA by computing AB and also BA for particular 2×2 matrices A and B.

5. Verify for particular matrices A and B, conformable for the multiplication AB, that $[cA]B = c[AB]$ and $A[cB] = c[AB]$ for any real number c.

6. Write a given system of n simultaneous linear equations in n unknowns as a matrix equation.

7. Use the definition of equality of $m \times n$ matrices.

8. Write the identity matrix I for a specified order.

9. Verify that $AI = A$ and $IA = A$ for a given square matrix A.

10. Verify that $AA^{-1} = I$ and $A^{-1}A = I$ for a given square matrix A and its inverse A^{-1}.

11. Determine whether or not a given square matrix is nonsingular.

12. Find the transpose of a given square matrix.

13. Find the inverse of a given nonsingular matrix.

14. Solve a given system of n simultaneous linear equations in n unknowns having a nonsingular coefficient matrix by writing the system as a matrix equation and using the inverse of the coefficient matrix.

SECTION QUIZ 3.4

1. If A is a 3×4 matrix and B is a 4×5 matrix, how many rows and how many columns are there in the matrix product AB?

Find the products in problems 2—5.

2. $\begin{pmatrix} 3 & 1 \\ -2 & 4 \end{pmatrix} \begin{pmatrix} -1 & 2 \\ 5 & -3 \end{pmatrix}$

3. $\begin{pmatrix} 2 & 3 & -1 \\ 5 & -6 & 4 \end{pmatrix} \begin{pmatrix} 3 & 2 \\ 0 & -1 \\ 1 & -2 \end{pmatrix}$

4. $\begin{pmatrix} 3 & 1 \\ -2 & 4 \end{pmatrix} \begin{pmatrix} x \\ y \end{pmatrix}$

5. $\dfrac{1}{5}\begin{pmatrix} 5 & 0 & 0 \\ 0 & 5 & 0 \\ 0 & 0 & 5 \end{pmatrix}$

6. Let $A = \begin{pmatrix} 5 & 2 \\ 1 & 4 \end{pmatrix}$ and $B = \begin{pmatrix} -2 & 1 \\ 0 & 5 \end{pmatrix}$. Find the product $A[3B]$.

7. Write the following system as a matrix equation.

$$2x - 3y - 5 = 0$$
$$4x - y + 1 = 0$$

8. Write the following as a system of equations.

$$\begin{pmatrix} 3x - 2y \\ x - y \end{pmatrix} = \begin{pmatrix} 5 \\ 4 \end{pmatrix}$$

9. Write the identity matrix for order 3.

10. Let $A = \begin{pmatrix} 3 & 4 \\ -2 & 5 \end{pmatrix}$ and let I be the identity matrix for order 2. Find the product AI and the product IA.

11. Verify by multiplication that $\dfrac{1}{17}\begin{pmatrix} 1 & -5 \\ 3 & 2 \end{pmatrix}$ is the inverse of $\begin{pmatrix} 2 & 5 \\ -3 & 1 \end{pmatrix}$.

12. Determine whether or not $\begin{pmatrix} 3 & 2 \\ 6 & 4 \end{pmatrix}$ is nonsingular.

13. Find the transpose of $\begin{pmatrix} 2 & 7 & 6 \\ 3 & 4 & 5 \\ 9 & 8 & 1 \end{pmatrix}$.

14. Find the inverse of $\begin{pmatrix} 1 & 3 & 0 \\ 0 & 0 & -1 \\ 4 & 0 & 0 \end{pmatrix}$.

15. Solve the following system of equations by writing the system as a matrix equation and using the inverse of the coefficient matrix.

$$5x - 2y = 1$$
$$x + 3y = -10$$

3.5 Partial Fractions

The purpose of this section is to present the partial fraction theorem and to demonstrate an application of methods of solving systems of simultaneous linear equations.

In basic algebra we learn how to express a sum of fractions as a single fraction. As we observed in Sections 1.6 and 1.10, it is sometimes necessary to work in reverse and express a fraction whose numerator and denominator are polynomials as a sum of "partial" fractions whose denominators are of lower degree than the denominator of the given fraction. This decomposition can be effected by applying the *partial fraction theorem*.

Before stating the partial fraction theorem, let us review some facts about polynomials.

1. Where a and b are real numbers, and $a \neq 0$, the expression $ax + b$ is a **linear** or *first-degree* polynomial.

Example: $3x + 7$

2. Where a, b, and c are real numbers, and $a \neq 0$, $ax^2 + bx + c$ is a **quadratic** or *second-degree* polynomial.

Example: $3x^2 + 5x - 2$

3. Any quadratic with real coefficients that cannot be factored into linear factors with real coefficients is called an **irreducible quadratic.**

Example: $3x^2 + 2x + 1$

4. If the quadratic equation $ax^2 + bx + c = 0$ has only imaginary roots, then $ax^2 + bx + c$ is an irreducible quadratic.

Example: $3x^2 + 2x + 1 = 0$

5. Any polynomial with real coefficients can be factored into a product of linear factors with real coefficients and irreducible quadratic factors. (See the factorization theorem. Section 3.1.)

Example: $x^4 + 3x^3 + 5x^2 + x - 10 = (x + 2)(x - 1)(x^2 + 2x + 5)$

6. In the factorization of a polynomial $P(x)$ into a product of linear factors and irreducible quadratic factors, one or more of the factors may be repeated. If k is a positive integer and $P(x)$ has $(ax + b)^k$ or $(ax^2 + bx + c)^k$ as a factor but has no higher power of $ax + b$ or of $ax^2 + bx + c$ as a factor, then the factor $(ax + b)$ or $(ax^2 + bx + c)$ is said to be of **multiplicity k.**

Example:

$$P(x) = (x + 2)^3(x - 1)(x^2 + 2x + 5)^2$$

$(x + 2)$ of multiplicity 3

$(x^2 + 2x + 5)$ of multiplicity 2

Now let us state the partial fraction theorem. (The proof will not be included in this text.)

Partial Fraction Theorem   If $N(x)$ and nonzero $D(x)$ are polynomials in x with real coefficients and with $N(x)$ of lower degree than $D(x)$, then $N(x)/D(x)$ can be expressed as a sum of partial fractions as follows:

1. For each unrepeated linear factor $(ax + b)$ of $D(x)$, there is a partial fraction of the form

$$\frac{A}{ax + b}$$

where A is a specific real number.

Example: For unrepeated factor $(2x + 7)$, there is a partial fraction $A/(2x + 7)$.

2. For each repeated linear factor $(ax + b)$ of multiplicity k, there is a sum of partial fractions of the form

$$\frac{A_1}{ax + b} + \frac{A_2}{(ax + b)^2} + \frac{A_3}{(ax + b)^3} + \cdots + \frac{A_k}{(ax + b)^k}$$

where $A_1, A_2, A_3, \ldots, A_k$ are specific real numbers.

Example: For factor $(2x + 7)$ of multiplicity 3, there is a sum of 3 partial fractions of the form

$$\frac{A_1}{2x + 7} + \frac{A_2}{(2x + 7)^2} + \frac{A_3}{(2x + 7)^3}.$$

3. For each unrepeated irreducible quadratic factor $(ax^2 + bx + c)$, there is a partial fraction of the form

$$\frac{Ax + B}{ax^2 + bx + c}$$

where A and B are specific real numbers.

Example: For unrepeated factor $(3x^2 + 2x + 1)$, there is a partial fraction

$$\frac{Ax + B}{3x^2 + 2x + 1}.$$

4. For each repeated irreducible quadratic factor $(ax^2 + bx + c)$ of multiplicity k, there is a sum of k partial fractions of the form

$$\frac{A_1x + B_1}{ax^2 + bx + c} + \frac{A_2x + B_2}{(ax^2 + bx + c)^2} + \cdots + \frac{A_kx + B_k}{(ax^2 + bx + c)^k}$$

where $A_1, B_1, A_2, B_2, \ldots, A_k, B_k$ are specific real numbers.

Example: For factor $(x^2 + 2x + 5)$ of multiplicity 3, there is a sum of 3 partial fractions of the form

$$\frac{A_1 x + B_1}{x^2 + 2x + 5} + \frac{A_2 x + B_2}{(x^2 + 2x + 5)^2} + \frac{A_3 x + B_3}{(x^2 + 2x + 5)^3}.$$

Note that the partial fraction theorem is stated only for the case in which the numerator is of lower degree than the denominator. If the numerator $N(x)$ is of the same or of higher degree than the denominator $D(x)$, we can divide $N(x)$ by $D(x)$ until the remainder $R(x)$ is of lower degree than $D(x)$; we can then apply the partial fraction theorem to the fraction $R(x)/D(x)$.

EXAMPLE 1

Write out the form of the partial-fraction decomposition of

$$\frac{2x^3 + 3x^2 - 24x - 23}{2x^2 - 5x - 7}.$$

Solution

Since the numerator is not of lower degree than the denominator, we must first divide.

$$
\begin{array}{r}
x + 4 \\
2x^2 - 5x - 7 \overline{)\, 2x^3 + 3x^2 - 24x - 23} \\
\underline{2x^3 - 5x^2 - 7x} \\
8x^2 - 17x - 23 \\
\underline{8x^2 - 20x - 28} \\
3x + 5
\end{array}
$$

$$\frac{2x^3 + 3x^2 - 24x - 23}{2x^2 - 5x - 7} = x + 4 + \frac{3x + 5}{2x^2 - 5x - 7}$$

$$= x + 4 + \frac{3x + 5}{(2x - 7)(x + 1)}$$

We may now express

$$\frac{3x + 5}{2x^2 - 5x - 7}$$

as a sum of partial fractions. The factors of the denominator are $(2x - 7)$ and $(x + 1)$.

For factor $(2x - 7)$ there is a partial fraction $A/(2x - 7)$.

For factor $(x + 1)$ there is a partial fraction $B/(x + 1)$.

The decomposition of the given fraction must be of the form

$$\frac{2x^3 + 3x^2 - 24x - 23}{2x^2 - 5x - 7} = x + 4 + \frac{A}{2x - 7} + \frac{B}{x + 1}.$$

EXAMPLE 2

Write out the form of the partial-fraction decomposition of

$$\frac{x^2 - 4x - 5}{(x - 3)(x - 1)^2(x^2 + 2x + 5)}.$$

Solution

The numerator is of lower degree than the denominator; the partial fraction theorem applies. The factors of the denominator are $(x - 3)$, $(x - 1)^2$, and $(x^2 + 2x + 5)$.

For factor $(x - 3)$ there is a partial fraction $A/(x - 3)$.

For factor $(x - 1)^2$ there is a sum of two partial fractions,

$$\frac{B}{x - 1} + \frac{C}{(x - 1)^2}.$$

For factor $(x^2 + 2x + 5)$ there is a partial fraction

$$\frac{Dx + E}{x^2 + 2x + 5}.$$

$$\frac{x^2 - 4x - 5}{(x - 3)(x - 1)^2(x^2 + 2x + 5)} = \frac{A}{x - 3} + \frac{B}{x - 1} + \frac{C}{(x - 1)^2}$$
$$+ \frac{Dx + E}{x^2 + 2x + 5}$$

EXAMPLE 3

Write out the form of the partial-fraction decomposition of

$$\frac{2x^2 - 5x - 3}{(x + 5)(x^2 + 9)^2}$$

Solution

The numerator is of lower degree than the denominator; the partial fraction theorem applies. The factors of the denominator are $(x + 5)$ and $(x^2 + 9)^2$.

For factor $(x + 5)$ there is a partial fraction $A/(x + 5)$.

For factor $(x^2 + 9)^2$ there is a sum of two partial fractions,

$$\frac{Bx + C}{x^2 + 9} + \frac{Dx + E}{(x^2 + 9)^2}.$$

Exercise 3.5, 1–5

$$\frac{2x^2 - 5x - 3}{(x + 5)(x^2 + 9)^2} = \frac{A}{x + 5} + \frac{Bx + C}{x^2 + 9} + \frac{Dx + E}{(x^2 + 9)^2}.$$

The next example illustrates an application of methods of solving systems of simultaneous linear equations. We have had several of these methods. Elementary methods were presented in Section 1.10, and methods using matrices and determinants were presented in Sections 3.2, 3.3, and 3.4.

EXAMPLE 4

Express the following as a sum of partial fractions.

$$\frac{6x^2 + 7x - 4}{x(x - 1)(x + 2)}$$

Solution

The numerator is of lower degree than the denominator; the partial-fraction theorem applies. The factors of the denominator are the unrepeated linear factors x, $x - 1$, and $x + 2$. The decomposition required is of the form

$$\frac{6x^2 + 7x - 4}{x(x - 1)(x + 2)} = \frac{A}{x} + \frac{B}{x - 1} + \frac{C}{x + 2}.$$

In Example 3 of Section 1.10, we assumed that this was the form of the decomposition of the given fraction, and we found A, B, and C by two methods. The first method, *equating coefficients of like powers*, led to a system of three equations in the three "unknowns" A, B, and C. We obtained that system by first combining the three partial fractions on the right and then equating coefficients of like terms in the numerators on the left and right of the equation. In particular, we obtained the system

$$\begin{aligned} A + B + C &= 6 \\ A + 2B - C &= 7 \\ -2A &= -4. \end{aligned}$$

We solved this system for A, B, and C by the methods of Section 1.10. We may now solve this system by using Gaussian elimination or by using Cramer's rule.

The second method presented in Section 1.10 for finding A, B, and C, *substituting convenient values,* is easier in this case. If repeated factors appear in the denominator, however, it might be easier to use the first method or a combination of the two methods.

Exercise 3.5, 6–15

EXERCISE 3.5 PARTIAL FRACTIONS

▶ Write out the form of the partial-fraction decomposition of each of the following.

EXAMPLES

(a) $\dfrac{3x^2 - 5x + 9}{(x - 3)^2(x^2 + 2x + 3)} = \dfrac{A}{x - 3} + \dfrac{B}{(x - 3)^2} + \dfrac{Cx + D}{x^2 + 2x + 3}$

(b) $\dfrac{2x^3 + 1}{(x - 5)(x^2 + 2x + 3)^2} = \dfrac{A}{x - 5} + \dfrac{Bx + C}{x^2 + 2x + 3} + \dfrac{Dx + E}{(x^2 + 2x + 3)^2}$

1. $\dfrac{x^2 - 7x + 5}{x(x - 3)(x + 7)}$

2. $\dfrac{3x^2 - 5x + 1}{(x - 2)^3(x^2 + 1)}$

3. $\dfrac{2x + 1}{x^2 - 6x + 9}$

4. $\dfrac{x^3 - 5x}{(x^2 + 1)^3(x - 1)}$

5. $\dfrac{3x^2 - 2}{(x - 1)^2(3x^2 + x + 1)}$

▶ Express each of the following as a sum of partial fractions. Unless otherwise specified by your instructor, use any method you wish.

EXAMPLE

$\dfrac{2x + 1}{(x - 1)^2}$

Solution

$$\frac{2x + 1}{(x - 1)^2} = \frac{A}{x - 1} + \frac{B}{(x - 1)^2}$$

$$= \frac{A(x - 1) + B}{(x - 1)^2}$$

$$2x + 1 = A(x - 1) + B$$

Let $x = 1$. $\qquad\qquad\qquad 3 = B$

Equate coefficients of x. $\qquad 2 = A$

$$\frac{2x + 1}{(x - 1)^2} = \frac{2}{x - 1} + \frac{3}{(x - 1)^2}$$

6. $\dfrac{4x - 5}{(x - 2)^2}$

7. $\dfrac{2x^2 - 11x + 3}{x(x - 1)(x - 3)}$

8. $\dfrac{-x^2 + 4x + 2}{(x - 1)(x^2 + x + 1)}$

9. $\dfrac{-5x^2 - 11x - 7}{(x - 2)(x^2 + x + 1)^2}$

10. $\dfrac{x^3 + 5x^2 + 13x + 7}{(x^2 + 3x + 4)^2}$

11. $\dfrac{x^3 + 7x^2 + 6x - 42}{x^2 + 2x - 8}$

12. $\dfrac{2x^3 - x^2 + 4x - 7}{x^2 + x + 2}$

13. $\dfrac{x^2 - 22x - 30}{x(x^2 + 3x - 10)}$

14. $\dfrac{x^2 - 2x + 2}{(x - 3)^3}$

15. $\dfrac{x^4 - 3x^3 + x^2 - 19x - 22}{(x - 2)(x^2 + 4)^2}$

CHECK LIST FOR THE STUDENT

If you have learned the material in Section 3.5, you should be able to do the following:

1. Give the form of the partial fraction(s) associated with each of the following types of factors of $D(x)$, where $N(x)$ and nonzero $D(x)$ are polynomials with real coefficients, and where the numerator of $N(x)/D(x)$ is of lower degree than the denominator.

    (a) Unrepeated linear: $ax + b$
    (b) Repeated linear of multiplicity k: $(ax + b)^k$
    (c) Unrepeated irreducible quadratic: $ax^2 + bx + c$
    (d) Repeated irreducible quadratic of multiplicity k: $(ax^2 + bx + c)^k$

2. Write out the form of the partial-fraction decomposition of a given $N(x)/D(x)$ where $N(x)$ and nonzero $D(x)$ are polynomials with real coefficients, where the degree of $N(x)$ is not necessarily lower than the degree of $D(x)$, and where $D(x)$ has been factored into a product of linear and irreducible quadratic factors.

3. Express a given fraction whose numerator and nonzero denominator are polynomials with real coefficients as a sum of partial fractions, and show how this could involve the solution of a system of simultaneous linear equations.

SECTION QUIZ 3.5

Write out the form of the partial-fraction decomposition of each of the following.

1. $\dfrac{2x^2 - 4x + 7}{(x - 4)(x + 5)^2}$

2. $\dfrac{5x^2 - 3x + 8}{(x - 3)(x^2 - 4x + 29)}$

3. $\dfrac{2x^4 + 1}{(x - 3)(x^2 - 4x + 29)^2}$

Express each of the following as a sum of partial fractions.

4. $\dfrac{2x - 3}{(x - 3)^2}$

5. $\dfrac{21x + 15}{x(x + 1)(x - 5)}$

3.6 Systems of Linear Inequalities in Two Variables

The purpose of this section is to consider linear inequalities in two variables, to show how to solve systems of simultaneous linear inequalities in two variables, and to suggest some applications.

Properties of inequalities and linear inequalities in one variable were considered in Section 1–9. In this section we shall consider linear inequalities in *two* variables. We shall consider inequalities such as

$$2x - y + 1 \geq 0.$$

A **linear inequality in two variables**, x and y, is an inequality that can be written in one of the forms

$$ax + by + c < 0 \qquad ax + by + c > 0$$
$$ax + by + c \leq 0 \qquad ax + by + c \geq 0$$

where $a \neq 0$, $b \neq 0$, and a, b, and c are specific real numbers. For example, $y \leq 2x + 1$ is a linear inequality in the two variables x and y because it can be written in one of the prescribed forms.

A **solution of an inequality in exactly two variables** with real numbers for coefficients is an ordered pair of real numbers that *satisfy* the inequality in the sense that the inequality becomes a true statement when the members of the ordered pair replace the variables. The *solution set* or simply *the solution* is the set of all ordered pairs of real numbers that satisfy the inequality. The pairs $(1, -1)$, $(3, 5)$, and $(4, 9)$ are some of the solutions (x, y) of

$$y \leq 2x + 1$$

because the members of each of these ordered pairs satisfy the inequality. The statements

$$-1 \leq 2(1) + 1, \quad 5 \leq 2(3) + 1, \quad \text{and} \quad 9 \leq 2(4) + 1$$

are all true.

The **graph** of the linear inequality

$$ax + by + c \leq 0 \quad \text{or of} \quad ax + by + c \geq 0$$

consists of all points on the line $ax + by + c = 0$, together with all points in one of the half-planes determined by the line. Likewise, the graph of the linear inequality

$$ax + by + c < 0 \quad \text{or of} \quad ax + by + c > 0$$

consists only of all points in one of the half-planes determined by the line $ax + by + c = 0$. To find out which half-plane, simply choose a convenient point in one half-plane and substitute its coordinates in the given inequality to determine whether or not they satisfy the inequality.

EXAMPLE 1

Solution

Sketch the graph of $y \leq 2x + 1$.

Every point on the line $y = 2x + 1$ has coordinates that satisfy the inequality, so the line $y = 2x + 1$ is part of the graph. It is shown in Figure 3.4. The rest of the graph consists of all the points on one "side" of the line $y = 2x + 1$. To determine which side, choose a convenient point and substitute its coordinates in the inequality. Let us choose the point $(0, 0)$ Since the inequality

$$0 \leq 2(0) + 1$$

is true, the point $(0, 0)$ is on the graph and so also are all the other points on that side of the line. This is shown by the shading in Figure 3.4.

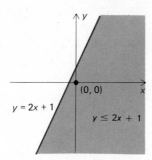

FIGURE 3.4   Graph of $y \leq 2x + 1$

Exercise 3.6, 1–15

Notice that the line $y = 2x + 1$ is very definitely included in the graph of $y \leq 2x + 1$. If the inequality had been $y < 2x + 1$, the line would not have been part of the graph. To show this, we would have drawn it as a dashed line instead of a solid line.

Sometimes it is necessary to consider a system of inequalities in two variables. The *solution set* or simply *the solution* of a system of inequalities in two variables with real-number coefficients is the set of all ordered pairs whose coordinates satisfy each and every one of the inequalities of the system. It is the *intersection* of the solution sets of all the individual inequalities in the system. One way to find the solution of a system graphically is to sketch the graphs of all the inequalities in the system on the same coordinate system, using a different color or different shading for each graph, and then note the intersection of all these graphs. This is illustrated in the next example.

EXAMPLE 2

Sketch the graph of the solution of the system

$$y \leq x - 3$$
$$y < -x + 5$$
$$y \geq 0.$$

Solution   We first sketch the graphs of the lines $y = x - 3$, $y = -x + 5$, and $y = 0$. They are shown in Figure 3.5. We drew the lines $y = x - 3$ and $y = 0$ solid because points on them are included on the graph; we drew the line $y = -x + 5$ dashed because no points on it are included on the graph.

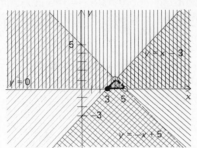

FIGURE 3.5   Graph of
$$\begin{cases} y \le x - 3 \\ y < -x + 5 \\ y \ge 0 \end{cases}$$

The graph of $y \le x - 3$ consists of all points on or below $y = x - 3$; the graph of $y < -x + 5$ consists of all points below $y = -x + 5$; the graph of $y \ge 0$ consists of all points on or above the x-axis. The set of all points common to all three of these graphs is the graph of the solution of the system. It is shown shaded in Figure 3.5. It consists of the interior, two sides, and one vertex of a triangle.

To show that the points of intersection of the excluded line $y = -x + 5$ and the lines $y = x - 3$ and $y = 0$ are not included, we put open dots around them. To show that the point of intersection of $y = x - 3$ and $y = 0$ *is* included, we put a solid dot there.

Exercise 3.6, 16–20

Systems of inequalities in two variables arise in many contexts. The next two examples suggest how they might arise in such diverse areas as manufacturing and archeological research. In each of these examples, two variables are subject to certain constraints that form a system of linear inequalities, and the problem is to find values of the variables that will make a certain linear expression in these variables take on its maximum value subject to the given constraints. In the mathematical subject called *linear programming*, it is shown that the maximum (or minimum) value of a linear expression that is defined subject to such constraints is taken on at the "corners" or vertices of the graph of the system of inequalities specifying the constraints. Accordingly, to solve problems such as those illustrated in the next two examples, we need only set up the system of inequalities specifying the constraints, sketch the graph of the system, locate the "corners," and then systematically compute the expression that is to be made a maximum (or minimum) at these corners.

EXAMPLE 3

A company makes flutes and piccolos. They can make no more than 500 instruments a year. They must make at least 200 flutes and at least 100 piccolos to satisfy their regular annual orders. They make a profit of $250 on each flute and $200 on each piccolo. If they sell all their instruments, how many of each kind must they make in a given year in order to realize the maximum profit?

Solution

Let x be the number of flutes.
Let y be the number of piccolos.

The annual profit in dollars is

$$P = 250x + 200y.$$

The constraints are

$$x + y \leq 500$$
$$x \geq 200$$
$$y \geq 100.$$

According to the theorem from linear programming that we referred to earlier, the maximum value of P will be taken on at one of the "corners" or vertices of the graph of this system of inequalities.

The graph of the system is shown in Figure 3.6.

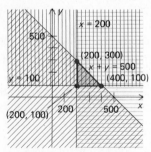

FIGURE 3.6 Graph of
$$\begin{cases} x + y \leq 500 \\ x \geq 200 \\ y \geq 100 \end{cases}$$

The vertices of the graph are

$$(200, 100), \quad (200, 300), \quad \text{and} \quad (400, 100).$$

The values of the annual profit P at these vertices are given in the following table.

Vertices	$P = 250x + 200y$
(200, 100)	$50{,}000 + 20{,}000 = 70{,}000$
(200, 300)	$50{,}000 + 60{,}000 = 110{,}000$
(400, 100)	$100{,}000 + 20{,}000 = 120{,}000$

The maximum value of P is taken on at the vertex (400, 100). To realize the maximum profit, the company should make 400 flutes and 100 piccolos.

(The solution to this problem might appear obvious, but the problem presents an application and serves as an illustration of a technique that can be used in more complicated problems whose solutions are not obvious.)

EXAMPLE 4

Professor Carberry, the psychoceramist, was granted $500 and 100 sections of a desert to search for archaic ceramics. He planned to use two research techniques: the new Smiley method, and the traditional shovel method. He estimated that the Smiley method would locate 1 pot per 5 sections at a cost of $50 per pot, and the value of each pot located this way would be $10. (These pots would probably not be cracked.) The traditional method locates 1 pot per section at a cost of only $2 per pot, but the pots located this way are in poorer condition and are worth only $1 each. On how many sections should Carberry use each of these methods in order for the collection of pots located to have the maximum value?

Solution

Let x be the number of sections researched by the Smiley method.

Let y be the number of sections researched by the traditional shovel method.

The number of pots located by each of these methods is $x/5$ and $y/1$, respectively. The value of the located pots in dollars is

$$V = 10\left(\frac{x}{5}\right) + \left(\frac{y}{1}\right)$$

$$= 2x + y.$$

The cost of the project is

$$50\left(\frac{x}{5}\right) + 2\left(\frac{y}{1}\right)$$

or
$$10x \times 2y.$$

The constraints of the problem are

$$x + y \leq 100$$
$$10x + 2y \leq 500$$
$$x \geq 0 \qquad \text{(Number of sections is nonnegative.)}$$
$$y \geq 0.$$

The linear programming theorem we have been using assures us that the maximum value of V will occur at one of the "corners" or vertices of the graph of this system of inequalities. The graph of the system is shown in Figure 3.7. Three of the vertices are at $(0, 0)$, $(0, 100)$, and $(50, 0)$. The fourth vertex is the point of intersection of the lines $x + y = 100$ and $10x + 2y = 500$ or $5x + y = 250$. Solving the equations of these lines simultaneously, we have

$$\begin{aligned} x + y &= 100 \\ 5x + y &= 250 \\ \hline -4x &= -150 \end{aligned}$$

$$x = \frac{150}{4} = \frac{75}{2}$$

$$y = 100 - \frac{75}{2} = \frac{125}{2}.$$

The fourth vertex is at $(\frac{75}{2}, \frac{125}{2})$.

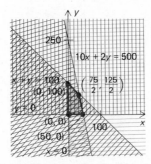

FIGURE 3.7 Graph of
$$\begin{aligned} x + y &\leq 100 \\ 10x + 2y &\leq 500 \\ x &\geq 0 \\ y &\geq 0 \end{aligned}$$

The values of $V = 2x + y$ at the four vertices are listed in the following table.

Vertices	$V = 2x + y$
(0, 0)	0
(0, 100)	\$100
(50, 0)	\$100
$(\frac{75}{2}, \frac{125}{2})$	\$137.50

The maximum value of V is at $(\frac{75}{2}, \frac{125}{2})$, that is, at $(37\frac{1}{2}, 62\frac{1}{2})$. In order to obtain the most valuable collection of pots within the given constraints, Carberry should use the Smiley method on $37\frac{1}{2}$ sections and the traditional shovel method on $62\frac{1}{2}$ sections.

Exercise 3.6, 21–25

EXERCISE 3.6 SYSTEMS OF LINEAR INEQUALITIES IN TWO VARIABLES

▶ Sketch the graph of each of the following.

EXAMPLE

$3x + 2y - 12 > 0$

Solution

The line $3x + 2y - 12 = 0$ is not part of the graph so it has been drawn dashed. Choose the point (0, 0) on one side. The statement $0 + 0 - 12 > 0$ is not true, so (0, 0) is not on the graph. The graph must consist of all points on the other side.

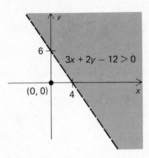

1. $3x + 2y - 12 < 0$ **2.** $3x + 2y - 12 \leq 0$

3. $3x + 2y - 12 \geq 0$ **4.** $x \leq 5$

5. $y > 3$ **6.** $y \geq 2x + 5$

7. $x > 0$ **8.** $y < 3x - 2$

9. $y \geq -1$ **10.** $x + y \leq 4$

11. $3x + 2y - 7 < 0$ **12.** $3x + 2y - 7 \leq 0$

13. $2x - 7y > 1$ **14.** $2x - 7y \geq 1$

15. $x - y \leq 1$

► Sketch the graph of the solution set of each of the following systems.

EXAMPLE

$x + y \leq 4$

$y \geq 0$

$y \leq x + 2$

Solution

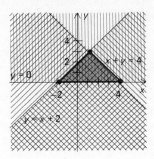

16. $x + y \leq 4$
$y \geq 0$
$x \geq 0$
$y \leq x + 2$

17. $x + y < 4$
$y > 2$
$y < x + 2$

18. $x + 2y \leq 4$
$x \geq 0$
$y > 1$

19. $3x + 2y \leq 12$
$y < x + 1$
$y \geq 0$

20. $4x - 3y \geq 12$
$x + y \leq 3$

21. A farmer has no more than 500 acres on which to plant corn and rye. He must use at least 100 acres for corn and at least 200 acres for rye. He can make a profit of $80 per acre on corn and a profit of $75 per acre on rye. If he sells all of his corn and rye, how many acres of each should he raise to make the greatest profit?

22. A 4-H club was granted $500 with which to experiment with two kinds of fertilizer on 100 acres of land. If they use Type X, they will need 2 acres of land for each carload of produce at a cost of $50 per carload, and the profit on this produce will be $10 per carload more than the average profit. If they use Type Y, they will need only 1 acre of land for.each carload of produce at a cost of only $5 per carload, but the profit on this produce would be only $2 per carload more than the average profit. On how many acres should they use each type of fertilizer in order to obtain the maximum profit more than the average profit?

23. A warehouse fire forced a manufacturer to sell at a loss at least enough boxes of product A and product B to make up 40,000 cubic meters of space. The manufacturer lost $1 on each box sold of product A and $2 on each box sold of product B. It was necessary to sell at least 1000 boxes of product A. If each box of product A takes 8 cubic meters and each box of product B takes 4 cubic meters, how many boxes of each product should he sell in order to minimize his loss?

24. A publisher has 4 square meters at a convention to display textbooks. She wants to display two kinds, school books and college books. She has no more than 8 square meters of school books to display and no more than 6 square meters of college books to display. She estimates that each square meter of school books on display attracts 4 requests per hour for sample copies, and each square meter of college books attracts 3 requests per hour for sample copies. How many square meters of each should she display in order to attract the maximum number of requests per hour for sample copies?

25. Dr. Trebbin can see no more than 24 of his patients per day. He must schedule at least 10 regular check-ups per day. He always schedules at least 3 emergencies per day. Suppose that he spends $\frac{1}{3}$ hour with each regular check-up and $\frac{1}{2}$ hour with each emergency. If the number of regular check-ups per day is at least 3 times the number of emergencies per day, how many of each kind of patient should he schedule a day in order for him to spend the maximum amount of time per day treating his patients? In that case, what would be the number of hours per day he would be working with patients?

CHECK LIST FOR THE STUDENT

If you have learned the material in Section 3.6, you should be able to do the following:

1. Sketch the graph of a given linear inequality in two variables.

2. Sketch the graph of a given system of linear inequalities in two variables.

3. Use a basic theorem from linear programming to find the maximum or minimum value of a given linear expression in two variables, subject to some given constraints expressed as a system of linear inequalities in the two variables.

SECTION QUIZ 3.6

Sketch the graph of each of the following.
1. $2x - 3y - 6 \leq 0$

2. $2x - 3y - 6 < 0$

3. $y < 2$

4. Sketch the graph of the solution of the following system.

$y \leq x + 3$
$y \geq 0$
$x \leq 1$

5. A family raises and sells lettuce and beans. They have room for no more than 20 rows of vegetables. In one summer they must raise at least 1 row of beans and 2 rows of lettuce to satisfy their regular customers. They make a profit of $22 on each row of lettuce and $25 on each row of beans.
   (a) Let x be the number of rows of lettuce, let y be the number of rows of beans, and set up a system of inequalities expressing the given constraints.
   (b) Graph the system of inequalities you have set up.
   (c) Assume that the family sells all their vegetables and compute their profit at each "corner" of the graph you have obtained.
   (d) How many rows of each vegetable must the family raise in a summer in order to realize the maximum profit?

3.7 Sequences and Series

The purpose of this section is to introduce sequences and series, to introduce the summation notation, to consider geometric and arithmetic sequences and series, and to suggest some applications.

INTRODUCTION TO SEQUENCES

The expression

$$\tfrac{1}{2}, \tfrac{1}{4}, \tfrac{1}{8}, \tfrac{1}{16}, \ldots, (\tfrac{1}{2})^n, \ldots$$

represents an **infinite sequence**. The letter n represents any positive integer. The numbers separated by the commas are *terms* of the sequence. The three dots at the end indicate that there is an unlimited number of terms formed according to the rule $(\tfrac{1}{2})^n$ by successively letting n take on the values 1, 2, 3, 4, etc. of all the positive integers.

It is convenient to represent an entire sequence simply by writing the rule of formation in braces. For instance,

$$\{(\tfrac{1}{2})^n\}$$

represents the sequence

$$\tfrac{1}{2}, \tfrac{1}{4}, \tfrac{1}{8}, \tfrac{1}{16}, \ldots, (\tfrac{1}{2})^n, \ldots.$$

Here are some more examples of expressions that represent infinite sequences.

$$2, 2^2, 2^3, 2^4, \ldots, 2^n, \ldots \qquad \{2^n\}$$
$$0.8, (0.8)^2, (0.8)^3, (0.8)^4, \ldots, (0.8)^n, \ldots \qquad \{(0.8)^n\}$$
$$3 - \tfrac{1}{2}, 3 - \tfrac{1}{4}, 3 - \tfrac{1}{8}, \ldots, 3 - (\tfrac{1}{2})^n, \ldots \qquad \{3 - (\tfrac{1}{2})^n\}$$
$$2, 5, 8, 11, \ldots, 2 + 3(n-1), \ldots \qquad \{2 + 3(n-1)\}$$
$$1, -1, 1, -1, \ldots, (-1)^{n+1}, \ldots \qquad \{(-1)^{n+1}\}$$

In each case, there is given a rule of formation expressed in terms of n with the understanding that the terms are formed by letting n successively take on the values 1, 2, 3, 4, etc. of all the positive integers.

Technically speaking, an infinite sequence is a function whose domain is the set of all positive integers and whose rule of assignment is the rule of formation of the terms. For example, the rule of assignment of the sequence

$$\tfrac{1}{2}, \tfrac{1}{4}, \tfrac{1}{8}, \tfrac{1}{16}, \ldots, (\tfrac{1}{2})^n, \ldots$$

assigns the number $(\tfrac{1}{2})^n$ to positive integer n. If we called this sequence the function f and used the usual function notation, we would write

$$f(n) = (\tfrac{1}{2})^n.$$

In working with infinite sequences, however, it is customary to let n represent any element in the domain (the positive integers) and to use the notation a_n for the rule of assignment rather than the notation $f(n)$. For our example, it is customary to write

$$a_n = (\tfrac{1}{2})^n.$$

The general expression

$$a_1, a_2, a_3, a_4, \ldots, a_n, \ldots$$

represents an infinite sequence whose rule assigns a_n to positive integer n. The a_1, a_2, a_3, a_4, etc. are called the *terms* of the sequence. The rule a_n is sometimes called the *general term* or the *nth* term of the sequence. The notation

$$\{a_n\}$$

represents the entire sequence.

In specifying an infinite sequence, it is not sufficient to give merely the first few terms of the sequence. In order to specify a sequence, *the rule of assignment* (the general term) *must be given*. For example, just knowing that

$$2, 4, 8$$

are the first three terms of a sequence is not enough to tell us the next term, much less the entire sequence. There is an unlimited number of infinite sequences starting out with these three terms. Some of them are

$$2, 4, 8, 16, \quad 32, \ldots, 2^n, \ldots$$
$$2, 4, 8, 22, \quad 56, \ldots, 2^n + (n - 1)(n - 2)(n - 3), \ldots$$
$$2, 4, 8, 34, 128, \ldots, 2^n + (n - 1)^2(n - 2)(n - 3), \ldots.$$

We observed in Part 2 of this text that there are *two* essential ingredients in the definition of any function, the domain and the rule of assignment. Both of these must be given in order for the function to be defined. An infinite sequence is a function whose domain is the set of all positive integers, and its rule of assignment (its general term) *must* be given or the sequence itself has not been given.

The expression

$$a_1, a_2, a_3, \ldots, a_n, \ldots a_k$$

represents a sequence with exactly k terms. Its domain is not the set of *all* positive integers but rather its domain is the set of all positive integers 1, 2, 3, etc. up to and including the positive integer k. Its terms are formed according to its rule of assignment, its general term a_n, by letting

n successively take on the values of the integers 1 through k. The expression

$$2, 2^2, 2^3, \ldots, 2^n, \ldots 2^8$$

represents the sequence

$$2, 2^2, 2^3, 2^4, 2^5, 2^6, 2^7, 2^8$$

whose domain is the set of positive integers 1, 2, 3, 4, 5, 6, 7, 8.

Sequences arise quite naturally in everyday life as well as in science and pure mathematics. The next few examples illustrate this.

EXAMPLE 1 A chemical evaporates in such a way that it loses half of its volume each day. If it initially has 1 unit of volume, during the first day it loses $\frac{1}{2}$ of that, leaving $\frac{1}{2}$ of a unit; during the second day it loses $\frac{1}{2}$ of the remains, leaving $\frac{1}{4}$ of a unit; the third day it loses $\frac{1}{2} \times \frac{1}{4}$, leaving $\frac{1}{8}$ of a unit, etc. The volume remaining at the end of the nth day is $(\frac{1}{2})^n$ of a unit. The volumes remaining at the ends of the successive days are the terms in the infinite sequence

$$\tfrac{1}{2}, \tfrac{1}{4}, \tfrac{1}{8}, \tfrac{1}{16}, \ldots, (\tfrac{1}{2})^n, \ldots$$

In this example, for that matter, the volumes *lost* each successive day are the terms in the same infinite sequence.

EXAMPLE 2 A concerned ecologist in Northern Alaska noticed that each year his community has twice as much nonrecyclable garbage as it had the previous year. He wondered how long the town dump could accommodate this. He started to make a study of the situation. The first year of the study, the community deposited 2 tons of the garbage, the second year it deposited $2 \times 2 = 2^2$ tons, the third year it deposited $2 \times 2^2 = 2^3$ tons, etc. At this rate, during the nth year 2^n tons would be deposited. If the dumping continues in this manner indefinitely, the number of tons deposited in succeeding years form the infinite sequence

$$2, 2^2, 2^3, 2^4, \ldots, 2^n, \ldots$$

EXAMPLE 3 A congressional committee proposed to give a rebate on federal income taxes in order to bolster the economy. An economist estimated that of each 1 million dollars in rebate, 80% would be spent in the United States (fed back into the local economy), and 20% would be either saved or spent abroad. Of each sum spent in the United States, 80% would again be spent in the United States and 20% would either be saved or spent abroad, etc. In other words, the economist estimated that for each million-dollar rebate, there would be a sequence of spending ''sprees,'' and that in millions of dollars the amounts fed back into the local economy would be

$$0.8 \quad \text{on the first spree,}$$
$$0.8(0.8) = (0.8)^2 \quad \text{on the second spree,}$$
$$0.8(0.8)^2 = (0.8)^3 \quad \text{on the third spree,}$$
$$\vdots \qquad \vdots$$
$$(0.8)^n \quad \text{on the } n\text{th spree.}$$

In millions of dollars, the successive amounts fed back into the local economy for each million-dollar rebate form the infinite sequence

$$0.8, \ (0.8)^2, \ (0.8)^3, \ \ldots, \ (0.8)^n, \ \ldots.$$

EXAMPLE 4

If a person deposited $100 in an account paying 6% interest compounded annually, at the end of the first year the number of dollars in the account would be the original 100 plus the interest.

$$100 + 100(0.06) = 100(1.06)$$

At the end of the second year, interest would be computed on the amount $100(1.06)$, and the number of dollars in the account would then be

$$100(1.06) + 100(1.06)(.06) = 100(1.06)(1.06)$$
$$= 100(1.06)^2.$$

Similarly, at the end of the third year, the number of dollars in the account would be

$$100(1.06)^3.$$

At the end of the nth year, the number of dollars in the account would be

$$100(1.06)^n.$$

The numbers of dollars in the account at the end of each successive year are the terms in the infinite sequence

$$100(1.06), \ 100(1.06)^2, \ 100(1.06)^3, \ \ldots, \ 100(1.06)^n, \ \ldots.$$

EXAMPLE 5

An organization established an annual prize with the agreement that the prize would be $2 the first year with an annual increase of $3 each year thereafter. In dollars, the prize the second year would be $2 + 3 = 5$, the prize the third year would be $2 + (2)(3) = 8$, the prize the fourth year would be $2 + (3)(3) = 11$, etc. The nth year the prize would be

$$2 + (n - 1)(3) = 2 + 3(n - 1).$$

The dollar values of the prize in succeeding years are the terms in the infinite sequence

$$2, \ 5, \ 8, \ 11, \ \ldots, \ 2 + 3(n - 1), \ \ldots.$$

EXAMPLE 6   As x takes on the values

$$\frac{\pi}{2}, \frac{3\pi}{2}, \frac{5\pi}{2}, \frac{7\pi}{2}, \ldots, (2n-1)\frac{\pi}{2}, \ldots,$$

sin x (defined in trigonometry) takes on the values

Exercise 3.7, 1–15 $$1, -1, 1, -1, \ldots, (-1)^{n+1}, \ldots.$$

CONVERGENCE OF SEQUENCES

As we go further and further out in some sequences (as n takes on larger and larger values), the terms get closer and closer in value to a particular number. For example, as we go further and further out in the sequence

$$\tfrac{1}{2}, \tfrac{1}{4}, \tfrac{1}{8}, \tfrac{1}{16}, \tfrac{1}{32}, \ldots, (\tfrac{1}{2})^n, \ldots,$$

the terms get closer and closer in value to the number 0. We might say that as n increases, the terms *approach* 0. Similarly, as we go further and further out in the sequence

$$3 - \tfrac{1}{2},\ 3 - \tfrac{1}{4},\ 3 - \tfrac{1}{8},\ 3 - \tfrac{1}{16},\ 3 - \tfrac{1}{32},\ \ldots,\ 3 - (\tfrac{1}{2})^n,\ \ldots,$$

the terms get closer and closer in value to the number 3. The terms of this sequence *approach* 3 as n increases. We say that the sequence $\{(\tfrac{1}{2})^n\}$ *converges* to 0; we say that the sequence $\{3 - (\tfrac{1}{2})^n\}$ *converges* to 3. We say that 0 is the *limit* of the sequence $\{(\tfrac{1}{2})^n\}$; we say that 3 is the *limit* of the sequence $\{3 - (\tfrac{1}{2})^n\}$. We express this symbolically by writing

$$\lim_{n \to \infty} (\tfrac{1}{2})^n = 0$$

and $$\lim_{n \to \infty} 3 - (\tfrac{1}{2})^n = 3.$$

For any general term a_n, the symbols

$$\lim_{n \to \infty} a_n$$

are read

          "the limit as n approaches infinity of a_n"

or   "the limit as n increases indefinitely of a_n."

   In general, the sequence $\{a_n\}$ is said to **converge** to L provided that as we go further and further out in the sequence, the terms get closer and closer in value to the number L. We say that L is the **limit** of the sequence. We may write

$\lim\limits_{n \to \infty} a_n = L$ provided that   we can find an n large enough so that each of the terms beyond a_n will be as close to L in value as we please.

More technically,

$\lim\limits_{n \to \infty} a_n = L$ provided that there exists a number N such that for all $n > N$, $|a_n - L|$ is less than any pre-assigned and arbitrarily small number.

If the terms of an infinite sequence $\{a_n\}$ do not approach a number L as n increases indefinitely (that is, if the sequence does not converge to a limit), then the sequence is said to **diverge**. The sequence

$$2, 2^2, 2^3, 2^4, \ldots, 2^n, \ldots$$

diverges. There is no number that the terms approach in value as n increases. The sequence

$$1, -1, 1, -1, \ldots, (-1)^{n-1}, \ldots$$

also diverges. As n increases, the terms just fluctuate in value from 1 to -1; they do not all approach a particular number L.

EXAMPLE 7

Write out the first four terms and the general term of $\{(\tfrac{1}{3})^n\}$ and of $\{7 - (\tfrac{1}{3})^n\}$. What is $\lim\limits_{n \to \infty} (\tfrac{1}{3})^n$, and what is $\lim\limits_{n \to \infty} 7 - (\tfrac{1}{3})^n$?

Solution

$\{(\tfrac{1}{3})^n\}$ $\qquad$ $\tfrac{1}{3}, \tfrac{1}{9}, \tfrac{1}{27}, \tfrac{1}{81}, \ldots, (\tfrac{1}{3})^n, \ldots$

$\{7 - (\tfrac{1}{3})^n\}$ $\qquad$ $7 - \tfrac{1}{3}, 7 - \tfrac{1}{9}, 7 - \tfrac{1}{27}, 7 - \tfrac{1}{81}, \ldots, 7 - (\tfrac{1}{3})^n, \ldots$

As n increases indefinitely, the terms of $\{(\tfrac{1}{3})^n\}$ get successively closer in value to 0, so

$$\lim\limits_{n \to \infty} (\tfrac{1}{3})^n = 0.$$

As n increases indefinitely, the terms of $\{7 - (\tfrac{1}{3})^n\}$ get successively closer in value to 7, so

Exercise 3.7, 16–20

$$\lim\limits_{n \to \infty} 7 - (\tfrac{1}{3})^n = 7.$$

GEOMETRIC SEQUENCES

Several of the sequences that we have used as examples are *geometric sequences*. In particular,

$$\tfrac{1}{2}, \tfrac{1}{4}, \tfrac{1}{8}, \tfrac{1}{16}, \ldots, (\tfrac{1}{2})^n, \ldots,$$

$$2, 2^2, 2^3, 2^4, \ldots, 2^n, \ldots,$$

$$\tfrac{1}{8}, \tfrac{1}{16}, \tfrac{1}{32}, \tfrac{1}{64}, \ldots, (\tfrac{1}{2})^{n+2}, \ldots,$$

and $\qquad$ $1, -1, 1, -1, \ldots, (-1)^{n+1}, \ldots$

are geometric sequences. After the first term, each of the terms of a geometric sequence can be formed by multiplying the preceding term by a nonzero constant called the *common ratio.* In other words, after the first term, the ratio of any term to the preceding term is a constant called the **common ratio.** The sequence $\{ar^{n-1}\}$,

$$a, \; ar, \; ar^2, \; ar^3, \; \ldots, \; ar^{n-1}, \; \ldots,$$

is a **geometric sequence** with first term $a \neq 0$ and common ratio r.

The first term of $\{(\frac{1}{2})^n\}$ is $\frac{1}{2}$, and the common ratio is $\frac{1}{2}$.

The first term of $\{2^n\}$ is 2, and the common ratio is 2.

The first term of $\{(\frac{1}{2})^{n+2}\}$ is $\frac{1}{8}$, and the common ratio is $\frac{1}{2}$.

The first term of $\{(-1)^{n+1}\}$ is 1, and the common ratio is -1.

Geometric sequences are sometimes called *geometric progressions.*

We saw that the geometric sequence $\{(\frac{1}{2})^n\}$ with common ratio $\frac{1}{2}$ converges to 0, whereas the geometric sequence $\{2^n\}$ with common ratio 2 does not converge, and the geometric sequence $\{(-1)^{n+1}\}$ with common ratio -1 does not converge. It is true that for any geometric sequence with common ratio r,

if $|r| < 1$, the sequence converges to 0, and

if $|r| > 1$, the sequence does not converge (diverges).

This is because

if $|r| < 1$, $\displaystyle\lim_{n \to \infty} r^n = 0$, and

if $|r| > 1$, $\displaystyle\lim_{n \to \infty} r^n$ fails to exist.

If $r = 1$ and the first term is a, the sequence is simply $a, a, a, a, \ldots, a, \ldots$, which converges to a. If $r = -1$ and the first term is a, the sequence is $a, -a, a, -a, \ldots, (-1)^{n+1}a, \ldots$, which does not converge.

EXAMPLE 8

State whether or not each of the following sequences converges. If it converges, to what does it converge?
(a) $\{(\frac{-1}{3})^{n-1}\}$ (b) $\{3(-1)^n\}$ (c) $\{(-3)^{n-1}\}$

Solution

(a) $\{(\frac{-1}{3})^{n-1}\}$ is a geometric sequence with common ratio $r = \frac{-1}{3}$. Since $|r| < 1$, the sequence converges to 0.
(b) $\{3(-1)^n\}$ is a geometric sequence with common ratio $r = -1$. Since $r = -1$, the sequence does not converge.
(c) $\{(-3)^{n-1}\}$ is a geometric sequence with common ratio $r = -3$. Since $|r| > 1$, the sequence does not converge.

We can derive a very useful formula for the sum of the first n terms of a geometric sequence with common ratio other than one. In fact, if S_n is the sum of the first n terms of the geometric sequence $\{ar^{n-1}\}$ with $r \neq 1$, that is, if

$$S_n = a + ar + ar^2 + ar^3 + \cdots + ar^{n-1},$$

and if $r \neq 1$ we can show that

$$S_n = \frac{a - ar^n}{1 - r}.$$

To show this, we first write out S_n, putting in the next to the last term of it for future reference. Next we multiply both sides of the equation defining S_n by the common ratio r. (In so doing, we use the fact that $ar^{n-2} \times r = ar^{n-1}$ and $ar^{n-1} \times r = ar^n$.)

$$S_n = a + ar + ar^2 + ar^3 + \cdots + ar^{n-2} + ar^{n-1}$$
$$S_n r = \quad\; ar + ar^2 + ar^3 + ar^4 + \cdots \quad\; + ar^{n-1} + ar^n$$

Now we subtract the second of these equations from the first. When we do this subtraction, all the terms on the right of the equals signs drop out except for the first term of S_n and the last term of $S_n r$. We get

$$S_n - S_n r = a - ar^n.$$

Finally we solve for S_n, thereby obtaining the formula we seek.

$$S_n(1 - r) = a - ar^n$$

$$S_n = \frac{a - ar^n}{1 - r} \qquad \text{if } r \neq 1$$

If $r = 1$, then $S_n = a + a \cdot 1 + a \cdot 1^2 + a \cdot 1^3 + \cdots + a \cdot 1^{n-1}$, so S_n is just the sum of n terms each of which is a. That is,

$$S_n = na, \qquad \text{if } r = 1.$$

EXAMPLE 9

(a) Write out a formula for the sum of the first n terms of $\{(\frac{1}{2})^n\}$. Use this formula to find the sum of the first 5 terms.

(b) Write a formula for the sum of the first n terms of $\{3 \times (1)^n\}$, and find the sum of the first 5 terms.

Solution

(a) In the sequence $\{(\frac{1}{2})^n\}$, $a = \frac{1}{2}$ and $r = \frac{1}{2}$.

$$S_n = \frac{a - ar^n}{1 - r}$$

$$= \frac{\frac{1}{2} - \frac{1}{2}(\frac{1}{2})^n}{1 - \frac{1}{2}}$$

$$= \frac{\frac{1}{2} - \frac{1}{2}(\frac{1}{2})^n}{\frac{1}{2}}$$

$$= 1 - (\tfrac{1}{2})^n$$

$$S_5 = 1 - (\tfrac{1}{2})^5$$

$$= 1 - \tfrac{1}{32}$$

$$= \tfrac{31}{32}$$

(b) In the sequence $\{3 \times (1)^n\}$, $a = 3$ and $r = 1$, so every term is 3.

$$S_n = 3n$$
$$S_5 = 15$$

Exercise 3.7, 21–40

ARITHMETIC SEQUENCES

The sequence

$$2, 5, 8, 11, \ldots, 2 + 3(n - 1), \ldots$$

is an *arithmetic sequence*. After the first term, each term of an arithmetic sequence can be formed by adding a nonzero constant called the *common difference* to the preceding term. The sequence

$$a, a + d, a + 2d, \ldots, a + (n - 1)d, \ldots$$

is an **arithmetic sequence** with first term a and common difference d. The first term of

$$2, 5, 8, 11, \ldots, 2 + 3(n - 1), \ldots$$

is 2, and the common difference is 3. Arithmetic sequences are sometimes called *arithmetic progressions*.

Every infinite arithmetic sequence diverges. There is no number that the terms approach in value as n increases indefinitely.

There is a very useful formula for the sum of the first n terms of an arithmetic sequence. To see this, let S_n be the sum of the first n terms of

$$a, a + d, a + 2d, \ldots, a + (n - 1)d, \ldots$$

so that

$$S_n = a + (a + d) + (a + 2d) + \cdots + a + (n - 1)d.$$

If we call the last term l, we have

$$l = a + (n - 1)d$$

and

$$S_n = a + (a + d) + (a + 2d) + \cdots + l$$

It can be shown that

$$S_n = \frac{n(a + l)}{2}.$$

The proof is outlined in Exercise 3.7.

EXAMPLE 10

Write out a formula for the sum of the first n terms of the following arithmetic sequence.

$$2, 5, 8, \ldots, 2 + 3(n - 1), \ldots.$$

Use this formula to find the sum of the first five terms.

Let S_n be the sum of the first n terms. The first term $a = 2$; the common difference $d = 3$; the last term (the nth term) of S_n is

$$l = a + (n - 1)d$$
$$= 2 + 3(n - 1).$$

$$S_n = \frac{n(a + l)}{2}$$

$$= \frac{n(2 + 2 + 3(n - 1))}{2}$$

$$= \frac{4n + 3n(n - 1)}{2}$$

$$S_5 = \frac{20 + 15(4)}{2}$$

$$= \frac{80}{2}$$

Exercise 3.7, 41–50

$$= 40$$

INTRODUCTION TO SERIES AND SUMMATION NOTATION

Sometimes we are more interested in the *sum* of the terms of a sequence than we are in the sequence itself. For instance, the economist in Example 3 might be more interested in knowing the total effect each million-dollar rebate would have on the local economy than in knowing the effect in any one particular year. That is, the economist might want to know the total amount

$$0.8 + (0.8)^2 + (0.8)^3 + \cdots + (0.8)^n + \cdots$$

fed back into the local economy if the spending sprees went on indefinitely as predicted. In order to provide this information, we must first know what is meant by the sum of an unlimited number of terms. We must consider *infinite series.*

If we replace the commas in the expression

$$a_1, a_2, a_3, \ldots, a_n \ldots$$

by plus signs and raise the dots to align with the plus signs, we have an expression for an **infinite series.**

$$a_1 + a_2 + a_3 + \cdots + a_n + \cdots \qquad \text{Infinite series}$$

Such a series can be written more compactly as

$$\sum_{n=1}^{\infty} a_n$$

where it is understood that this means the sum of terms formed by successively letting n take on the values 1, 2, 3, 4, etc. of all the positive integers. The general term a_n is displayed in the **summation notation** $\sum_{n=1}^{\infty} a_n$. When this sum is written out, the general term must be written in so that the rule of formation of the terms will be known.

$$\sum_{n=1}^{\infty} a_n = a_1 + a_2 + a_3 + \cdots + a_n + \cdots$$

EXAMPLE 11   Write out $\displaystyle\sum_{n=1}^{\infty} 2^n$.

Solution $\displaystyle\sum_{n=1}^{\infty} 2^n = 2 + 2^2 + 2^3 + 2^4 + \cdots + 2^n + \cdots$

We may use summation notation to indicate the sum of the first k terms of a series, where k is a particular positive integer. For example, the sum of the first five terms of $\displaystyle\sum_{n=1}^{\infty} a_n$ may be written $\displaystyle\sum_{n=1}^{5} a_n$.

$$\sum_{n=1}^{5} a_n = a_1 + a_2 + a_3 + a_4 + a_5$$

We may also use summation notation to indicate a sum of k successive terms that does not begin with the first term. For example, the sum of the five terms beginning with a_3 may be written $\displaystyle\sum_{n=3}^{7} a_n$.

$$\sum_{n=3}^{7} a_n = a_3 + a_4 + a_5 + a_6 + a_7$$

EXAMPLE 12   Write out $\displaystyle\sum_{n=1}^{5} 2^n$. Write out $\displaystyle\sum_{n=4}^{6} 2^n$.

Solution $\displaystyle\sum_{n=1}^{5} 2^n = 2 + 2^2 + 2^3 + 2^4 + 2^5$

$$\sum_{n=4}^{6} 2^n = 2^4 + 2^5 + 2^6$$

In summation notation, there is nothing sacred about the letter n. We may use any letter instead of n. Many authors use i and write $\sum_{i=1}^{\infty} a_i$ instead of $\sum_{n=1}^{\infty} a_n$. Other authors use k and write $\sum_{k=1}^{\infty} a_k$. They all represent

Exercise 3.7, 51–65 the same series.

PARTIAL SUMS, CONVERGENCE, AND SUMS OF INFINITE SERIES

To assign a meaning to the sum of an unlimited number of terms, we first set up a sequence of so-called *partial sums* of the terms. The first partial sum S_1 of

$$\sum_{n=1}^{\infty} a_n = a_1 + a_2 + a_3 + \cdots + a_n + \cdots$$

is just the first term.

$$S_1 = a_1$$

The second partial sum S_2 is the sum of the first two terms.

$$S_2 = a_1 + a_2$$

The third partial sum S_3 is the sum of the first three terms.

$$S_3 = a_1 + a_2 + a_3$$

The nth partial sum S_n is the sum of the first n terms.

$$S_n = a_1 + a_2 + a_3 + \cdots + a_n$$

If the sequence $\{S_n\}$ of partial sums converges to some number L, the infinite series $\sum_{n=1}^{\infty} a_n$ is said to converge to L; the number L is defined to be the **sum** of the infinite series. If the sequence $\{S_n\}$ of partial sums diverges, then the series $\sum_{n=1}^{\infty} a_n$ is said to diverge.

If $\{S_n\}$ converges to L, then $\sum_{n=1}^{\infty} a_n$ converges to L.

If $\{S_n\}$ diverges, then $\sum_{n=1}^{\infty} a_n$ diverges.

EXAMPLE 13   Write out the first three terms and write the nth term of the sequence of

partial sums of $\sum\limits_{n=1}^{\infty} (\frac{1}{2})^n$. Does the series $\sum\limits_{n=1}^{\infty} (\frac{1}{2})^n$ converge? If so, what is

its sum?

Solution $S_1 = \frac{1}{2}$

$S_2 = \frac{1}{2} + (\frac{1}{2})^2$

$S_3 = \frac{1}{2} + (\frac{1}{2})^2 + (\frac{1}{2})^3$

$\vdots$

$S_n = \frac{1}{2} + (\frac{1}{2})^2 + (\frac{1}{2})^3 + \cdots + (\frac{1}{2})^n$

Here S_n is the sum of the first n terms of the geometric sequence $\{(\frac{1}{2})^n\}$
with first term $a = \frac{1}{2}$ and common ratio $r = \frac{1}{2}$.

$$S_n = \frac{a - ar^n}{1 - r}$$

$$= \frac{\frac{1}{2} - \frac{1}{2}(\frac{1}{2})^n}{1 - \frac{1}{2}}$$

$$= 1 - (\frac{1}{2})^n$$

The series $\sum\limits_{n=1}^{\infty} (\frac{1}{2})^n$ converges to the number L provided that the
sequence of partial sums $\{1 - (\frac{1}{2})^n\}$ converges to L. The sequence of
partial sums $\{1 - (\frac{1}{2})^n\}$ converges to 1 because

$$\lim_{n \to \infty} 1 - (\frac{1}{2})^n = 1.$$

Hence the series $\sum\limits_{n=1}^{\infty} (\frac{1}{2})^n$ converges to 1. We say that the sum

$$\frac{1}{2} + (\frac{1}{2})^2 + (\frac{1}{2})^3 + \cdots + (\frac{1}{2})^n + \cdots$$

Exercise 3.7, 66–70   is 1.

GEOMETRIC AND ARITHMETIC SERIES

Any series such as $\sum\limits_{n=1}^{\infty} (\frac{1}{2})^n$ that is the sum of the terms of a geometric

sequence is called a *geometric series*. The series

$$\sum_{n=1}^{\infty} ar^{n-1} = a + ar + ar^2 + ar^3 + \cdots + ar^{n-1} + \cdots$$

is the **geometric series** with first term a and common ratio r. The nth term S_n of its sequence of partial sums—that is, the sum S_n of its first n terms—is

$$S_n = \frac{a - ar^n}{1 - r}$$

$$= \frac{a}{1 - r} - \frac{ar^n}{1 - r}.$$

If $|r| < 1$, this nth term S_n of the sequence $\{S_n\}$ approaches $a/(1-r)$ as n increases, because r^n approaches 0. That is, if $|r| < 1$, the sequence $\{S_n\}$ of partial sums converges to $a/(1-r)$. By definition, this means that if $|r| < 1$, the geometric series itself converges to $a/(1-r)$. It can be shown that the series diverges for $|r| \geq 1$.

If $|r| < 1$, the geometric series $\sum_{n=1}^{\infty} ar^{n-1}$ converges to $a/(1-r)$.

If $|r| \geq 1$, the geometric series $\sum_{n=1}^{\infty} ar^{n-1}$ diverges.

EXAMPLE 14   State whether or not the series

$$\sum_{n=1}^{\infty} \left(\tfrac{-1}{2}\right)^{n-1} = 1 - \tfrac{1}{2} + \tfrac{1}{4} - \tfrac{1}{8} + \tfrac{1}{16} + \cdots + \left(\tfrac{-1}{2}\right)^{n-1} + \cdots$$

converges. If it converges, what is its sum?

Solution   This is a geometric series with first term $a = 1$ and common ratio $r = \tfrac{-1}{2}$. Since $|r| < 1$, the series converges. It converges to

$$\frac{a}{1 - r} = \frac{1}{1 - \left(\tfrac{-1}{2}\right)}$$

$$= \frac{1}{1 + \tfrac{1}{2}}$$

$$= \tfrac{2}{3}.$$

EXAMPLE 15   State whether or not the series $\sum_{n=1}^{\infty} 2^n$ converges.

Solution   The series $\sum_{n=1}^{\infty} 2^n$ is a geometric series with common ratio $r = 2$. Since $|r| \geq 1$, the series diverges (does not converge).

EXAMPLE 16   Find the total amount fed into the local economy for each million-dollar tax rebate as predicted by the economist in Example 3.

Solution

The total amount will be the initial 1 million plus the amount fed *back* into the economy. In millions of dollars, that will be 1 plus

$$\sum_{n=1}^{\infty} (0.8)^n = 0.8 + (0.8)^2 + (0.8)^3 + \cdots + (0.8)^n + \cdots .$$

The series $\displaystyle\sum_{n=1}^{\infty} (0.8)^n$ is a geometric series with first term $a = 0.8$ and common ratio $r = 0.8$. Since $|r| < 1$, the series converges. Its sum is

$$\frac{a}{1 - r} = \frac{0.8}{1 - 0.8}$$

$$= \frac{0.8}{0.2}$$

$$= 4.$$

The total amount fed into the economy by the million-dollar rebate is therefore

$$1 + 4 = 5 \text{ million dollars.}$$

Any series such as

$$2 + 5 + 8 + 11 + \cdots + [2 + 3(n - 1)] + \cdots$$

whose terms are the terms of an arithmetic sequence is called an *arithmetic series.* The nth term S_n of the sequence of partial sums of the **arithmetic series**

$$a + (a + d) + (a + 2d) + \cdots + [a + (n - 1)d] + \cdots$$

is

$$S_n = \frac{n(a + \ell)}{2}$$

where

$$\ell = a + (n - 1)d.$$

Every arithmetic series diverges because its sequence of partial sums $\{S_n\}$ fails to converge. (As n increases indefinitely, S_n fails to approach a particular number.)

The series

$$2 + 2^2 + 2^3 + \cdots + 2^n + \cdots$$

and

$$3 + 5 + 8 + \cdots + [2 + 3(n - 1)] + \cdots$$

are both series of positive terms that diverge. In each of these series, the general term increases indefinitely as n increases indefinitely. In particular, as n increases indefinitely, neither of these general terms approaches zero. In calculus, it is shown that in order for a series of positive

terms to converge, the *general term a_n must approach zero* as n increases
indefinitely. Many tests for convergence and divergence of various types
of series are developed in calculus.

Exercise 3.7, 71–82

EXERCISE 3.7 SEQUENCES AND SERIES

INTRODUCTION TO SEQUENCES

▶ Using braces, rewrite each of the following sequences. Call each sequence f,
and write its rule of assignment with the usual function notation.

EXAMPLE

3, 5, 7, 9, . . . , $2n + 1$, . . .

Solution

$\{2n + 1\}$; $f(n) = 2n + 1$

1. 3, 6, 9, . . . , $3n$, . . .   **2.** 4, 7, 10, . . . , $3n + 1$, . . .

3. 2, 5, 10, . . . , $n^2 + 1$, . . .   **4.** 1, 8, 27, . . . , n^3, . . .

5. 3, 9, 19, . . . , $2n^2 + 1$, . . .

▶ For each of the following, write an expression that does not use braces and in-
cludes at least the first three terms.

EXAMPLE

$\{5n + 1\}$

Solution

6, 11, 16, . . . , $5n + 1$, . . .

6. $\{6n - 1\}$ **7.** $\{2n^3 + 1\}$ **8.** $\{n^2 + 2n\}$

9. $\{n(n + 1)\}$ **10.** $\{2n - 1\}$

11. If the chemical of Example 1 lost one quarter of its volume each day, what
would be the sequence of volumes remaining at the end of successive days?

12. If the community in Example 2 deposited 3 tons of nonrecyclable garbage
the first year and $3 \times 2^{n-1}$ tons the nth year, what is the sequence whose
terms are the number of tons deposited in succeeding years?

13. Suppose that the economist in Example 3 is wrong, and of each 1 million
dollars in rebate, only 75% would be spent in the States. Write an infinite
sequence whose terms are the successive amounts, in millions of dollars, fed
back into the local economy for each million-dollar rebate.

14. Suppose that the person in Example 4 deposited $300 in an account paying
5% interest compounded annually. Write out an infinite sequence whose
terms are the numbers of dollars in the account at the end of each successive
year.

15. Suppose that the prize in Example 5 was $5 the first year with an annual increase of $3 each year thereafter. Write out an infinite sequence whose terms are the dollar values of the prize in succeeding years.

CONVERGENCE OF SEQUENCES

▶ Write out the first four terms and the general term of each of the following, and find $\lim\limits_{n \to \infty}$ of the general term.

EXAMPLES

(a) $\{(\frac{1}{4})^n\}$ (b) $\{8 + (\frac{1}{4})^n\}$

Solutions

(a) $\frac{1}{4}, \frac{1}{16}, \frac{2}{64}, \frac{1}{256}, \ldots, (\frac{1}{4})^n, \ldots$

$$\lim_{n \to \infty} (\tfrac{1}{4})^n = 0$$

(b) $8 + \frac{1}{4}, 8 + \frac{1}{16}, 8 + \frac{1}{64}, 8 + \frac{1}{256}, \ldots, 8 + (\frac{1}{4})^n, \ldots$

$$\lim_{n \to \infty} 8 + (\tfrac{1}{4})^n = 8$$

16. $\{(\frac{1}{5})^n\}$ **17.** $\{3 + (\frac{1}{5})^n\}$ **18.** $\{(\frac{1}{6})^n\}$

19. $\{1 + (\frac{1}{6})^n\}$ **20.** $\{(\frac{1}{2})^n + (\frac{1}{3})^n\}$

GEOMETRIC SEQUENCES

▶ Find the common ratio of each of the following geometric sequences and state whether or not the sequence converges. If it converges, to what does it converge?

EXAMPLES

(a) $-3, 9, -27, 81, \ldots, (-3)^n, \ldots$

(b) $1, \frac{-1}{3}, \frac{1}{9}, \frac{-1}{27}, \ldots, (\frac{-1}{3})^{n-1}, \ldots$

Solutions

(a) $r = \frac{9}{-3} = -3;$ $|r| > 1$, diverges

(b) $r = \frac{-1}{3};$ $|r| < 1$, converges to 0.

21. $-2, 4, -8, 16, \ldots, (-2)^n, \ldots$

22. $1, \frac{1}{3}, \frac{1}{9}, \frac{1}{27}, \ldots, (\frac{1}{3})^{n-1}, \ldots$

23. $3, -3, 3, -3, \ldots, 3 \times (-1)^{n+1}, \ldots$

24. $3, 6, 12, 24, \ldots, 3 \times 2^{n-1}, \ldots$

25. $4, 2, 1, \frac{1}{2}, \ldots, 4 \times (\frac{1}{2})^{n-1}, \ldots$

26. $-1, 1, -1, 1, \ldots, (-1)(-1)^{n+1}, \ldots$

27. $5, -5, 5, -5, \ldots, 5 \times (-1)^{n+1}, \ldots$

28. $\frac{1}{4}, \frac{-1}{16}, \frac{1}{64}, \frac{-1}{256}, \ldots, \frac{1}{4} \times (\frac{-1}{4})^{n-1}, \ldots$

29. 1, 2, 4, 8, . . . , 2^{n-1}, . . .

30. 0.1, 0.01, 0.001, 0.0001, . . . , $(0.1)^n$, . . .

▶ Write out a formula for the sum of the first n terms of each of the following. Use this formula to find the sum of the first n terms for the given particular value of n.

EXAMPLE

Solution

$\{5^{n-1}\}$; $n = 4$

$a = 1, r = 5$, $\quad S_n = \dfrac{1 - 5^n}{1 - 5} = \dfrac{1 - 5^n}{-4}$

$$S_4 = \dfrac{1 - 5^4}{-4} = \dfrac{1 - 625}{-4} = 156$$

31. $\{4^{n-1}\}$, $n = 5$ **32.** $\{(\frac{-1}{3})^{n-1}\}$; $n = 2$

33. $\{(\frac{1}{3})^n\}$; $n = 5$ **34.** $\{(\frac{-1}{3})^n\}$; $n = 5$

35. $\{(-2)^n\}$; $n = 5$ **36.** $\{3 \times (-1)^{n-1}\}$; $n = 4$

37. $\{(-1)^{n-1}\}$; $n = 6$ **38.** $\{5 \times (-1)^{n+1}\}$; $n = 4$

39. $\{4 \times (\frac{1}{2})^{n-1}\}$; $n = 3$ **40.** $\{(0.1)^n\}$; $n = 5$

ARITHMETIC SEQUENCES

▶ Find the common difference of each of the following arithmetic sequences. Write out a formula for the sum of the first n terms. Use this formula to find the sum of the first five terms.

EXAMPLE

Solution

1, 3, 5, 7, . . . , $1 + 2(n - 1)$, . . .

$d = 3 - 1 = 2, a = 1, \ell = a + (n - 1)d$

$\qquad\qquad = 1 + 2(n - 1)$

$$S_n = \dfrac{n(a + \ell)}{2}$$

$$= \dfrac{n[2 + 2(n - 1)]}{2}$$

$$= \dfrac{2n[1 + (n - 1)]}{2}$$

$$= n^2$$

$$S_5 = 5^2 = 25$$

41. 2, 4, 6, 8, . . . , $2 + 2(n - 1)$, . . .

42. 3, 7, 11, 15, . . . , $3 + 4(n - 1)$, . . .

43. $1, -1, -3, -5, \ldots, 1 + (-2)(n - 1), \ldots$

44. $5, 10, 15, 20, \ldots, 5 + 5(n - 1), \ldots$

45. $3, 0, -3, -6, \ldots, 3 + (-3)(n - 1), \ldots$

▶ Classify each of the following as either geometric or arithmetic. If it is geometric, give the common ratio. If it is arithmetic, give the common difference.

EXAMPLES

(a) $3, 6, 9, 12, \ldots, 3 + 3(n - 1), \ldots$
(b) $3, 6, 12, 24, \ldots, 3 \times 2^{n-1}, \ldots$

Solutions

(a) Arithmetic, $d = 3$
(b) Geometric, $r = 2$

46. $4, 8, 12, 16, \ldots, 4 + 4(n - 1), \ldots$

47. $4, 8, 16, 32, \ldots, 4 \times 2^{n-1}, \ldots$

48. $\frac{1}{2}, 1, \frac{3}{2}, 2, \ldots, \frac{1}{2} + \frac{1}{2}(n - 1), \ldots$

49. $\frac{1}{2}, 2, 8, 32, \ldots, \frac{1}{2} \times (4)^{n-1}, \ldots$

50. $5, 10, 20, 40, \ldots, 5 \times 2^{n-1}, \ldots$

INTRODUCTION TO SERIES AND SUMMATION NOTATION

▶ Use summation notation to write each of the following.

EXAMPLE

$2 + 6 + 12 + 20 + \cdots + n(n + 1) + \cdots$

Solution

$$\sum_{n \to 1}^{\infty} n(n + 1)$$

51. $3 + 6 + 12 + 24 + \cdots + 3 \times 2^{n-1} + \cdots$

52. $4 + 8 + 12 + 16 + \cdots + [4 + 4(n - 1)] + \cdots$

53. $1 + 4 + 9 + 16 + \cdots + n^2 + \cdots$

54. $1 + 8 + 27 + 64 + \cdots + n^3 + \cdots$

55. $4 + 10 + 18 + 28 + \cdots + n(n + 3) + \cdots$

▶ Write out each of the following.

EXAMPLE

$$\sum_{n=1}^{\infty} 3^n$$

Solution

$3 + 3^2 + 3^3 + 3^4 + \cdots + 3^n + \cdots$

56. $\displaystyle\sum_{n=1}^{5} 3^n$ **57.** $\displaystyle\sum_{n=4}^{9} 3^n$ **58.** $\displaystyle\sum_{n=1}^{\infty} \frac{n}{n.+1}$

59. $\displaystyle\sum_{n=1}^{\infty} \frac{1}{2n}$ **60.** $\displaystyle\sum_{n=1}^{\infty} \frac{n}{n^3+2}$ **61.** $\displaystyle\sum_{n=3}^{7} \frac{n}{n^2+1}$

62. $\displaystyle\sum_{n=1}^{5} n^4$ **63.** $\displaystyle\sum_{n=1}^{\infty} \frac{n^3}{3^n}$ **64.** $\displaystyle\sum_{n=1}^{\infty} \frac{2^n}{n}$

65. $\displaystyle\sum_{n=1}^{9} \sqrt{n}$

PARTIAL SUMS, CONVERGENCE, AND SUMS OF INFINITE SERIES

▶ Write out the first three terms and write the nth term S_n of the sequence of partial sums of each of the following. Find the sum of the series by finding $\displaystyle\lim_{n\to\infty} S_n$.

EXAMPLE

$\displaystyle\sum_{n=1}^{\infty} (\tfrac{1}{5})^{n-1}$

Solution

$S_1 = 1,\ S_2 = 1 + \tfrac{1}{5},\ S_3 = 1 + \tfrac{1}{5} + (\tfrac{1}{5})^2$

$S_n = 1 + \tfrac{1}{5} + (\tfrac{1}{5})^2 + \cdots + (\tfrac{1}{5})^{n-1}$

$S_n = \dfrac{1 - (\tfrac{1}{5})^n}{1 - \tfrac{1}{5}}$

$\quad = \tfrac{5}{4}[1 - (\tfrac{1}{5})^n]$

$\displaystyle\lim_{n\to\infty} S_n = \tfrac{5}{4}$

66. $\displaystyle\sum_{n=1}^{\infty} (\tfrac{1}{4})^{n-1}$ **67.** $\displaystyle\sum_{n=1}^{\infty} (\tfrac{-1}{3})^{n-1}$ **68.** $\displaystyle\sum_{n=1}^{\infty} (\tfrac{1}{3})^n$

69. $\displaystyle\sum_{n=1}^{\infty} (\tfrac{1}{6})^{n-1}$ **70.** $\displaystyle\sum_{n=1}^{\infty} (\tfrac{-1}{4})^{n-1}$

GEOMETRIC AND ARITHMETIC SERIES

▶ Tell whether or not each of the following converges. If it converges, find its sum.

EXAMPLES

(a) $\displaystyle\sum_{n=1}^{\infty} (\tfrac{1}{5})^{n-1}$ (b) $\displaystyle\sum_{n=1}^{\infty} 5^{n-1}$

Solutions

(a) $\sum_{n=1}^{\infty} (\frac{1}{5})^{n-1}$ is a geometric series with $r = \frac{1}{5}$.

$r < 1$, series converges to $\dfrac{a}{1-r} = \dfrac{1}{\frac{4}{5}} = \frac{5}{4}$

(b) $\sum_{n=1}^{\infty} 5^{n-1}$ is a geometric series with $r = 5$.

$r > 1$, series diverges

71. $\sum_{n=1}^{\infty} (\frac{1}{4})^{n-1}$ **72.** $\sum_{n=1}^{\infty} 4^{n-1}$ **73.** $\sum_{n=1}^{\infty} (\frac{-1}{3})^{n-1}$

74. $\sum_{n=1}^{\infty} (-2)^{n}$ **75.** $\sum_{n=1}^{\infty} (-1)^{n-1}$

76. Assume that the evaporation of the chemical in Example 1 continues indefinitely, and find the sum of the volumes lost each day.

77. How many tons of nonrecyclable garbage would the community of Example 2 deposit in the first 10 years of the ecologist's study?

78. If the economist in Example 3 estimated that only 70% of money fed into the local economy would be spent locally, and if this situation continued indefinitely, what would be the total amount fed back into the local economy for each million-dollar tax rebate?

79. How much would the person of Example 4 have in the account at the end of 10 years?

80. Suppose that the dump in Example 2 can accommodate no more than 2044 tons of nonrecyclable garbage, and suppose that the people decided they would wait until the dump was half full before removing any garbage. How many years would they have to wait before the dump would be half full? How many years would they then have to remove the garbage before the dump was full? If they waited until the dump was half full, and if they can remove only 500 tons of garbage per year, within how many years from then would they have more garbage than the dump can accommodate?

81. A farmer noticed that 1 square meter of his 512-square meter pond was covered with weeds that were doubling the area they covered each day. He decided to wait until the weeds covered half the area of his pond before removing them. How many days would he have to wait for the weeds to cover half of the pond? How many days would he then have before they covered the entire pond?

82. Let S_n be the sum of the first n terms of the arithmetic sequence

$$a, a + d, a + 2d, \ldots, a + (n - 1)d, \ldots,$$

and call the nth term l so that $S_n = a + (a + d) + (a + 2d) + \cdots + (l - 2d) + (l - d) + l$. (Note that the next to the last term is $l - d$, and the term before that is $l - 2d$.) Derive the formula

$$S_n = \frac{n(a + l)}{2}.$$

Hint   Rewrite the equation defining S_n by reversing the order of the terms on the right so that

$$S_n = l + (l - d) + (l - 2d) + \cdots + (a + 2d) + (a + d) + a.$$

Next add the two equations defining S_n; note that the terms involving d drop out, and note that

$$2S_n = n(a + l).$$

CHECK LIST FOR THE STUDENT

If you have learned the material in Section 3.7, you should be able to do the following:

1. Write one expression for an infinite sequence using braces and one without using braces.

2. Recognize that a sequence is a function, and give the domain and rule of assignment of a given sequence.

3. Tell what is meant by the terms and the general term of a sequence.

4. Tell why it is not sufficient to give merely the first few terms in specifying a sequence.

5. Give instances in which sequences arise in everyday life, in science, or in mathematics.

6. Write out the first four terms and the general term of a sequence written with braces.

7. Give an example of a convergent sequence, and tell what is meant by its limit.

8. Tell what is meant when a sequence is said to diverge.

9. Find $\lim_{n \to \infty}$ of the general term of some given sequence.

10. Tell how the second term and the succeeding terms of a geometric sequence can be formed.

11. Find the common ratio of a given geometric sequence.

12. Complete the statements

If $|r| < 1$, a geometric sequence with common ratio r _____.
If $|r| > 1$, a geometric sequence with common ratio r _____.
If $r = 1$, a geometric sequence with common ratio r _____.
If $r = -1$, a geometric sequence with common ratio r _____.

13. Complete the statements.

If $|r| < 1$, $\lim_{n \to \infty} r^n =$ _____.
If $|r| > 1$, $\lim_{n \to \infty} r^n =$ _____.

14. State whether or not a given geometric sequence converges, and tell to what it converges if it *does* converge.

15. Write and derive a formula for the sum S_n of the first n terms of a geometric sequence with first term a and common ratio $r \neq 1$; write a formula for S_n if $r = 1$.

16. Find the sum of the first n terms of a given geometric sequence for a particular given n.

17. Tell how the second term and the succeeding terms of an arithmetic sequence can be formed.

18. Find the common difference of a given arithmetic sequence.

19. Write a formula for the sum of the first n terms of an arithmetic sequence with first term a, with nth term l, and with common difference d.

20. Find the sum of the first n terms of a given arithmetic sequence for a particular n.

21. Write one expression for an infinite series using summation notation and one without using summation notation.

22. Write out an expression for a series written with summation notation.

23. Tell what is meant by the nth partial sum of an infinite series.

24. Write the first three terms and the nth term of the sequence of partial sums of a given infinite series.

25. Tell what is meant when an infinite series is said to converge and when an infinite series is said to diverge.

26. Tell what is meant by the sum of an infinite series.

27. Complete the statements, where $\{S_n\}$ is the sequence of partial sums of $\sum\limits_{n=1}^{\infty} a_n$.

If $\{S_n\}$ converges to L, then $\sum\limits_{n=1}^{\infty} a_n$ _____.

If $\{S_n\}$ diverges, then $\sum\limits_{n=1}^{\infty} a_n$ _____.

28. Give the definition of a geometric series.

29. Find the common ratio of a given geometric series.

30. Complete the statements.
If $|r| < 1$, the geometric series with first term a and common ratio r converges to _____.
If $|r| \geq 1$, a geometric series with common ratio r _____.

31. State whether or not a given geometric series converges, and find its sum if it *does* converge.

32. Give the definition of an arithmetic series.

33. Tell whether or not an arithmetic series converges.

SECTION QUIZ 3.7

1. (a) Using braces, rewrite the sequence 5, 7, 9, 11, . . . , $2n + 3$, ...
   (b) Without using braces, write an expression for $\{3n - 2\}$. Include at least the first three terms.

2. Write the first four terms and the general term a_n of $\{(\frac{1}{3})^n\}$, and give $\lim\limits_{n \to \infty} a_n$.

Classify each of the following as either a geometric sequence or an arithmetic sequence. If it is a geometric sequence, give its common ratio; if it is an arithmetic sequence, give its common difference.

3. 1, 5, 9, 13, . . . , $1 + 4(n - 1)$, . . .

4. 1, 3, 9, 27, . . . , 3^{n-1}, . . .

Write out a formula for the sum of the first n terms of each of the following.

5. $5, 5^2, 5^3, \ldots, 5^n, \ldots$

6. $3, 6, 9, \ldots, 3n, \ldots$

State whether or not each of the following converges. If it converges, find its sum.

7. $1 + \frac{1}{3} + \frac{1}{9} + \frac{1}{27} + \cdots + (\frac{1}{3})^{n-1} + \cdots$

8. $1 + 3 + 9 + 27 + \cdots + (3)^{n-1} + \cdots$

9. (a) Write out the first four terms, and write the nth term of

$$\sum_{n=1}^{\infty} \frac{n}{n+2}.$$

(b) Write the first, second, and third partial sums of

$$\sum_{n=1}^{\infty} \frac{n}{n+2}.$$

10. A pollution-free energy source is being studied. It is found that in one application a measure of the energy lost during the first hour is 0.1, during the second hour it is $(0.1)^2$, during the third hour it is $(0.1)^3$, etc. Assume that the energy lost during the nth hour is $(0.1)^n$.

(a) What is a formula for the sum of the energy lost during the first n hours?

(b) Use this formula to find the sum of the energy lost during the first 10 hours.

(c) If the energy loss continued indefinitely as assumed here, what would be the total sum of the energy lost?

3.8 Mathematical Induction

The purpose of this section is to present the principle of mathematical induction and to illustrate its use.

The *principle of mathematical induction* is used to prove various statements concerning integers, usually positive integers. Here are some examples of statements that can be proved by using this principle.

1. The sum of the squares of the first n positive integers is

$$\frac{n(n + 1)(2n + 1)}{6}.$$

$$1^2 + 2^2 + \cdots + n^2 = \frac{n(n + 1)(2n + 1)}{6} \qquad \text{for all integers } n \geq 1$$

2. The sum of the first n odd positive integers is n^2.

$$1 + 3 + 5 + \cdots + (2n - 1) = n^2 \qquad \text{for all integers } n \geq 1$$

3. For all integers $n \geq 5$, 2^n is greater than n^2.

$$2^n > n^2 \qquad \text{for all integers } n \geq 5$$

4. For real numbers x and y and for any positive integer n, $(xy)^n = x^n y^n$.

$$(xy)^n = x^n y^n \qquad \text{for all integers } n \geq 1$$

5. For real numbers x and y, and for any positive integer n, $x - y$ is a factor of $x^n - y^n$.

$$(x - y) \text{ is a factor of } (x^n - y^n) \qquad \text{for all integers } n \geq 1$$

We shall prove each of these statements after we have formally stated the principle of mathematical induction.

The *principle of mathematical induction* may be thought of as being the *domino theory*. If dominoes are lined up so that if any one of them falls, then the next one will fall, and if some one of the dominoes *does* fall, then it follows that all of the dominoes in line after that one will also fall. In politics, the domino theory refers to countries rather than dominoes. The theory is that certain countries are "lined up" so that if any one of them falls under unsavory influence, then the next one will also fall; it follows that if the first of these countries *does* fall, then all of the others after it will fall. In mathematics, statements concerning integers take the place of dominoes or countries. Roughly speaking, the principle of mathematical induction asserts that if a statement being true (falling) for any one integer makes the statement true (fall) for the next integer, and if the statement *is* true for some one integer, then it is true for all larger integers.

To prove a statement true for all positive integers n, we must prove two things.

1. If the statement is true for any positive integer k, then it is true for the next positive integer $k + 1$.

2. The statement *is* true for the first positive integer $n = 1$.

To prove a statement true for all integers n greater than or equal to some particular integer q, we must also prove two things.

1. If the statement is true for any integer $k \geq q$, then it is true for the next integer $k + 1$.

2. The statement *is* true for integer q.

These two parts of a "proof by induction" are usually proved in reverse order because it is usually easier to prove or disprove that a statement is true for a particular integer than it is to prove that truth for any one integer forces truth for the next integer.

The principle of mathematical induction is stated formally below. In the formal statement of the principle, the symbol P_n is used for a statement concerning integer n.

$$P_n \text{ means a statement concerning integer } n.$$

Principle of Mathematical Induction If a statement P_n concerning integer n satisfies the following two conditions, then P_n is true for all integers n greater than or equal to a particular integer q. The two conditions are

1. P_n is true for $n = q$.

2. If P_n is true for any integer $k \geq q$, then P_n is true for the next integer $k + 1$.

(Note that if $q = 1$, and if P_n satisfies the two conditions of the principle of mathematical induction, then P_n is true for all positive integers.)

The two conditions of the principle of mathematical induction can be stated symbolically as

1. P_q is true.

2. P_k true $\rightarrow P_{k+1}$ true, for integer $k \geq q$.

The first part of the symbolic form of the second condition is read, "If P_k is true, then P_{k+1} is true." Or it is read, "P_k true implies P_{k+1} true."

In using the principle of mathematical induction and letting q represent a particular integer, the proof that a statement P_n is true for all integers $n \geq q$ consists of two parts.

Part I Prove: Statement P_n is true for $n = q$.

Part II Prove: If P_n is true for $n = k$, then P_n is true for $n = k + 1$, where k is any integer greater than or equal to q.

Such a proof can be summarized as follows:

Part I Prove: P_q is true.

Part II Prove: P_k true $\rightarrow P_{k+1}$ true   for any integer $k \geq q$.

Conclusion If P_q is true, and if P_k true $\rightarrow P_{k+1}$ true, for any integer $k \geq q$, then P_n is true for all integers $n \geq q$.

EXAMPLE 1   Prove that

$$1^2 + 2^2 + \cdots + n^2 = \frac{n(n + 1)(2n + 1)}{6}$$

for all positive integers n.

Solution   Let P_n be the statement

$$1^2 + 2^2 + \cdots + n^2 = \frac{n(n + 1)(2n + 1)}{6}. \qquad P_n$$

Part I Prove: P_n is true for $n = 1$. (Prove: P_1 is true.)

When $n = 1$, the statement P_n becomes P_1.

$$1^2 = \frac{1(1 + 1)(2 \times 1 + 1)}{6} \qquad P_1$$

That is,

$$1 = \frac{(2)(3)}{6}. \qquad \text{This is true.}$$

Part II Prove: If P_n is true for $n = k$ where k is any integer ≥ 1, then P_n is true for $n = k + 1$. (Prove: If P_k is true, then P_{k+1} is true, for any integer $k \geq 1$.)

Suppose P_n is true for $n = k$. That is, suppose

$$1^2 + 2^2 + \cdots + k^2 = \frac{k(k + 1)(2k + 1)}{6}. \qquad P_k$$

$$(P_n \text{ for } n = k)$$

On the basis of this supposition, we must prove that P_n is true for $n = k + 1$ and $k \geq 1$. That is, we must prove that P_k true implies that the following is true for $k \geq 1$.

$$1^2 + 2^2 + \cdots + (k + 1)^2 = \frac{(k + 1)(k + 1 + 1)(2(k + 1) + 1)}{6}. \qquad P_{k+1}$$

$$(P_n \text{ for } n = k + 1)$$

This statement, P_{k+1}, was obtained from P_n by replacing n by $k + 1$. In this type of problem (a certain indicated sum is given by a certain formula), it is usually helpful to write in the next to the last term of the indicated sum on the left of P_{k+1}. If we do that here, and if we also collect terms in the factors on the right, P_{k+1} becomes

$$1^2 + 2^2 + \cdots + k^2 + (k+1)^2 = \frac{(k+1)(k+2)(2k+3)}{6}. \qquad P_{k+1}$$

Proceeding to prove that for integer $k \geq 1$, P_k true implies that P_{k+1} is true, we start with the left side of P_{k+1} and use the fact that if P_k is true, then $1^2 + 2^2 + \cdots + k^2$ can be replaced by

$$\frac{k(k+1)(2k+1)}{6}.$$

$$1^2 + 2^2 + \cdots + k^2 + (k+1)^2 = \frac{k(k+1)(2k+1)}{6} + (k+1)^2, \text{ if } P_k \text{ is true}$$

$$= \frac{k(k+1)(2k+1)}{6} + \frac{6(k+1)^2}{6}$$

$$= \frac{(k+1)[k(2k+1) + 6(k+1)]}{6}$$

$$= \frac{(k+1)[2k^2 + k + 6k + 6]}{6}$$

$$= \frac{(k+1)(2k^2 + 7k + 6)}{6}$$

$$= \frac{(k+1)(k+2)(2k+3)}{6}$$

That is,

$$1^2 + 2^2 + \cdots + k^2 + (k+1)^2 = \frac{(k+1)(k+2)(2k+3)}{6}, \text{ if } P_k \text{ is true,}$$

which says P_{k+1} is true if P_k is true.

Conclusion   Since P_1 is true, and since if P_k is true, then P_{k+1} is true for any integer $k \geq 1$, it follows that P_n is true for all positive integers n.

EXAMPLE 2   Use mathematical induction to prove that

$$1 + 3 + \cdots + (2n - 1) = n^2$$

for all integers $n \geq 1$.

Solution   Let P_n be the statement

$$1 + 3 + \cdots + (2n - 1) = n^2. \qquad P_n$$

Part I   Prove: P_1 is true.

When $n = 1$, P_n becomes P_1.

$$1 = 1^2. \qquad P_1$$

This is true.

Part II Prove: If P_k is true, then P_{k+1} is true, for any integer $k \geq 1$.

When $n = k$, P_n becomes P_k.

$$1 + 3 + \cdots + (2k - 1) = k^2 \qquad P_k$$

When $n = k + 1$, P_n becomes P_{k+1}.

$$1 + 3 + \cdots + [2(k + 1) - 1] = (k + 1)^2. \qquad P_{k+1}$$

If we simplify the last term on the left side of P_{k+1}, and if we write in the next to the last term, P_{k+1} becomes

$$1 + 3 + \cdots + (2k - 1) + (2k + 1) = (k + 1)^2. \qquad P_{k+1}$$

We must prove that for $k \geq 1$, if P_k is true, then P_{k+1} is true. To prove this, we start with the left side of P_{k+1} and note that if P_k is true, we can replace $1 + 3 + \cdots + (2k - 1)$ by k^2.

$$1 + 3 + \cdots + (2k - 1) + (2k + 1) = k^2 + (2k + 1), \text{ if } P_k \text{ is true}$$
$$= k^2 + 2k + 1$$
$$= (k + 1)^2$$

That is,

$$1 + 3 + \cdots + (2k - 1) + (2k + 1) = (k + 1)^2, \text{ if } P_k \text{ is true.}$$

In other words, P_{k+1} is true if P_k is true.

Conclusion Since P_1 is true, and since P_k true $\rightarrow P_{k+1}$ true for any integer $k \geq 1$, it follows that P_n is true for all integers $n \geq 1$.

Note that in Example 2, the left side of P_n is the sum of the first n terms of an arithmetic sequence. In this case, we can therefore check the assertion that P_n is true by using the formula for the sum of the first n terms of an arithmetic sequence. Here the first term $a = 1$ and the last term $l = 2n - 1$, so according to that formula,

$$1 + 3 + \cdots + (2n - 1) = \frac{n(1 + 2n - 1)}{2}$$
$$= \frac{2n^2}{2}$$
$$= n^2.$$

Exercise 3.8, 1–20 This does check.

EXAMPLE 3   Use mathematical induction to prove that $2^n > n^2$ for all integers $n \geq 5$. (You may use the fact that $2^{k+1} = 2^k \times 2$ for any positive integer k.)

Solution   Let P_n be the statement

$$2^n > n^2. \qquad P_n$$

Part I   Prove: P_5 is true.

When $n = 5$, P_n becomes P_5.

$$2^5 > 5^2 \qquad P_5$$
$$32 > 25 \qquad \text{This is true.}$$

Part II   Prove: If P_k is true, then P_{k+1} is true, for any integer $k \geq 5$.

When $n = k$, P_n becomes P_k.

$$2^k > k^2 \qquad P_k$$

When $n = k + 1$, P_n becomes P_{k+1}.

$$2^{k+1} > (k + 1)^2 \qquad P_{k+1}$$

We must prove that for any integer $k \geq 5$, P_{k+1} is true if P_k is true. To prove this, we shall start with the left side of P_{k+1} and show that if P_k is true and $k \geq 5$, then the left side of P_{k+1} *is* greater than $(k + 1)^2$ so that P_{k+1} *is* true. In doing this, we shall use the fact that if P_k is true, $2^k > k^2$. We shall also use the fact that $k \geq 5$ so that $k > 3$ and $k > 1$. Here is the proof.

Left side
of P_{k+1}

$$
\begin{aligned}
2^{k+1} &= 2^k \times 2 \\
&> k^2 \times 2, &&\text{if } P_k \text{ is true} \\
&= k^2 + k^2 \\
&= k^2 + k \times k \\
&> k^2 + 3k &&k \geq 5 \text{ so } k > 3 \\
&= k^2 + 2k + k \\
&> k^2 + 2k + 1 &&k \geq 5 \text{ so } k > 1 \\
&= (k + 1)^2
\end{aligned}
$$

Right side
of P_{k+1}

This chain of inequalities and equalities says that $2^{k+1} > (k + 1)^2$, if P_k is true and integer $k \geq 5$. In other words, P_{k+1} is true if P_k is true and integer $k \geq 5$.

Conclusion Since P_5 is true, and since P_k true $\rightarrow P_{k+1}$ true for any integer $k \geq 5$, then P_n is true for all integers $n \geq 5$. (Note that this P_n is *not* true for *all* integers n. In particular, it is not true for $n = 2$, 3, or 4.)

Exercise 3.8, 21–25

EXAMPLE 4 Use mathematical induction to prove that for real numbers x and y and for any positive integer n, $(xy)^n = x^n y^n$. (You may assume that for any real number a, $a^1 = a$, and also that $a^{k+1} = a^k \cdot a$ for any positive integer k.)

Solution Let P_n be the statement

$$(xy)^n = x^n y^n. \qquad P_n$$

Part I Prove: P_1 is true.

When $n = 1$, P_n becomes P_1.

$$(xy)^1 = x^1 y^1 \qquad P_1$$
$$xy = xy \qquad \text{Since } a^1 = a \text{ for any real number } a.$$

This is true.

Part II Prove: If P_k is true, then P_{k+1} is true, for any integer $k \geq 1$.

When $n = k$, P_n becomes P_k.

$$(xy)^k = x^k y^k \qquad P_k$$

When $n = k + 1$, P_n becomes P_{k+1}.

$$(xy)^{k+1} = x^{k+1} y^{k+1} \qquad P_{k+1}$$

We must prove that for any integer $k \geq 1$, P_{k+1} is true if P_k is true. To prove this, we shall start with the left side of P_{k+1} and show that if k is a positive integer and P_k is true, then the left side of P_{k+1} *is* equal to $x^{k+1} y^{k+1}$ so that P_{k+1} *is* true. In doing this we shall use the fact that if P_k is true, then $(xy)^k = x^k y^k$.

We shall also use the fact that $(xy)^{k+1} = (xy)^k \cdot (xy)$. Here is the proof.

Left side
of P_{k+1}

$$(xy)^{k+1} = (xy)^k \cdot (xy) \qquad \text{since } a^{k+1} = a^k \cdot a$$
for real number a
and integer $k \geq 1$

$$= (x^k y^k) \cdot (xy) \qquad \text{if } P_k \text{ is true}$$
$$= x^k \cdot x \cdot y^k \cdot y$$
$$= x^{k+1} \cdot y^{k+1} \qquad \text{since } a^{k+1} = a^k \cdot a$$

Right side
of P_{k+1}

This chain of equalities says that

$$(xy)^{k+1} = x^{k+1} y^{k+1}, \text{ if } P_k \text{ is true and integer } k \geq 1$$

In other words, P_{k+1} is true if P_k is true and integer $k \geq 1$.

Conclusion
Exercise 3.8, 26–29

Since P_1 is true, and since P_k true $\rightarrow P_{k+1}$ true for any integer $k \geq 1$, then P_n is true for all positive integers n.

Some proofs by mathematical induction are quite straightforward; others are not. The next example demonstrates a proof that depends on the use of a bit of ingenuity.

EXAMPLE 5

Use mathematical induction to prove that for real numbers x and y and for any positive integer n, $(x - y)$ is a factor of $x^n - y^n$. (You may use the fact that for any real number a, $a^1 = a$, and that $a^{k+1} = a^k \cdot a$ for any positive integer k.)

Solution

Let P_n be the statement

$$(x - y) \text{ is a factor of } (x^n - y^n). \qquad P_n$$

Part I Prove: P_1 is true.

When $n = 1$, P_n becomes P_1.

$$(x - y) \text{ is a factor of } (x^1 - y^1). \qquad P_1$$

That is,

$$(x - y) \text{ is a factor of } (x - y). \qquad P_1$$

This is true because $(x - y) = 1 \cdot (x - y)$.

Part II Prove: If P_k is true, then P_{k+1} is true for any integer $k \geq 1$.

When $n = k$, P_n becomes P_k.

$$(x - y) \text{ is a factor of } (x^k - y^k). \qquad P_k$$

When $n = k + 1$, P_n becomes P_{k+1}.

$$(x - y) \text{ is a factor of } (x^{k+1} - y^{k+1}). \qquad P_{k+1}$$

We must prove that for any integer $k \geq 1$, P_{k+1} is true if P_k is true. To prove this, we shall consider $(x^{k+1} - y^{k+1})$. We shall add and subtract xy^k to $x^{k+1} - y^{k+1}$ so that we can group terms and factor. We shall then use the fact that $(x - y)$ is a factor of $(x^k - y^k)$ if P_k is true. Here is the proof.

$$x^{k+1} - y^{k+1} = x^{k+1} - xy^k + xy^k - y^{k+1}$$

$$= x^k \cdot x - xy^k + xy^k - y^k \cdot y \qquad \text{Since } a^{k+1} = a^k \cdot a$$
$$\text{for real number } a$$
$$\text{and positive integer}$$
$$k$$

$$= x(x^k - y^k) + y^k(x - y)$$

PART THREE | HIGHER ALGEBRA

This says that

$$x^{k+1} - y^{k+1} = x(x^k - y^k) + y^k(x - y).$$

Note that $(x - y)$ is a factor of the second term on the right. Now if P_k is true, then $(x - y)$ is also a factor of $(x^k - y^k)$, and hence is a factor of both terms on the right of the expression for $x^{k+1} - y^{k+1}$. It follows that $(x - y)$ is a factor of $x^{k+1} - y^{k+1}$.

Conclusion

Exercise 3.8, 30–32 Since P_1 is true, and since P_k true $\rightarrow P_{k+1}$ is true for any integer $k \geq 1$, the statement P_n is true for all positive integers n.

The next example illustrates that it is not sufficient merely to prove Part II of an induction proof without proving Part I. There do exist false statements P_n for which Part II can be carried out successfully.

EXAMPLE 6

Suppose that we were trying to use induction to prove that $n > n + 2$ for all positive integers n. Show that Part II of the induction proof can be carried out successfully, but Part I cannot.

Solution

Let P_n be the statement

$$n > n + 2. \qquad P_n$$

To show that Part II of the induction proof can be carried out successfully, we must show that if P_k is true, then P_{k+1} is true, for any integer $k \geq 1$. When $n = k$, P_n becomes P_k.

$$k > k + 2 \qquad P_k$$

When $n = k + 1$, P_n becomes P_{k+1}.

$$k + 1 > k + 1 + 2 \qquad P_{k+1}$$

We can show that P_{k+1} is true if P_k is true, by noting that if P_k is true, then the statement obtained by adding the number 1 to both sides of P_k is also true, and the statement so obtained is P_{k+1}. Indeed, if P_k is true, and if we add 1 to both sides of it, we obtain the true statement

$$k + 1 > k + 1 + 2,$$

which is P_{k+1}.

Part I of the induction proof cannot be done because P_1 is *not* true. The statement P_1 says that

$$1 > 1 + 2,$$

which is false. The statement P_n is not true for $n = 1$; in fact, it is not true for any positive integer.

Example 6 showed that for an induction proof, it is not sufficient merely to prove Part II. Likewise, merely proving Part I is not sufficient to guarantee that statement P_n is true for all integers $n \geq q$. Even verifying that P_n is true for several values of n is not sufficient. For example, the statement

$$1 + 2 + \cdots + n = \frac{n(n + 1)}{2} + (n - 1)(n - 2)(n - 3)$$

is true for $n = 1$, for $n = 2$, and for $n = 3$, but it is not true for all positive integers n. (Try it for $n = 4$.) In presenting a proof by mathematical induction, both Part I *and* Part II must be done.

Exercise 3.8, 33

Many theorems in mathematics are proved by using mathematical induction. The formula for the sum of the first n terms of a geometric or an arithmetic sequence can be established in this manner.

Exercise 3.8, 34–35

EXERCISE 3.8 MATHEMATICAL INDUCTION

▶   Use mathematical induction to prove that each of the following is true for all the prescribed positive integers n. If possible, check the truth of P_n by using the formula for the sum of the first n terms of an arithmetic sequence or a geometric sequence.

EXAMPLE $2 + 2^2 + \cdots + 2^n = 2(2^n - 1)$, $n \geq 1$

Solution   Let P_n be: $2 + 2^2 + \cdots + 2^n = 2(2^n - 1)$.

Part I P_1 is: $2 = 2(2 - 1)$. This is true.

Part II P_k is: $2 + 2^2 + \cdots + 2^k = 2(2^k - 1)$.

P_{k+1} is: $2 + 2^2 + \cdots + 2^k + 2^{k+1} = 2(2^{k+1} - 1)$.

Start with the left of P_{k+1}.

$$\begin{aligned}
2 + 2^2 + \cdots + 2^k + 2^{k+1} &= 2(2^k - 1) + 2^{k+1}, \text{ if } P_k \text{ is true} \\
&= 2^{k+1} - 2 + 2^{k+1} \\
&= 2 \times 2^{k+1} - 2 \\
&= 2(2^{k+1} - 1)
\end{aligned}$$

That is,

$$2 + 2^2 + \cdots + 2^k + 2^{k+1} = 2(2^{k+1} - 1), \text{ if } P_k \text{ is true.}$$

In other words,

$$P_{k+1} \text{ is true if } P_k \text{ is true.}$$

Conclusion P_n is true for all integers $n \geq 1$.

Check $2, 2^2, \ldots, 2^n$ is geometric with first term $a = 2$ and common ratio $r = 2$.

$$\begin{aligned}
2 + 2^2 + \cdots + 2^n &= \frac{2(1 - 2^n)}{1 - 2} \\
&= -2(1 - 2^n) \\
&= 2(2^n - 1) \qquad \text{It checks.}
\end{aligned}$$

1. $3 + 3^2 + \cdots + 3^n = \frac{3}{2}(3^n - 1), \; n \geq 1$

2. $1 + 2 + \cdots + n = \dfrac{n(n + 1)}{2}, \; n \geq 1$

3. $1^3 + 2^3 + \cdots + n^3 = \dfrac{n^2(n + 1)^2}{4}, \; n \geq 1$

4. $2 + 4 + 6 + \cdots + 2n = n(n + 1), \; n \geq 1$

5. $3 + 9 + 15 + \cdots + 3(2n - 1) = 3n^2, \; n \geq 1$

6. $4 + 8 + \cdots + 4n = 2n(n + 1), \; n \geq 1$

7. $2 + 8 + 18 + \cdots + 2n^2 = \dfrac{n(n + 1)(2n + 1)}{3}, \; n \geq 1$

8. $\frac{1}{2} + \frac{1}{4} + \cdots + \dfrac{1}{2^n} = 1 - (\frac{1}{2})^n, \; n \geq 1$

9. $2 + 16 + \cdots + 2n^3 = \dfrac{n^2(n + 1)^2}{2}, \; n \geq 1$

10. $2 + 5 + 8 + \cdots + [2 + 3(n - 1)] = \dfrac{4n + 3n(n - 1)}{2}, \; n \geq 1$

11. $1 + 7 + 19 + \cdots + [n^3 - (n - 1)^3] = n^3, \; n \geq 1$

12. $0.8 + (0.8)^2 + (0.8)^3 + \cdots + (0.8)^n = 4[1 - (0.8)^n], \; n \geq 1$

13. $1 + 15 + 66 + \cdots + [n^4 - (n - 1)^4] = n^4, \; n \geq 1$

14. $1 + \frac{1}{3} + \frac{1}{9} + \cdots + (\frac{1}{3})^{n-1} = \frac{3}{2}[1 - (\frac{1}{3})^n], \; n \geq 1$

15. $1 + 5 + 5^2 + \cdots + 5^{n-1} = \dfrac{5^n - 1}{4}, \; n \geq 1$

16. $\dfrac{1}{1 \times 2} + \dfrac{1}{2 \times 3} + \dfrac{1}{3 \times 4} + \cdots + \dfrac{1}{n(n + 1)} = \dfrac{n}{n + 1}, \; n \geq 1$

17. $3 + 7 + 11 + \cdots + [3 + 4(n - 1)] = n(2n + 1), \; n \geq 1$

18. $0.1 + (0.01) + (0.001) + \cdots + (0.1)^n = \dfrac{1 - (0.1)^n}{9}, \; n \geq 1$

19. $\dfrac{1}{6} + \dfrac{1}{6 \times 11} + \dfrac{1}{11 \times 16} + \cdots + \dfrac{1}{(5n - 4)(5n + 1)} = \dfrac{n}{5n + 1},$ $n \geq 1$

20. $5 + 10 + 15 + \cdots + [5 + 5(n - 1)] = \dfrac{5n(n + 1)}{2}, \; n \geq 1$

21. $n^3 > n^2, \; n \geq 2$

22. $2^n > 2n^2, \; n \geq 7$ (You may use the fact that $2^{k+1} = 2^k \times 2$ for any positive integer k.)

23. $n^3 > n, \; n \geq 2$

24. $n^2 > n$, $n \geq 2$

25. $3^n > 6n$, $n \geq 3$ (You may use the fact that $3^{k+1} = 3^k \times 3$ for any positive integer k.)

26. For real numbers x and y and for $n \geq 1$, $(x/y)^n = x^n/y^n$. (You may assume that for any real number a, $a^1 = a$, and also that $a^{k+1} = a^k \cdot a$ for any positive integer k.)

27. For real number x and $n \geq 1$, $x^2 \cdot x^n = x^{2+n}$. (You may assume that for any real number a, $a^1 = a$, and also that $a^{m+1} = a^m \cdot a$ for any positive integer m.)

28. For real number x and $n \geq 1$, $(x^2)^n = x^{2n}$. (You may assume that for any real number a, $a^1 = a$, $a^{m+1} = a^m \cdot a^1$ and also that $a^2 \cdot a^m = a^{2+m}$ for any positive integer m.)

29. For real number x and $n \geq 3$, $x^n/x^2 = x^{n-2}$. (You may assume that for any real number a, $a^1 = a$, and also that $a^{m+1} = a^m \cdot a$ for any positive integer m.)

30. For real numbers x and y and $n \geq 1$, $(x + y)$ is a factor of $x^{2n-1} + y^{2n-1}$. (You may use the fact that for any real number a and positive integer m, $a^{m+2} = a^m \cdot a^2$.)

31. For real numbers x and y and $n \geq 1$, $(x - y)$ is a factor of $(x^{2n} - y^{2n})$. (You may use the fact that for any real number a and positive integer m, $a^{m+2} = a^m \cdot a^2$.)

32. For $n \geq 1$, 2 is a factor of $n^2 + n$.

33. Suppose we were trying to use induction to prove that $n > n + 3$ for all positive integers n. Show that Part II of the induction proof can be carried out successfully, but Part I cannot.

34. For real number $r \neq 1$ and for $n \geq 1$, prove that

$$1 + r + r^2 + \cdots + r^{n-1} = \frac{1 - r^n}{1 - r}.$$

35. For real number d and for $n \geq 1$, prove that

$$1 + (1 + d) + (1 + 2d) + \cdots + [1 + (n - 1)d] = \frac{n[2 + (n - 1)d]}{2}.$$

CHECK LIST FOR THE STUDENT

If you have learned the material in Section 3.8, you should be able to do the following:

1. Tell what is meant by the domino theory.

2. Tell what two things must be proved in order to prove that a statement is true for all positive integers n or for all integers n greater than or equal to some particular integer q.

3. Complete the following statement of the principle of mathematical induction: If a statement P_n concerning integer n satisfies the following two conditions:

1. _____

2. _____

then P_n is true for all integers $n \geq q$.

4. Complete the following: The proof that a statement P_n is true for all integers $n \geq q$ consists of two parts.

Part I   Prove: _____

Part II   Prove: _____

5. Use mathematical induction to prove that certain given statements concerning integers are true; if possible, check the truth of a given statement by using the formula for the sum of the first n terms of an arithmetic or a geometric sequence.

6. Show by presenting examples that in a proof by mathematical induction, both Part I *and* Part II of the proof must be done.

SECTION QUIZ 3.8

Use mathematical induction to prove that each of the following is true for all the prescribed positive integers n. If possible, check the truth by using the formula for the sum of the first n terms of an arithmetic sequence or a geometric sequence.

1. $5 + 5^2 + 5^3 + \cdots + 5^n = \frac{5}{4}(5^n - 1)$, $n \geq 1$

2. $3 + 6 + 9 + \cdots + 3n = \frac{3}{2}(n + n^2)$, $n \geq 1$

3. $2^n > n + 2$, $n \geq 3$

4. $3 + 12 + \cdots + 3n^2 = \dfrac{n(n + 1)(2n + 1)}{2}$, $n \geq 1$

5. 2 is a factor of $n^2 + 3n$, $n \geq 1$

3.9 Binomial Theorem

The purpose of this section is to state and illustrate the binomial theorem and to present Pascal's triangle.

The binomial theorem provides us with a formula for expanding—that is, for writing out as a sum—the power of a binomial such as $(a + b)^3$, or $(a + b)^{25}$, or $(a + b)^n$ for any positive integer n. It enables us to write out any particular term of such an expansion without writing out the entire expansion. It can be extended to provide series representations for expressions such as $(1 + x)^r$ where r is any real number and $|x| < 1$.

As an introduction, let us write out $(a + b)^n$ for a few specific values of n so that we can see the pattern of the terms. We already know that $(a + b)^2 = a^2 + 2ab + b^2$. We can find successively higher powers of $(a + b)$ by successively multiplying by $a + b$.

$$(a + b)^3 = (a + b)^2(a + b),$$
$$(a + b)^4 = (a + b)^3(a + b),$$
$$(a + b)^5 = (a + b)^4(a + b), \text{ etc.}$$

Here are some powers of $(a + b)$ written out.

$$(a + b)^1 = 1 \cdot a + 1 \cdot b$$
$$(a + b)^2 = 1 \cdot a^2 + 2ab + 1 \cdot b^2$$
$$(a + b)^3 = 1 \cdot a^3 + 3a^2b + 3ab^2 + 1 \cdot b^3$$
$$(a + b)^4 = 1 \cdot a^4 + 4a^3b + 6a^2b^2 + 4ab^3 + 1 \cdot b^4$$
$$(a + b)^5 = 1 \cdot a^5 + 5a^4b + 10a^3b^2 + 10a^2b^3 + 5ab^4 + 1 \cdot b^5$$

To begin to see how the terms are formed, look at $(a + b)^4$ as an example. Note that the coefficient of a^4 is 1 and that the coefficient of b^4 is 1. The degree of each term is 4. We may think of the last term as being a^0b^4, and we may think of the first term as being a^4b^0. Thus proceeding left to right, the exponents of the a are successively 4, 3, 2, 1, 0, and the exponents of the b are successively 0, 1, 2, 3, 4.

Now look at $(a + b)^5$ as an example. The coefficient of a^5 is 1; the coefficient of b^5 is 1. The degree of each term is 5. Proceeding left to right, the exponents of the a are successively 5, 4, 3, 2, 1, 0; the exponents of the b are successively 0, 1, 2, 3, 4, 5.

The patterns that we observe in the expansions of $(a + b)^4$ and $(a + b)^5$ lead us to expect similar patterns in the expansion of $(a + b)^n$ for any positive integer n. As part of the binomial theorem, we can prove by mathematical induction that in the expansion of $(a + b)^n$,

1. The coefficient of a^n is 1; the coefficient of b^n is 1.

2. The degree of each term is n.

3. The exponents of the a are successively n $(n-1)$, $(n-2)$, . . . , 2, 1, 0.

4. The exponents of the b are successively 0, 1, 2, . . . , $(n-2)$, $(n-1)$, n.

Knowing this, to write out $(a+b)^n$ for any positive integer n, we only need a formula for computing the coefficients.

EXAMPLE 1

Leaving blank parentheses for the coefficients of other than a^6 and b^6, write out $(a+b)^6$.

Solution

The coefficients of a^6 and b^6 are both 1. The degree of each term is 6; the exponents of a are successively 6, 5, 4, 3, 2, 1, 0; the exponents of b are successively 0, 1, 2, 3, 4, 5, 6.

$$(a+b)^6 = a^6 + (\ )a^5b + (\ )a^4b^2 + (\ )a^3b^3 + (\ )a^2b^4 + (\ )ab^5 + b^6$$

One way to compute the coefficients of $(a+b)^n$ for positive integer n was observed by the French mathematician Blaise Pascal in the seventeenth century and continues to be useful for small values of n. His method becomes apparent when the coefficients of $(a+b)^n$ for the first few positive integers n are written in an array called *Pascal's triangle*. Pascal's triangle for $n = 1, 2, 3, 4, 5$ is shown in Figure 3.8. Each row consists of the coefficients in the expansion of the expression written in the left margin. Except for the ones, each entry is the sum of the entry just to its left and the entry just to its right in the preceding row. For example, the 6 in the row of coefficients of $(a+b)^4$ is $3+3$, and the first 5 in the row of coefficients of $(a+b)^5$ is $1+4$. Notice that the array of coefficients in any row is symmetric in that the second one in from the right in each row is the same as the second one in from the left in that row, etc.

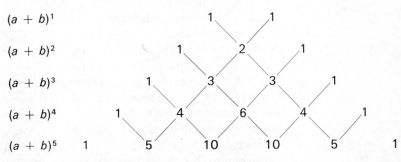

FIGURE 3.8   Pascal's triangle for $n = 1, 2, 3, 4, 5$

EXAMPLE 2

Use Pascal's triangle to find the coefficients and expand $(a+b)^6$.

Solution

Starting with the fifth row of Pascal's triangle as written in Figure 3.8 we can compute the sixth row.

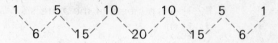

$$1 \quad\quad 5 \quad\quad 10 \quad\quad 10 \quad\quad 5 \quad\quad 1$$
$$6 \quad\quad 15 \quad\quad 20 \quad\quad 15 \quad\quad 6$$

Knowing the pattern of the terms, we can now write the expansion.

$$(a + b)^6 = a^6 + 6a^5b + 15a^4b^2 + 20a^3b^3 + 15a^2b^4 + 6ab^5 + b^6$$

Exercise 3.9, 1–10

Pascal's triangle is impractical to use for large values of n, so we are led to consider the binomial tneorem. We shall see that according to the binomial theorem, we can find the coefficient of any term in the expansion of $(a + b)^n$ by considering the preceding term in *that* expansion rather than considering two coefficients in the expansion of $(a + b)^{n-1}$, as one has to do in Pascal's triangle. In fact, using the binomial theorem, we shall see that the following rule is true.

Binomial Coefficient Rule

To find the coefficient of any term after the first in the expansion $(a + b)^n$,

1. Consider the *preceding* term; multiply its coefficient by the exponent of a.

2. Divide this product by the number of the *preceding* term.

Let us check this rule in the expansion of $(a + b)^4$.

Number of term 1   2   3   4   5

$$1 \cdot a^4 + 4a^3b + 6a^2b^2 + 4a^1b^3 + 1 \cdot b^4$$

$$\frac{1 \times 4}{1} \quad \frac{4 \times 3}{2} \quad \frac{6 \times 2}{3} \quad \frac{4 \times 1}{4}$$

The rule checks. The student should verify that the rule also checks in the expansion of $(a + b)^n$ for 1, 2, 3, 5, and 6.

As a preliminary to stating the binomial theorem itself, Table 3.1 shows how the coefficients are formed when the binomial coefficient rule is applied to $(a + b)^n$ for any positive integer n. Note that in each term the exponent of b is one less than the number of the term. Where r is less than n, the chart gives a general formula for the term numbered $r + 1$. That general formula is based on the rth term and follows the patterns that may be observed in the formation of the first four terms. We shall see that the binomial theorem simply states that the formula shown in the chart is correct.

The entries in Table 3.1 can be written more concisely by employing *factorial* notation. For any positive integer n, the symbol $n!$, read "n factorial" represents the product $1 \times 2 \times 3 \times \cdots \times (n - 1) \times n$.

$$n! = 1 \times 2 \times 3 \times \cdots \times (n - 1) \times n$$
$$5! = 1 \times 2 \times 3 \times 4 \times 5$$

TABLE 3.1   FORMATION OF COEFFICIENTS OF $(a + b)^n$

Number of term	Coefficient	Term
1	1	$1 \times a^n$
2	$\dfrac{1 \times n}{1}$	$\dfrac{n}{1} a^{n-1} b$
3	$\dfrac{n}{1}(n - 1) \div 2$	$\dfrac{n(n - 1)}{1 \times 2} a^{n-2} b^2$
4	$\dfrac{n(n - 1)}{1 \times 2}(n - 2) \div 3$	$\dfrac{n(n - 1)(n - 2)}{1 \times 2 \times 3} a^{n-3} b^3$
$\vdots$	$\vdots$	$\vdots$
$r + 1$	$\dfrac{n(n - 1)(n - 2) \cdots (n - (r - 1))}{1 \times 2 \times 3 \cdots \times (r - 1)} \div r$	$\dfrac{n(n - 1)(n - 2) \cdots (n - (r - 1))}{1 \times 2 \times 3 \times \cdots \times (r - 1) \times r} a^{n-r} b^r$
$\vdots$	$\vdots$	$\vdots$
$n - 1$	n	nab^{n-1}
n	1	$1 \times b^n$

Note that $(n - 1)! \times n = n!$. Note also that

$$1! = 1$$

and

$$2! = 1 \times 2 = 2.$$

By definition,

$$0! = 1.$$

Using this notation in Table 3.1, we see that the denominator of the coefficient of term number 3 is 2!. The denominator of the coefficient of term number 4 is 3!, and the denominator of the coefficient of term number $r + 1$ is $r!$.

Let us now formally state the binomial theorem.

Binomial Theorem   If a and b represent real numbers, if n is a positive integer, and if positive integer r is less than n, then $(a + b)^n$ is the sum of $(n + 1)$ terms as follows:

$$(a + b)^n = a^n + na^{n-1}b + \frac{n(n - 1)}{2!} a^{n-2} b^2$$

$$+ \frac{n(n - 1)(n - 2)}{3!} a^{n-3} b^3 + \cdots$$

$$+ \frac{n(n - 1)(n - 2) \cdots (n - (r - 1))}{r!} a^{n-r} b^r + \cdots$$

$$+ \frac{n(n - 1)(n - 2) \cdots 1}{(n - 1)!} ab^{n-1} + b^n.$$

(The next to the last term can also be written nab^{n-1} because

$$\frac{n(n-1)(n-2)\cdots 1}{(n-1)!} = \frac{n\cdot(n-1)!}{(n-1)!} = n.)$$

Let us take a second look at the statement of this theorem. It does say that the coefficients are formed according to the rule used in Table 3.1 In fact, the binomial theorem states that in the expansion of $(a+b)^n$,

the coefficient of $a^{n-r}b^r$ is $\dfrac{\overbrace{n(n-1)(n-2)\cdots(n-(r-1))}^{r \text{ factors}}}{r!}$.

Note that this is the coefficient of term number $r+1$; note that the exponent of the b is one less than the number of the term; note that there are r factors starting with the number n in the numerator of this coefficient, and r factorial is the denominator.

Exercise 3.9, 11–20

The binomial theorem is proved by mathematical induction.

Proof   Let P_n be the statement

$$(a+b)^n = a^n + na^{n-1}b + \frac{n(n-1)}{2!}a^{n-2}b^2 + \cdots$$

$$+ \frac{n(n-1)\cdots(n-(r-1))}{r!}a^{n-r}b^r + \cdots$$

$$+ nab^{n-1} + b^n. \qquad P_n$$

(For conciseness, we have not written in the fourth term, and we have written the next to the last term in simplified form. It is understood that there are $n+1$ terms on the right side of P_n; the terms are formed according to the rule indicated until b^n is obtained.)

Part I   Prove: P_1 is true.

When $n = 1$, there are $1 + 1 = 2$ terms on the right of P_n, and P_n becomes P_1:

$$(a+b)^1 = a^1 + b^1 \qquad P_1$$

This is true.

Part II   Prove: If P_k is true, then P_{k+1} is true, for any integer $k \geq 1$.

When $n = k$, P_n becomes P_k.

$$(a+b)^k = a^k + ka^{k-1}b + \frac{k(k-1)}{2!}a^{k-2}b^2 + \cdots$$

$$+ \frac{k(k-1)\cdots(k-(r-1))}{r!}a^{k-r}b^r$$

$$+ \cdots + kab^{k-1} + b^k. \qquad P_k$$

When $n = k + 1$, P_n becomes P_{k+1}:

$$(a+b)^{k+1} = a^{k+1} + (k+1)a^{k+1-1}b$$

$$+ \frac{(k+1)(k+1-1)}{2!}a^{k+1-2}b^2 + \cdots$$

$$+ \frac{(k+1)(k+1-1)\cdots(k+1-(r-1))}{r!}a^{k+1-r}b^r + \cdots$$

$$+ (k+1)ab^{k+1-1} + b^{k+1} \qquad P_{k+1}$$

If P_k is true, the statement obtained by multiplying both sides of P_k by $(a + b)$ is also true. The statement so obtained is actually P_{k+1}. **To see how this works out without writing out all the algebraic details, let us use abbreviated and simplified expressions for P_k and for P_{k+1}.** Let us use the expression

$$(a+b)^k = a^k + ka^{k-1}b + \frac{k(k-1)}{2!}a^{k-2}b^2 + \cdots + kab^{k-1} + b^k \qquad P_k$$

for P_k, and let us use the expression

$$(a+b)^{k+1} = a^{k+1} + (k+1)a^k b + \frac{(k+1)k}{2!}a^{k-1}b^2 + \cdots$$

$$+ (k+1)ab^k + b^{k+1} \qquad P_{k+1}$$

for P_{k+1}. (For future reference, we have written in the next to the last terms of P_k and P_{k+1}.) Let us now multiply both sides of P_k by $(a + b)$.

$$(a+b)(a+b)^k = a\left[a^k + ka^{k-1}b + \frac{k(k-1)}{2!}a^{k-2}b^2 + \cdots + b^k\right]$$

$$+ b\left[a^k + ka^{k-1}b + \cdots + kab^{k-1} + b^k\right]$$

(To simplify the presentation, we did not write the next to the last term in the first bracket or the third term in the second bracket.) Multiplying this out, we have

$$(a+b)^{k+1} = a^{k+1} + ka^k b + \frac{k(k-1)}{2!}a^{k-1}b^2 + \cdots + ab^k$$

$$+ a^k b + ka^{k-1}b^2 + \cdots + kab^k + b^{k+1}.$$

Now collecting like terms we have

$$(a + b)^{k+1} = a^{k+1} + (k + 1)a^k b + \left[\frac{k(k - 1)}{2!} + k \right] a^{k-1}b^2$$

$$+ \cdots + (k + 1)ab^k + b^{k+1}$$

The coefficient of the third term on the right can be simplified as follows.

$$\frac{k(k - 1)}{2!} + k = \frac{k^2 - k + 2k}{2}$$

$$= \frac{k^2 + k}{2}$$

$$= \frac{(k + 1)k}{2!}$$

Therefore we have

$$(a + b)^{k+1} = a^{k+1} + (k + 1)a^k b + \frac{(k + 1)k}{2!}a^{k-1}b^2 + \cdots$$

$$+ (k + 1)ab^k + b^{k+1}$$

which *is* P_{k+1}.

Conclusion   Since P_1 is true, and since P_k true $\longrightarrow$ P_{k+1} true, for any integer $k \geq 1$, it follows that P_n is true for all integers $n \geq 1$.

The symbol $\binom{n}{r}$ is commonly used for the coefficient of $a^{n-r}b^r$ in the expansion of $(a + b)^n$. The number $\binom{n}{r}$ is called the **binomial coefficient** of $a^{n-r}b^r$ in $(a + b)^n$. It is read "n over r."

$$\binom{n}{r} = \text{coefficient of } a^{n-r}b^r \text{ in } (a + b)^n$$

$$\binom{n}{r} = \frac{\overbrace{n(n - 1)(n - 2) \cdots (n - (r - 1))}^{r \text{ factors}}}{\underbrace{r!}_{r \text{ factorial}}} \quad \begin{array}{l} r \text{ factors} \\ \text{starting with } n \end{array}$$

For example, the expansion of $(a + b)^7$, the coefficient of a^4b^3 is $\binom{7}{3}$.

$$\binom{7}{3} = \text{coefficient of } a^4b^3 \text{ in } (a + b)^7$$

$$\binom{7}{3} = \frac{7 \times 6 \times 5}{1 \times 2 \times 3} \qquad \begin{array}{l} \text{3 factors} \\ \text{starting with 7} \end{array}$$

3 factors

3 factorial

Using this notation, in $(a + b)^n$,

1 factor

the coefficient of $a^{n-1}b$ is $\binom{n}{1} = \dfrac{n}{1}$,

1 factorial

n factors

and the coefficient of b^n is $\binom{n}{n} = \dfrac{n(n - 1)(n - 2) \cdots 1}{n!}$

n factorial

$$= \frac{n!}{n!}$$

$$= 1.$$

$$\binom{n}{n} = 1$$

Since the coefficient of $a^n = a^n b^0$ is also 1, for consistency we define $\binom{n}{0}$ to be 1.

$$\binom{n}{0} = 1$$

EXAMPLE 3

Find the value of the binomial coefficient $\binom{9}{6}$, and write the term having $\binom{9}{6}$ for its coefficient in the expansion of $(x + y)^9$.

Solution

$$\binom{9}{6} = \frac{9 \times 8 \times 7 \times 6 \times 5 \times 4}{1 \times 2 \times 3 \times 4 \times 5 \times 6}$$

$$= 84$$

$$\binom{9}{6} x^3 y^6 = 84x^3 y^6$$

EXAMPLE 4   Use the notation $\binom{n}{r}$ for the coefficient of $a^{n-r}b^r$ in $(a + b)^n$, and write the first three rows of Pascal's triangle.

Solution $n = 1$: $\binom{1}{0}$ $\binom{1}{1}$

$n = 2$: $\binom{2}{0}$ $\binom{2}{1}$ $\binom{2}{2}$

$n = 3$: $\binom{3}{0}$ $\binom{3}{1}$ $\binom{3}{2}$ $\binom{3}{3}$

EXAMPLE 5   Show that the binomial coefficient $\binom{n}{r}$ can also be written in the form

$$\frac{n!}{r!\,(n - r)!}.$$

Solution $$\binom{n}{r} = \frac{n(n - 1) \cdots (n - (r - 1))}{r!}$$

Multiplying numerator and denominator by $(n - r)!$, we have

$$\binom{n}{r} = \frac{n(n - 1) \cdots (n - r + 1)(n - r)!}{r!\,(n - r)!}.$$

Since $(n - r)! = (n - r)(n - r - 1) \cdots 1$, the numerator is simply $n!$. It follows that

Exercise 3.9, 21–55

$$\binom{n}{r} = \frac{n!}{r!\,(n - r)!}.$$

In calculus it is shown that for *any* number n (not just for *positive integer n*), and for real number x such that $|x| < 1$, the expression $(1 + x)^n$ can be represented by an infinite series. It is shown that the terms of the series representing $(1 + x)^n$ are formed according to the same rule as that given by the binomial theorem. This is illustrated for $(1 + x)^{1/2}$ in the next example.

EXAMPLE 6   Assume that there is an infinite series representing $(1 + x)^{1/2}$ and that the terms of the series are formed according to the rule given in the binomial theorem. Find the first four terms.

Solution   According to the binomial theorem,

$$(a + b)^n = a^n + na^{n-1}b + \frac{n(n - 1)}{2!}a^{n-2}b^2$$

$$+ \frac{n(n - 1)(n - 2)}{3!}a^{n-3}b^3 + \cdots .$$

If we take $a = 1$, $b = x$, and $n = \frac{1}{2}$, we have

$$(1 + x)^{1/2} = 1^{1/2} + \tfrac{1}{2}(1)^{(1/2)-1}x + \frac{\frac{1}{2}(\frac{1}{2} - 1)}{2!}(1)^{(1/2)-2}x^2$$

$$+ \frac{\frac{1}{2}(\frac{1}{2} - 1)(\frac{1}{2} - 2)}{3!}(1)^{(1/2)-3}x^3 + \cdots .$$

Simplifying, we have

$$(1 + x)^{1/2} = 1 + \frac{1}{2}x + \frac{1}{2}\left(\frac{-1}{2}\right)\left(\frac{1}{2}\right)x^2 + \left(\frac{1}{2}\right)\left(\frac{-1}{2}\right)\left(\frac{-3}{2}\right)\left(\frac{1}{2 \times 3}\right)x^3 + \cdots$$

Exercise 3.9, 56–60
$$= 1 + \frac{x}{2} - \frac{x^2}{2^3} + \frac{x^3}{2^4} + \cdots .$$

EXERCISE 3.9 BINOMIAL THEOREM

▶ Use Pascal's triangle to find the coefficients and expand each of the following.

EXAMPLES   (a) $(1 + x)^7$ (b) $(x - y)^4$

Solutions   (a) Sixth row: 1   6   15   20   15   6   1

Seventh row: 1   7   21   35   35   21   7   1

Since 1^7, 1^6, 1^5, 1^4, 1^3, and 1^2 are all equal to 1,

$(1 + x)^7 = 1 + 7x + 21x^2 + 35x^3 + 35x^4 + 21x^5 + 7x^6 + x^7.$

(b) Fourth row: 1   4   6   4   1

$$(x - y)^4 = x^4 + 4x^3(-y) + 6x^2(-y)^2 + 4x(-y)^3 + (-y)^4$$
$$= x^4 - 4x^3y + 6x^2y^2 - 4xy^3 + y^4$$

1. $(1 + x)^8$ **2.** $(x - y)^6$ **3.** $(x + 2y)^4$ **4.** $(a + b)^9$

5. $(1 + x)^6$ **6.** $(a - b)^{10}$ **7.** $(x - y)^5$ **8.** $(x - 1)^7$

9. $(2a + 3b)^4$ **10.** $(a - 2b)^5$

▶ Use the binomial theorem and write out the first four terms in the expansion of each of the following.

EXAMPLES   (a) $(a + b)^{12}$ (b) $(x - 2y)^5$

Solutions   (a) $(a + b)^{12} = a^{12} + 12a^{11}b + \dfrac{12 \cdot 11}{2!}a^{10}b^2 + \dfrac{12 \cdot 11 \cdot 10}{3!}a^9b^3 + \cdots$

$$= a^{12} + 12a^{11}b + 66a^{10}b^2 + 220a^9b^3 + \cdots$$

(b) $(x - 2y)^5 = x^5 + 5x^4(-2y) + \dfrac{5 \cdot 4}{2!} x^3(-2y)^2$

$$+ \dfrac{5 \cdot 4 \cdot 3}{3!} x^2(-2y)^3 + \cdots$$

$$= x^5 - 10x^4y + 40x^3y^2 - 80x^2y^3 + \cdots$$

11. $(a + b)^{20}$ **12.** $(a - b)^{20}$ **13.** $(1 + x)^{12}$ **14.** $(1 - x)^{12}$

15. $(3x - y)^6$ **16.** $(2x + y)^5$ **17.** $(x - y)^{10}$ **18.** $(1 + 2x)^7$

19. $(x + y)^{30}$ **20.** $(x - y)^{18}$

▶ Find the numerical value of each of the given binomial coefficients, and write the term having that binomial coefficient in the expansion indicated.

EXAMPLE $\dbinom{8}{3}, \quad (x + y)^8$

Solution $\dbinom{8}{3} = \dfrac{8 \times 7 \times 6}{1 \times 2 \times 3} = 56, \quad \dbinom{8}{3} x^5 y^3 = 56x^5 y^3$

21. $\dbinom{8}{5}, \quad (x + y)^8$ **22.** $\dbinom{9}{4}, \quad (a + b)^9$

23. $\dbinom{10}{6}, \quad (x + y)^{10}$ **24.** $\dbinom{11}{4}, \quad (x - y)^{11}$

25. $\dbinom{30}{2}, \quad (x + 2y)^{30}$ **26.** $\dbinom{15}{15}, \quad (a + b)^{15}$

27. $\dbinom{12}{10}, \quad (a - b)^{12}$ **28.** $\dbinom{23}{1}, \quad (x + y)^{23}$

29. $\dbinom{16}{4}, \quad (x - y)^{16}$ **30.** $\dbinom{25}{24}, \quad (2x + y)^{25}$

▶ Find the coefficient of each of the following in the expansion indicated.

EXAMPLE $x^7 y^3, \quad (x - 2y)^{10}$

Solution $\dbinom{10}{3} (x)^7 (-2y)^3 = \dfrac{10 \times 9 \times 8}{1 \times 2 \times 3} x^7 (-8y^3)$

$$= -960 x^7 y^3$$

31. $x^2 y^5, \quad (3x + 2y)^7$ **32.** $x^5 y^5, \quad (2x - y)^{10}$

33. a^9b^2, $(a - 5b)^{11}$ **34.** a^3b^9, $(3a + b)^{12}$

35. x^7y^8, $(2x - y)^{15}$

▶ Use the notation $\binom{n}{r}$ for the coefficient of $a^{n-r}b^r$ in $(a + b)^n$, and write the indicated row of Pascal's triangle.

EXAMPLE   Row 4

Solution

$$\binom{4}{0} \quad \binom{4}{1} \quad \binom{4}{2} \quad \binom{4}{3} \quad \binom{4}{4}$$

36. Row 5   **37.** Row 6   **38.** Row 7   **39.** Row 8   **40.** Row 9

▶ Verify each of the following by finding the values of the binomial coefficients.

EXAMPLE

$$\binom{3}{1} = \binom{2}{0} + \binom{2}{1}$$

Solution

$$\binom{3}{1} = \frac{3}{1} = 3, \quad \binom{2}{0} = 1, \quad \binom{2}{1} = \frac{2}{1} = 2$$

$$3 = 1 + 2$$

41. $\binom{4}{1} = \binom{3}{0} + \binom{3}{1}$ **42.** $\binom{6}{1} = \binom{5}{0} + \binom{5}{1}$ **43.** $\binom{6}{2} = \binom{5}{1} + \binom{5}{2}$

44. $\binom{6}{4} = \binom{5}{3} + \binom{5}{4}$ **45.** $\binom{7}{1} = \binom{6}{0} + \binom{6}{1}$ **46.** $\binom{7}{6} = \binom{6}{5} + \binom{6}{6}$

47. $\binom{4}{1} = \binom{4}{3}$ **48.** $\binom{5}{2} = \binom{5}{3}$ **49.** $\binom{6}{1} = \binom{6}{5}$

50. $\binom{8}{3} = \binom{8}{5}$

▶ For the given n and r, show that $n(n - 1) \cdots (n - r + 1)(n - r)!$ is $n!$.

EXAMPLE $n = 8, r = 3$

Solution $8 \times 7 \times 6 \times 5! = 8 \times 7 \times 6 \times 5 \times 4 \times 3 \times 2 \times 1 = 8!$

51. $n = 9, r = 3$ **52.** $n = 7, r = 2$ **53.** $n = 10, r = 4$

54. $n = 8, r = 4$ **55.** $n = 11, r = 4$

▶ Assume that there is an infinite series representing each of the following and that the terms of the series are formed according to the rule given in the binomial theorem. Find the first four terms.

EXAMPLE

$(1 + x)^{-2}$

Solution

$$(1 + x)^{-2} = 1 + (-2)(1)^{-3}x + \frac{(-2)(-3)}{2!}(1)^{-4}x^2$$

$$+ \frac{(-2)(-3)(-4)}{3!}(1)^{-5}x^3 + \cdots$$

$$= 1 - 2x + 3x^2 - 4x^3 + \cdots$$

56. $(1 + x)^{-1}$ **57.** $(1 + x)^{1/3}$ **58.** $(1 + x)^{-1/2}$

59. $(1 + x)^{3/2}$ **60.** $(1 + x)^{-3}$

CHECK LIST FOR THE STUDENT

If you have learned the material in Section 3.9, you should be able to do the following:

1. Give the coefficients of a^n and b^n, and give the degree of each term in the expansion $(a + b)^n$, for any positive integer n.

2. Proceeding left to right, tell what the exponents of the a and of the b are in successive terms of the expansion of $(a + b)^n$, for any positive integer n.

3. Show how to form the rows of Pascal's triangle.

4. Use Pascal's triangle and expand $(a + b)^n$ for some given positive integers n.

5. Use factorial notation. Give the values of 1! and 0!. Show that

$$\frac{n!}{(n - 1)!} = n.$$

6. State the binomial theorem.

7. Use the binomial theorem and expand $(a + b)^n$ for some given positive integers n.

8. Give the coefficient of $a^{n-r}b^r$ in the expansion of $(a + b)^n$.

9. Give the number of the term having $a^{n-r}b^r$ as a factor in the expansion of $(a + b)^n$.

10. Outline the proof of the binomial theorem.

11. Tell what coefficient it is that $\binom{n}{r}$ represents in $(a + b)^n$.

12. Find the value of $\binom{n}{r}$ for given n and r.

13. Give the values of $\binom{n}{1}$, $\binom{n}{n}$, and $\binom{n}{0}$, for any positive integer n.

14. Use the notation $\binom{n}{r}$ and write the first few rows of Pascal's triangle.

15. Show that

$$\binom{n}{r} = \frac{n(n - 1) \cdots (n - r + 1)}{r!}$$

can also be written

$$\frac{n!}{r!(n - r)!}.$$

16. Assume there is an infinite series representing $(1 + x)^r$ for a given real number r, and use the rule given in the binomial theorem to find the first few terms.

SECTION QUIZ 3.9

1. Leave blank parentheses for the coefficients of other than a^7 and b^7, and write out $(a + b)^7$.

2. Use Pascal's triangle to find the coefficients and expand $(x + y)^5$.

3. Use the binomial theorem and write out the first four terms in the expansion of $(a + b)^{10}$.

4. Find the numerical value of the binomial coefficient $\binom{12}{4}$, and write the term having that binomial coefficient in the expansion of $(x + y)^{12}$.

5. Find the coefficient of x^2y^4 in the expansion of $(2x + y)^6$.

3.10 Permutations, Combinations, and Counting

The purpose of this section is to introduce and illustrate permutations, combinations, and the fundamental counting principle.

A **permutation** is an ordered arrangement of a collection or set of things such as symbols, routes, or events. For example,

$$pqr$$

is a permutation of the three letters p, q, and r. Another permutation of these same letters is

$$prq.$$

Unless otherwise specified, by a permutation we mean a *linear* permutation, an ordered arrangement in a line.

Sometimes it is necessary to know the *number* of possible permutations of some given things. It is easy to determine the number of permutations of the three letters p, q, and r because we can write these permutations out. They are

pqr	qpr	qrp
prq	rpq	rqp

There are six of them. If we had more than just three things, it would not be so easy to count the number of permutations by writing them all out. In that case, we could use the *fundamental counting principle.*

Exercise 3.10, 1–5

Fundamental Counting Principle

If there are exactly m ways of doing a first thing, and if there are then exactly n ways of doing a second thing, the number of ways of doing the first thing and then doing the second thing is $m \times n$. If there are exactly w ways of doing a third thing after the first thing and then the second thing have been done, the number of ways of doing the first, and then the second, and then the third is $m \times n \times w$, etc.

EXAMPLE 1

Use the fundamental counting principle to determine the number of permutations of the three letters p, q, and r.

Solution

In writing a permutation of the three letters p, q, and r, we have to choose a first letter, then a second letter, and then a third. We have to fill three "blanks" in order. There are *three* ways of choosing a first letter, and then after it has been chosen, there are *two* ways of choosing a second letter, and then finally there is only *one* way of choosing the third letter. By the fundamental counting principle, there are

$$3 \times 2 \times 1 = 6$$

ways to choose a first letter, and then a second letter, and then a third letter. There are six permutations of the three letters p, q, r.

The number of permutations of n things is $n!$. That is because there are n ways of choosing the first thing, and then $(n - 1)$ ways of choosing the second thing, etc., until finally there is only one way of choosing the nth thing. By the fundamental counting principle, there are

$$n \times (n - 1) \times (n - 2) \times \cdots \times 1 = n!$$

ways of choosing the n things and taking order into consideration. We use the symbol $_nP_n$ for the number of permutations of n things. The symbol

$$_nP_n$$

is read "the number of permutations of n things taken all at a time."

$$_nP_n = n!$$

The number of permutations of the three letters p, q, and r is

$$_3P_3 = 3! = 3 \times 2 \times 1 = 6.$$

Sometimes it is necessary to know the number of possible permutations of r things chosen from a collection of more than r things. The number of ordered arrangements of r things chosen from n things is called the number of permutations of n things taken r at a time.

EXAMPLE 2

Use the fundamental counting principle to determine the number of permutations of the four letters p, q, r, and s, taking the letters two at a time. Write out all these permutations.

Solution

To write a permutation of the four letters p, q, r, and s taken two at a time, we must choose a first letter, and then choose a second letter. There are four ways of choosing a first letter, and then there are three ways of choosing a second letter. By the fundamental counting principle, there are

$$4 \times 3 = 12$$

ways of choosing a first letter and then a second letter. There are twelve permutations of the four letters p, q, r, and s taken two at a time. These twelve permutations are

pq	*qp*	*rp*	*sp*
pr	*qr*	*rq*	*sq*
ps	*qs*	*rs*	*sr*

We use the symbol $_nP_r$ for the number of permutations of n things taken r at a time. (Some authors use the symbol $P(n, r)$; others use P_r^n.) By the fundamental counting principle, the number $_nP_r$ is equal to the product of r factors $n(n - 1)(n - 2) \cdots (n - (r - 1))$ because there are n ways of choosing the first thing, and then there are $(n - 1)$ ways of choosing the second thing, etc., until finally there are only $(n - (r - 1)) = (n - r + 1)$

ways of choosing the rth thing. The symbol

$$_nP_r$$

is read "the number of permutations of n things taken r at a time."

r factors, starting with n

$$_nP_r = n(n - 1)(n - 2) \cdots (n - (r - 1))$$

The number of permutations of the four letters p, q, r, and s taken two at a time is

2 factors, starting with 4

Exercise 3.10, 6–20

$$_4P_2 = 4 \times 3 = 12.$$

If a collection consists of n things that are not all different, there are still $n!$ permutations of the n things taken all at a time, but the permutations are not all distinguishable. For example, suppose we have the three letters p, q, and q. There are $3! = 6$ permutations of these three letters, but the permutations are not all distinguishable. We can obtain a list of these six permutations by writing out the permutations of the three different letters p, q, and r and then replacing each r by q. Doing this, we see that we obtain the six permutations

$$
\begin{array}{ccc}
q & q & q \\
p\,q\,\cancel{r} & q\,p\,\cancel{r} & q\,\cancel{r}\,p \\
q & q & q \\
p\,\cancel{r}\,q & \cancel{r}\,p\,q & \cancel{r}\,q\,p,
\end{array}
$$

but only *three* of these are distinguishable. The *distinguishable* permutations of p, q, and q are

$$pqq, \qquad qpq, \quad \text{and} \quad qqp.$$

The next two examples show how to compute the number of distinguishable permutations of n things that are not all different.

EXAMPLE 3

Find the number of distinguishable permutations of the letters of the word *nutty*.

Solution

The total number of permutations of the five letters is $5!$, but these permutations are not all distinguishable because two of the given letters are alike. Let P be the number of *distinguishable* permutations. Within each of these P permutations there are two t's, and there are $2!$ ways of arranging the two t's. Thus $P \times 2!$ must be the total number of permutations of the five letters. That is,

$$P \times 2! = 5!$$

$$P = \frac{5!}{2!}$$

$$= \frac{1 \times 2 \times 3 \times 4 \times 5}{1 \times 2}$$

$$= 60.$$

EXAMPLE 4

Find the number of distinguishable permutations of the seven letter grades A, A, A, B, B, C, and D.

Solution

The total number of permutations of the seven letter grades is 7!, but these permutations are not all distinguishable because three of the given letters are alike and two others are also alike. Let P be the number of distinguishable permutations. Within each of these P permutations there are three A's and two B's. There are 3! ways of arranging the A's, and for each of these ways there are 2! ways of arranging the B's. By the fundamental counting principle, $P \times 3! \times 2!$ must be the number of permutations of the seven letter grades.

$$P \times 3! \times 2! = 7!$$

$$P = \frac{7!}{3!\ 2!}$$

$$= \frac{1 \times 2 \times 3 \times 4 \times 5 \times 6 \times 7}{1 \times 2 \times 3 \times 1 \times 2}$$

Exercise 3.10, 21–25

$$= 420$$

The next two examples look quite similar, but they illustrate two totally different types of counting problems. In Example 5, we must count the number of ways of choosing *and* ordering two people, whereas in Example 6 we must simply count the number of ways of *choosing* two people.

EXAMPLE 5

In how many ways can a pilot and copilot be chosen from five astronauts? Call the astronauts *a*, *b*, *c*, *d*, and *e*, and list all possible choices.

Solution

There are five ways of choosing a pilot, and after that is done there are four ways of choosing a copilot. By the fundamental counting principle, there are $5 \times 4 = 20$ ways of choosing a pilot and a copilot. These twenty ways are as follows:

Pilot	Copilot	Pilot	Copilot	Pilot	Copilot	Pilot	Copilot	Pilot	Copilot
a	b	b	a	c	a	d	a	e	a
a	c	b	c	c	b	d	b	e	b
a	d	b	d	c	d	d	c	e	c
a	e	b	e	c	e	d	e	e	d

Each of these choices is a permutation. Order must be taken into consideration, because the choice of *a* for pilot and *b* for copilot is different from the choice of *a* for copilot and *b* for pilot.

EXAMPLE 6

In how many ways can a crew of two be chosen, without regard to order, from five astronauts? Call the astronauts *a*, *b*, *c*, *d*, and *e*, and list all possible choices.

Solution

Here we simply need to choose two things from five things without regard to order. Let *C* be the number of ways we can make this choice of things. Then since there are 2! ways of arranging two things once they have been chosen, the number of ways of choosing *and* arranging the two things must be

$$C \times 2!.$$

But the number of ways of choosing two things from five *and* ordering them is $_5P_2 = 5 \times 4$. Therefore,

$$C \times 2! = 5 \times 4$$

$$C = \frac{5 \times 4}{2!} = 10.$$

There are ten ways of choosing a crew of two from five astronauts. These ten ways are

ab	bc	cd	de
ac	bd	ce	
ad	be		
ae			

The order of listing the members in any one choice does not matter here. For example, the crew consisting of *a* and *b* is the same regardless of whether we list *a* first or *b* first.

None of the choices listed in Example 6 is a permutation because the order of arrangement does not matter. Each is simply a selection of two things from five things. A selection of things from a given collection made without regard to order is called a **combination**. Each of the choices in Example 6 is a combination.

Exercise 3.10, 26–30

We use the symbol $_nC_r$ for the number of possible collections of r things each, that can be chosen from a collection of n things. (Some authors use the symbol $C(n, r)$; others use C_r^n.) The symbol

$$_nC_r$$

is read "the number of combinations of n things taken r at a time."

We can show that $_nC_r = {}_nP_r/r!$. To see this, first note that $_nP_r$ is the number of ways of choosing r things from n things *and ordering* the r things. Since $_nC_r$ is the number of ways of choosing r things, and $r!$ is the number of ways of ordering the r things, it follows from the fundamental counting principle that $_nC_r \times r!$ must be $_nP_r$.

$$_nC_r \times r! = {}_nP_r$$

$$_nC_r = \frac{_nP_r}{r!}$$

Since $_nP_r = n(n - 1)(n - 2) \cdots (n - (r - 1))$, we may also write

$$_nC_r = \frac{n(n - 1)(n - 2) \cdots (n - (r - 1))}{r!}.$$

The number of ways of choosing a crew of two from five astronauts without regard to order is

$$_5C_2 = \frac{5 \times 4}{2!}$$

$$= 10.$$

The formula we just derived for $_nC_r$ is precisely the same as the formula we derived in Section 3.9 for the coefficient of $a^{n-r}b^r$ in the expansion of $(a + b)^n$. The notation we used for the binomial coefficient is $\binom{n}{r}$. It follows that

$$_nC_r = \binom{n}{r}.$$

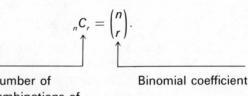

Number of
combinations of
n things taken
r at a time

Binomial coefficient

We saw in Section 3.9 that $\binom{n}{r}$ can also be written

$$\binom{n}{r} = \frac{n!}{r!\,(n-r)!}.$$

Since $_nC_r = \binom{n}{r}$, $_nC_r$ is also equal to $\dfrac{n!}{r!\,(n-r)!}$. Many authors use

$$_nC_r = \frac{n!}{r!\,(n-r)!}$$

as their basic formula for $_nC_r$. Note that $_nC_n = 1$. We define $_nC_0$ to be 1.

$$_7C_3 = \frac{7 \times 6 \times 5}{1 \times 2 \times 3} \quad \text{or} \quad {_7C_3} = \frac{7!}{3!\,(7-3)!}$$

$$= \frac{1 \times 2 \times 3 \times 4 \times 5 \times 6 \times 7}{1 \times 2 \times 3 \times 1 \times 2 \times 3 \times 4}$$

$$= 35$$

Exercise 3.10, 31–50

Many counting problems involve decks of playing cards. A standard deck of 52 cards consists of 13 *spades,* 13 *clubs,* 13 *hearts,* and 13 *diamonds.* Each of these sets of 13 cards is called a *suit.* The spades and clubs are black; the hearts and diamonds are red. Within each suit the 13 cards consist of 9 cards numbered 2 through 10, together with a *Jack,* a *Queen,* a *King,* and an *ace.* The Jack, Queen, and King are called *face cards.*

EXAMPLE 7

Two cards are drawn in succession (that is, order is taken into consideration) and without replacement from a standard deck of 52 cards. What is the number of ways of drawing two spades?

Solution

There are 13 ways of drawing the first spade, and after that is done there are 12 ways of drawing the second spade. By the fundamental counting principle, the number of ways of drawing first one spade and then another is

$$13 \times 12 = 156.$$

(Notice that here we have counted *permutations.* For example, the ace on the first draw with the Jack on the second draw is one of the 156 ways, and the Jack on the first draw with the ace on the second draw is another of the 156 ways.)

EXAMPLE 8

Two cards are drawn simultaneously (a selection is made without regard to order) from a standard deck of 52 cards. What is the number of different pairs of spades that can be drawn?

Solution	Here we simply want to know the number of ways of choosing a pair of spades from the 13 spades in the deck without regard to order. That number is $_{13}C_2$.

$$_{13}C_2 = \frac{13 \times 12}{1 \times 2} = 78$$

<table>
<tr><td>EXAMPLE 9</td><td>How many different pairs of cards can be dealt from a standard deck of 52 cards without regard to order? (How many ways can 2 cards be drawn simultaneously?)</td></tr>
<tr><td>Solution</td><td>$_{52}C_2 = \dfrac{52 \times 51}{1 \times 2} = 1326$</td></tr>
</table>

Many counting problems involve dice and coins. The word *die* is the singular of the word *dice*. A die is a cube whose six sides are numbered 1 through 6. A coin has two sides; one side is called *heads;* the other side is called *tails.*

EXAMPLE 10	A die and a coin are tossed together. What is the total number of possible outcomes? List these outcomes.
Solution	There are six possible outcomes from the toss of the die, and there are two possible outcomes from the toss of the coin. By the fundamental counting principle, there are

$$6 \times 2 = 12$$

possible outcomes.

Using H for heads and T for tails, and using 1, 2, 3, 4, 5, 6 for the six possible outcomes on the die, the 12 possible outcomes on the die and coin together are

1H	2H	3H	4H	5H	6H
1T	2T	3T	4T	5T	6T

EXAMPLE 11	(a) Two dice are thrown. What is the number of possible outcomes? (b) List the number of ways of getting a total or sum of 7 on the dice. How many of these ways are there?
Solution	(a) There are six possible outcomes on one die, and for each of these there are six possible outcomes on the other die. There are

$$6 \times 6 = 36$$

possible outcomes when the two dice are thrown together.
(b) Let us call the dice Die 1 and Die 2. A total of 7 can be gotten on the two dice in the following ways:

Die 1	1	6	2	5	3	4
Die 2	6	1	5	2	4	3

There are six ways to get a total of 7 on a toss of two dice together.

Counting problems appear in many different contexts. The next examples illustrate two more of these contexts.

EXAMPLE 12

A traffic engineer has to determine the total number of possible routes taken by commuters from Town N to University U. (Figure 3.9). He knows that all these commuters must cross Bridge B between N and U. He knows that the commuters have a choice of three different routes from N to the bridge, and then they have a choice of five different routes from the bridge to the university. What is the total number of possible routes from N to U via the bridge? Label these routes and list them.

Solution

There are three ways to go from N to B, and after that is done there are five ways to go from B to U. By the fundamental counting principle, there are $3 \times 5 = 15$ ways to go from N to B and hence to U.

The routes are labeled in Figure 3.9. Using these labels, the 15 routes are

$$
\begin{array}{ccc}
1-4 & 2-4 & 3-4 \\
1-5 & 2-5 & 3-5 \\
1-6 & 2-6 & 3-6 \\
1-7 & 2-7 & 3-7 \\
1-8 & 2-8 & 3-8 \\
\end{array}
$$

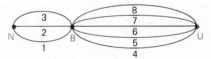

FIGURE 3.9   Routes from N to B to U

EXAMPLE 13

A basket contains five red balls, three black balls, and no others. Two balls are drawn.
(a) If the balls are drawn in succession, without replacement, what is the number of ways of getting a black on the second draw *after* getting a red on the first draw?

Solution

3

(b) If the balls are drawn in succession, without replacement, what is the number of ways of getting a red on the first draw and a black on the second draw? List these ways.

Solution

$5 \times 3 = 15$

Let us label the red balls R_1, R_2, R_3, R_4, and R_5 and label the black balls B_1, B_2, and B_3. Now we can list the 15 ways of getting a red on the first draw and a black on the second draw.

$$R_1B_1 \qquad R_2B_1 \qquad R_3B_1 \qquad R_4B_1 \qquad R_5B_1$$
$$R_1B_2 \qquad R_2B_2 \qquad R_3B_2 \qquad R_4B_2 \qquad R_5B_2$$
$$R_1B_3 \qquad R_2B_3 \qquad R_3B_3 \qquad R_4B_3 \qquad R_5B_3$$

(c) If the balls are drawn in succession, without replacement, what is the number of ways of getting a black on the first draw and also a black on the second draw? List these ways.

Solution

$3 \times 2 = 6$

Label the black balls B_1, B_2, and B_3. The 6 ways of getting first a black and then another black are:

$$B_1B_2 \qquad B_2B_1 \qquad B_3B_1$$
$$B_1B_3 \qquad B_2B_3 \qquad B_3B_2$$

(d) If the balls are drawn in succession, without replacement, what is the number of ways of getting a black on the second draw?

Solution

In part (b) we saw that there are 15 ways of getting a black on the second draw provided that the first ball drawn is red, and in part (c) we saw that there are 6 ways of getting a black on the second draw provided that the first ball drawn is black. Since the first ball drawn must be *either* red or black, the number of ways of getting a black on the second draw must be the number of ways of getting a red and then a black *plus* the number of ways of getting a black and then a black, or

$$15 + 6 = 21.$$

(e) If the balls are drawn in succession, without replacement, what is the number of ways of drawing two balls?

Solution

There are eight ways of drawing the first ball, and after that is done there are seven ways of drawing the second ball. There must be

$$8 \times 7 = 56$$

ways of drawing two balls. (This is the number of ways of getting a black on the second plus the number of ways of getting a red on the second. We have seen that the number of ways of getting a black on the second is $(5 \times 3) + (3 \times 2) = 21$. The number of ways of getting a red on the second is the number of ways of getting a red after a black plus the number of ways of getting a red after a red, or $(3 \times 5) + (5 \times 4) = 35$. Note that $21 + 35$ *is* equal to 56.)

(f) If the balls are drawn simultaneously (without regard to order), what is the number of ways of drawing two balls?

Solution

$$_8C_2 = \frac{8 \times 7}{2!} = 28$$

Sometimes it is necessary to count arrangements in which each thing is adjacent to *two* other things, as they would be if they were arranged in a circle. Such ordered arrangements are called *cyclic* permutations. They are illustrated in the next example.

EXAMPLE 14

The four representatives at an international conference must sit around a circular table. Find the number of different ways of seating them around the table. Call the representatives *a*, *b*, *c*, and *d*, and make diagrams showing all possible arrangements.

Solution

We can find the number of arrangements by thinking of one of the representatives as fixed and then finding the number of ways of seating the other three as if they were sitting in a line. If one is fixed, there are 3! ways of arranging the other three, so there are

$$3! = 6$$

Exercise 3.10, 51–70 different arrangements. These are shown in the diagrams in Figure 3.10

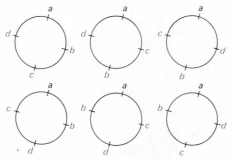

FIGURE 3.10   Cyclic permutations of *a*, *b*, *c*, and *d*

EXERCISE 3.10 PERMUTATIONS, COMBINATIONS, AND COUNTING

▶   In each of the following, list all the permutations of the given symbols.

EXAMPLE

1, 2, 3

Solution

123, 132, 213, 231, 312, 321

1. *a, b, c*   **2.** *p, q*   **3.** *p, q, r, s*   **4.** 2, 7   **5.** 1, 2, 3, 4

▶ In each of the following, find the number of permutations of the given symbols taken all at a time and taken r at a time.

EXAMPLE $\qquad$ $a, b, c, d, e;\ r = 3$

Solution $\qquad$ $_5P_5 = 5 \cdot 4 \cdot 3 \cdot 2 \cdot 1 = 120$

$_5P_3 = 5 \cdot 4 \cdot 3 = 60$

6. $a, b, c, d;\ r = 3$ **7.** $a, b, c, d, e, f;\ r = 5$

8. $a, b, c, d, e, f;\ r = 4$ **9.** $a, b, c, d, e;\ r = 4$

10. $a, b, c, d, e;\ r = 2$

▶ Compute the value of each of the following.

EXAMPLES $\qquad$ (a) $_4P_4 = 4! = 24$ (b) $_4P_2 = 4 \times 3 = 12$

11. $_5P_5$ **12.** $_5P_2$ **13.** $_5P_3$ **14.** $_6P_6$ **15.** $_7P_5$

▶ In each of the following, list all the permutations of the given symbols taken r at a time.

EXAMPLE $\qquad$ $a, b, c;\ r = 2$

Solution $\qquad$ ab, ac, bc, ba, ca, cb

16. $a, b, c, d;\ r = 3$ **17.** $a, b, c, d, e;\ r = 2$

18. $1, 2, 3, 4;\ r = 2$ **19.** $1, 2, 3;\ r = 2$

20. $a, b, c, d, e, f;\ r = 2$

▶ Find the number of distinguishable permutations of the letters of each of the following words. List the distinguishable permutations where requested.

EXAMPLE $\qquad$ gig   List distinguishable permutations.

Solution $\qquad$ $\dfrac{3!}{2!} = 3$; gig, ggi, igg

21. bib   List distinguishable permutations.

22. rabbit   **23.** coffee   **24.** bikini   **25.** Mississippi

▶ In each of the following, list all the combinations of the given symbols taken r at a time.

EXAMPLE $\qquad$ $a, b, c;\ r = 2$

Solution ab, ac, bc

26. a, b, c, d; $r = 2$ **27.** a, b, c, d; $r = 3$

28. a, b, c, d, e; $r = 4$ **29.** a, b, c, d, e; $r = 3$

30. a, b, c, d, e, f; $r = 5$

▶ In each of the following, find the number of combinations of the given symbols taken r at a time.

EXAMPLE a, b, c, d, e; $r = 3$

Solution

$$_5C_3 = \frac{5 \times 4 \times 3}{1 \times 2 \times 3} = 10$$

31. a, b, c, d, e; $r = 2$ **32.** a, b, c, d, e, f; $r = 4$

33. a, b, c, d, e, f; $r = 3$ **34.** a, b, c, d, e, f; $r = 2$

35. 1, 2, 3, 4, 5, 6, 7, 8; $r = 5$

▶ Compute the value of each of the following.

EXAMPLE

$$_8C_3 = \frac{8 \times 7 \times 6}{1 \times 2 \times 3} = 56$$

36. $_8C_4$ **37.** $_7C_4$ **38.** $_6C_2$ **39.** $_9C_6$ **40.** $_{10}C_9$

▶ Verify each of the following by finding the values.

EXAMPLE $_3C_1 = {_2C_0} + {_2C_1}$

Solution $_3C_1 = \dfrac{3}{1} = 3$, $_2C_0 = 1$, $_2C_1 = \dfrac{2}{1} = 2$

$3 = 1 + 2$

41. $_4C_1 = {_3C_0} + {_3C_1}$ **42.** $_6C_2 = {_5C_1} + {_5C_2}$

43. $_7C_6 = {_6C_5} + {_6C_6}$ **44.** $_4C_1 = {_4C_3}$ **45.** $_8C_3 = {_8C_5}$

▶ Use the notation $_nC_r$ for the binomial coefficient $\binom{n}{r}$, and write the indicated row of Pascal's triangle.

EXAMPLE   Row 4

Solution $_4C_0$ $_4C_1$ $_4C_2$ $_4C_3$ $_4C_4$

46. Row 5   **47.** Row 6   **48.** Row 7   **49.** Row 8   **50.** Row 9

EXAMPLE

How many different 5-card hands can be dealt from a standard deck of 52 cards without regard to order?

Solution

$$_{52}C_5 = \frac{52 \times 51 \times 50 \times 49 \times 48}{1 \times 2 \times 3 \times 4 \times 5}$$

$$= 2,598,960$$

51. How many different 3-card hands can be dealt from a standard deck of 52 cards without regard to order?

52. Four cards are drawn in succession and without replacement from a standard deck of 52 cards.
(a) What is the number of ways of drawing four aces in order?
(b) What is the number of ways of drawing four face cards in order?

53. Four cards are drawn in succession and without replacement from a standard deck of 52 cards.
(a) What is the number of ways of drawing four aces without regard to order?
(b) What is the number of ways of drawing four face cards without regard to order?

54. Four cards are drawn in succession from a standard deck of 52 cards. What is the number of ways of getting an ace, a King, a Queen, and a Jack in that order, if each card is replaced before the next one is drawn?

55. A pair of dice and a coin are tossed together. What is the total number of outcomes?

56. Two dice are thrown. List the number of ways of getting totals of 2, 3, 4, 5, and 6.

57. Two dice are thrown. List the number of ways of getting totals of 8, 9, 10, 11, and 12.

58. If Professor Carberry has four ways (car, bicycle, camel, or skate board) to get from Brown University to the airport, and if he then has three other ways (plane, eagle, or broomstick) to get to his camp, in how many different ways can he get from Brown to his camp via the airport?

59. How many different committees of 3 people each can be chosen from 15 people?

60. How many different committees of 3 people each can be chosen from 15 people, if the first person chosen chairs the committee, the second is secretary, and the third is treasurer?

61. Two coins are tossed. List all the possible outcomes.

62. Three coins are tossed. List all the possible outcomes.

63. If Mary and Roland have five routes to get to New Jersey, and if they then have three routes to get to Florida, how many routes do they have to get to Florida via New Jersey?

64. A basket contains four red balls, five black balls, and no others. Two balls are drawn.
   (a) If the balls are drawn in succession, without replacement, what is the number of ways of drawing two red balls? Label the red balls R_1, R_2, R_3, and R_4, and list these ways.
   (b) If the balls are drawn simultaneously (without regard to order), what is the number of ways of drawing two red balls? Label the red balls R_1, R_2, R_3, and R_4, and list these ways.
   (c) If the balls are drawn in succession, without replacement, what is the number of ways of drawing a red and a black?
   (d) If the balls are drawn in succession, without replacement, what is the number of ways of drawing two balls?
   (e) If the balls are drawn simultaneously (without regard to order), what is the number of ways of drawing two balls?
   (f) What is the number of unordered pairs of balls that can be drawn?

65. Ida sells five kinds of submarine sandwiches, and she sells halves. Tom wants to order two different halves to make up a whole submarine. How many different whole submarines can Tom order?

66. Bell's Pizza House sells 15 different kinds of pizza, and these can be combined in any way to make up a special order. Fred wants to order a pizza in which each of the different kinds is put on as a layer, but he wants the bottom two layers to be two specific kinds in either order. How many different pizzas can Fred order of the type he wants?

67. The "combination" locks at a Swim Club all have tick marks for 40 numbers on them. To open any of these locks, one must turn first left, then right, and then left to three given numbers respectively. The "combination" for one lock is 35-14-21. Is this a combination in the mathematical sense of the word? Explain. How many different "combinations" could possibly be given for locks of the type at this Swim Club?

68. What is the number of different ways of seating five people around a circular table?

69. What is the number of different ways of seating five people around a circular table if two of the people must not sit in adjacent seats?

70. Use the fact that $_nC_r = \binom{n}{r}$, and show that $_nC_1 + {_nC_2} + \cdots + {_nC_r} + \cdots + {_nC_n} = 2^n - 1$. (Hint: Write an expression for the expansion of $(a + b)^n$ using $_nC_r$ for $\binom{n}{r}$, and then let $a = 1$ and $b = 1$.)

71. Show that if $r \leq n - 1$, then $_nC_r = {_{n-1}C_{r-1}} + {_{n-1}C_r}$. How is this related to Pascal's triangle?

72. Show that for $k \leq n$, $_nC_k = {_nC_{n-k}}$. What does this say about the coefficients in the expansion of $(a + b)^n$?

CHECK LIST FOR THE STUDENT

If you have learned the material in Section 3.10, you should be able to do the following:

1. Tell what is meant by a permutation of n things.

2. List all the permutations of three different letters.

3. State and use the fundamental counting principle.

4. Give the number of permutations of n things.

5. Tell what is meant by the number of permutations of n things taken r at a time.

6. List all the permutations of five different letters taken two at a time.

7. Find $_nP_n$ and $_nP_r$ for given n and r.

8. List all the distinguishable permutations of three letters exactly two of which are alike.

9. Find the number of distinguishable permutations of n given letters that are not all different.

10. Tell what is meant by a combination of n things taken r at a time.

11. Find the number of combinations of n things taken r at a time.

12. List all the combinations of five different letters taken two at a time.

13. Find $_nC_r$ for given n and r.

14. Give the relationship between $_nC_r$ and $\binom{n}{r}$.

15. Give two basic formulas for $_nC_r$.

16. Do some counting problems including some involving dice, playing cards, coins, routes, and cyclic permutations.

SECTION QUIZ 3.10

1. List all the permutations of the letters q, r, s, t.

2. (a) Find the number of permutations of the letters a, b, c, d, e, f taken all at a time.
   (b) Find the number of permutations of the letters a, b, c, d, e, f, g taken two at a time.

3. (a) List all the permutations of the letters a, b, c, d, e taken two at a time.
   (b) List all the combinations of the letters a, b, c, d, e taken two at a time.

4. (a) Find the number of combinations of the letters a, b, c, d, e, f taken all at a time.
   (b) Find the number of combinations of the letters a, b, c, d, e, f, g taken two at a time.

5. Compute the value of each of the following.
   (a) $_9P_4$ (b) $_9C_4$

6. Find the numbers of distinguishable permutations of the letters of the word *funny*.

7. Use the notation $_nC_r$ for the binomial coefficient $\binom{n}{r}$, and write the first three rows of Pascal's triangle.

8. (a) How many different four-card hands can be dealt from a standard deck of 52 cards without regard to order?
   (b) If four cards are drawn in succession and without replacement from a standard deck of 52 cards, what is the number of ways of drawing four Jacks in order?

9. A pair of dice and two coins are tossed together. What is the number of ways of getting a total of five on the dice and exactly one head on the coins?

10. (a) How many different committees of 5 people each can be chosen from 12 people?
    (b) What is the number of different ways of seating five people and a dog around a circular table, if one particular person and the dog must sit in adjacent seats?

3.11 Probability

The purpose of this section is to introduce probability and to show how to find the probability that some particular events will occur.

Some processes, such as the process of throwing two dice, tossing a coin, or drawing a card from a standard deck, have a finite number (not an unlimited number) of outcomes each of which is equally likely to occur. For such processes it is of interest to consider the *probability* that a particular outcome or set of outcomes will occur. If a process has a finite number of equally likely outcomes, any single possible outcome or any set of possible outcomes is called an *event.* We define the *probability that an event E will occur* as the ratio of the number of ways the event can occur to the total number of possible outcomes. We call this the **probability of event E.**

$$\frac{\text{The probability}}{\text{of event E}} = \frac{\text{Number of ways E can occur}}{\text{Number of possible outcomes}}$$

The probability of an event E is denoted by $P(E)$. The number of ways E can occur is really the number of outcomes producing E. The set of all possible outcomes of a process is sometimes called the *sample space* of the process. As an example, consider the process of tossing a coin once. (We assume that all coins are two-sided and uniformly made so that a coin is equally likely to fall heads as to fall tails.) There are two possible outcomes, heads or tails. *Getting heads* is an event. There is one way of getting heads. The probability of getting heads is $\frac{1}{2}$.

As another example, consider the process of throwing a single die once. (We assume that all dice are uniformly made and unloaded so that a die is equally likely to fall on any one of its sides.) The number of possible outcomes is six. These possible outcomes are getting a one, two, three, four, five, or six. *Getting a three* is an event. There is one way of getting a three. The probability of getting a three is $\frac{1}{6}$.

As still another example, consider the process of tossing a pair of dice, call them red and green. There are $6 \times 6 = 36$ ways the dice may fall, so there are 36 possible outcomes. *Getting a sum of five* on the two dice is an event. The number of ways of getting a sum of five is four. They are as follows:

Red	1	4	2	3
Green	4	1	3	2

The probability of getting a sum of five on a single throw of two dice is $\frac{4}{36}$.

Some properties of probability can be readily deduced from the definition given. For instance, the probability of an event that is certain to occur is 1, because the number of ways that such an event can occur is equal to the number of possible outcomes. For example, consider the process of drawing one ball from a basket containing five red balls and no others. *Getting a red ball* is an event that is certain to occur. The number of ways of getting a red ball is 5; the number of possible outcomes is 5 (getting any one of the five balls is an outcome). The probability of getting a red ball is $\frac{5}{5} = 1$.

$$P(\text{Event certain to occur}) = 1$$

Similarly, the probability of an event certain *not* to occur must be 0, because the number of ways that such an event can occur is 0.

$$P(\text{Event certain not to occur}) = 0$$

Another property of probability that is apparent from the definition given is that the probability of any event must be nonnegative and less than or equal to 1. It must be nonnegative because the numerator and denominator of the defining ratio are both nonnegative integers. It must be less than or equal to 1 because the number of ways that an event E can occur (that is, the number of outcomes producing E) is less than or equal to the number of possible outcomes. Thus the numerator of the defining ratio is less than or equal to the denominator, and hence the ratio itself is less than or equal to 1.

$$P(E) \geq 0 \quad \text{and} \quad P(E) \leq 1$$

Otherwise written,

$$0 \leq P(E) \leq 1.$$

Still another useful property of probability is that the probability that an event E either *will* occur or *will not* occur is 1. This is because the *event* A that either E will occur or will not occur is itself certain, and so the probability of A is 1. For example, in the single toss of a coin the probability of either getting heads or not getting heads is 1. For another example, consider the process of throwing a pair of dice once. Either getting a sum of five or not getting a sum of five is an event that is certain to occur. The probability of either getting a sum of five or not getting a sum of five is 1. If a single card is drawn from a standard 52-card deck, the probability that it is either red or not red is 1. For any event E,

Exercise 3.11, 1–8

$$P(\text{either E or not E}) = 1.$$

Sometimes it is necessary to compute the probability of having first one event and then a second event. This is illustrated in the next two examples.

EXAMPLE 1

A basket contains five red balls, three black balls, and no others. Two balls are drawn in succession, without replacement. What is the probability of getting a red on the first draw and a black on the second?

Solution   The number of ways of getting first a red and then a black is

$$5 \times 3.$$

The number of possible outcomes is the number of ways of drawing two balls in succession, without replacement. That number is

$$8 \times 7.$$

The probability of getting first a red and then a black is

$$\frac{5 \times 3}{8 \times 7} = \frac{15}{56}.$$

(Note that this is the product of the probability of getting a red on the first draw, $\frac{5}{8}$, and the probability of getting a black on the second draw *after getting a red on the first draw*, namely $\frac{3}{7}$.)

EXAMPLE 2   A basket contains five red balls, three black balls, and no others. One ball is drawn, and then it is replaced. Again one ball is drawn. What is the probability of getting a red on the first draw and a black on the second draw?

Solution   The number of ways of getting first a red and then a black is

$$5 \times 3.$$

The number of possible outcomes is the number of ways of drawing two balls in order, replacing the first before the second is drawn. That number is

$$8 \times 8.$$

The probability of getting first a red and then a black is

$$\frac{5 \times 3}{8 \times 8} = \frac{15}{64}.$$

(Note that this is the product of the probability of getting a red on the first draw, $\frac{5}{8}$, and the probability of getting a black on the second draw *after getting a red on the first draw*, namely $\frac{3}{8}$.)

Let us take a second look at Examples 1 and 2 to see what they suggest. Let E_1 be the event of getting a red on the first draw, and let E_2 be the event of getting a black on the second draw. In Example 1, the ball was not replaced after the first draw, so the probability of event E_2 was affected by the occurrence of E_1, whereas in Example 2 the ball *was* replaced after the first draw, so the probability of event E_2 was in no way affected by the occurrence of E_1. In both cases, however, the probability of E_1 *and* E_2 turned out to be the product of the probability of E_1 and the probability of E_2 *after E_1 has occurred*. This suggests the following theorem.

Theorem 1   The probability of first event E_1 and then event E_2 is the product of the probability of E_1 and the probability of E_2 after E_1.

$$P(E_1 \text{ and } E_2) = P(E_1) \times P(E_2 \text{ after } E_1)$$

This theorem is true. It can be extended to cover any finite number of events occurring in order. The reasoning in the proof is a generalization of the reasoning used in Examples 1 and 2.

Two events E_1 and E_2 are said to be **independent** provided that the probability of any one of them is in no way affected by the occurrence of the other. In Example 2, the event E_1 of getting a red on the first draw and the event E_2 of getting a black on the second draw are independent; the first ball drawn is replaced before the second ball is drawn, so that $P(E_2)$ is in no way affected by the occurrence of E_1. If events E_1 and E_2 are independent, the probability of E_2 after E_1 is the same as the probability of E_2 regardless of whether or not E_1 has occurred. For instance, in Example 2 the probability of getting a black on the second draw is $\frac{3}{8}$ regardless of whether or not a red was drawn on the first draw.

If E_1 and E_2 are independent events,

$$P(E_2 \text{ after } E_1) = P(E_2).$$

According to Theorem 1, the probability of E_1 *and* E_2 is the product $P(E_1) \times P(E_2 \text{ after } E_1)$. It follows that if E_1 and E_2 are independent events, the probability of E_1 and E_2 is the product $P(E_1) \times P(E_2)$. We shall call this Theorem 2.

Theorem 2   If E_1 and E_2 are independent events,

$$P(E_1 \text{ and } E_2) = P(E_1) \times P(E_2).$$

Again use Example 2 as an illustration. There the probability of E_1 (getting a red on the first draw) is $\frac{5}{8}$, the probability of E_2 (getting a black on the second draw) is $\frac{3}{8}$, and the probability of E_1 *and* E_2 is the product $\frac{5}{8} \times \frac{3}{8}$. Theorem 2 can be extended and applied to any finite number of independent events.

We have noted that if P_1 and P_2 are independent events,

$$P(E_2 \text{ after } E_1) = P(E_2)$$
$$\text{and}$$
$$P(E_1 \text{ and } E_2) = P(E_1) \times P(E_2).$$

Events E_1 and E_2 may be *defined* to be independent provided that either of these conditions is true.

Events E_1 and E_2 are said to be *dependent* if they are not independent. If two events are **dependent**, the probability of any one of them *is* affected by the occurrence of the other. In Example 1, the event E_1 of getting a red on the first draw and the event E_2 of getting a black on the second draw are dependent; the first ball is not replaced before the second ball is drawn, so that $P(E_2)$ *is* affected by the occurrence of E_1. We have just seen that if events E_1 and E_2 are independent, $P(E_2$ after $E_1) = P(E_2)$ and also the probability of E_1 and E_2 is simply the product of $P(E_1)$ and $P(E_2)$. If events E_1 and E_2 are *dependent*, however, $P(E_2$ after $E_1)$ is *not* equal to $P(E_2)$, and the probability of E_1 and E_2 is *not* simply the product of $P(E_1)$ and $P(E_2)$. This is emphasized in the next example.

EXAMPLE 3

A basket contains five red balls, three black balls, and no others. Two balls are drawn in succession, without replacement. Note that the event E_1 of getting a red on the first draw and the event E_2 of getting a black on the second draw are *dependent*.
(a) Show that $P(E_2$ after $E_1) \neq P(E_2)$.
(b) Show that $P(E_1$ and $E_2) = P(E_1) \times P(E_2$ after $E_1)$, but $P(E_1$ and $E_2) \neq P(E_1) \times P(E_2)$.

Solution

(a) Let us first compute $P(E_2$ after $E_1)$. The number of ways E_2 after E_1 can occur (the number of ways of getting a black on the second draw *after* getting a red on the first draw) is 3. The number of possible outcomes is 7, because after getting a red on the first draw we have seven balls left from which to draw the second.

$$P(E_2 \text{ after } E_1) = \tfrac{3}{7}.$$

Now let us compute $P(E_2)$. The number of ways E_2 (getting a black on the second draw) can occur is the number of ways of getting a black on the second draw if the first ball drawn was red, together with the number of ways of getting a black on the second draw if the first ball drawn was black. That is,

$$(5 \times 3) + (3 \times 2) = 21.$$

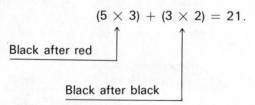

Black after red

Black after black

The number of possible outcomes is

$$8 \times 7 = 56.$$

$$P(E_2) = \frac{21}{56}$$

$$= \frac{3}{8}$$

But $P(E_2$ after $E_1) = \frac{3}{7}$, so for dependent events E_1 and E_2,

$$P(E_2 \text{ after } E_1) \neq P(E_2).$$

(b) Let us list the various probabilities involved in this problem for dependent events E_1 and E_2.

$P(E) = \dfrac{5}{8}$ 　　　　　On the first draw there are 5 ways to choose a red ball from the 8 balls.

$P(E_2) = \dfrac{3}{8}$ 　　　　　Computed in Part (a).

$P(E_2 \text{ after } E_1) = \dfrac{3}{7}$ 　　　　　Computed in Part (a).

$P(E_1 \text{ and } E_2) = \dfrac{5 \times 3}{8 \times 7} = \dfrac{15}{56}$ 　　　　　Computed in Example 1.

For *dependent* events E_1 and E_2,

$$P(E_1 \text{ and } E_2) = P(E_1) \times P(E_2 \text{ after } E_1), \text{ but}$$

Exercise 3.11, 9–20

$$P(E_1 \text{ and } E_2) \neq P(E_1) \times P(E_2).$$

Two events E_1 and E_2 are said to be **mutually exclusive** provided that the occurrence of either one of them *precludes* the occurrence of the other. If one of them occurs, the other *cannot* occur. For example, consider the process of throwing a single die once. The event E_1 of getting a four on that one throw and the event E_2 of getting a five on that one throw are mutually exclusive; it is impossible for both to occur. As another example, consider the process of drawing one ball from a basket containing 5 red balls, 3 black balls, 9 white balls, and no others. The event E_1 of getting a red ball on that one draw and the event E_2 of getting a black ball on that one draw are mutually exclusive. Probabilities of mutually exclusive events are computed in the next two examples.

EXAMPLE 4

A single die is thrown once. Find the probability of getting a four; find the probability of getting a five; find the probability of getting *either* a four or a five.

Solution

$$P(\text{getting a four}) = \frac{1}{6}$$

$$P(\text{getting a five}) = \frac{1}{6}$$

$$P(\text{getting either a four or a five}) = \frac{1+1}{6} = \frac{2}{6}$$

EXAMPLE 5

A basket contains five red balls, three black balls, nine white balls, and no others. One ball is drawn. Find the probability that it is red; find the probability that it is black; find the probability that it is either red or black.

Solution

$$P(\text{getting a red}) = \frac{5}{17}$$

$$P(\text{getting a black}) = \frac{3}{17}$$

$$P(\text{getting either a red or a black}) = \frac{5 + 3}{17} = \frac{8}{17}$$

Let us take a second look at Examples 4 and 5. In each example we had to find the probability of either one or the other of two mutually exclusive events, call them E_1 and E_2. In both examples, that probability turned out to be the *sum* of the probability of E_1 and the probability of E_2. That suggests the following theorem.

Theorem 3

If E_1 and E_2 are mutually exclusive events,

$$P(\text{either } E_1 \text{ or } E_2) = P(E_1) + P(E_2).$$

This theorem is true. It can be extended and applied to any finite number of mutually exclusive events.

EXAMPLE 6

Two dice are thrown together once. Find the probability of getting a sum of two; find the probability of getting a sum of three; find the probability of getting a sum of four; find the probability of getting either a sum of two or a sum of three or a sum of four.

Solution

The events of getting a sum of two, getting a sum of three, and getting a sum of four on a single throw of two dice are mutually exclusive.

The number of ways of getting a sum of two is 1.

The number of ways of getting a sum of three is 2.

The number of ways of getting a sum of four is 3.

The number of possible outcomes if $6 \times 6 = 36$.

$$P(\text{getting sum of two}) = \frac{1}{36}$$

$$P(\text{getting sum of three}) = \frac{2}{36}$$

$$P(\text{getting sum of four}) = \frac{3}{36}$$

$$P(\text{getting sum of either two, three, or four}) = \frac{1}{36} + \frac{2}{36} + \frac{3}{36} = \frac{6}{36}$$

$$= \frac{1}{6}$$

The next example uses the words "at random." If we say that a selection is made *at random,* we mean that the possible outcomes are equally likely to occur.

EXAMPLE 7

Doug, Fred, and Tom all keep their socks in the same drawer. They have 18 brown socks, 23 black socks, and no others.
(a) If Tom pulls out 2 socks at random, what is the probability that he will have a pair of socks of the same color?
(b) If Fred pulls out 3 socks at random, what is the probability that he will have a pair of the same color?
(c) If Doug dumps out all the socks except for 8 brown and 10 black, and if he throws in 6 green socks, what is the probability that he will then get a pair of the same color by pulling out 2 socks at random?

Solution

(a) Tom will have a pair of the same color if he has either 2 brown socks or 2 black socks. Note that in the beginning there are 41 socks in the drawer.

$$P(\text{pulling a pair of brown}) = \frac{_{18}C_2}{_{41}C_2} = \frac{\dfrac{18 \times 17}{2!}}{\dfrac{41 \times 40}{2!}} = \frac{18 \times 17}{41 \times 40}$$

$$P(\text{pulling a pair of black}) = \frac{_{23}C_2}{_{41}C_2} = \frac{\dfrac{23 \times 22}{2!}}{\dfrac{41 \times 40}{2!}} = \frac{23 \times 22}{41 \times 40}$$

The event of pulling a pair of brown and the event of pulling a pair of black are mutually exclusive. By Theorem 3, the probability of pulling *either* a pair of brown or a pair of black is the sum of the probability of pulling a pair of brown and the probability of pulling a pair of black.

$$\frac{18 \times 17}{41 \times 40} + \frac{23 \times 22}{41 \times 40} = \frac{(18 \times 17) + (23 \times 22)}{41 \times 40}$$

$$= \frac{812}{1640}$$

$$= \frac{203}{410}$$

(b) If Fred pulls out three socks at random, he is certain to get a pair of the same color because the three socks must either be all brown, all black, two brown and one black, or two black and one brown. The probability that he will have a pair of the same color is 1.

(c) Doug will have a pair of the same color if he has either a pair of brown, a pair of black, or a pair of green. He has 24 socks consisting of 8 brown, 10 black, and 6 green from which to pull.

$$P\text{(pulling a pair of brown)} = \frac{{}_8C_2}{{}_{24}C_2} = \frac{\dfrac{8 \times 7}{2!}}{\dfrac{24 \times 23}{2!}} = \frac{8 \times 7}{24 \times 23}$$

$$P\text{(pulling a pair of black)} = \frac{{}_{10}C_2}{{}_{24}C_2} = \frac{\dfrac{10 \times 9}{2!}}{\dfrac{24 \times 23}{2!}} = \frac{10 \times 9}{24 \times 23}$$

$$P\text{(pulling a pair of green)} = \frac{{}_6C_2}{{}_{24}C_2} = \frac{\dfrac{6 \times 5}{2!}}{\dfrac{24 \times 23}{2!}} = \frac{6 \times 5}{24 \times 23}$$

By an extension of Theorem 3, the probability of pulling either a pair of brown or a pair of black or a pair of green is the sum of these three probabilities.

$$\frac{8 \times 7}{24 \times 23} + \frac{10 \times 9}{24 \times 23} + \frac{6 \times 5}{24 \times 23} = \frac{176}{552}$$
$$= \frac{22}{69}$$

Exercise 3.11, 21–24

EXAMPLE 8

A basket contains five red balls, three black balls, and no others. Two balls are drawn in succession, without replacement. What is the probability of getting at least one black ball?

Solution

Method 1   If we draw two balls, we will get at least one black ball provided that we get either exactly one black or exactly two blacks. The events of getting exactly one black and getting exactly two blacks are mutually exclusive. By Theorem 3,

P(at least one black) = P(exactly one black) + P(exactly two blacks).

Let us first compute the probability of getting exactly one black. The number of ways of getting exactly one black is the number of ways of getting it on the first draw plus the number of ways of getting it on the second draw. That is,

$$(3 \times 5) + (5 \times 3) = 30.$$

The number of ways of drawing two balls is

$$8 \times 7 = 56.$$

$$P(\text{exactly one black}) = \frac{30}{56}$$

Now let us compute the probability of getting exactly two blacks. The number of ways of getting exactly two blacks is

$$3 \times 2 = 6.$$

The number of ways of drawing two balls is

$$8 \times 7 = 56.$$

$$P(\text{exactly two blacks}) = \frac{6}{56}$$

Therefore,

$$P(\text{at least one black}) = \frac{30}{56} + \frac{6}{56}$$

$$= \frac{36}{56}$$

$$= \frac{9}{14}.$$

Method 2   If we draw two balls, it is certain that we will get either at least one black or no blacks (both reds). Therefore the probability of getting either at least one black or no blacks is 1.

$$P(\text{either at least one black or no blacks}) = 1$$

The event of getting at least one black and the event of getting no blacks are mutually exclusive. Therefore

$$P\binom{\text{either at least one black}}{\text{or no blacks}} = P\binom{\text{at least one}}{\text{black}} + P\binom{\text{no}}{\text{black}}.$$

But this probability is 1, so that

$$1 = P(\text{at least one black}) + P(\text{no blacks}).$$

Solving this equation for $P(\text{at least one black})$, we have

$$P(\text{at least one black}) = 1 - P(\text{no blacks}).$$

Now let us compute the probability of getting no blacks. That is, let us compute the probability of getting both reds. The number of ways of getting both reds is

$$5 \times 4 = 20.$$

The number of ways of drawing two balls is

$$8 \times 7 = 56.$$

The probability of getting both reds (no blacks) is

$$\frac{20}{56}.$$

Therefore,

$$P(\text{at least one black}) = 1 - \frac{20}{56}$$

$$= \frac{36}{56}$$

$$= \frac{9}{14},$$

which agrees with the result of Method 1.

In the second method in Example 8, we noted that

$$P(\text{at least one black}) = 1 - P(\text{no blacks}).$$

The same reasoning can be used to establish the general fact that the probability of some event E happening at least once is 1 *minus* the probability of it not happening at all.

$$P(\text{at least one occurrence of E}) = 1 - P(\text{no occurrence of E})$$

EXAMPLE 9   A die is thrown three times. What is the probability of getting at least one "snake eye" (a one)?

Solution $P(\text{at least one snake eye}) = 1 - P(\text{no snake eyes})$

Let us compute the probability of getting no snake eyes. There are five ways of getting something other than a snake eye, and there are six possible outcomes. The probability of getting no snake eyes on the first throw is

$$\frac{5}{6}.$$

In fact, the probability of getting no snake eyes on any throw is $\frac{5}{6}$. By

Theorem 1, the probability of getting *no* snake eyes on the first, and on the second, and on the third throw is

$$\frac{5}{6} \times \frac{5}{6} \times \frac{5}{6} = \left(\frac{5}{6}\right)^3.$$

Therefore,

$$P(\text{at least one snake eye}) = 1 - \left(\frac{5}{6}\right)^3$$

$$= 1 - \frac{125}{216}$$

$$= \frac{91}{216}.$$

Exercise 3.11, 25–28

The *odds in favor of an event E occurring* are defined as the ratio of the probability that E will occur to the probability that E will not occur.

$$\text{Odds in favor of E occurring} = \frac{P(E)}{P(\text{not } E)}$$

The odds *against* E occurring are defined as the reciprocal of this.

$$\text{Odds against E occurring} = \frac{P(\text{not } E)}{P(E)}$$

For example, consider the event E of getting a snake eye on a single throw of a die. The probability of E is $\frac{1}{6}$; the probability of not E is $\frac{5}{6}$. The odds in favor of getting a snake eye are

$$\frac{\frac{1}{6}}{\frac{5}{6}} = \frac{1}{5}.$$

This is usually read "one to five." Likewise, the odds *against* getting a snake eye are "five to one."

EXAMPLE 10

What are the odds in favor of getting an ace if one card is drawn from a standard 52-card deck? What are the odds against getting an ace?

Solution

There are 4 aces.

The probability of getting an ace is $\frac{4}{52}$.

The probability of *not* getting an ace is $\frac{48}{52}$.

Since

$$\frac{P(\text{ace})}{P(\text{not an ace})} = \frac{\frac{4}{52}}{\frac{48}{52}} = \frac{4}{48} = \frac{1}{12},$$

Exercise 3.11, 29–32
the odds in favor of getting an ace are 1 to 12. The odds against getting an ace are 12 to 1.

Exercise 3.11, 33–45
Probability problems appear in many different contexts.

EXERCISE 3.11 PROBABILITY

EXAMPLE A die is thrown once. What is the probability of getting a four?

Solution $P(\text{getting a four}) = \frac{1}{6}$

EXAMPLE A pair of dice are thrown. What is the probability of getting a sum of three?

Solution There are two ways of getting a sum of three. There are $6 \times 6 = 36$ possible outcomes.

$$P(\text{sum of three}) = \frac{2}{36}$$

EXAMPLE A card is drawn from a standard 52-card deck. What is the probability that it is a face card?

Solution There are 12 ways of getting a face card. There are 52 possible outcomes.

$$P(\text{getting a face card}) = \frac{12}{52}$$

EXAMPLE A card is drawn from a standard 52-card deck. What is the probability that the card is either an ace or is not an ace?

Solution It is certain to be either an ace or not an ace.

$$P(\text{either ace or not ace}) = 1$$

1. A coin is tossed once. What is the probability of getting tails?

2. A die is thrown once.
   (a) What is the probability of getting a five?
   (b) What is the probability of getting a six?

3. A pair of dice are thrown. Find the probability of getting
   (a) a sum of six   (b) a sum of seven
   (c) a sum of eight   (d) a sum of nine

4. A card is drawn from a standard 52-card deck. Find the probability that
   (a) the card is a heart   (b) the card is black
   (c) the card is an ace   (d) the card is a red Queen

5. Two balls are drawn from a basket containing eight red balls and no others. What is the probability that the balls drawn are both red?

6. A card is drawn from a standard 52-card deck. What is the probability that the card is either a face card or not a face card?

7. A stable contains 17 Morgans and five horses of other breeds. If Debbie selects three mounts at random, what is the probability that she will have at least a pair of Morgans?

8. Jeff is choreographing a number for a show. He knows ten samba steps and six polka steps. If he chooses one of these steps at random for the chorus, what is the probability that it will be a samba step?

EXAMPLE   Two cards are drawn in succession, without replacement, from a standard 52-card deck. What is the probability that they both are red?

Solution

Method 1   The number of ways of getting both red is 26×25. The number of possible outcomes is 52×51.

$$P(\text{both red}) = \frac{26 \times 25}{52 \times 51}$$

Method 2 $P(\text{first red}) = \frac{26}{52}$

$P(\text{second red after first red}) = \frac{25}{51}$

$P(\text{both red}) = \frac{26}{52} \times \frac{25}{51} = \frac{26 \times 25}{52 \times 51}$

EXAMPLE   A card is drawn from a standard 52-card deck. The card is replaced, and again a card is drawn. What is the probability that they both are red?

Solution $P(\text{first red}) = \frac{26}{52}$

$P(\text{second red}) = \frac{26}{52}$

$P(\text{both red}) = \frac{26}{52} \times \frac{26}{52} = \frac{26 \times 26}{52 \times 52}$

9. A box contains 15 new pennies, nine "wheat-ear" pennies, and no others. If Marianne draws two pennies in succession, without replacement, what is the probability of her getting a wheat-ear on the first draw and also on the second draw?

10. A box contains 15 new pennies, nine "wheat-ear" pennies and no others.
   (a) If Fred draws a penny, replaces it, and then again draws a penny, what is the probability of his getting a new penny on the first draw and a wheat-ear on the second draw?
   (b) If Fred drew 2 pennies from the box in succession, without replacement, what would be the probability of his getting a new penny on the first draw and a wheat-ear on the second draw?

11. One room in an apple storage contains 32 boxes of Macintosh, 21 boxes of Delicious, and 8 boxes of Baldwins. Three boxes are selected at random.

(a) If the boxes are selected in succession, without replacement, what is the probability that the first is a box of Macintosh, the second is a box of Delicious, and the third is a box of Baldwins?

(b) If the boxes are selected without regard to order, what is the probability that all three boxes will be Macintosh?

12. A Vermonter has 25 maple sap buckets thrown together in his truck. Of these sap buckets, 15 are wood and 10 are metal.

(a) If he selects 2 buckets at random, without replacement, what is the probability that they are both wood?

(b) If he selects a bucket at random, replaces that bucket, and then again selects a bucket, what is the probability that the bucket selected first is wood and the bucket selected second is also wood?

13. A die is thrown twice. What is the probability of getting a five on the first throw and a six on the second throw?

14. Two cards are drawn in succession, without replacement, from a standard 52-card deck. What is the probability that they are both black?

15. A die and a coin are thrown once. What is the probability of getting a four on the die and heads on the coin?

16. Two cards are drawn in succession, without replacement, from a standard 52-card deck. What is the probability of getting first an ace and then a Jack?

17. Two cards are drawn in succession, without replacement, from a standard 52-card deck. What is the probability of getting first an ace and then another ace?

18. One card is drawn from a standard 52-card deck and then replaced. A second card is drawn and then replaced. A third card is drawn. What is the probability that they are all hearts?

19. One card is drawn from a standard 52-card deck and then replaced. A second card is drawn and then replaced. A third card is drawn. What is the probability that the first is an ace, the second is a King, and the third is a Queen?

20. A toy piano has ten keys on which the first four notes of *How Dry I Am* can be played in only one way. A monkey is equally likely to strike one key as another. If the monkey strikes four keys in succession, what is the probability that it will play the first four notes of this song?

EXAMPLE

One card is drawn from a standard 52-card deck. What is the probability that it is either a red Queen or a Jack?

Solution

Method 1   There are 6 ways for it to be a red Queen or a Jack. There are 52 possible outcomes.

$$P(\text{either red Queen or Jack}) = \frac{6}{52}$$

Method 2 $P(\text{red Queen}) = \dfrac{2}{52}$

$P(\text{Jack}) = \dfrac{4}{52}$

$P(\text{either red Queen or Jack}) = \dfrac{2}{52} + \dfrac{4}{52} = \dfrac{6}{52}$

21. One die is thrown. What is the probability of getting either a five or a six?

22. Two coins are tossed. What is the probability of getting either both heads or both tails?

23. A pair of dice is thrown. What is the probability of getting either a sum of ten, or a sum of eleven, or a sum of twelve?

24. A hutch contains five rabbits, four hamsters, six guinea pigs, and nothing else. If a magician pulls out one animal at random, what is the probability that it is either a rabbit or a hamster?

EXAMPLE

Two dice are thrown. What is the probability of getting a sum of at least three?

Solution

$P(\text{sum of at least three}) = 1 - P(\text{sum of two})$

$$= 1 - \frac{1}{36}$$

$$= \frac{35}{36}$$

25. Two dice are thrown. What is the probability of getting a sum of at least four?

26. A box of candy contains 2 caramels, 48 creams, and no others. If Tom selects 2 pieces at random, what is the probability that he will get at least 1 cream?

27. Five cards are drawn in succession, without replacement, from a standard 52-card deck. What is the probability that at least three are spades?

28. A manufacturer makes five different kinds of parts for automobiles. The front end of an automobile is made up of one each of the five parts. A study claims that in any box of 50 parts, there are two defective ones. If the study is correct, what is the probability that a front end will contain at least one defective part?

EXAMPLE

What are the odds in favor of getting a four on a single throw of a die? What are the odds against getting a four?

Odds in favor of a four $= \dfrac{P(\text{getting a four})}{P(\text{not getting a four})}$

$$= \dfrac{\frac{1}{6}}{\frac{5}{6}}$$

$$= \tfrac{1}{5}$$

The odds in favor of a four are "one to five."
The odds against a four are "five to one."

29. Two dice are thrown.
    (a) What are the odds in favor of getting a sum of seven?
    (b) What are the odds against getting a sum of seven?

30. A 5-card hand is dealt at random from a standard 52-card deck.
    (a) What are the odds in favor of the hand consisting of four aces and one King?
    (b) What are the odds against it consisting of four aces and one King?

31. A 13-card hand is dealt at random from a standard 52-card deck.
    (a) What are the odds in favor of it consisting of all spades?
    (b) What are the odds against it consisting of all spades?

32. A carnival wheel is marked with exactly 32 spaces numbered 1 through 32. If the wheel is equally likely to stop on any of these spaces, what are the odds in favor of it stopping on either 1 or 7?

33. A crew of two is to be chosen at random from eight astronauts. If three of the astronauts are women and five are men, what is the probability that the crew will consist of two women?

34. In a dormitory-room drawing there are 1365 tickets numbered 1 through 1365. If Curt and Fred are the first ones to draw, and if they each draw one ticket, what is the probability that at least one of them will draw a ticket numbered 10 or under?

35. Doug, Fred, and Tom found all their socks in the same laundry basket. They had 7 green socks, 16 brown socks, and 21 black socks.
    (a) If Tom pulled out 2 socks at random, what is the probability that he had a pair of the same color?
    (b) If Fred pulled out 4 socks at random, what is the probability that he had at least a pair of the same color?
    (c) If Doug pulled out 3 socks at random, what is the probability that he had at least a pair of the same color?

36. When Doug was married, his belongings included a box of 99 dirty socks. There were 79 army socks and 20 civilian socks. If somebody washed just four socks drawn at random from the box, what is the probability that at least one pair of army socks was washed?

37. A typewriter has 44 keys. If a monkey strikes 5 keys at random, what is the probability that it will type the word *Bruno*?

38. What are the odds in favor of getting a four or higher on a single throw of a die?

39. Tom is auditioning for an orchestra. He will be asked to play a classical piece chosen at random from 3 on a list and a jazz piece chosen at random from 25 on a list. He knows 2 of the classical pieces well, and he knows 8 of the jazz pieces well. What is the probability that he will be asked to play pieces that he knows well?

40. Roland has a basket of fish to give away to the neighbors. He has ten blues, eight weakfish, and nine flounders. If the first neighbor takes three fish at random, what is the probability that he takes either three blues or three flounders?

41. Mary found a box of spools of thread. There were five black, eight white, six brown, and seven red. If she pulls out two spools at random, what is the probability that she will have at least one spool other than white?

42. There are 20 questions on a multiple-choice examination with 5 choices for the answer to each question. If Mary-Jane chooses all her answers at random, what is the probability that she will get at least 2 questions right?

43. A weather forecaster has the choice of predicting that the temperature will be warmer, or colder, or the same. He also has the choice of predicting rain, snow, sleet, or sunshine.
    (a) If he makes his choices at random, what is the probability that he predicts a warmer temperature and snow?
    (b) If he makes his choices at random, what is the probability that he predicts either warmer and sunshine or colder and snow?

44. Each of the members of a 4-member ski team must draw a card at random from a box of cards numbered 1 through 23 to determine the order in which they run the downhill. What is the probability that at least one member of the team will draw a card numbered less than 5?

45. In cross-pollinating a particular vegetable, a seed gets one color-characteristic gene from each of two sources. There are two types of these genes, A and B. Assume that a seed is equally likely to get type A as it is to get type B from each of its sources. If the seed gets two genes of type A, it is blue; otherwise it is green. What is the probability that the seed is green?

CHECK LIST FOR THE STUDENT

If you have learned the material in Section 3.11, you should be able to do the following:

1. State the given definition of the probability of event E.

2. Use the definition to find the probability $P(E)$ of some event E, such as the event of getting a sum of five on a single throw of two dice.

3. Give the probability of an event certain to occur.

4. Give the probability of an event certain not to occur.

5. Explain why $P(E) \geq 0$ and $P(E) \leq 1$ for any event E.

6. Give the probability of either E or not E for any event E.

7. State the conclusion of the theorem concerning the probability of first event E_1 and then event E_2 by completing the statement: $P(E_1 \text{ and } E_2) = P(E_1) \times$ _____ .

8. Tell what is meant by *independent* events.

9. Complete the statement. If E_1 and E_2 are independent events, $P(E_2 \text{ after } E_1) =$ _____ .

10. Complete the statement: If E_1 and E_2 are independent events, $P(E_1 \text{ and } E_2) =$ _____ .

11. Tell what is meant by *dependent* events.

12. Show with an example that for dependent events E_1 and E_2, $P(E_2 \text{ after } E_1) \neq P(E_2)$.

13. Tell what is meant by *mutually exclusive* events.

14. Complete the statement: If E_1 and E_2 are mutually exclusive events, $P(\text{either } E_1 \text{ or } E_2) =$ _____ .

15. Complete the statement.

$$P(\text{at least one occurence of E}) = \underline{\hspace{2cm}} .$$

16. Complete the statement: The odds in favor of E occurring = _____ .

17. Complete the statement: The odds against E occurring = _____ .

SECTION QUIZ 3.11

1. A die is thrown once. What is the probability of getting a two?

2. A card is drawn from a standard 52-card deck. What is the probability of getting a Jack?

3. Two coins are tossed. What is the probability of getting either two heads or at least one tail?

4. A basket contains seven red balls, five black balls, and no others. Two balls are drawn in succession, without replacement. What is the probability of getting a red on the first draw and a black on the second?

5. A basket contains seven red balls, five black balls, and no others. One ball is drawn, and then it is replaced. Again one ball is drawn. What is the probability of getting a red on the first draw and a black on the second draw?

6. A basket contains seven red balls, five black balls, and no others. Two balls are drawn in succession, without replacement. Let E_1 be the event of getting a red on the first draw, and let E_2 be the event of getting a black on the second draw. Show that $P(E_2 \text{ after } E_1) \neq P(E_2)$.

7. Two dice are thrown. Find the probability of getting either a sum of three or a sum of five.

8. Three dice are thrown. What is the probability of getting at least one four?

9. Two dice are thrown.
   (a) What are the odds in favor of getting a sum of two?
   (b) What are the odds against getting a sum of two?

10. Jo has six business books, five psychology books, four marketing books, and no others. If she chooses three books at random to set up a display, what is the probability that she will have one book of each kind?

Part 3 Review Exercises

1. What are the roots of the equation $(x^2 + 3x)(x - 2)(x + 5) = 0$?

2. What are the roots, and what is the multiplicity of each root of the equation $x^3(x + 2)^5(x - 3) = 0$?

3. What is the greatest number of distinct complex number roots that the equation $x^4 - 3x^2 + 2x - 5 = 0$ may have?

4. Use the factor theorem to construct the lowest possible degree equation with the roots 5, -4, 0, and 3.

5. Use the remainder theorem to determine whether or not 2 is a root of the equation $x^3 - 2x^2 - 7x + 14 = 0$.

6. Find all the rational roots of $7x^4 - 45x^3 + 81x^2 - 41x + 6 = 0$.

7. If $2 - 7i$ is a root of a polynomial equation $P(x) = 0$, what is an irreducible quadratic factor of $P(x)$?

8. Use the location of roots theorem to show that there is at least one root of $x^3 - 2x^2 - 4x + 6 = 0$ between 2 and 3.

9. Show that there is a root of $9x^4 - 36x^3 + 112x^2 + 20x - 65 = 0$ between 0 and 1, and use linear interpolation to find a first approximation to it.

If possible, solve each of the following systems with the Gaussian elimination method. If there is no solution, or if there is an unlimited number of solutions, say so.

10. $2x - 5y + z = 16$
    $3x + y - 2z = 5$
    $x - 3y - z = 4$

11. $3x + 2y = 14$
    $6x + 4y = 7$

12. What is the minor of 7 in the following determinant?

$$\begin{vmatrix} 3 & 7 & -2 \\ 4 & 5 & 1 \\ -1 & -3 & 6 \end{vmatrix}$$

13. Find the value of $\begin{vmatrix} 3 & 7 \\ -2 & 4 \end{vmatrix}$.

14. Use expansion by minors of elements in the second column, and find the value of the following determinant.

$$\begin{vmatrix} 3 & 4 & -2 \\ 2 & 0 & 5 \\ -7 & 1 & -1 \end{vmatrix}$$

Use Cramer's rule to find the solution of each of the following systems.

15. $3x + 7y = 15$
  $2x + 5y = 11$

16. $2x - y + z = 6$
  $-x - y + 3z = -4$
  $x + y - z = 6$

Using properties of determinants, fill in the blank parentheses in each of the following.

17. $\begin{vmatrix} 4 & 5 & -1 \\ 0 & 0 & 0 \\ 2 & 7 & 6 \end{vmatrix} = (\qquad)$

18. $\begin{vmatrix} 4 & 4 & 2 \\ 1 & 1 & 5 \\ 3 & 3 & 8 \end{vmatrix} = (\qquad)$

19. $\begin{vmatrix} 4 & 1 & 3 \\ 2 & 5 & 8 \\ 6 & 2 & 9 \end{vmatrix} = (\qquad) \begin{vmatrix} 4 & 2 & 6 \\ 1 & 5 & 2 \\ 3 & 8 & 9 \end{vmatrix}$

20. $\begin{vmatrix} 4 & 1 & 3 \\ 4 & 10 & 16 \\ 6 & 2 & 9 \end{vmatrix} = (\qquad) \begin{vmatrix} 4 & 1 & 3 \\ 2 & 5 & 8 \\ 6 & 2 & 9 \end{vmatrix}$

21. $\begin{vmatrix} 2 & 4 & 7 \\ 9 & 8 & 1 \\ 3 & 5 & 6 \end{vmatrix} = (\qquad) \begin{vmatrix} 2 & 4 & 7 \\ 3 & 5 & 6 \\ 9 & 8 & 1 \end{vmatrix}$

22. $\begin{vmatrix} 2 & 4 & 7 \\ 1 & 3 & -1 \\ 6 & 5 & 9 \end{vmatrix} = (\qquad) \begin{vmatrix} 0 & -2 & 9 \\ 1 & 3 & -1 \\ 6 & 5 & 9 \end{vmatrix}$

Find each of the following products.

23. $\begin{pmatrix} 2 & 5 \\ -1 & 6 \end{pmatrix} \begin{pmatrix} 3 & 7 \\ 6 & 1 \end{pmatrix}$

24. $\begin{pmatrix} 3 & 5 & 7 \\ 2 & -1 & 0 \end{pmatrix} \begin{pmatrix} 8 & 1 \\ 7 & 2 \\ 0 & 3 \end{pmatrix}$

25. $\begin{pmatrix} 3 & 6 & 2 \\ 1 & 4 & 8 \\ 3 & 9 & -7 \end{pmatrix} \begin{pmatrix} 1 & 0 & 0 \\ 0 & 1 & 0 \\ 0 & 0 & 1 \end{pmatrix}$

26. $\begin{pmatrix} 3 & 5 \\ -2 & 9 \end{pmatrix} \begin{pmatrix} x \\ y \end{pmatrix}$

27. $\dfrac{-1}{2} \begin{pmatrix} -2 & 0 & 0 \\ 0 & -2 & 0 \\ 0 & 0 & -2 \end{pmatrix}$

28. (a) $\begin{pmatrix} 3 & 6 \\ 2 & 7 \end{pmatrix} \begin{pmatrix} -1 & 1 \\ 4 & -3 \end{pmatrix}$ **(b)** $\begin{pmatrix} -1 & 1 \\ 4 & -3 \end{pmatrix} \begin{pmatrix} 3 & 6 \\ 2 & 7 \end{pmatrix}$

29. Write the following as a system of equations.

$$\begin{pmatrix} 2x - 7y \\ 3x + y \end{pmatrix} = \begin{pmatrix} 5 \\ 2 \end{pmatrix}$$

30. Write the identity matrix of order 4.

31. Determine whether or not $\begin{pmatrix} 6 & 3 \\ 8 & 4 \end{pmatrix}$ is nonsingular.

32. Verify by multiplication that $\dfrac{1}{31} \begin{pmatrix} 1 & 7 \\ -4 & 3 \end{pmatrix}$ is the inverse of $\begin{pmatrix} 3 & -7 \\ 4 & 1 \end{pmatrix}$.

33. Find the transpose of $\begin{pmatrix} 6 & 1 & 3 \\ 4 & 2 & 7 \\ 1 & 9 & 5 \end{pmatrix}$.

34. Find the inverse of $\begin{pmatrix} 2 & 5 \\ 3 & -1 \end{pmatrix}$.

35. Find the inverse of $\begin{pmatrix} -1 & 0 & 2 \\ 0 & 0 & 1 \\ -3 & 1 & 0 \end{pmatrix}$.

36. Solve the following system of equations by writing the system as a matrix equation and using the inverse of the coefficient matrix.

$$5x - y = 6$$
$$3x + y = 10$$

Write out the form of the partial fraction decomposition of each of the following.

37. $\dfrac{3x^2 - 5x + 8}{(x - 3)(x + 2)^2}$

38. $\dfrac{4x^3 - 2x^2 + 1}{(x - 1)(x + 2)(x^2 - 2x + 15)}$

39. $\dfrac{2x^3 + 2x + 1}{(x - 4)(x^2 - 2x + 15)^2}$

Express each of the following as a sum of partial fractions.

40. $\dfrac{x^3 - 6x^2 + 12x - 2}{(x - 1)^2}$ **41.** $\dfrac{3x^3 - 3x^2 + 13x + 17}{(x - 1)^2(x^2 + 9)}$

Sketch the graph of each of the following.

42. $x \geq 3$

43. $x + y - 5 < 0$

44. $x + y \leq 16$

45. $y > -x$

46. $y \geq 2x$

Sketch the graph of the solution of each of the following systems.

47.
$$\begin{cases} x + y \leq 9 \\ y \geq 0 \\ x \geq 0 \end{cases}$$

48.
$$\begin{cases} x + y - 2 \geq 0 \\ y \leq 2 \\ y \geq x - 2 \end{cases}$$

49. Write out the first four terms and the general term a_n of $\{(\frac{2}{3})^n\}$, and give $\lim_{n \to \infty} a_n$.

Classify each of the following as either geometric or arithmetic. If it is geometric, give its common ratio. If it is arithmetic, give its common difference.

50. $2, 5, 8, \ldots, 2 + 3(n - 1), \ldots$

51. $3, 6, 12, \ldots, 3 \times 2^{n-1}, \ldots$

Write out a formula for the sum of the first n terms of each of the following.

52. $2, 6, 10, \ldots, 2 + 4(n - 1), \ldots$

53. $2, \frac{2}{3}, \frac{2}{9}, \ldots, 2(\frac{1}{3})^{n-1}, \ldots$

State whether or not each of the following converges. If it converges, find its sum.

54. $1 + \frac{1}{6} + \frac{1}{36} + \frac{1}{216} + \cdots + (\frac{1}{6})^{n-1} + \cdots$

55. $1 + 6 + 36 + 216 + \cdots + 6^{n-1} + \cdots$

56. Write out the first four terms and write the nth term of $\sum_{n=1}^{\infty} n^2 + 1$.

57. Write out the first, second, and third partial sum of $\sum_{n=1}^{\infty} \frac{1}{n}$.

58. In a New England town dump the number of nonreturnable bottles is increasing at such a rate that each year there are three times as many as the year before. (a) If none are removed, and if there were 500 the first year, how many would there be at the end of four years? (b) Use summation notation and write an expression for the total number at the end of 10 years.

Use mathematical induction to prove that each of the following is true for all the prescribed positive integers n.

59. $6 + 12 + 18 + \cdots + 6n = 3(n + n^2), n \geq 1$

60. $2^n > n + 4, n \geq 3$

61. 4 is a factor of $2n^2 + 6n$.

62. Leave blank parentheses for the coefficients of other than x^{10} and y^{10}, and write out $(x + y)^{10}$.

63. Use Pascal's triangle in determining the coefficients, and expand $(2x + y)^5$.

64. Use the binomial theorem and write out the first four terms in the expansion of $(x + y)^{12}$.

65. Find the numerical value of the binomial coefficient $\binom{10}{3}$, and write the term having that coefficient in the expansion of $(x + y)^{10}$.

66. Find the coefficient of x^4y^2 in the expansion of $(2x + y)^6$.

67. Verify that $\binom{7}{2} = \binom{7}{5}$.

Find the value of each of the following.

68. (a) $\binom{17}{17}$ (b) $\binom{18}{17}$

69. (a) $\binom{17}{1}$ (b) $\binom{17}{0}$

70. A club plans to sell candles and flowers. They must agree to sell at least 100 boxes of candles and at least 150 boxes of flowers. They are allowed to order no more than a total of 400 boxes of merchandise. They will make a profit of $1 on each box of candles and of $2 on each box of flowers they sell. How many boxes of each should they order to make the maximum profit, provided that they sell all they order?

71. List all the permutations of the symbols a, b, c, d.

72. List all the combinations of the symbols a, b, c, d, e taken four at a time.

Compute the value of each of the following.

73. $_9P_9$ **74.** $_9P_6$

75. $_9C_9$ **76.** $_9C_6$

77. List all the distinguishable permutations of the letters of the word *tat*.

78. Find the number of distinguishable permutations of the letters of the words *Silly Putty*.

79. Use the notation $_nC_r$ for the binomial coefficient $\binom{n}{r}$ and write the fourth and fifth rows of Pascal's triangle.

80. A merry-go-round has 20 different horses for seats. (a) In how many ways can 20 people be seated on these horses? (b) In how many ways can 20 people be arranged in a circle?

81. Five cards are drawn in succession and without replacement from a standard deck of 52 cards. (a) What is the number of ways of drawing 5 spades that can be arranged in consecutive order? (b) What is the number of ways of drawing 5 spades?

82. Two dice are thrown. What is the probability of getting either a total of 4 or a total of 5?

83. Five coins are tossed. What is the probability of getting exactly 3 heads?

84. There are 10 ways to go from A to B and then there are 9 ways to go from B to C. How many ways are there to go from A to B and then to C?

85. (a) How many different groups of 4 people each can be chosen from a chorus of 30 people?
    (b) How many different quartets can be chosen from a chorus of 30 people, if the 4 people chosen are assigned bass, tenor, alto, and soprano in the order chosen, regardless of the parts they usually sing?

In each of the following a basket contains 12 red balls, 10 green balls, and no others. Two balls are drawn.

86. If the balls are drawn in succession, without replacement, what is the probability of drawing 2 red?

87. If a ball is drawn and replaced, and if again a ball is drawn, what is the probability of drawing 2 red?

88. If the balls are drawn simultaneously (without regard to order), what is the probability of drawing 2 red?

89. If the balls are drawn in succession, without replacement, what is the probability of drawing a red and a green?

90. Two cards are drawn in succession, without replacement, from a standard 52-card deck. What is the probability that they are both red face cards?

91. A basket contains 18 red balls, 12 green balls, and no others. A ball is drawn, its color recorded, and it is replaced. If this is done 5 times, what is the probability of getting at least 1 green ball?

92. A die is thrown 5 times. What is the probability of getting at least 1 snake eye?

93. A card is drawn from a standard 52-card deck. What are the odds in favor of its being an ace?

94. Five coins are tossed. What are the odds against getting all heads?

95. Two dice are thrown. What are the odds against getting a total of 2 on the 2 dice?

96. At a department picnic, there was a tub containing 8 bottles of beer and 10 bottles of soft drinks. If Don Catlin pulled out 2 bottles at random, what is the probability that he pulled out 2 bottles of beer?

97. Marianne made 2 dozen cookies and 18 brownies for roommates Curt and Fred. If Curt first ate 2 dozen cookies and brownies chosen at random, what is the probability that there was at least 1 cookie left for Fred?

98. Tom knows 11 hustle steps. How many different dance routines can he do of 5 different hustle steps each? What is the probability that one of these routines consists of the first 5 hustle steps that he ever learned?

99. Mary bought 4 cans of house paint at a sale. On the store shelf there were 8 cans of white paint and 10 cans of paint of various colors. If she selected 4 cans at random, what is the probability that she got at least 3 cans of white?

100. In his garage, Roland had 2 cans of marine varnish, 3 cans of yacht-white, and 9 cans of other assorted paints. If he selected 1 can at random, what is the probability that he selected a can of marine varnish or a can of yacht-white?

Part 3 Final Test

1. Use the remainder theorem to determine whether or not 4 is a root of the equation $x^3 - 4x^2 - 2x + 8 = 0$.

2. Use the factor theorem to construct an equation with the lowest possible degree having the roots 2, -3, 0, and -1.

3. Consider the equation $12x^4 - 5x^3 - 26x^2 + 10x + 4 = 0$.
   (a) What is the greatest number of distinct complex-number roots that this equation may have?
   (b) Find all the rational roots.

4. If $2 + 3i$ is a root of polynomial equation $P(x) = 0$, what is an irreducible quadratic factor of $P(x)$?

5. Use the location-of-roots theorem to show that there is at least one root of $x^3 - 6x^2 + 6x + 8 = 0$ between 2 and 3.

6. Show that there is a root of $4x^4 - 8x^3 + 17x^2 + 6x - 15 = 0$ between 0 and 1, and use linear interpolation to find a first approximation to it.

If possible, solve each of the following systems with the Gaussian elimination method. If there is not exactly one solution, say so.

7. $3x - 2y + 5z = -7$
   $2x + y - z = 6$
   $x - 4y + z = 3$

8. $5x - 3y = 10$
   $10x - 6y = 5$

9. (a) Find the value of $\begin{vmatrix} 2 & 5 \\ -3 & 4 \end{vmatrix}$.

   (b) Use expansion by minors of elements in the first row and find the value of
   $$\begin{vmatrix} 2 & 1 & 0 \\ 4 & 2 & -6 \\ -1 & 3 & 0 \end{vmatrix}.$$

10. Use Cramer's rule to find the solution of the following system.
$$x + y - z = -1$$
$$-x - y + 3z = 3$$
$$2x + y + z = 1$$

11. Find each of the following products:

   (a) $\begin{pmatrix} 3 & 7 & -1 \\ 2 & -3 & 1 \end{pmatrix} \begin{pmatrix} 2 & 3 \\ 0 & 1 \\ -2 & 4 \end{pmatrix}$

   (b) $\begin{pmatrix} 2 & 3 \\ -4 & 1 \end{pmatrix} \begin{pmatrix} x \\ y \end{pmatrix}$

(c) $\frac{1}{7}\begin{pmatrix} 7 & 0 & 0 \\ 0 & 7 & 0 \\ 0 & 0 & 7 \end{pmatrix}$

12. Find the inverse of $\begin{pmatrix} 1 & 0 & 2 \\ -1 & 3 & 0 \\ 0 & 0 & 1 \end{pmatrix}$.

13. Solve the following system of equations by writing the system as a matrix equation and using the inverse of the coefficient matrix.

$$3x - 2y = 5$$
$$-x + y = -1$$

14. Express the following as a sum of partial fractions.

$$\frac{x^2 - 9x + 20}{x(x - 1)(x + 5)}$$

15. Write out the form of the partial-fraction decomposition of the following.

$$\frac{3x^2 - 7x + 2}{(x + 1)^2(x^2 + 5)^2}$$

16. Sketch the graph of $2x + y + 1 > 0$.

17. Sketch the graph of the solution of the following system.

$$y \le x + 2$$
$$x + y - 5 \ge 0$$
$$x \le 4$$

18. Write out the first four terms and the general term a_n of $\{(\frac{1}{5})^n\}$, and give $\lim\limits_{n \to \infty} a_n$.

19. Classify each of the following as either geometric or arithmetic and write out a formula for the sum of the first n terms.
   (a) $7, 7^2, 7^3, \ldots, 7^n, \ldots$
   (b) $7, 14, 21, \ldots, 7n, \ldots$

20. (a) Write out $\sum\limits_{n=1}^{5} \dfrac{n}{n + 3}$.

   (b) Write out the first four terms and write the nth term of $\sum\limits_{n=1}^{\infty} n^3$.

21. (a) A substance decays at such a rate that the amount of decay during the nth hour is $(\frac{1}{2})^n$ for any positive integer n. Write an expression for the sum of the total amount decayed during the first n hours.
   (b) State whether or not the series $1 + \frac{1}{5} + \frac{1}{25} + \cdots + (\frac{1}{5})^{n-1} + \cdots$ converges. If it converges, find its sum.

Use mathematical induction to prove that each of the following is true for all the prescribed positive integers n.

22. $4 + 32 + \cdots + 4n^3 = n^2(n + 1)^2, \quad n \geq 1$

23. $5^n > n + 5, \quad n \geq 2$

24. (a) Leave blank parentheses for the coefficients of other than a^8 and b^8, and write out $(a + b)^8$.

    (b) Use Pascal's triangle in determining the coefficients and expand $(x + 2y)^4$.

25. (a) Use the binomial theorem and write out the first four terms in the expansion $(a + b)^9$.

    (b) Find the numerical value of the binomial coefficient $\binom{15}{3}$, and write the term having that coefficient in the expansion of $(x + y)^{15}$.

26. (a) What is the number of permutations of ten different things taken five at a time?

    (b) What is the number of combinations of ten different things taken five at a time?

    (c) What is the number of distinguishable permutations of the letters in the words *funny bunny*?

27. (a) How many different teams of 6 players each can be chosen from 30 players?

    (b) How many different teams of 6 players each can be chosen from 30 players, if the first person chosen is the goalie, the second is center, the third is left wing, the fourth is right wing, the fifth is left defense, and the sixth is right defense?

28. A bus driver must go from Town A to Town B and then to Town M. There are four routes from A to B, two of which are approved. There are seven routes from B to M, three of which are approved.

    (a) If he chooses his routes at random, what is the probability that he chooses approved routes for the entire trip from A to B to M?

    (b) What are the odds in favor of his choosing an approved route for the entire trip?

29. (a) Four cards are drawn at random from a standard 52-card deck. What is the probability of getting all red face cards?

    (b) A pair of dice and four coins are tossed together. What is the probability of getting a total of 6 on the dice and exactly one head on the coins?

30. A basket contains nine red balls, five black balls, and no others. Three balls are drawn in succession without replacement. What is the probability of getting at least one red ball?

Appendix A Properties of the Real Numbers

PROPERTIES OF THE REAL NUMBERS

The following properties are true for all real numbers a, b, and c.

1. $a + b$ is a real number   Closure law for addition

    ab is a real number   Closure law for multiplication

2. $a + b = b + a$ Commutative law for addition

    $ab = ba$ Commutative law for multiplication

3. $(a + b) + c = a + (b + c) = a + b + c$ Associative law for addition

    $(ab)c = a(bc) = abc$ Associative law for multiplication

4. $a(b + c) = ab + ac$ Distributive law

5. $\left.\begin{array}{l} a + 0 = a \\ 0 + a = a \end{array}\right\}$ 0 is the additive identity

6. $\left.\begin{array}{l} a \cdot 0 = 0 \\ 0 \cdot a = 0 \end{array}\right\}$

7. $\left.\begin{array}{l} a \cdot 1 = a \\ 1 \cdot a = a \end{array}\right\}$ 1 is the multiplicative identity

8. $\left.\begin{array}{l} a + (-a) = 0 \\ (-a) + a = 0 \end{array}\right\}$ $-a$ is the additive inverse of a.

9. $\left.\begin{array}{l} a \cdot \left(\dfrac{1}{a}\right) = 1 \\[2em] \left(\dfrac{1}{a}\right) \cdot a = 1 \end{array}\right\}$ provided $a \neq 0$ $\dfrac{1}{a}$ is the multiplicative inverse of a

Appendix B Tables

TABLE 1 | COMMON LOGARITHMS

x	0	1	2	3	4	5	6	7	8	9
1.0	.0000	.0043	.0086	.0128	.0170	.0212	.0253	.0294	.0334	.0374
1.1	.0414	.0453	.0492	.0531	.0569	.0607	.0645	.0682	.0719	.0755
1.2	.0792	.0828	.0864	.0899	.0934	.0969	.1004	.1038	.1072	.1106
1.3	.1139	.1173	.1206	.1239	.1271	.1303	.1335	.1367	.1399	.1430
1.4	.1461	.1492	.1523	.1553	.1584	.1614	.1644	.1673	.1703	.1732
1.5	.1761	.1790	.1818	.1847	.1875	.1903	.1931	.1959	.1987	.2014
1.6	.2041	.2068	.2095	.2122	.2148	.2175	.2201	.2227	.2253	.2279
1.7	.2304	.2330	.2355	.2380	.2405	.2430	.2455	.2480	.2504	.2529
1.8	.2553	.2577	.2601	.2625	.2648	.2672	.2695	.2718	.2742	.2765
1.9	.2788	.2810	.2833	.2856	.2878	.2900	.2923	.2945	.2967	.2989
2.0	.3010	.3032	.3054	.3075	.3096	.3118	.3139	.3160	.3181	.3201
2.1	.3222	.3243	.3263	.3284	.3304	.3324	.3345	.3365	.3385	.3404
2.2	.3424	.3444	.3464	.3483	.3502	.3522	.3541	.3560	.3579	.3598
2.3	.3617	.3636	.3655	.3674	.3692	.3711	.3729	.3747	.3766	.3784
2.4	.3802	.3820	.3838	.3856	.3874	.3892	.3909	.3927	.3945	.3962
2.5	.3979	.3997	.4014	.4031	.4048	.4065	.4082	.4099	.4116	.4133
2.6	.4150	.4166	.4183	.4200	.4216	.4232	.4249	.4265	.4281	.4298
2.7	.4314	.4330	.4346	.4362	.4378	.4393	.4409	.4425	.4440	.4456
2.8	.4472	.4487	.4502	.4518	.4533	.4548	.4564	.4579	.4594	.4609
2.9	.4624	.4639	.4654	.4669	.4683	.4698	.4713	.4728	.4742	.4757
3.0	.4771	.4786	.4800	.4814	.4829	.4843	.4857	.4871	.4886	.4900
3.1	.4914	.4928	.4942	.4955	.4969	.4983	.4997	.5011	.5024	.5038
3.2	.5051	.5065	.5079	.5092	.5105	.5119	.5132	.5145	.5159	.5172
3.3	.5185	.5198	.5211	.5224	.5237	.5250	.5263	.5276	.5289	.5302
3.4	.5315	.5328	.5340	.5353	.5366	.5378	.5391	.5403	.5416	.5428
3.5	.5441	.5453	.5465	.5478	.5490	.5502	.5514	.5527	.5539	.5551
3.6	.5563	.5575	.5587	.5599	.5611	.5623	.5635	.5647	.5658	.5670
3.7	.5682	.5694	.5705	.5717	.5729	.5740	.5752	.5763	.5775	.5786
3.8	.5798	.5809	.5821	.5832	.5843	.5855	.5866	.5877	.5888	.5899
3.9	.5911	.5922	.5933	.5944	.5955	.5966	.5977	.5988	.5999	.6010
4.0	.6021	.6031	.6042	.6053	.6064	.6075	.6085	.6096	.6107	.6117
4.1	.6128	.6138	.6149	.6160	.6170	.6180	.6191	.6201	.6212	.6222
4.2	.6232	.6243	.6253	.6263	.6274	.6284	.6294	.6304	.6314	.6325
4.3	.6335	.6345	.6355	.6365	.6375	.6385	.6395	.6405	.6415	.6425
4.4	.6435	.6444	.6454	.6464	.6474	.6484	.6493	.6503	.6513	.6522
4.5	.6532	.6542	.6551	.6561	.6571	.6580	.6590	.6599	.6609	.6618
4.6	.6628	.6637	.6646	.6656	.6665	.6675	.6684	.6693	.6702	.6712
4.7	.6721	.6730	.6739	.6749	.6758	.6767	.6776	.6785	.6794	.6803
4.8	.6812	.6821	.6830	.6839	.6848	.6857	.6866	.6875	.6884	.6893
4.9	.6902	.6911	.6920	.6928	.6937	.6946	.6955	.6964	.6972	.6981
5.0	.6990	.6998	.7007	.7016	.7024	.7033	.7042	.7050	.7059	.7067
5.1	.7076	.7084	.7093	.7101	.7110	.7118	.7126	.7135	.7143	.7152
5.2	.7160	.7168	.7177	.7185	.7193	.7202	.7210	.7218	.7226	.7235
5.3	.7243	.7251	.7259	.7267	.7275	.7284	.7292	.7300	.7308	.7316
5.4	.7324	.7332	.7340	.7348	.7356	.7364	.7372	.7380	.7388	.7396
x	0	1	2	3	4	5	6	7	8	9

TABLE 1 (*continued*)

x	0	1	2	3	4	5	6	7	8	9
5.5	.7404	.7412	.7419	.7427	.7435	.7443	.7451	.7459	.7466	.7474
5.6	.7482	.7490	.7497	.7505	.7513	.7520	.7528	.7536	.7543	.7551
5.7	.7559	.7566	.7574	.7582	.7589	.7597	.7604	.7612	.7619	.7627
5.8	.7634	.7642	.7649	.7657	.7664	.7672	.7679	.7686	.7694	.7701
5.9	.7709	.7716	.7723	.7731	.7738	.7745	.7752	.7760	.7767	.7774
6.0	.7782	.7789	.7796	.7803	.7810	.7818	.7825	.7832	.7839	.7846
6.1	.7853	.7860	.7868	.7875	.7882	.7889	.7896	.7903	.7910	.7917
6.2	.7924	.7931	.7938	.7945	.7952	.7959	.7966	.7973	.7980	.7987
6.3	.7993	.8000	.8007	.8014	.8021	.8028	.8035	.8041	.8048	.8055
6.4	.8062	.8069	.8075	.8082	.8089	.8096	.8102	.8109	.8116	.8122
6.5	.8129	.8136	.8142	.8149	.8156	.8162	.8169	.8176	.8182	.8189
6.6	.8195	.8202	.8209	.8215	.8222	.8228	.8235	.8241	.8248	.8254
6.7	.8261	.8267	.8274	.8280	.8287	.8293	.8299	.8306	.8312	.8319
6.8	.8325	.8331	.8338	.8344	.8351	.8357	.8363	.8370	.8376	.8382
6.9	.8388	.8395	.8401	.8407	.8414	.8420	.8426	.8432	.8439	.8445
7.0	.8451	.8457	.8463	.8470	.8476	.8482	.8488	.8494	.8500	.8506
7.1	.8513	.8519	.8525	.8531	.8537	.8543	.8549	.8555	.8561	.8567
7.2	.8573	.8579	.8585	.8591	.8597	.8603	.8609	.8615	.8621	.8627
7.3	.8633	.8639	.8645	.8651	.8657	.8663	.8669	.8675	.8681	.8686
7.4	.8692	.8698	.8704	.8710	.8716	.8722	.8727	.8733	.8739	.8745
7.5	.8751	.8756	.8762	.8768	.8774	.8779	.8785	.8791	.8797	.8802
7.6	.8808	.8814	.8820	.8825	.8831	.8837	.8842	.8848	.8854	.8859
7.7	.8865	.8871	.8876	.8882	.8887	.8893	.8899	.8904	.8910	.8915
7.8	.8921	.8927	.8932	.8938	.8943	.8949	.8954	.8960	.8965	.8971
7.9	.8976	.8982	.8987	.8993	.8998	.9004	.9009	.9015	.9020	.9025
8.0	.9031	.9036	.9042	.9047	.9053	.9058	.9063	.9069	.9074	.9079
8.1	.9085	.9090	.9096	.9101	.9106	.9112	.9117	.9122	.9128	.9133
8.2	.9138	.9143	.9149	.9154	.9159	.9165	.9170	.9175	.9180	.9186
8.3	.9191	.9196	.9201	.9206	.9212	.9217	.9222	.9227	.9232	.9238
8.4	.9243	.9248	.9253	.9258	.9263	.9269	.9274	.9279	.9284	.9289
8.5	.9294	.9299	.9304	.9309	.9315	.9320	.9325	.9330	.9335	.9340
8.6	.9345	.9350	.9355	.9360	.9365	.9370	.9375	.9380	.9385	.9390
8.7	.9395	.9400	.9405	.9410	.9415	.9420	.9425	.9430	.9435	.9440
8.8	.9445	.9450	.9455	.9460	.9465	.9469	.9474	.9479	.9484	.9489
8.9	.9494	.9499	.9504	.9509	.9513	.9518	.9523	.9528	.9533	.9538
9.0	.9542	.9547	.9552	.9557	.9562	.9566	.9571	.9576	.9581	.9586
9.1	.9590	.9595	.9600	.9605	.9609	.9614	.9619	.9624	.9628	.9633
9.2	.9638	.9643	.9647	.9652	.9657	.9661	.9666	.9671	.9675	.9680
9.3	.9685	.9689	.9694	.9699	.9703	.9708	.9713	.9717	.9722	.9727
9.4	.9731	.9736	.9741	.9745	.9750	.9754	.9759	.9763	.9768	.9773
9.5	.9777	.9782	.9786	.9791	.9795	.9800	.9805	.9809	.9814	.9818
9.6	.9823	.9827	.9832	.9836	.9841	.9845	.9850	.9854	.9859	.9863
9.7	.9868	.9872	.9877	.9881	.9886	.9890	.9894	.9899	.9903	.9908
9.8	.9912	.9917	.9921	.9926	.9930	.9934	.9939	.9943	.9948	.9952
9.9	.9956	.9961	.9965	.9969	.9974	.9978	.9983	.9987	.9991	.9996
x	0	1	2	3	4	5	6	7	8	9

TABLE 2 | NATURAL LOGARITHMS

n	$\log_e n$	n	$\log_e n$	n	$\log_e n$
	*	4.5	1.5041	9.0	2.1972
0.1	7.6974 ⌉	4.6	1.5261	9.1	2.2083
0.2	8.3906	4.7	1.5476	9.2	2.2192
0.3	8.7960	4.8	1.5686	9.3	2.2300
0.4	9.0837	4.9	1.5892	9.4	2.2407
0.5	9.3069	5.0	1.6094	9.5	2.2513
0.6	9.4892	5.1	1.6292	9.6	2.2618
0.7	9.6433	5.2	1.6487	9.7	2.2721
0.8	9.7769	5.3	1.6677	9.8	2.2824
0.9	9.8946 ⌋	5.4	1.6864	9.9	2.2925
1.0	0.0000	5.5	1.7047	10	2.3026
1.1	0.0953	5.6	1.7228	11	2.3979
1.2	0.1823	5.7	1.7405	12	2.4849
1.3	0.2624	5.8	1.7579	13	2.5649
1.4	0.3365	5.9	1.7750	14	2.6391
1.5	0.4055	6.0	1.7918	15	2.7081
1.6	0.4700	6.1	1.8083	16	2.7726
1.7	0.5306	6.2	1.8245	17	2.8332
1.8	0.5878	6.3	1.8405	18	2.8904
1.9	0.6419	6.4	1.8563	19	2.9444
2.0	0.6931	6.5	1.8718	20	2.9957
2.1	0.7419	6.6	1.8871	25	3.2189
2.2	0.7885	6.7	1.9021	30	3.4012
2.3	0.8329	6.8	1.9169	35	3.5553
2.4	0.8755	6.9	1.9315	40	3.6889
2.5	0.9163	7.0	1.9459	45	3.8067
2.6	0.9555	7.1	1.9601	50	3.9120
2.7	0.9933	7.2	1.9741	55	4.0073
2.8	1.0296	7.3	1.9879	60	4.0943
2.9	1.0647	7.4	2.0015	65	4.1744
3.0	1.0986	7.5	2.0149	70	4.2485
3.1	1.1314	7.6	2.0281	75	4.3175
3.2	1.1632	7.7	2.0412	80	4.3820
3.3	1.1939	7.8	2.0541	85	4.4427
3.4	1.2238	7.9	2.0669	90	4.4998
3.5	1.2528	8.0	2.0794	100	4.6052
3.6	1.2809	8.1	2.0919	110	4.7005
3.7	1.3083	8.2	2.1041	120	4.7875
3.8	1.3350	8.3	2.1163	130	4.8676
3.9	1.3610	8.4	2.1282	140	4.9416
4.0	1.3863	8.5	2.1401	150	5.0106
4.1	1.4110	8.6	2.1518	160	5.0752
4.2	1.4351	8.7	2.1633	170	5.1358
4.3	1.4586	8.8	2.1748	180	5.1930
4.4	1.4816	8.9	2.1861	190	5.2470

*Subtract 10 for $n < 1$. Thus $\log_e 0.1 = 7.6974 - 10 = -2.3026$.

Answers to Quizzes, Tests, and Review Exercises

Section Quiz 1.1

1. (a) Positive integers, integers, rational numbers, real numbers, complex numbers (b) Integers, rational numbers, real numbers, complex numbers **2.** (a) Negative integers, integers, rational numbers, real numbers, complex numbers (b) Integers, rational, real, and complex numbers **3.** (a) Positive integers, integers, rational numbers, real numbers, complex numbers (b) Rational numbers, real numbers, complex numbers
4. (a) Irrational numbers, real numbers, complex numbers (b) Complex numbers, imaginary numbers
5. -3, -2, 0, 2, $\frac{3}{5}$, 2.3, $2.\overline{3}$, 9, $\frac{22}{7}$, 3.14, $\frac{-2}{3}$, 1.732 **6.** $\sqrt{3}$, π **7.** (a) Irrational, nonterminating, and nonrepeating (b) Rational, repeating (c) Rational, terminating **8.** (a) $\{x \mid x \geq 0\}$ ———————•——— 0

(b) $\{x \mid x < 0\}$ ———————◦——— 0 **9.** (a) $\{x \mid x \leq 0\}$ ———————•——— 0 (b) $\{x \mid x > 0\}$ ———————◦——— 0

10. -4

Section Quiz 1.2

1. (a) 125 (b) 5 (c) $\frac{1}{125}$ (d) 1 (e) 0 **2.** (a) 4 (b) -4 (c) 3 (d) -3 (e) -5 **3.** (a) 5 (b) 0 (c) 3 (d) 3 (e) $2\frac{5}{9}$ **4.** (a) 5 (b) 0 (c) Not defined (d) Not defined (e) Not real **5.** (a) 7 (b) 3 (c) 10
(d) $5, 5$ (e) $5, 5$ **6.** $\dfrac{y}{x^2}$, $x \neq 0$, $y \neq 0$ **7.** $\dfrac{1}{x^2 y^5}$, $x \neq 0$, $y \neq 0$ **8.** $\dfrac{y^2}{x^3}$, $x \neq 0$ **9.** $\dfrac{9x^6}{16y^8}$, $y \neq 0$
10. (a) 1 (b) $-i$

Section Quiz 1.3

1. (a) $\sqrt[4]{5^3}$, $(\sqrt[4]{5})^3$ (b) $5^{2/3}$ **2.** (a) 16 (b) $\frac{1}{16}$ (c) 16 (d) 8 (e) 0 **3.** (a) $\dfrac{y^2}{x^{2/3}}$ (b) $\dfrac{1}{x^{1/4}y^2}$ **4.** (a) 12

(b) $\dfrac{-3}{5}$ **5.** (a) $2xy^2$ (b) $15xy$ (c) $2y$

Section Quiz 1.4

1. (a) No (b) No (c) Yes (d) Yes (e) Yes **2.** (a) 4 (b) $-5x^3 - 2x^2 - 3x + 7$ **3.** (a) -9 (b) 4
(c) -9 **4.** $2x^3 + 4x^2 - 9x - 13$ **5.** $2x^3 + 2x^2 - x + 1$ **6.** $2x^5 - 5x^4 - 31x^3 - 7x^2 + 59x + 42$
7. $2x + 11 + \dfrac{53x + 71}{x^2 - 4x - 7}$ **8.** $3x^2 + 6xh + 3h^2 - 2x - 2h - 5$ **9.** $2x^2 + 5x - \dfrac{6}{x - 1}$
10. (a) $5 - 2i$ (b) $-1 + 12i$ (c) $41 + i$

Section Quiz 1.5

1. (a) $x^2 - y^2$ (b) $x^2 + 2xy + y^2$ (c) $x^3 - y^3$ (d) $3x^2 - 19x - 14$ (e) $-6x^3 + 2x^2 - 2x$
2. $(3x - 2)^2$ **3.** $(3x + 2)(3x - 2)$ **4.** $(3x + 2)(9x^2 - 6x + 4)$ **5.** $(x + 3)^2$ **6.** Not factorable
7. $(3x + 4)(3x - 2)$ **8.** $(a + b)(2 - 3a)$ **9.** $2a^2 b(a^3 - 3ab + 7)$ **10.** 36, $(x - 6)^2$

Section Quiz 1.6

1. (a) -1 (b) $a^2 - 4a + 3$ 2. $\dfrac{x + 3}{2(x + 2)}$ 3. $(x - 5)^3(x + 5)^2(x + 1)$ 4. $\dfrac{5x + 10}{7x^2}$ 5. $\dfrac{x - 3}{7x + 7}$

6. $\dfrac{5x^3 + 12x^2 - 2x}{x^3 + 4x^2 + x - 6}$ 7. $\dfrac{3x + 7}{x^2 + 2x - 3}$ 8. $3x^4 - 2x^3$ 9. $\dfrac{3x^2 - 5x}{7x^3 - 1}$ 10. (a) $\dfrac{3(\sqrt{5} + 1)}{4}$

(b) $\dfrac{-29}{53} + \dfrac{31}{53}i$

Section Quiz 1.7

1. $\dfrac{9}{8}$ 2. $\dfrac{-2}{13}$ 3. $0, 1, -3$ 4. $B = \dfrac{A - CA}{C}$ 5. (a) 2 (b) 0 6. (a) 1 (b) 2 7. (a) 0, 3

(b) $5, -2$ 8. $\dfrac{2 \pm \sqrt{5}}{3}$ 9. $2, \dfrac{-1}{5}$ 10. $\sqrt{9}\ \sqrt{(x - 1)^2 - 2^2}, h = 1, k = 2$

Section Quiz 1.8

1. $8, -3$ 2. $\dfrac{x^2}{2^2} + \dfrac{y^2}{1^2} = 1; a = 2, b = 1$ 3. 3 4. $5, \dfrac{-1}{3}$ ⟶ $-\frac{1}{3}\ 0 \quad\quad 5$

5. (a) $\pm\sqrt{5}$ ⟶ $-\sqrt{5}\ 0\ \sqrt{5}$ (b) $-4, 3$ ⟶ $-4\quad 0\quad 3$

Section Quiz 1.9

1. $x \geq -2$ ⟶ $-2\ 0$ 2. $\{x \mid -1 \leq x \leq 4\}$ ⟶ $-1 \quad\quad 4$

3. $\{x \mid x > 2 + \sqrt{3} \text{ or } x < 2 - \sqrt{3}\}.$ ⟶ $2 - \sqrt{3}\quad 2 + \sqrt{3}$ 4. $-1 < x < 6$ ⟶ $-1\ 0 \quad\quad 6$

5. $x \geq 3 \text{ or } x \leq \dfrac{5}{3}$ ⟶ $0\ \frac{5}{3}\ 3$

Section Quiz 1.10

1. $x = 0, y = 2; x = \dfrac{-4}{3}, y = 0; x = 1, y = \dfrac{7}{2}; x = \dfrac{-2}{3}, y = 1$ 2. $x = -2, y = -1$ 3. $x = 3, y = 4$

4. $x = -1, y = 2, z = 3$ 5. $A = 3, B = 4, C = 0$

Section Quiz 2.1

1. $\{12, 3, 4, 7\}$ 2. $5, 3, a^2 - 3a + 5$ 3. It is a function; domain of $f = \{\frac{1}{3}, 1, 0, 2\}; \frac{1}{3} \to -4, 1 \to -2,$
$0 \to -5, 2 \to -2$; range $= \{-4, -2, -5\}$ 4. (a) $\{x \mid x \neq -5\}$ (b) $\{x \mid x \geq -5\}$

5. (a) $-, -$ (b) 6. 7.

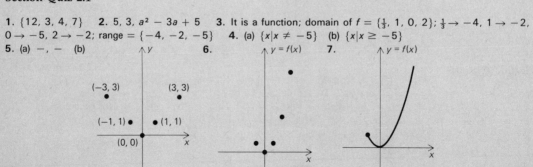

8. (a) Function   (b) Not a function

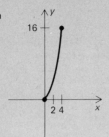

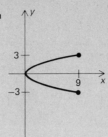

9. (a) Yes (b) No   **10.** $2x + h - 5$

Section Quiz 2.2

1.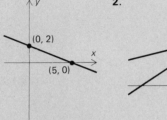

2.

3. (a)    (b)

4. (a) $\dfrac{-7}{4}$ (b) $\dfrac{f(a + h) - f(a)}{h}$ **5.**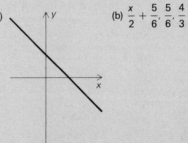

6. $2x - 5y - 11 = 0$ **7.** $y = \dfrac{3x}{2} + 5, \dfrac{3}{2}, 5$

8. (a) $\dfrac{2}{9}$ (b) $\dfrac{-9}{2}$ **9.** $4x - y - 7 = 0$ **10.** (a)   (b) $\dfrac{x}{2} + \dfrac{5}{6}, \dfrac{5}{6}, \dfrac{4}{3}$

Section Quiz 2.3

1. $(0, 2)$, $(0, -2)$, $(5, 0)$, $(-5, 0)$ **2.** (a) Yes (b) No (c) Yes (d) No   **3.** (a) No (b) Yes (c) Yes (d) No
4. (a) No (b) No (c) Yes (d) Yes   **5.** (a) $x = 3$, $x = -3$ (b) No vertical asymptotes (c) $x = 9$
6. (a) $y = 0$ (b) $y = \frac{2}{5}$ (c) No horizontal asymptotes   **7.** $x > 4$ and $y < 0$, $-1 < x < 4$ and $y > 0$,
$-1 > x$ and $y < 0$ **8.** $y < -2$, $y > 2$, $x < -5$, $x > 5$ **9.** Intercepts: $(0, 5)$, $(0, -5)$, $(-2, 0)$, $(2, 0)$;
Symmetric with respect to the x-axis, the y-axis, and the origin; No asymptotes; Excluded areas are $y > 5$,
$y < -5$, $x > 2$, $x < -2$.

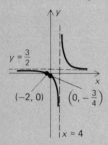

10. Intercepts: $(0, \frac{-3}{4})$, $(-2, 0)$; Asymptotes: $x = 4$, $y = \frac{3}{2}$; Excluded areas: $x > 4$ and $y \le 0$, $-2 < x < 4$ and
$y \ge 0$, $-2 > x$ and $y \le 0$, $y > \frac{3}{2}$ and $x \le 0$, $\frac{-3}{4} < y < \frac{3}{2}$ and $x \ge 0$, $\frac{-3}{4} > y$ and $x \le 0$; Not symmetric with
respect to the origin, the x-axis, or the y-axis.

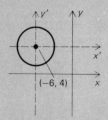

Section Quiz 2.4

1. (a) 5 (b) $(4, 3)$ **2.** $(x - 2)^2 + (y + 5)^2 = 49$ **3.** Center $(-4, 1)$, radius 6
4. (a) $(x - 7)^2 + (y + 2)^2 = 106$ (b) $(x - 1)^2 + (y + 2)^2 = -9$; no graph

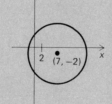

5. $(x')^2 + (y')^2 = 9$

Section Quiz 2.5

1. $(0, 0)$, $x = 0$, $(0, 5)$, $y = -5$

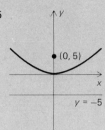

$(0, 5)$

$y = -5$

2. $(0, 0)$, $y = 0$, $(-5, 0)$, $x = 5$

$(-5, 0)$

$x = 5$

3.

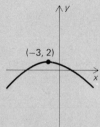

$(-3, 2)$

4.

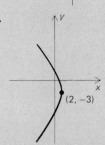

$(2, -3)$

5. $(y - 3)^2 = 20x$

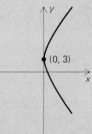

$(0, 3)$

6. $(x - 1)^2 = 8(y + 1)$

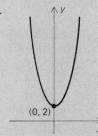

$(1, -1)$

7. (a) Parabola, up   (b) Parabola, down

8.

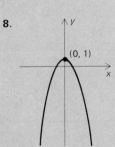

$(0, 1)$

9.

$(0, 2)$

10. 7 seconds

Section Quiz 2.6

1. Center $(0, 0)$

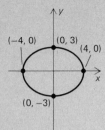

$(-4, 0)$ $(0, 3)$ $(4, 0)$ $(0, -3)$

2. $\dfrac{x'^2}{9} + \dfrac{y'^2}{25} = 1$

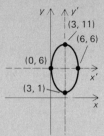

$(3, 11)$ $(6, 6)$ $(0, 6)$ $(3, 1)$

3. $\dfrac{(x - 3)^2}{9} + \dfrac{(y + 4)^2}{4} = 1$; ellipse **4.** $\dfrac{(x - 3)^2}{9} - \dfrac{(y + 4)^2}{9} = 1$; hyperbola **5.** Center $(0, 0)$; vertices:

$(-5, 0)$, $(5, 0)$; asymptotes: $y = \pm \dfrac{3x}{5}$

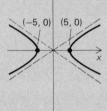

$(-5, 0)$ $(5, 0)$

6. $\dfrac{(y')^2}{25} - \dfrac{(x')^2}{9} = 1$; asymptotes: $y' = \dfrac{-5}{3}x'$, $y' = \dfrac{5}{3}x'$; asymptotes: $y - 6 = \dfrac{5}{3}(x - 3)$, $y - 6 = \dfrac{-5}{3}(x - 3)$

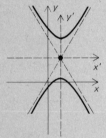

7. (a) Ellipse (b) Circle **8.** (a) Hyperbola (b) Two intersecting lines **9.** (a) No graph (b) Parabola
10. (a) Hyperbola (b) Parabola

Section Quiz 2.7

1. (a) Not a function

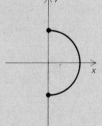

(b) Function

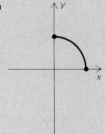

2. (a) $\{(x, y)|x = 2y + 7\}$, function   (b) $\{(x, y)|x = 8y^2\}$, not a function   **3.** (a) Yes   (b) No   **4.** (a) No
(b) Yes   (c) No   **5.** Domain of $f^{-1} = \{x|x \geq 3\}$, range of $f^{-1} = \{y|y \geq 0\}$, $f^{-1}(x) = \dfrac{x}{2} - \dfrac{3}{2}$

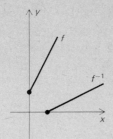

Section Quiz 2.8

1.

2.

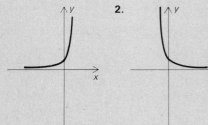

3. Domain = $\{x|x > 0\}$, range = set of all real numbers, $x = 3^y$

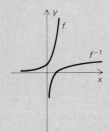

4. (a) 6   (b) 0   (c) 1   (d) -1 (e) -3
5.

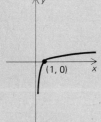

(1, 0)

Section Quiz 2.9

1. (a) $2^b = a$ (b) $\log_3 y = 5$ (c) $3 = e^{\ln 3}$ (d) $\dfrac{\ln 37}{\ln 10}$ **2.** (a) 0 (b) 1 (c) 5 (e) 2 **3.** (a) 2.8451

(b) $0.8451 - 1$ **4.** (a) 1.5441 (b) 0.1461 **5.** (a) 2.5353 (b) 0.2330

Section Quiz 2.10

1. (a) $\{x \mid 0 \le x \le 3\}$ (b) $\{x \mid 0 \le x < 3\}$ **2.** (a) 33 (b) -29 (c) 62 **3.** (a) $\dfrac{2}{31}$ (b) $2\sqrt[3]{4}$ (c) 6

4. (a) $(f + g)(x) = \sqrt{x} + x^2 + 9$, $\{x \mid x \ge 0\}$ (b) $(f - g)(x) = \sqrt{x} - x^2 - 9$, $\{x \mid x \ge 0\}$

(c) $fg(x) = x^2\sqrt{x} + 9\sqrt{x}$, $\{x \mid x \ge 0\}$ **5.** (a) $\dfrac{f}{g}(x) = \dfrac{\sqrt{x}}{x^2 + 9}$, $\{x \mid x \ge 0\}$ (b) $f \circ g(x) = \sqrt{x^2 + 9}$, set of all real

numbers (c) $g \circ f(x) = x + 9$, set of all real numbers

Section Quiz 3.1

1. 6 **2.** $x = \dfrac{-b \pm \sqrt{b^2 - 4ac}}{2a}$ **3.** $4x^3 - 3x^2 + 5x - 30 = (x - 2)(4x^2 + 5x + 15) + 0$, so 2 *is* a root.

4. $x(x - 1)(x - 3)(x + 4) = 0$ **5.** 0, multiplicity 2; 3, multiplicity 4; -1, multiplicity 1

6. $x^3(x + 4)^2(x - 6) = 0$ **7.** 4 **8.** 5 **9.** $(2 - i)^2 - 4(2 - i) + 5 = 4 - 4i - 1 - 8 + 4i + 5 = 0$

$(2 + i)^2 - 4(2 + i) + 5 = 4 + 4i - 1 - 8 - 4i + 5 = 0$ **10.** $(x - (3 - 4i))(x - (3 + 4i)) = x^2 - 6x + 25$

11. $1^3 - 5(1)^2 + 11(1) - 7 = 1 - 5 + 11 - 7 = 0$; $x^3 - 5x^2 + 11x - 7 = (x - 1)(x^2 - 4x + 7)$, so the

roots are 1, $2 + \sqrt{3}i$, $2 - \sqrt{3}i$. **12.** ± 1, $\pm\frac{1}{7}$, $\pm\frac{3}{7}$, $\pm\frac{5}{7}$, $\pm\frac{15}{7}$, ± 3, ± 5, ± 15 **13.** 3, -2, $\frac{1}{2}$

14. 2, $3i$, $-3i$ **15.** Let $P(x) = 5x^3 + 2x^2 - 16x + 4 = 0$. $P(0) = 4$ (positive), and $P(1) = -5$ (negative),

so there is at least one root between 0 and 1. First approximation, 0.4.

Section Quiz 3.2

1. 3 rows; 4 columns **2.** 9 **3.** 2, 7, 5

4. $\begin{pmatrix} 1 & 0 & 6 \\ 0 & 1 & 4 \\ 0 & 0 & 5 \end{pmatrix}$ **5.** $\begin{pmatrix} 4 & -3 & 7 \\ -4 & 4 & -6 \\ 5 & 9 & -8 \end{pmatrix}$ **6.** $\begin{pmatrix} 1 & -1 & 2 \\ 0 & 7 & 9 \\ 0 & 1 & 8 \end{pmatrix}$

7. $x = 2$, $y = -3$ **8.** $x = 1$, $y = -2$, $z = -3$ **9.** Not exactly one solution (no solution)

10. Not exactly one solution (unlimited number of solutions)

Section Quiz 3.3

1. Row 3, column 2 **2.** Row 2, column 3; minus position **3.** $\begin{vmatrix} 3 & 4 \\ 0 & 8 \end{vmatrix}$ **4.** 31

5. $3(-7 - 30) - (-4 - 10) - 2(24 - 14) = 105$ **6.** (a) 0 (b) 0 **7.** (a) 1 (b) 2 **8.** (a) 1 (b) 1

9. $x = -2$, $y = -4$ **10.** $x = \dfrac{8}{7}$, $y = \dfrac{19}{7}$, $z = \dfrac{4}{7}$

Section Quiz 3.4

1. 3 rows, 5 columns **2.** $\begin{pmatrix} 2 & 3 \\ 22 & -16 \end{pmatrix}$ **3.** $\begin{pmatrix} 5 & 3 \\ 19 & 8 \end{pmatrix}$ **4.** $\begin{pmatrix} 3x + y \\ -2x + 4y \end{pmatrix}$

5. $\begin{pmatrix} 1 & 0 & 0 \\ 0 & 1 & 0 \\ 0 & 0 & 1 \end{pmatrix}$ **6.** $\begin{pmatrix} -30 & 45 \\ -6 & 63 \end{pmatrix}$ **7.** $\begin{pmatrix} 2 & -3 \\ 4 & -1 \end{pmatrix}\begin{pmatrix} x \\ y \end{pmatrix} = \begin{pmatrix} 5 \\ -1 \end{pmatrix}$

8. $3x - 2y = 5$
$\quad\;\; x - \;\; y = 4$

9. $\begin{pmatrix} 1 & 0 & 0 \\ 0 & 1 & 0 \\ 0 & 0 & 1 \end{pmatrix}$

10. $\begin{pmatrix} 3 & 4 \\ -2 & 5 \end{pmatrix}, \begin{pmatrix} 3 & 4 \\ -2 & 5 \end{pmatrix}$

11. $\dfrac{1}{17}\begin{pmatrix} 1 & -5 \\ 3 & 2 \end{pmatrix}\begin{pmatrix} 2 & 5 \\ -3 & 1 \end{pmatrix} = I, \quad \begin{pmatrix} 2 & 5 \\ -3 & 1 \end{pmatrix}\dfrac{1}{17}\begin{pmatrix} 1 & -5 \\ 3 & 2 \end{pmatrix} = I$

12. $\begin{vmatrix} 3 & 2 \\ 6 & 4 \end{vmatrix} = 0$, not nonsingular

13. $\begin{pmatrix} 2 & 3 & 9 \\ 7 & 4 & 8 \\ 6 & 5 & 1 \end{pmatrix}$

14. $\dfrac{1}{-12}\begin{pmatrix} 0 & 0 & -3 \\ -4 & 0 & 1 \\ 0 & 12 & 0 \end{pmatrix}$

15. $\begin{pmatrix} 5 & -2 \\ 1 & 3 \end{pmatrix}\begin{pmatrix} x \\ y \end{pmatrix} = \begin{pmatrix} 1 \\ -10 \end{pmatrix}$

$\begin{pmatrix} x \\ y \end{pmatrix} = \dfrac{1}{17}\begin{pmatrix} 3 & 2 \\ -1 & 5 \end{pmatrix}\begin{pmatrix} 1 \\ -10 \end{pmatrix}$

$\begin{pmatrix} x \\ y \end{pmatrix} = \begin{pmatrix} -1 \\ -3 \end{pmatrix}$

$x = -1, y = -3$

Section Quiz 3.5

1. $\dfrac{A}{x - 4} + \dfrac{B}{x + 5} + \dfrac{C}{(x + 5)^2}$

2. $\dfrac{A}{x - 3} + \dfrac{Bx + C}{x^2 - 4x + 29}$

3. $\dfrac{A}{x - 3} + \dfrac{Bx + C}{x^2 - 4x + 29} + \dfrac{Dx + E}{(x^2 - 4x + 29)^2}$

4. $\dfrac{2}{x - 3} + \dfrac{3}{(x - 3)^2}$

5. $\dfrac{-3}{x} - \dfrac{1}{(x + 1)} + \dfrac{4}{x - 5}$

Section Quiz 3.6

1. 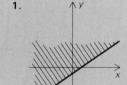   **2.** 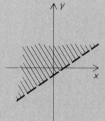   **3.** 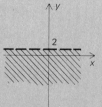   **4.**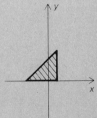

5. (a) $x \geq 2, y \geq 1, x + y \leq 20$ (b)

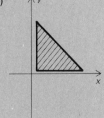

(c) $69, $443, $494   (d) 2 rows of lettuce, 18 rows of beans

Section Quiz 3.7

1. (a) $\{2n + 3\}$ (b) $1, 4, 7, \ldots, 3n - 2, \ldots$ **2.** $\dfrac{1}{3}, \dfrac{1}{9}, \dfrac{1}{27}, \dfrac{1}{81}; \dfrac{1}{3^n}; 0$ **3.** Arithmetic, 4

4. Geometric, 3 **5.** $\dfrac{5^{n+1} - 5}{4}$ **6.** $\dfrac{3n(n + 1)}{2}$ **7.** Converges, $\frac{3}{2}$ **8.** Does not converge

9. (a) $\dfrac{1}{3}, \dfrac{1}{2}, \dfrac{3}{5}, \dfrac{2}{3}; \dfrac{n}{n + 2}$ (b) $\dfrac{1}{3}, \dfrac{1}{3} + \dfrac{1}{2}, \dfrac{1}{3} + \dfrac{1}{2} + \dfrac{3}{5}$ **10.** (a) $\dfrac{1 - (0.1)^n}{9}$ (b) $\dfrac{1 - (0.1)^{10}}{9}$ (c) $\dfrac{1}{9}$

Section Quiz 3.8

1. Part 1: $5 = \frac{5}{4}(5^1 - 1)$ since $5 = \dfrac{5 \cdot 4}{4}$

Part 2: Suppose $5 + 5^2 + 5^3 + \cdots + 5^k = \frac{5}{4}(5^k - 1)$. Then

$$5 + 5^2 + 5^3 + \cdots + 5^k + 5^{k+1} = \frac{5}{4}(5^k - 1) + 5^{k+1}$$
$$= \frac{5(5^k - 1) + 4 \cdot 5^{k+1}}{4}$$
$$= \frac{5(5^k - 1 + 4 \cdot 5^k)}{4}$$
$$= \frac{5(5 \cdot 5^k - 1)}{4}$$
$$= \tfrac{5}{4}(5^{k+1} - 1)$$

Check: $S_n = \dfrac{5 - 5 \cdot 5^n}{1 - 5} = \dfrac{5(1 - 5^n)}{-4} = \frac{5}{4}(5^n - 1)$

2. Part 1: $3 = \frac{3}{2}(1 + 1^2)$ since $3 = \frac{3}{2} \cdot 2$

Part 2: Suppose $3 + 6 + 9 + \cdots + 3k = \frac{3}{2}(k + k^2)$. Then

$$3 + 6 + 9 + \cdots + 3k + 3(k + 1) = \tfrac{3}{2}(k + k^2) + 3(k + 1)$$
$$= \frac{3(k + k^2) + 6(k + 1)}{2}$$
$$= \frac{3k + 3k^2 + 6k + 6}{2}$$
$$= \frac{3(k^2 + 3k + 2)}{2}$$
$$= \frac{3((k^2 + 2k + 1) + (k + 1))}{2}$$
$$= \frac{3[(k + 1)^2 + (k + 1)]}{2}$$
$$= \frac{3[(k + 1) + (k + 1)^2]}{2}$$

Check: $\ell = 3 + (n - 1)3$

$$S_n = \frac{n(3 + 3 + (n - 1)3)}{2} = \frac{3(n + n^2)}{2}$$

3. Part 1: $2^3 > 3 + 2$ since $8 > 5$.

Part 2: $2^{k+1} = 2^k \cdot 2$
$$> (k + 2) \cdot 2$$
$$= 2k + 4$$
$$= k + k + 4$$
$$\geq k + 3 + 4$$
$$> k + 3$$
$$= (k + 1) + 2$$
$$2^{k+1} > (k + 1) + 2$$

4. Part 1: $3 = \dfrac{1(1 + 1)(2 \cdot 1 + 1)}{2}$ since $3 = \dfrac{1(2)(3)}{3}$

Part 2: If $3 + 12 + \cdots + 3k^2 = \dfrac{k(k + 1)(2k + 1)}{2}$, then

$$3 + 12 + \cdots + 3k^2 + 3(k + 1)^2 = \frac{k(k + 1)(2k + 1)}{2} + 3(k + 1)^2$$

$$= \frac{(k + 1)[k(2k + 1) + 6(k + 1)]}{2} = \frac{(k + 1)(2k^2 + 7k + 6)}{2}$$

$$= \frac{(k + 1)(k + 2)(2k + 3)}{2} = \frac{(k + 1)(k + 2)[2(k + 1) + 1]}{2}.$$

5. Part 1: 2 is a factor of $1^2 + 3 \cdot 1$ since 2 is a factor of 4

Part 2: $(k + 1)^2 + 3(k + 1) = k^2 + 5k + 4$
$$= k^2 + 5k + 4 + [k^2 + 3k - (k^2 + 3k)]$$
$$= 2k^2 + 8k + 4 - (k^2 + 3k)$$
$$= 2(k^2 + 4k + 2) - (k^2 + 3k)$$

2 is a factor of $2(k^2 + 4k + 2)$, so if 2 is a factor of $k^2 + 3k$, then 2 is a factor of $(k + 1)^2 + 3(k + 1)$.

Section Quiz 3.9

1. $a^7 + (\ \ )a^6b + (\ \ )a^5b^2 + (\ \ )a^4b^3 + (\ \ )a^3b^4 + (\ \ )a^2b^5 + (\ \ )ab^6 + b^7$
2. $x^5 + 5x^4y + 10x^3y^2 + 10x^2y^3 + 5xy^4 + y^5$ **3.** $a^{10} + 10a^9b + 45a^8b^2 + 120a^7b^3$ **4.** 495, $495x^8y^4$ **5.** 60

Section Quiz 3.10

1. *qrst, qrts, qstr, qsrt, qtrs, qtsr, rqst, rqts, rsqt, rstq, rtqs, rtsq, srqt, srtq, sqrt, sqtr, srqt, srtq, tqrs, tqsr, tsqr, tsrq, trqs, trsq*
2. (a) 720 (b) 42
3. (a) *ab, ac, ad, ae, ba, bc, bd, be, ca, cb, cd, ce, da, db, dc, de, ea, eb, ec, ed*
(b) *ab, ac, ad, ae, bc, bd, be, cd, ce, de*
4. (a) 1 (b) 21 **5.** (a) 3024 (b) 126 **6.** 60 **7.** $_1C_1$; $_2C_1$, $_2C_2$; $_3C_1$, $_3C_2$, $_3C_3$
8. (a) 270725 (b) 24 **9.** 8 **10.** (a) 792 (b) 48

Section Quiz 3.11

1. $\frac{1}{6}$ **2.** $\frac{1}{13}$ **3.** 1 **4.** $\frac{35}{132}$ **5.** $\frac{35}{144}$ **6.** $P(E_2 \text{ after } E_1) = \frac{5}{11}$ $P(E_2) = \frac{5}{12}$ **7.** $\frac{1}{6}$ **8.** $\frac{7}{216}$
9. (a) 1 to 35 (b) 35 to 1 **10.** $\frac{24}{91}$

Part 1 Final Test

1. (a) $\{0, 1, 2, 3, \ldots\}$ (b) $\{x \mid x > 0\}$

2. (a) $-5, 0, |-5|, 0.5, \dfrac{1}{5}, 2.\overline{51}, \dfrac{22}{7}, 3.14, \dfrac{-4}{5}$ (b) $\sqrt{5}, \pi, \sqrt{2}$ **3.** (a) 243 (b) 3 (c) 1 (d) $\dfrac{1}{243}$ (e) 0

4. (a) 3 (b) Not defined   (c) Not real   (d) 5 (e) $\dfrac{3}{2}$ **5.** (a) $\dfrac{x^6}{y^7}, y \neq 0$ (b) $\dfrac{9x^4}{y^4}, y \neq 0, x \neq 0$

6. (a) 16 (b) 5 (c) $\frac{1}{2}$ (d) 9 (e) 0 **7.** (a) y (b) xy **8.** (a) $2x^3y$ (b) $\dfrac{x^2}{2y^2}$

9. (a) $-2x^3 - 4x^2 + 3x - 4$ (b) $3x^3 - 23x^2 + 15x - 7$ **10.** $4x^2 - 14x + 43 - \dfrac{134}{x + 3}$

11. $5x^2 + 10xh + 5h^2 - 3x - 3h - 7$ **12.** (a) $(5x - 6)(5x + 6)$ (b) $(5x + 2)^2$
(c) $(3x - 2)(9x^2 + 6x + 4)$ (d) $(5x + 2)(5x - 4)$ **13.** $169; (x - 13)^2$ **14.** (a) $(x - 2)(x - 3)$
(b) $\dfrac{3x - 6}{x^2 - 3x}$ **15.** (a) $\dfrac{-2x^2 - 2x - 3}{x(x - 1)(x + 3)}$ (b) $\dfrac{x^2 - 5x + 4}{x + 4}$ **16.** $\dfrac{3x^2 - 2x}{2 + x^3}$ **17.** (a) $x = \dfrac{7}{9}$ (b) $x = 0, -2, 5$

18. (a) $x = 0, \dfrac{5}{3}$ (b) $x = \dfrac{5 \pm \sqrt{2}}{3}$ **19.** $x = -2$ **20.** $x = 5$ **21.** (a) $x = 3, \dfrac{-3}{5}$

(b) $x = 0, 3, 1, 2$ **22.** $x \leq \dfrac{-8}{9}$

23. $\{x \mid -2 \leq x \leq 5\}$ **24.** $\dfrac{-1}{8} \leq x \leq \dfrac{7}{8}$

25. $x = \dfrac{-1}{5}, y = 2$

Part 2 Final Test

1. (a) $\{1, 3, 4, 6\}$ (b)   **2.** (a) $\{x \mid x \geq 4\}$ (b) $\{x \mid x \neq 0\}$

3.   **4.**   **5.** (a) $\dfrac{-3}{5}$ (b) $\dfrac{5}{3}$ (c) $3x + 5y - 26 = 0$

6. (a) $3x - y - 1 = 0$ (b) $\dfrac{5}{3}; \left(0, \dfrac{-7}{3}\right)$ (c)

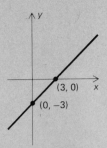

(3, 0) x
(0, −3)

7. (a) (1, 0), (−1, 0); Symmetric with respect to the x-axis, the y-axis, and the origin; No vertical or horizontal asymptotes (b) (0, 0); Not symmetric with respect to the x-axis or the y-axis, symmetric with respect to the origin; Vertical asymptotes: $x = 2$, $x = -2$, horizontal asymptote: $y = 0$

8.

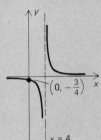

$\left(0, -\dfrac{3}{4}\right)$ x
$x = 4$

9. (a) 5 (b) $\left(\dfrac{-1}{2}, 5\right)$ **10.** (a) $(x - 3)^2 + (y + 7)^2 = 36$ (b) Center (5, −2); radius 9

11. $x'^2 + y'^2 = 4$

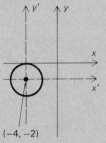

x
x'
(−4, −2)

12. (a)

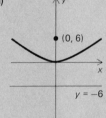

(0, 6)
x
$y = -6$

(b)

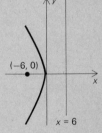

(−6, 0)
x
$x = 6$

13. $(y - 3)^2 = 24(x + 2)$

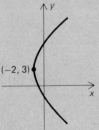

(−2, 3)
x

14.

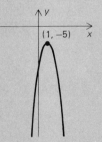

(1, −5) x

15. Major axis 6, minor axis 4

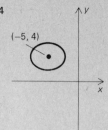

16. Asymptotes $y = \pm \dfrac{3}{2}(x - 5)$

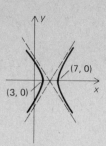

17. (a) Line (b) Parabola (c) No graph (d) Circle (e) Point   **18.** (a) Not one-to-one (b) One-to-one

19. Domain of $f^{-1} = \{x \mid -1 \le x \le 5\}$; $f^{-1}(x) = \dfrac{x + 1}{3}$

20. **21.**

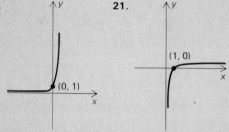

22. (a) 0 (b) 1 (c) 9 (d) -4 (e) -2 **23.** (a) 2.4771 (b) 0.6021 $- 3$ (c) 1.0792 (d) 0.1250
(e) 0.2007 (f) 1.9084   **24.** (a) $\sqrt{x - 3}$ (b) $\sqrt{x} - 3$ **25.** (a) $\{x \mid x \ge 0\}$ (b) $\{x \mid x \ge 0 \text{ and } x \ne 3\}$
(c) $\{x \mid x \ge 3\}$ (d) $\{x \mid x \ge 0\}$

Part 3 Final Test

1. Remainder $= 0$; 4 is a root   **2.** $x(x - 2)(x + 3)(x + 1) = 0$ **3.** (a) 4 (b) $\dfrac{2}{3}, \dfrac{-1}{4}$ **4.** $x^2 - 4x + 13$

5. $P(2) = 4 > 0$, $P(3) = -1 < 0$ **6.** Let $P(x) = 4x^4 - 8x^3 + 17x^2 + 6x - 15$. $P(0) = -15$ (negative),
and $P(1) = 4$ (positive), so there is at least one root between 0 and 1. First approximation, $\frac{15}{19}$

7. $x = 2, y = -1, z = -3$ **8.** Not exactly one solution (no solution)   **9.** (a) 23 (b) 42

10. $x = 0, y = 0, z = 1$ **11.** (a) $\begin{pmatrix} 8 & 12 \\ 2 & 7 \end{pmatrix}$ (b) $\begin{pmatrix} 2x + 3y \\ -4x + y \end{pmatrix}$ (c) $\begin{pmatrix} 1 & 0 & 0 \\ 0 & 1 & 0 \\ 0 & 0 & 1 \end{pmatrix}$ **12.** $\dfrac{1}{3}\begin{pmatrix} 3 & 0 & -6 \\ 1 & 1 & -2 \\ 0 & 0 & 3 \end{pmatrix}$

13. $\begin{pmatrix} 3 & -2 \\ -1 & 1 \end{pmatrix}\begin{pmatrix} x \\ y \end{pmatrix} = \begin{pmatrix} 5 \\ -1 \end{pmatrix}$

$\begin{pmatrix} x \\ y \end{pmatrix} = \begin{pmatrix} 1 & 2 \\ 1 & 3 \end{pmatrix}\begin{pmatrix} 5 \\ -1 \end{pmatrix}$

$\begin{pmatrix} x \\ y \end{pmatrix} = \begin{pmatrix} 3 \\ 2 \end{pmatrix}$, $x = 3, y = 2$

14. $\dfrac{-4}{x} + \dfrac{2}{x - 1} + \dfrac{3}{x + 5}$ **15.** $\dfrac{A}{x + 1} + \dfrac{B}{(x + 1)^2} + \dfrac{Cx + D}{x^2 + 5} + \dfrac{Ex + F}{(x^2 + 5)^2}$

16.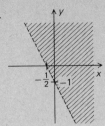

17.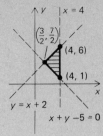

18. $\dfrac{1}{5}, \dfrac{1}{25}, \dfrac{1}{125}, \dfrac{1}{625}, \ldots, \left(\dfrac{1}{5}\right)^n, \ldots; 0$ **19.** (a) Geometric, $S_n = \dfrac{7 - 7^{n+1}}{-6}$

(b) Arithmetic, $S_n = \dfrac{7n + 7n^2}{2}$

20. (a) $\dfrac{1}{4} + \dfrac{2}{5} + \dfrac{1}{2} + \dfrac{4}{7} + \dfrac{5}{8}$ (b) $1 + 8 + 27 + 64 + \cdots + n^3 + \cdots$ **21.** (a) $1 - \dfrac{1}{2^n}$ (b) Converges; $\dfrac{5}{4}$

22. *Part I* $4 = 1^2(1 + 1)^2$ since $4 = 1 \times 4$.

Part II If $4 + 32 + \cdots + 4k^3 = k^2(k + 1)^2$, then
$$4 + 32 + \cdots + 4k^3 + 4(k + 1)^3 = k^2(k + 1)^2 + 4(k + 1)^3$$
$$= (k + 1)^2(k^2 + 4k + 4)$$
$$= (k + 1)^2(k + 2)^2$$
$$= (k + 1)^2(k + 1 + 1)^2$$

23. *Part I* $5^2 > 2 + 7$ since $25 > 9$.

Part II If $5^k > k + 5$, then
$$5 \times 5^k > 5(k + 5)$$
$$5^{k+1} > 5k + 25$$
$$= k + 1 + 5 + 4k + 19$$
$$> k + 1 + 5$$

24. (a) $a^8 + (\ \ )a^7b + (\ \ )a^6b^2 + (\ \ )a^5b^3 + (\ \ )a^4b^4 + (\ \ )a^3b^5 + (\ \ )a^2b^6 + (\ \ )ab^7 + b^8$
(b) $x^4 + 8x^3y + 24x^2y^2 + 32xy^3 + 16y^4$ **25.** (a) $a^9 + 9a^8b + 36a^7b^2 + 84a^6b^3$
(b) $455, 455x^{12}y^3$ **26.** (a) 30,240   (b) 252   (c) 37,800   **27.** (a) 593,775   (b) 427,518,000

28. (a) $\dfrac{3}{14}$ (b) 6 to 22   **29.** (a) $\dfrac{3}{54145}$ (b) $\dfrac{5}{144}$ **30.** $\dfrac{177}{182}$

Part 1 Review Exercises

1. (a) $-6, 0, 1, 5, \frac{2}{7}, 7, \frac{1}{3}, \frac{-3}{4}, 2.\overline{7}, \frac{22}{7}, 3.14$ (b) $\pi, \sqrt{2}, \sqrt{7}$ (c) $-6, 0, 1, 5, \frac{2}{7}, 7, \frac{1}{3}, \frac{-3}{4}, 2.\overline{7}, \pi, \frac{22}{7}, 3.14,$
$\sqrt{2}, \sqrt{7}$ (d) $-6, 0, 1, 5, 7$ (e) $0, 1, 5, 7$ (f) $-6, 0$ (g) $5, 7$ **2.** (a) Irrational, nonterminating and
nonrepeating   (b) Irrational; nonterminating and nonrepeating   (c) Rational; repeating   (d) Rational, terminating
3. (a) 8   (b) 8   (c) 0   (d) -6 **4.** (a) 216   (b) $\frac{1}{8}$ (c) 1   (d) 1   **5.** (a) 6   (b) 6   (c) -6 (d) 2
6. (a) $\frac{5}{3}$ (b) 1   (c) 0   (d) 1   **7.** (a) 2   (b) $\frac{1}{4}$ (c) 7   (d) 0   **8.** $\sqrt[5]{23^4}, \left(\sqrt[5]{23}\right)^4$ **9.** $2^{3/5}$ **10.** a^{10} **11.** $\dfrac{1}{a^4}$

12. $\dfrac{1}{a^{21}}$ **13.** $\dfrac{1}{x^9y^{12}}$ **14.** $\dfrac{x^{1/5}}{y^{1/5}}$ **15.** (a) i (b) -1 **16.** (a) -10 (b) $\frac{4}{3}$ **17.** (a) $3x^2y^3$ (b) $\dfrac{2x}{7y}$ (c) y

18. (a) 1   (b) 4   **19.** (a) $2x^3 + 3x - 6$ (b) $2x^3 - 2x^2 + 5x - 4$ (c) $2x^5 - 3x^4 + 3x^3 - 8x^2 + x + 5$

(d) $2x + 1 + \dfrac{7x - 4}{x^2 - x - 1}$ **20.** (a) $5x^2 + 10xh + 5h^2 - 3x - 3h - 7$ (b) $10xh + 5h^2 - 3h$

21.
$$-2\overline{)\begin{array}{cccc} 1 & -3 & 4 & -2 \\ & -2 & 10 & -28 \end{array}}$$
$$\begin{array}{cccc} \hline 1 & -5 & 14 & -30 \end{array}$$
$$x^2 - 5x + 14 + \frac{-30}{x+2}$$

22. (a) $2 + 13i$ (b) $4 + i$ (c) $-45 + 11i$ **23.** $(5a - 4b)(5a + 4b)$ **24.** $(x - 3y)(x^2 + 3xy + 9y^2)$
25. $(3x - 5)^2$ **26.** $(3x - 5)(3x + 2)$ **27.** $(x + y)(y + 3)$
28. $(x + a + b)(x^2 - ax - bx + a^2 + 2ab + b^2)$ **29.** Not factorable   **30.** $\frac{225}{4}, (x - \frac{15}{2})^2$ **31.** (a) $+1$
(b) $x^2 - 4$

32. $\dfrac{3(x + 2)}{(x - 3)^2}$ **33.** $\dfrac{2(x - 3)}{x^2(x + 2)}$ **34.** $\dfrac{x + 3}{2x^4(x - 2)}$ **35.** $\dfrac{3x^2 - 3x - 2}{x(x - 1)^2}$ **36.** $\dfrac{2x^2y - 3x}{x^4y^3 + 5y^5}$ **37.** $\dfrac{5\sqrt{5} - 5\sqrt{3}}{2}$

38. $\dfrac{-16}{13} + \dfrac{37}{13}i$ **39.** $P_1 = \dfrac{PV}{V_1}$ **40.** $x = -\frac{2}{23}$ **41.** $x = 0, -4, 7$ **42.** $x = 0, 7$ **43.** $x = 7, -1$

44. $x = 7 \pm \sqrt{2}$ **45.** $x = \pm 3, \pm 1$ **46.** $x = 0, 3, 2, 1$ **47.** $x = -7, 1, -3 \pm 2\sqrt{2}$ **48.** $x = -1$

49. $x = 9$ **50.** $x = 7$ **51.** $x = 4, -3$ **52.** $x = \dfrac{1 \pm \sqrt{161}}{2}$ **53.** $x = 0, \pm \sqrt{21}$ **54.** $x = \frac{23}{3}$

55. $x = \pm 4$ **56.** $x = 4, -1$ **57.** $x = \frac{17}{5}, -\frac{1}{5}$ **58.** $x \le -\frac{3}{2}$ ———————●——————
$-\frac{3}{2}$

59. $-1 \le x \le 8$ ——●————————●——
-1 8

60. $x < -9$ or $x > 5$ ————⊕—————————⊕————
-9 5

61. $-\frac{10}{7} \le x \le 2$ ————●——●————
$-\frac{10}{7}$ 2

62. $x \ge \frac{11}{3}$ or $x \le -7$ ————————●————————●————————
-7 $\frac{11}{3}$

63. $x = \frac{1}{7}, y = 3$ **64.** $\dfrac{x^2}{1} - \dfrac{y^2}{2} = 1, a = 1, b = \sqrt{2}$ **65.** $\dfrac{x^2}{9} - \dfrac{y^2}{1} = 1, a = 3, b = 1$ **66.** $\dfrac{x^2}{9} + \dfrac{y^2}{4} = 1,$
$a = 3, b = 2$ **67.** $\dfrac{x^2}{4} + \dfrac{y^2}{1} = 1, a = 2, b = 1$ **68.** $y^2 = 16x, p = 8$ **69.** $x^2 = 24y, p = 12$

70. $y^2 = 8x, p = 4$

Part 2 Review Exercises

1. $\{1, 7\}$

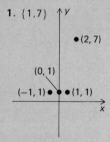

A22

2. 1, 1, 3, $a^2 - a + 1$, $(a + h)^2 - (a + h) + 1$, $a^2 + 2ah + h^2 - a - h + 1$ **3.** (a) Function; domain $= \{1, 2, -2, 3\}$; $1 \to 1$, $2 \to 32$, $-2 \to -32$, $3 \to 243$; range $= \{1, 32, -32, 243\}$ (b) Not a function
4. (a) $\{x \mid x \neq 2 \text{ or } x \neq -2\}$ (b) $\{x \mid x \geq 4\}$ (c) $\{x \mid x \geq 2 \text{ or } x \leq -2\}$ (d) $\{x \mid x > 7\}$ (e) All real numbers
(f) All real numbers

5.

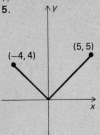

$(-4, 4)$ $(5, 5)$

6.

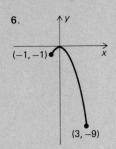

$(-1, -1)$ $(3, -9)$

7.

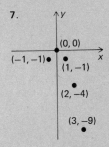

$(0, 0)$ $(-1, -1)$ $(1, -1)$ $(2, -4)$ $(3, -9)$

8. (a) Not a function   (b) Function   (c) Not a function   (d) Function   (e) Function   (f) Not a function
9. $6x + 3h - 1$ **10.** $2x_1 + \Delta x - 3$

11. (a)

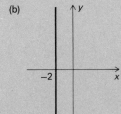

4 $-\frac{8}{3}$

(b)

-2

(c)

4

(d)

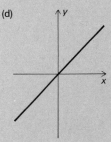

(e)

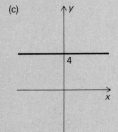

-1 $\frac{1}{3}$

12. $(-\frac{1}{5}, 1)$ **13.** $\dfrac{f(x + h) - f(x)}{h}$ **14.** $-\frac{8}{13}$ **15.** $3x - 8y - 62 = 0$ **16.** $11x + 4y - 13 = 0$
17. $5x + 3y - 15 = 0$ **18.** $2x - 3y - 15 = 0$ **19.** $4x + 3y = 0$ **20.** $y + 1 = 0$ **21.** $x + 1 = 0$
22. $x - y + 2 = 0$ **23.** $x - y - 2 = 0$ **24.** Slope $= \frac{7}{9}$, y-intercept $= \frac{11}{9}$

25.

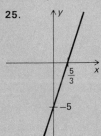

$\frac{5}{3}$ -5

26. (a) $(1, 0), (-1, 0), (0, 7), (0, -7)$ (b) $(1, 0), (-1, 0)$ **27.** (a) Not symmetric with respect to x-axis, symmetric with respect to y-axis, not symmetric with respect to origin   (b) Not symmetric with respect to either axis, symmetric with respect to origin   (c) Not symmetric with respect to x-axis, symmetric with respect to y-axis, not symmetric with respect to origin   (d) Symmetric with respect to both axes and the origin   (e) Not symmetric with respect to either axis, symmetric with respect to origin

28. (a) $x = 7, y = 0$ (b) $x = 7, y = 3$ (c) $x = 7$ (d) $y = 0$ **29.** (a) y positive   (b) y negative   (c) y positive   **30.** $x < 1$

31.

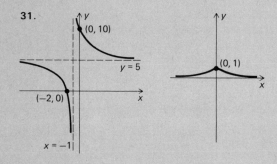

32. (a) 13 (b) $(0, \frac{13}{2})$ (c) $24x + 10y - 65 = 0$ (d) $x^2 + (y - \frac{13}{2})^2 = \frac{169}{4}$ **33.** $(x - 2)^2 + (y - 3)^2 = 13$
34. $(-\frac{3}{2}, -4), \sqrt{11}$ **35.** $\dfrac{x'^2}{25} + \dfrac{y'^2}{36} = 1$

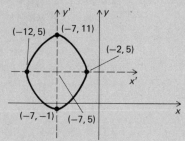

36. (a) Vertex $(0, 0)$; axis $x = 0$; focus $(0, 7)$; directrix $y = -7$

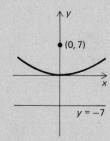

(b) Vertex (0, 0); axis $x = 0$; focus $(0, -7)$; directrix $y = 7$

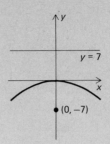

$y = 7$

$(0, -7)$

(c) Vertex (0, 0); axis $y = 0$; focus $(-7, 0)$; directrix $x = 7$

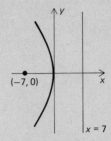

$(-7, 0)$

$x = 7$

(d) Vertex (0, 0); axis $y = 0$; focus $(7, 0)$; directrix $x = -7$

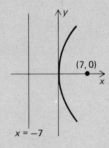

$(7, 0)$

$x = -7$

(e) Vertex $(1, -4)$; axis $x = 1$; focus $(1, 4)$; directrix $y = -12$

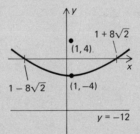

$1 + 8\sqrt{2}$

$(1, 4)$

$1 - 8\sqrt{2}$ $(1, -4)$

$y = -12$

(f) Vertex (1, −4); axis $y = -4$; focus $(-7, -4)$, directrix $x = 9$

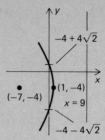

(g) Vertex (2, −1); axis $x = 2$; focus $(2, -8)$; directrix $y = 6$

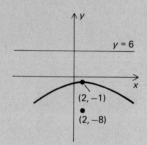

(h) Vertex (2, −1); axis $y = -1$; focus $(9, -1)$; directrix $x = -5$

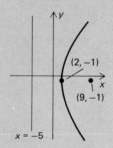

37. (a)

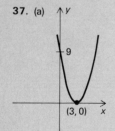

(b)

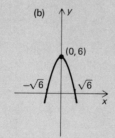

38. (a)

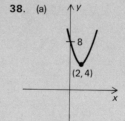

(b)

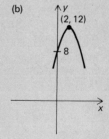

39. 9 seconds; 1296 feet   **40.** 3 years; 9000   **41.** Ellipse, center $(2, -5)$

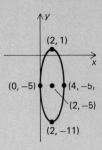

42. Hyperbola, asymptotes $3x + y - 1 = 0$ and $3x - y - 11 = 0$

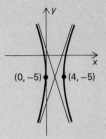

43. Circle, center $(2, -5)$, radius 1

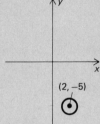

44. Parabola, vertex $(2, 1)$

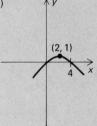

45. Parabola, vertex $(1, -5)$

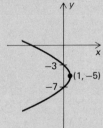

46. Circle   **47.** Hyperbola   **48.** No graph   **49.** Two intersecting lines   **50.** Hyperbola   **51.** Ellipse
52. Parabola   **53.** Parabola

54. Not a function

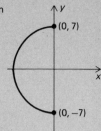

(0, 7)

(0, −7)

55. Function

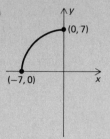

(0, 7)

(−7, 0)

56. One-to-one   **57.** One-to-one   **58.** One-to-one   **59.** Not one-to-one   **60.** $\{x \mid -3 \le x \le 5\}$;
$f^{-1}(x) = \dfrac{x - 3}{-2}$ **61.** $\{x \mid x \ge 1\}$; $f^{-1}(x) = \sqrt{x - 1}$

62.

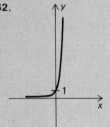

63.

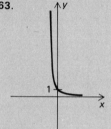

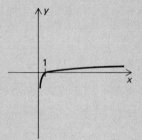

65. 2   **66.** −2   **67.** 8   **68.** 25   **69.** 3.6877   **70.** 0.1913 − 3 or −2.8087   **71.** 0.3741

72. 0.5879   **73.** 9.395   **74.** (a) $\{x \mid x \ge 8\}$ (b) $\sqrt[3]{x} + \sqrt{x - 2}$ (c) $\{x \mid x \ge 8\}$ (d) $\dfrac{\sqrt[3]{x}}{\sqrt{x - 2}}$

75. (a) $\{x \mid x \ge 66\}$ (b) $\sqrt[3]{\sqrt{x - 2}} = \sqrt[6]{x - 2}$ (c) $\{x \mid x \ge 8\}$ (d) $\sqrt{\sqrt[3]{x} - 2}$

Part 3 Review Exercises

1. $x = 0, -3, 2, -5$ **2.** $x = 0$, multiplicity 3; $x = -2$, multiplicity 5; $x = 3$, multiplicity 1   **3.** 4

4. $x(x - 5)(x + 4)(x - 3) = 0$ **5.** $\dfrac{x^3 - 2x^2 - 7x + 14}{x - 2} = x^2 + 2x - 3$. The remainder is 0, so 2 is a root.

6. $x = \frac{3}{7}, 2$ **7.** $x^2 - 4x + 53$ **8.** Let $P(x) = x^3 - 2x^2 - 4x + 6$. $P(2) = -2$ (negative), $P(3) = 3$
(positive), so there is at least one root between 2 and 3.   **9.** Let $P(x) = 9x^4 - 36x^3 + 112x^2 + 20x - 65$.
$P(0) = -65$ (negative), $P(1) = 40$ (positive), so there is at least one root between 0 and 1. First approximation, $\frac{13}{21}$.

10. $x = 4, y = -1, z = 3$ **11.** No solution   **12.** $\begin{vmatrix} 4 & 1 \\ -1 & 6 \end{vmatrix} = 25$ **13.** 26   **14.** $-|5|$

15. $x = -2, y = 3$ **16.** $x = 4, y = 3, z = 1$ **17.** 0   **18.** 0   **19.** +1   **20.** 2   **21.** −1   **22.** +1

23. $\begin{pmatrix} 36 & 19 \\ 33 & -1 \end{pmatrix}$ **24.** $\begin{pmatrix} 59 & 34 \\ 9 & 0 \end{pmatrix}$ **25.** $\begin{pmatrix} 3 & 6 & 2 \\ 1 & 4 & 8 \\ 3 & 9 & -7 \end{pmatrix}$ **26.** $\begin{pmatrix} 3x + 5y \\ -2x + 9y \end{pmatrix}$ **27.** $\begin{pmatrix} 1 & 0 & 0 \\ 0 & 1 & 0 \\ 0 & 0 & 1 \end{pmatrix}$

28. (a) $\begin{pmatrix} 21 & -15 \\ 26 & -19 \end{pmatrix}$ (b) $\begin{pmatrix} -1 & 1 \\ 6 & 3 \end{pmatrix}$ **29.** $\begin{aligned} 2x - 7y &= 5 \\ 3x + y &= 2 \end{aligned}$ **30.** $\begin{pmatrix} 1 & 0 & 0 & 0 \\ 0 & 1 & 0 & 0 \\ 0 & 0 & 1 & 0 \\ 0 & 0 & 0 & 1 \end{pmatrix}$

31. $\begin{vmatrix} 6 & 3 \\ 8 & 4 \end{vmatrix} = 0$, not nonsingular   **32.** $\dfrac{1}{31}\begin{pmatrix} 1 & 7 \\ -4 & 3 \end{pmatrix}\begin{pmatrix} 3 & -7 \\ 4 & 1 \end{pmatrix} = \begin{pmatrix} 1 & 0 \\ 0 & 1 \end{pmatrix}, \begin{pmatrix} 3 & -7 \\ 4 & 1 \end{pmatrix} \cdot$
$\dfrac{1}{31}\begin{pmatrix} 1 & 7 \\ -4 & 3 \end{pmatrix} = \begin{pmatrix} 1 & 0 \\ 0 & 1 \end{pmatrix}$

33. $\begin{pmatrix} 6 & 4 & 1 \\ 1 & 2 & 9 \\ 3 & 7 & 5 \end{pmatrix}$ **34.** $-\dfrac{1}{17}\begin{pmatrix} -1 & -5 \\ -3 & 2 \end{pmatrix}$ **35.** $\begin{pmatrix} -1 & 2 & 0 \\ -3 & 6 & 1 \\ 0 & 1 & 0 \end{pmatrix}$ **36.** $\begin{pmatrix} 5 & -1 \\ 3 & 1 \end{pmatrix}\begin{pmatrix} x \\ y \end{pmatrix} = \begin{pmatrix} 6 \\ 10 \end{pmatrix}$

$$\begin{pmatrix} x \\ y \end{pmatrix} = \frac{1}{8}\begin{pmatrix} 1 & 1 \\ -3 & 5 \end{pmatrix}\begin{pmatrix} 6 \\ 10 \end{pmatrix}$$

$$= \frac{1}{8}\begin{pmatrix} 16 \\ 32 \end{pmatrix}$$

$$= \begin{pmatrix} 2 \\ 4 \end{pmatrix},$$

$$x = 2, \; y = 4$$

37. $\dfrac{A}{x-3} + \dfrac{B}{x+2} + \dfrac{C}{(x+2)^2}$ **38.** $\dfrac{A}{x-1} + \dfrac{B}{x+2} + \dfrac{Cx+D}{x^2-2x+15}$

39. $\dfrac{A}{x-4} + \dfrac{Bx+C}{x^2-2x+15} + \dfrac{DX+E}{(x^2-2x+15)^2}$ **40.** $x - 4 + \dfrac{3}{x-1} + \dfrac{5}{(x-1)^2}$

41. $\dfrac{1}{x-1} + \dfrac{3}{(x-1)^2} + \dfrac{2x-1}{x^2+9}$

42.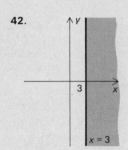

43.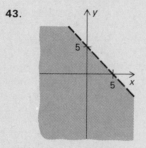

44.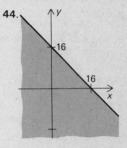

45.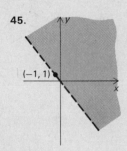

46.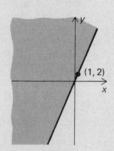

47.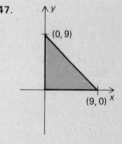

48.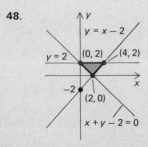

49. $\frac{2}{3}, \frac{4}{9}, \frac{8}{27}, \frac{16}{81}, (\frac{2}{3})^n; 0$ **50.** Arithmetic; 3   **51.** Geometric; 2   **52.** $2n^2$ **53.** $3 - (\frac{1}{3})^{n-1}$

54. Converges; $\frac{6}{5}$ **55.** Does not converge   **56.** $2 + 5 + 10 + 17 + \cdots + (n^2 + 1) + \cdots$

57. $1, 1 + \frac{1}{2}, 1 + \frac{1}{2} + \frac{1}{3}$ **58.** (a) 20,000   (b) $\sum_{n=1}^{10} 3^{n-1}(500)$

59. *Part I* $6 = 3(1 + 1^2)$ since $6 = 3 \times 2$
   Part II   If $6 + 12 + 18 + \cdots + 6k = 3(k + k^2)$, then
$$6 + 12 + 18 + \cdots + 6k + 6(k + 1) = 3(k + k^2) + 6(k + 1)$$
$$= 3(k + k^2 + 2k + 2)$$
$$= 3(k + 1 + k^2 + 2k + 1)$$
$$= 3((k + 1) + (k + 1)^2)$$

60. *Part I* $2^3 > 3 + 4$ since $8 > 7$
   Part II   If $2^k > k + 4$, then
$$2^{k+1} > 2(k + 4)$$
$$= 2k + 8$$
$$= k + 1 + 4 + k + 3$$
$$> (k + 1) + 4 \text{ for } k \geq 3$$

61. *Part I*   4 is a factor of $2(1^2) + 6(1)$ since 4 is a factor of 8.
   Part II $2(k + 1)^2 + 6(k + 1) = 2(k^2 + 2k + 1) + 6k + 6$
$$= (2k^2 + 6k) + (4k + 8)$$
Note that 4 is a factor of $4k + 8$, so if 4 is a factor of $2k^2 + 6k$, then 4 is a factor of $2(k + 1)^2 + 6(k + 1)$.

62. $x^{10} + (\quad)x^9y + (\quad)x^8y^2 + (\quad)x^7y^3 + (\quad)x^6y^4 + (\quad)x^5y^5 + (\quad)x^4y^6 + (\quad)x^3y^7 + (\quad)x^2y^8$
   $+ (\quad)xy^9 + y^{10}$ **63.** $32x^5 + 80x^4y + 80x^3y^2 + 40x^2y^3 + 10xy^4 + y^5$

64. $x^{12} + 12x^{11}y + 66x^{10}y^2 + 220x^{11}y^3$ **65.** $120; 120x^7y^3$ **66.** 240

67. $\binom{7}{2} = \frac{7 \times 6}{2} = 21$

$\binom{7}{5} = \frac{7 \times 6 \times 5 \times 4 \times 3}{2 \times 3 \times 4 \times 5} = 21$

68. (a) 1   (b) 18   **69.** (a) 17   (b) 1   **70.** 100 boxes candles, 300 boxes flowers   **71.** *abcd, abdc, acbd, acdb, adbc, adcb, bacd, badc, bcad, bcda, bdac, bdca, cabd, cadb, cbad, cbda, cdab, cdba, dabc, dacb, dbac, dbca, dcab, dcba*   **72.** *abcd, abce, abde, acde, bcde*   **73.** 362,880   **74.** 60,480   **75.** 1   **76.** 84

77. *tta, tat, att*   **78.** 453,600   **79.** $_4C_0$ $_4C_1$ $_4C_2$ $_4C_3$ $_4C_4$, $_5C_0$ $_5C_1$ $_5C_2$ $_5C_3$ $_5C_4$ $_5C_5$

80. (a) 20!   (b) 19!   **81.** (a) 1080   (b) 154,440   **82.** $\frac{7}{36}$ **83.** $\frac{5}{16}$ **84.** 90   **85.** (a) 27,405

(b) 657,720   **86.** $\frac{2}{7}$ **87.** $\frac{36}{121}$ **88.** $\frac{2}{7}$ **89.** $\frac{49}{77}$ **90.** $\frac{5}{442}$ **91.** $1 - (\frac{3}{5})^5 = \frac{2882}{3125}$ **92.** $1 - (\frac{5}{6})^5 = \frac{4651}{7776}$

93. 1 to 12   **94.** 31 to one   **95.** 35 to one   **96.** $\frac{56}{306}$ **97.** $\frac{353,697,121,049}{353,697,121,050}$ **98.** 55,440, $\frac{120}{55,440}$ **99.** $\frac{7}{34}$

100. $\frac{5}{14}$

Odd-Numbered Exercise Answers; Selected Graphs and Proofs

Exercise 1.1

1. 0, 1, 4, 7, 11   **3.** -5, -1 **5.** 0, 4   **7.** -5, -1, 0, 1, 4, $\frac{3}{4}$, 7, 11, $\frac{22}{7}$, -1.5, $1.\overline{5}$

9. -5, -1, 0, 1, 4, $\frac{3}{4}$, π, 7, 11, $\sqrt{3}$, e, $\frac{22}{7}$, -1.5, $1.\overline{5}$, $1.020020002...$ **11.** 0, 1, 4, $\frac{3}{4}$, π, 7, 11, $\sqrt{3}$,

e, $\frac{22}{7}$, $1.\overline{5}$, $1.020020002...$ **13.** $-3i$, $4 + 3i$ **15.** -5, -1, 0, -1.5

17. Real numbers, positive real numbers, rational numbers, nonnegative real numbers, complex numbers
19. Real numbers, positive real numbers, irrational numbers, nonnegative real numbers, complex numbers
21. Real numbers, negative integers, integers, odd integers, rational numbers, negative real numbers, nonpositive real numbers, complex numbers
23. Real numbers, positive real numbers, rational numbers, nonnegative real numbers, complex numbers
25. Real numbers, positive real numbers, irrational numbers, complex numbers, nonnegative real numbers
27. Real numbers, negative integers, integers, odd integers, rational numbers, negative real numbers, nonpositive real numbers, complex numbers
29. Real numbers, positive real numbers, irrational numbers, nonnegative real numbers, complex numbers
31. Real numbers, positive real numbers, irrational numbers, nonnegative real numbers, complex numbers
33. Complex numbers
35. Imaginary numbers, complex numbers
37. Real numbers, irrational numbers, negative real numbers, nonpositive real numbers, complex numbers
39. Real numbers, negative real numbers, nonpositive real numbers, irrational numbers, complex numbers
41. Complex numbers
43. Imaginary numbers, complex numbers
45. Real numbers, positive real numbers, rational numbers, nonnegative real numbers, complex numbers
47. Repeating; rational   **49.** Repeating; rational   **51.** Terminating; rational   **53.** Terminating; rational
55. Nonterminating, nonrepeating; irrational   **57.** Terminating; rational   **59.** Terminating; rational
61. $\{x|x > 2\}$ **63.** $\{x|2 < x < 5\}$
65. $\{x|-2 \le x \le 2\}$ **67.** $\{x|0 \le x \le 9$ and x is an integer$\}$
69. $\{x| |x| < 5$ and x is an integer$\}$
71. $\{x| |x| \ge 5\}$ **73.** $\{x|0 \le x < 7$ and x is an integer$\}$
75. $\{x|x \ge -3\}$ **77.** 0   **79.** -8 **81.** -2 **83.** 6   **85.** -10

Exercise 1.2

1. $\frac{9}{25}$ **3.** $\frac{25}{9}$ **5.** 0   **7.** -3 **9.** $\frac{1}{5}$ **11.** 17   **13.** -8 **15.** 2   **17.** 2   **19.** $\frac{1}{2}$ **21.** 0   **23.** -2
25. 11   **27.** 7   **29.** 7   **31.** 4   **33.** -4 **35.** 9   **37.** 7   **39.** 7   **41.** 0   **43.** Not defined   **45.** 0

47. Not defined　**49.** 0　**51.** Not defined　**53.** 0　**55.** Not defined　**57.** Not defined　**59.** 0
61. Not defined　**63.** Not defined　**65.** 0　**67.** 36　**69.** Not real　**71.** Not real　**73.** 1　**75.** -1
77. x^8y^9　**79.** $2x^2y^2$　**81.** $2x$　**83.** $\dfrac{1}{x^3y^3};\ x \neq 0,\ y \neq 0$　**85.** $\dfrac{1}{y};\ y \neq 0$　**87.** x^4y^{10}
89. $\dfrac{25x^8}{2y^{12}};\ x \neq 0,\ y \neq 0$　**91.** $\dfrac{y}{x};\ x \neq 0,\ y \neq 0$　**93.** $\dfrac{2}{x^2y^4};\ x \neq 0,\ y \neq 0$　**95.** $\dfrac{x^2}{3}$　**97.** $+i$　**99.** $+1$
101. $-i$　**103.** -1　**105.** $+1$　**107.** -1　**109.** $+i$

Exercise 1.3

1. $\sqrt[5]{8^4},\ (\sqrt[5]{8})^4$　**3.** $\sqrt[3]{25},\ (\sqrt[3]{5})^2$　**5.** $\sqrt{\pi^{-1}},\ (\sqrt{\pi})^{-1}$　**7.** $\sqrt{8},\ (\sqrt{2})^3$　**9.** $\sqrt[8]{6^7},\ (\sqrt[8]{6})^7$　**11.** $7^{2/3}$
13. $3^{1/2}$　**15.** $7^{3/2}$　**17.** $5^{1/2}$　**19.** $2^{3/2}$　**21.** 343　**23.** 9　**25.** $\frac{1}{9}$　**27.** $\frac{1}{64}$　**29.** 10　**31.** $\frac{2}{5}$
33. 100　**35.** -2　**37.** $\frac{1}{12}$　**39.** 9　**41.** 0.3　**43.** 0.2　**45.** 0.1　**47.** 2　**49.** 16
51. $\dfrac{y^{7/2}}{x}$　**53.** $\dfrac{x^2}{y^{1/2}}$　**55.** $x^{10}y^2$　**57.** $x^{3/2}y^2$　**59.** $a^{7/3}b^{7/2}$　**61.** $\dfrac{21x^{1/2}}{2\sqrt{7}}$　**63.** $\dfrac{8x^{1/3}}{3\sqrt[3]{4}}$　**65.** $\dfrac{-21}{2\sqrt{7^3}}x^{-5/2}$
67. $\dfrac{-15}{\sqrt{6^3}}x^{-7/2}$　**69.** $\dfrac{\sqrt{2}}{6}x^{-5/6}$　**71.** 50　**73.** 6　**75.** $\frac{1}{45}$　**77.** 4×10^5　**79.** 20　**81.** $2xy\sqrt[3]{y}$
83. $9x^3y\sqrt{y}$　**85.** $-2xy^2\sqrt[3]{2y}$　**87.** $2x^2y^2\sqrt[5]{2y}$　**89.** $xz^2\sqrt[5]{-81x^4y^3z^2}$　**91.** $\dfrac{\sqrt{6x}}{3x}$　**93.** $\dfrac{\sqrt[3]{20x^2y}}{2x}$
95. $\dfrac{\sqrt{5}}{5}$　**97.** $\dfrac{\sqrt[4]{3xy^2}}{y}$　**99.** $\dfrac{\sqrt[3]{4}}{2}$

Exercise 1.4

1. $2x^2y,\ 5x^2y;\ 3x,\ -7x;\ 7y,\ 8y$　**3.** $x^5y^2,\ -x^5y^2;\ 2x^2y^5,\ -x^2y^5$　**5.** $-x^5y,\ 7x^5y;\ 7x^4y^2,\ -x^4y^2;\ 3y,\ 19y$
7. $x^3y,\ 2x^3y;\ xy^3,\ 2xy^3;\ 2xy,\ xy$　**9.** $\sqrt{2}x^2y,\ 3x^2y;\ \sqrt{2}xy,\ 3xy$　**11.** Polynomial, but not over the integers
13. Not polynomial　**15.** Polynomial over the integers　**17.** Polynomial over the integers
19. Polynomial, but not over the integers　**21.** Polynomial, but not over the integers　**23.** Not polynomial
25. Polynomial over the integers　**27.** Polynomial, but not over the integers　**29.** Not polynomial
31. Polynomial　**33.** Not polynomial　**35.** Polynomial　**37.** Polynomial　**39.** Not polynomial
41. Tenth; fourth in x; seventh in y　**43.** Fifth; fifth in x; second in y　**45.** First; first in x; first in y
47. Second; second in x; first in y　**49.** First; first in x; first in y　**51.** $-5x^4 + 11x^3 - 7$
53. $x^2 - 5xy + y^2 + 7$　**55.** $3h$　**57.** $5h$　**59.** $3x^2h + 3xh^2 + h^3$　**61.** $-3,\ -2,\ -7,\ 4,\ -36,\ 90$
63. $0,\ 21,\ 18,\ 33,\ 990,\ 936$　**65.** $3a^2 - 2a,\ 2a + 7,\ 12a^2 + 80a + 133,\ 6a^2 - 4a + 7$
67. $x^3 + x^2 - 1$　**69.** $4x^4 - 2x^3 + 10x^2 - x + 8$　**71.** 0　**73.** $-5x^5 + 5x^4 + 7x^3 + 6x^2 - 16$
75. $3x^3 + 5x^2 - 12x - 1$　**77.** $48x^5 - 40x^4 - 12x^3 + 18x^2 - 2$　**79.** $5x^5 - 17x^4 + 16x^3 + x^2 - 2x$
81. $9x^2 - 3x - 56$　**83.** $x^2 - 3x - 28$　**85.** $x^2 - 4$　**87.** $x^3 + 8$　**89.** $x^3 - y^3$　**91.** $x^3 + y^3$
93. $x^2 - 10x + 25$　**95.** $a^2 - 2ab + b^2$　**97.** 7　**99.** -1　**101.** $x - 6\sqrt{x} + 9$
103. $x - 2\sqrt{xy} + y$　**105.** $x + 4\sqrt{x - 2} + 2$　**107.** $5x + 4\sqrt{x^2 - 3x} - 3$
109. $53 - 14\sqrt{x + 4} + x$　**111.** $2a^2 + 4ah + 2h^2 - 3a - 3h + 5,\ 2x^2 + 4xh + 2h^2 - 3x - 3h + 5$
113. $a^2 + 2ah + h^2 + 5a + 5h - 3,\ x^2 + 2xh + h^2 + 5x + 5h - 3$
115. $a^3 + 3a^2h + 3ah^2 + h^3 - a^2 - 2ah - h^2 + 3a + 3h + 2,$
$x^3 + 3x^2h + 3xh^2 + h^3 - x^2 - 2xh - h^2 + 3x + 3h + 2$　**117.** $7a + 7h + 15,\ 7x + 7h + 15$
119. $9a^2 + 18ah + 9h^2 + 1,\ 9x^2 + 18xh + 9h^2 + 1$　**121.** $6xh + 3h^2 - 7h$
123. $6x^2h + 6xh^2 + 2h^3 - 2xh - h^2$　**125.** $12x^2h + 12xh^2 + 4h^3$　**127.** $-2xh - h^2 + h$
129. $6x^2h + 6xh^2 + 2h^3 - 5h$　**131.** $45x^2 + 69x + 26$　**133.** $x^3 + 6x^2 + 12x + 8$
135. $9x^4 - 21x^2 + 10$　**137.** 3　**139.** $2x^2 + x$　**141.** $5x + 13 + \dfrac{10x - 80}{x^2 - 3x + 6}$
143. $2x - 4 + \dfrac{11}{2x^2 + x + 3}$　**145.** $2x + 9 + \dfrac{48x^2 - 12x - 59}{x^3 - 5x^2 + 6}$　**147.** $2x + 9$, Remainder 0

149. $x + 3$, Remainder 0 **151.** $6; 2x + 5 + \dfrac{6}{x-3}$ **153.** $-8; 3x^2 + 2x - 1 + \dfrac{-8}{x-1}$

155. $-6; 2x^2 + 3x - 1 + \dfrac{-6}{x-2}$ **157.** $-10; 5x^2 - 3x + \dfrac{-10}{x+4}$ **159.** $-4; 7x^2 - 2x + 3 + \dfrac{-4}{x+5}$

161. $x^4 + 3x^2 + 5x + 12 + \dfrac{25}{x-2}$ **163.** $3x^3 - 11x^2 + 33x - 92 + \dfrac{272}{x+3}$

165. $5x + 18 + \dfrac{79}{x-4}$ **167.** $2x^4 + 2x^3 - x^2 + 4x + 4 + \dfrac{2}{x-1}$

169. $5x^3 + 35x^2 + 242x + 1690 + \dfrac{11836}{x-7}$ **171.** 28 **173.** -23 **175.** -237 **177.** 304

179. -234 **181.** $7 - i$ **183.** $54 - 10i$ **185.** $15 - 35i$ **187.** $-2 - 7i$ **189.** $59 - 26i$

191. $5 + 0i$ **193.** $0 + 0i$ **195.** $\frac{34}{25} + 0i$ **197.** 0, 0 **199.** 2, 2

Exercise 1.5

1. $5^2 \times 3^2$ **3.** $2^3 \times 5^2 \times 7$ **5.** 2×13^2 **7.** $2 \times 3 \times 7 \times 17$ **9.** $2^2 \times 5 \times 11 \times 41$

11. $a^2 - b^2$ **13.** $a^2 - 2ab + b^2$ **15.** $a^3 - b^3$ **17.** $a^5 - b^5$ **19.** $x^2 - 2x - 15$

21. $x^2 + 2x - 24$ **23.** $6x^2 - 11x - 7$ **25.** $12x^2 + 10x - 12$ **27.** $21x^2 + 22x - 8$

29. $21x^7 - 6x^6 + 27x^5$ **31.** $5x^5 - 4x^3 + 2x^2 - 7x$ **33.** $-5x^6 - 15x^4 + 10x^3 + 45x^2$

35. $3x^6 - 2x^4 + 17x^3 - x + 8$ **37.** $3x^8 + 7x^7 - 6x^6 - 14x^5 + 18x + 42$

39. $x^2 - 4y^2$ **41.** $9x^2 - 24x + 16$ **43.** $x^2 + 4xy + 4y^2$ **45.** $x^6 - y^6$ **47.** $x^6 + 2x^3y^3 + y^6$

49. $x^2 - 4x + 4$ **51.** $9x^2 - 4$ **53.** $x^6 - 4$ **55.** $x^2 + 2xh + h^2$ **57.** $x^2 - 2x + 1$ **59.** $9x^2 - 49$

61. $9x^2 - 18x - 7$ **63.** $x^3 - 1$ **65.** $27x^3 + 343$ **67.** $x^3 + 8y^3$ **69.** $16 - 81x^4$

71. $(5x + 4)(5x - 4)$ **73.** $x^2y(3z - 2x^3y + xyz)$ **75.** $(2w + 3z)(4w^2 - 6wz + 9z^2)$

77. $(3x + 7y)(3x - 7y)$ **79.** $xy(3x - 4y)(9x^2 + 12xy + 16y^2)$ **81.** $(x^2 + y^2)^2(x + y)^2(x - y)^2$

83. $(x - y)(x^2 + xy + y^2)(x^6 + x^3y^3 + y^6)$ **85.** $(x - 11)(x + 4)$

87. $(2x - y)(2x + y)(4x^2 + 2xy + y^2)(4x^2 - 2xy + y^2)$ **89.** $(2x + 5)(x - 4)$ **91.** $4(3x + y)(3x - y)$

93. $2a^2b^2(a + b)^2$ **95.** $(11a + 13y)(11a - 13y)$ **97.** $(3w + 8z)^2$ **99.** Not factorable

101. $b(3a + 2b)(9a^2 - 6ab + 4b^2)$ **103.** $x(4xy + 5z)(4xy - 5z)$ **105.** $x^2y(x - 8)(x + 7)$

107. Not factorable **109.** $6(x - 2)(x + 1)$ **111.** $(x^2 + 4y^2)(x + 2y)(x - 2y)$

113. $(a - b)(a^4 + a^3b + a^2b^2 + ab^3 + b^4)(a + b)(a^4 - a^3b + a^2b^2 - ab^3 + b^4)$

115. $(x - y)(x^2 + xy + y^2)(x + y)(x^2 - xy + y^2)(x^2 + y^2)(x^4 - x^2y^2 + y^4)$

117. $\dfrac{49}{4}; \left(x + \dfrac{7}{2}\right)^2$ **119.** $\dfrac{81}{4}; \left(x - \dfrac{9}{2}\right)^2$ **121.** $1; (x - 1)^2$ **123.** $1; (x + 1)^2$

125. $\dfrac{b^2}{4a^2}; \left(x + \dfrac{b}{2a}\right)^2$ **127.** $(y^7 - 3)^{l}x + 2)$ **129.** $a^3(3a^2 - 2)(a - 1)$

Exercise 1.6

1. Minus **3.** Plus **5.** Plus **7.** Plus **9.** Minus **11.** Minus **13.** Minus **15.** Plus **17.** $\dfrac{2}{9}$ **19.** $\dfrac{7}{4}$

21. In lowest terms **23.** In lowest terms **25.** In lowest terms **27.** In lowest terms **29.** $\dfrac{6(x + 1)}{x + 4}$

31. $\dfrac{x + 7}{x + 4}$ **33.** $\dfrac{2x^2 - x + 5}{x(x - 5)}$ **35.** In lowest terms **37.** $\dfrac{x - 5}{x - 7}$ **39.** $\dfrac{x^2 + x + 1}{x - 1}$ **41.** 110 **43.** 75

45. $19(x + 2)$ **47.** $5(x + 2)$ **49.** $x^2(x^2 + 2x + 4)$ **51.** 180 **53.** $567x^5$ **55.** $(x - 1)^3(x^2 + x + 1)$

57. $x^3(x - 2)^7(x + 1)^3(x - 1)$ **59.** $(x - y)^2(x + y)$ **61.** $\dfrac{35}{88}$ **63.** $\dfrac{111}{88}$ **65.** $\dfrac{24}{5}$ **67.** $\dfrac{-101}{630}$

69. $\dfrac{269}{420}$ **71.** $\dfrac{153}{40}$ **73.** $\dfrac{1427}{1302}$ **75.** $\dfrac{253}{2808}$ **77.** $\dfrac{1}{3}$ **79.** $\dfrac{1}{\sqrt{3}}$ **81.** $\dfrac{10x^3}{(x^2-9)(x-3)}$

83. $\dfrac{(x+2)(x+1)^2}{(x-2)(x+3)(x-3)}$ **85.** $\dfrac{16a^2b^2y}{3x}$ **87.** $\dfrac{4x^2a^3b^3c^3}{9yz^3}$ **89.** $\dfrac{7(x+2)}{x-3}$ **91.** $\dfrac{5x^2+3x}{x^2-9}$

93. $\dfrac{7x^2+6x}{x^2-9}$ **95.** $\dfrac{2x^2-27x+64}{(x-2)(x^2-10x+25)}$ **97.** $\dfrac{2x^2+4}{(x-1)(x+2)}$ **99.** $\dfrac{7x^2+3x+9}{5}$

101. $\dfrac{2}{3x-1}+\dfrac{5}{x+2}=\dfrac{2(x+2)+5(3x-1)}{(3x-1)(x+2)}=\dfrac{17x-1}{3x^2+5x-2}$

103. $\dfrac{3}{x-5}+\dfrac{6}{x+5}+\dfrac{2}{x}=\dfrac{3x(x+5)+6x(x-5)+2(x-5)(x+5)}{x(x+5)(x-5)}$

$$=\dfrac{11x^2-15x-50}{x^3-25x}$$

105. $\dfrac{7}{x+3}+\dfrac{1}{(x+3)^2}=\dfrac{7(x+3)+1}{(x+3)^2}=\dfrac{7x+22}{x^2+6x+9}$

107. $\dfrac{3x+2}{x^2+x+1}+\dfrac{1}{x-1}=\dfrac{(3x+2)(x-1)+(x^2+x+1)}{(x-1)(x^2+x+1)}=\dfrac{4x^2-1}{x^3-1}$

109. $\dfrac{7}{x}+\dfrac{1}{x+3}+\dfrac{4}{(x+3)^2}=\dfrac{7(x+3)^2+x(x+3)+4x}{x(x+3)^2}$

$$=\dfrac{8x^2+49x+63}{x^3+6x^2+9x}$$

111. $\dfrac{10x^3-6}{6x^6+x^4}$ **113.** $\dfrac{3x^2-14x}{4x^2-6x-4}$ **115.** $\dfrac{-x^2+2x+1}{4x^2-3}$ **117.** $\dfrac{20-4\sqrt{2}}{23}$ **119.** $\sqrt{5}-\sqrt{3}$

121. $\dfrac{22}{41}+\dfrac{7}{41}i$ **123.** $\dfrac{17}{10}+\dfrac{1}{10}i$ **125.** $9-7i$

Exercise 1.7

1. Any solution of (1) is a solution of (2), but (2) may have additional solutions.
3. The solutions of (1) and (2) are the same
5. Any solution of (1) is a solution of (2), but (2) may have additional solutions.
7. Any solution of (1) is a solution of (2), but (2) may have additional solutions.
9. Any solution of (1) is a solution of (2), but (2) may have additional solutions.

11. $x=3$ **13.** $x=8$ **15.** $x=\dfrac{7}{11}$ **17.** $x=\dfrac{-4}{3}$ **19.** $x=-19$ **21.** $h=\dfrac{V}{2\pi r}$ **23.** $h=\dfrac{3V}{\pi r^2}$

25. $P=\dfrac{A}{1+ni}$ **27.** $y'=\dfrac{-2x-y}{x+2},\ x\neq-2$ **29.** $y'=\dfrac{-x}{y},\ y\neq0$ **31.** 2 real solutions

33. 2 real solutions **35.** 2 imaginary solutions **37.** 1 real solution **39.** 2 imaginary solutions
41. $x=8,-4,7$ **43.** $x=-4,-3,2$ **45.** $x=-4,3$ **47.** $x=0,3$ **49.** $x=4,\pm2$

51. $x=9,-2$ **53.** $x=7,-5$ **55.** $x=\dfrac{1}{3},\dfrac{-1}{2}$ **57.** $x=\dfrac{-1}{8},\dfrac{1}{7}$ **59.** $x=\dfrac{-5}{3}$

61. $x=\dfrac{1}{3},-2$ **63.** $x=1\pm\sqrt{11}$ **65.** $x=\dfrac{8\pm6i}{5}$ **67.** $x=\pm3$ **69.** $x=\dfrac{-1}{2}\pm\dfrac{5}{2}i$

71. $x=\dfrac{-1}{2},\dfrac{5}{3}$ **73.** $x=\dfrac{-1}{3},\dfrac{1}{5}$ **75.** $x=5$ **77.** $x=7,\dfrac{1}{3}$ **79.** $x=1,\dfrac{-5}{2}$ **81.** $x=\pm1,\pm4i$

83. $x=\pm3i,\pm2$ **85.** $x=3,1,\dfrac{-3\pm3\sqrt{3}i}{2},\dfrac{-1\pm\sqrt{3}i}{2}$

87. $\dfrac{(x-1)^2}{9}-\dfrac{(y+2)^2}{4}=1;\ h=1,\ a=3,\ k=-2,\ b=2$

89. $(y - 3)^2 = 8(x - (-1)); k = 3, p = 4, h = -1$

91. $\dfrac{(y - 3)^2}{4} - \dfrac{(x + 1)^2}{9} = 1; k = 3, a = 2, h = -1, b = 3$

93. $\dfrac{(x + 5)^2}{4} + (y - 4)^2 = 1; h = -5, a = 2, k = 4, b = 1$

95. $\dfrac{(x - 3)^2}{16} + \dfrac{(y + 1)^2}{4} = 1; h = 3, a = 4, k = -1, b = 2$ **97.** $3\sqrt{1 - (x - 2)^2}; h = 2, k = 1$

97. $3\sqrt{1 - (x - 2)^2}; h = 2, k = 1$

99. $2\sqrt{\left(x - \dfrac{3}{2}\right)^2 - 36}; h = \dfrac{3}{2}, k = 6$ **101.** $\sqrt{(x + 5)^2 + 9}; h = -5, k = 3$

103. $2\sqrt{\left(x - \dfrac{5}{2}\right)^2 - 9}; h = \dfrac{5}{2}, k = 3$ **105.** $2\sqrt{\left(x + \dfrac{1}{2}\right)^2 + 4}; h = \dfrac{-1}{2}, k = 2$

107. $3250   **109.** 10 seconds, 20 seconds   **111.** $\frac{1}{2}$ m

Exercise 1.8

1. 3   **3.** No solution   **5.** No solution   **7.** $x = 20$ **9.** $x = 12$ **11.** $\dfrac{x^2}{25} + \dfrac{y^2}{9} = 1; a = 5, b = 3$

13. $\dfrac{x^2}{169} + \dfrac{y^2}{25} = 1; a = 13, b = 5$ **15.** $\dfrac{x^2}{100} + \dfrac{y^2}{64} = 1; a = 10, b = 8$

17. $\dfrac{x^2}{144} - \dfrac{y^2}{25} = 1; a = 12, b = 5$ **19.** $\dfrac{x^2}{4} - \dfrac{y^2}{9} = 1; a = 2, b = 3$ **21.** $y^2 = 20x; \dfrac{p}{2} = 5$

23. $x^2 = -20y; \dfrac{p}{2} = -5$ **25.** $y^2 = -8x; \dfrac{p}{2} = -2$ **27.** No solution   **29.** $x = 4$ **31.** $x = 1, -5$

33. $x = 2$ **35.** $x = 4$ **37.** $x = \dfrac{3}{2}, \dfrac{-7}{2}$ **39.** $x = \dfrac{10}{3}, \dfrac{-2}{3}$

41. $x = 1, -10$ **43.** $x = -1, -7$

45. $x = 2, -2$

47. $x = -2, -1, \dfrac{-3 \pm \sqrt{97}}{2}$

49. $x = \pm 3, \pm \sqrt{3}$ **51.** $\dfrac{(x - 2)^2}{25} + \dfrac{(y - 1)^2}{16} = 1; h = 2, a = 5, k = 1, b = 4$

53. $\dfrac{(x + 1)^2}{16} - \dfrac{(y + 2)^2}{9} = 1; h = -1, a = 4, k = -2, b = 3$

55. $(y - 2)^2 = 16x, \dfrac{p}{2} = 4$

Exercise 1.9

1. $x \geq \dfrac{-8}{5}$ **3.** $x \leq 0$ **5.** $x \leq 5$

7. $x > -1$ **9.** $x \geq 3$

11. $\{x \mid x \geq 2 \text{ or } x \leq -3\}$

13. $\{x \mid -3 \leq x \leq 1\}$ **15.** $\{x \mid -2 \leq x \leq 3\}$

17. $\{x \mid -7 < x < 4\}$ **19.** $\{x \mid x > -1 \text{ or } x < -2\}$

21. $\{x \mid 1 - \sqrt{2} \leq x \leq 1 + \sqrt{2}\}$

23. $\{x \mid x > 1 + \sqrt{3} \text{ or } x < 1 - \sqrt{3}\}$

25. $\{x \mid -1 - \sqrt{2} \leq x \leq -1 + \sqrt{2}\}$

27. $\{x \mid 1 - \sqrt{5} \leq x \leq 1 + \sqrt{5}\}$

29. $\{x \mid x > -1 + \sqrt{3} \text{ or } x < -1 - \sqrt{3}\}$

31. $x \leq -3 \text{ or } x \geq \dfrac{-5}{3}$ **33.** $x \leq -1 \text{ or } x \geq 2$

35. $-2 < x < 2$ **37.** $-1 \leq x \leq \dfrac{-1}{2}$

39. $x \leq 1 \text{ or } x \geq 4$ **41.** $x > -2 \text{ or } x < -3$ **43.** $x \geq 11 \text{ or } x < 5$

45. $x > 3 \text{ or } x < 0$ **47.** 89   **49.** Between 26 and 36 inclusive   **51.** 17   **53.** 24,000
55. More than 15,625

Exercise 1.10

1. $x = 1, y = -2$ **3.** $A = \dfrac{-72}{25}, B = \dfrac{102}{25}$ **5.** $A = -1, B = 1$ **7.** $x = 8, y = -9$ **9.** $x = \dfrac{3}{2}, y = -1$

11. $A = -4, B = -1$ **13.** $x = \dfrac{29}{107}, y = \dfrac{268}{107}, z = \dfrac{-300}{107}$ **15.** $A = -1, B = -1, C = 3$

17. $x = -2, y = 3, z = 4$ **19.** $x = 1, y = -1, z = 3, w = 2$ **21.** $A = 7, B = 4, C = 5, D = 2$
23. $A = 1, B = -1, C = 2, D = -2, E = 3$ **25.** $x = 1, y = 0, z = -1, w = 2, q = -2$
27. $A = -3, B = -5$ **29.** $A = 3, B = -5, C = 7$ **31.** $A = 3, B = -3, C = 4$
33. $A = 1, B = 2, C = 1, D = 2$ **35.** $A = 1, B = 4, C = 5, D = -1, E = -2$ **37.** Corn, 5; rice, 8
39. Carberry, 0; Smiley, 50   **41.** 12   **43.** Tuba, 33; piccolo, 6   **45.** Silver, 7; new pennies, 8; wheat-ears, 6

Exercise 2.1

1. $-3 \to 10, \dfrac{-1}{2} \to \dfrac{5}{4}, 0 \to 1, \dfrac{1}{2} \to \dfrac{5}{4}, 3 \to 10; \left\{10, \dfrac{5}{4}, 1\right\}$

3. $0 \to 0, 1 \to 1, \dfrac{9}{4} \to \dfrac{3}{2}, 4 \to 2, 9 \to 3; \left\{0, 1, \dfrac{3}{2}, 2, 3\right\}$

5. $\dfrac{-1}{8} \to \dfrac{-1}{2}, -1 \to -1, 0 \to 0, 1 \to 1, 8 \to 2; \left\{\dfrac{-1}{2}, -1, 0, 1, 2\right\}$

7. $1, -1, 1, a^3 - a^2 + 1, a^3 + 3a^2h + 3ah^2 + h^3 - a^2 - 2ah - h^2 + 1$

9. $\dfrac{1}{2}, \dfrac{7}{6}, 2, \dfrac{1}{2a} + 1, \dfrac{1}{2(a+h)} + 1$ **11.** $x_1^2 + 2x_1(\Delta x) + (\Delta x)^2 - 5x_1 - 5(\Delta x) + 3$

13. $4x_1^2 + 8x_1(\Delta x) + 4(\Delta x)^2 - 2x_1 - 2(\Delta x) + 1$ **15.** $2x_1^3 + 6x_1^2(\Delta x) + 6x_1(\Delta x)^2 + 2(\Delta x)^3$

17. $2x^2 + 4xh + 2h^2 + 3x + 3h - 4$ **19.** $5x^2 + 10xh + 5h^2 - x - h + 7$

21. Function; domain $= \left\{\dfrac{-1}{2}, \dfrac{1}{2}, \dfrac{1}{3}, \dfrac{3}{2}\right\}; \dfrac{-1}{2} \to -1, \dfrac{1}{2} \to 0, \dfrac{1}{3} \to 0, \dfrac{3}{2} \to 1$; range $= \{-1, 0, 1\}$

23. Function; domain $= \{-2, 0, -1, 2\}; -2 \to -8, 0 \to 0, -1 \to -1, 2 \to 8$; range $= \{-8, 0, -1, 8\}$

25. Not a function   **27.** $\{x \mid x \ge 1\}$ **29.** $\{x \mid x \ge -2\}$

31. 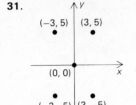   **33.** 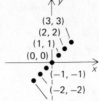   **35.** 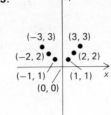   **37.** $(-5, -2)$ **39.** $(-2, 0)$

41. 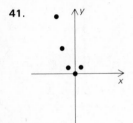   **43.** 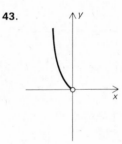   **45.** 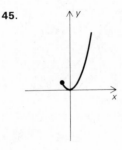   **47.**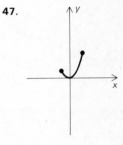

49. 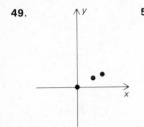   **51.** 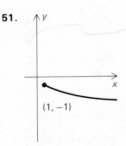   **53.** 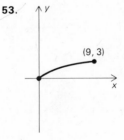   **55.**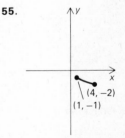

57.

x	y
-3	2
-2	1
-1	0
0	1
1	2

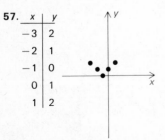

59.

x	y
-3	-1
-2	0
-1	1
0	2
1	3
2	4
3	5

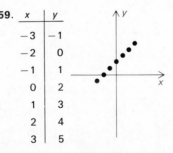

61. Not a function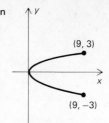

63. Function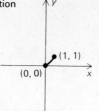

65. Not a function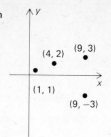

67. Not a function   **69.** Not a function   **71.** Function   **73.** Not a function   **75.** Function
77. Not a function   **79.** Function   **81.** $3h + 6x - 2$ **83.** $4h + 8x + 3$ **85.** $7h + 14x - 4$
87. $3x_1^2 + 3x_1(\Delta x) + (\Delta x)^2$ **89.** $3x_1^2 + 3x_1(\Delta x) + (\Delta x)^2 - 2$
91. $A(x) = (50 - 2x)x$; 132 square meters   **93.** $P(x) = 60x - 180,000$; 3000; $\$120,000$

Exercise 2.2

1.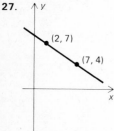

7.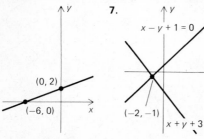

11.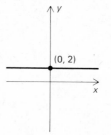

13.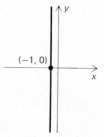

17. $\dfrac{-1}{5}$ **19.** $\dfrac{-4}{3}$ **21.** $(1) - (2) + 1 = 0$; $(3) - (4) + 1 = 0$; 1

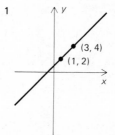

23. $3(2) + 5(1) - 11 = 0$; $3(-3) + 5(4) - 11 = 0$; $\dfrac{-3}{5}$ **25.** $(-3) + 4(2) - 5 = 0$; $1 + 4(1) - 5 = 0$; $\dfrac{-1}{4}$

27.

31. $2x + 5y - 26 = 0$ **33.** $x + y + 1 = 0$ **35.** $3x - 8y + 43 = 0$ **37.** $3x + 4y + 20 = 0$

39. $x - y + 3 = 0$ **41.** $y = \dfrac{2}{7}x + 4$; $\dfrac{2}{7}$; $(0, 4)$ **43.** $y = -5x + 11$; -5; $(0, 11)$

45. $y = \dfrac{5}{8}x + \dfrac{3}{2}$; $\dfrac{5}{8}$; $\left(0, \dfrac{3}{2}\right)$ **47.** $\dfrac{-1}{2}$; 2 **49.** -2; $\dfrac{1}{2}$ **51.** $5x + y - 11 = 0$ **53.** $x - 5y - 14 = 0$

55. $5x - 9y + 64 = 0$ **57.** $x + y - 1 = 0$ **59.** $3x + 2y - 6 = 0$ **61.** $3x - 4y = 0$

63. $x - 2y = 0$ **65.** $x - y + 2 = 0$

67.

77. $2x + 18$; 20; 18 **79.** $-x + 6$; 5; 6 **81.** $4x + 2h + 4$ **83.** $14x + 7h + 3$ **85.** $8x + 4h - 3$

87. 0; 20π meters; The half-line emanating from the origin with slope 2π.   **89.** $A(x) = 0.4x - 4000$; $10,000$; The salary which would be totally used in taxes and insurance.

Exercise 2.3

1. Intercept: $(1, 0)$; symmetric with respect to the x-axis; excluded area: $x < 1$

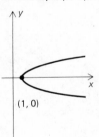

(1, 0)

3. Intercept: $(0, 0)$; symmetric with respect to the x-axis; excluded area: $x < 0$

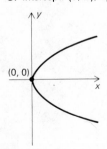

(0, 0)

5. Intercept: (1, 0); symmetric with respect to the x-axis; excluded area: $x < 1$ **7.** Intercept: (0, -1);
symmetric with respect to the y-axis; excluded area: $y < -1$ **9.** Intercept: (0, 0); symmetric with respect to the
y-axis; excluded area: $y < 0$ **11.** Intercept: (0, 0); symmetric with respect to the origin   **13.** Intercept: (0, 0);
symmetric with respect to the origin   **15.** Intercept: (0, 0); symmetric with respect to the origin
17. Intercept: (0, 0); symmetric with respect to the origin   **19.** Intercept: (0, 1); asymptotes: $y = 0$, $x = -1$;
excluded areas: $x > -1$ and $y < 0$, $x < -1$ and $y > 0$

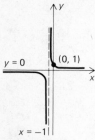

21. Intercept: (0, $\frac{1}{3}$); asymptotes: $x = -3$, $y = 0$; excluded areas: $x > -3$ and $y < 0$, $x < -3$ and $y > 0$
23. Intercept: (0, -3); asymptotes: $x = 1$, $y = 0$; excluded areas: $x > 1$ and $y < 0$, $x < 1$ and $y > 0$
25. Symmetric with respect to the origin; asymptotes: $x = 0$, $y = 0$; excluded areas: $x > 0$ and $y > 0$, $x < 0$
and $y < 0$

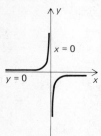

27. Intercepts: (0, -6), (2, 0); asymptotes: $x = \dfrac{-1}{2}$, $y = \dfrac{3}{2}$; excluded areas: $x > 2$ and $y < 0$, $\dfrac{-1}{2} < x < 2$
and $y > 0$, $y < \dfrac{-1}{2}$ and $y < 0$

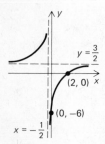

29. Intercepts: $\left(0, \dfrac{-10}{3}\right)$, ($-2$, 0); asymptotes: $x = 3$, $y = 5$; excluded areas: $x > 3$ and $y < 0$, $-2 < x < 3$
and $y > 0$, $x < -2$ and $y < 0$ **31.** Intercepts: $\left(0, \dfrac{-4}{3}\right)$, (4, 0); asymptotes: $x = -2$, $y = \dfrac{2}{3}$; excluded areas:
$x > 4$ and $y < 0$, $-2 < x < 4$ and $y > 0$, $x < -2$ and $y < 0$

33. Intercept: (0, 0); asymptotes: $x = 2$, $x = -2$, $y = \frac{1}{3}$; excluded areas: $x > 2$ and $y < 0$, $-2 < x < 2$ and $y > 0$, $x < -2$ and $y < 0$

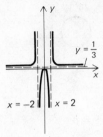

35. Intercept: (0, 0); asymptotes: $x = 2$, $x = -2$, $y = 3$; excluded areas: $x > 2$ and $y < 0$, $-2 < x < 2$ and $y > 0$, $x < -2$ and $y < 0$ **37.** Intercept: (0, 0); asymptote: $x = 1$; excluded areas: $x > 1$ and $y < 0$, $x < 1$ and $y > 0$ **39.** Intercept: (0, 0); asymptote: $x = -1$; excluded areas: $x > -1$ and $y < 0$, $x < -1$ and $y > 0$ **41.** Intercepts: (0, 3), (0, −3), (2, 0), (−2, 0); excluded areas: $x > 2$, $x < -2$, $y > 3$, $y < -3$

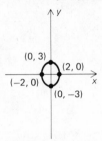

43. Intercepts: (0, 3), (0, −3) (1, 0), (−1, 0); excluded areas: $y > 3$, $y < -3$, $x > 1$, $x < -1$
45. Intercepts: (0, 1), (0, −1), (3, 0), (−3, 0); excluded areas: $x > 3$, $x < -3$, $y > 1$, $y < -1$
47. Intercepts: (3, 0), (−3, 0); excluded area: $-3 < x < 3$

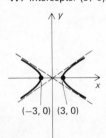

49. Intercepts: (0, 4), (0, −4), (5, 0), (−5, 0); excluded areas: $y > 4$, $y < -4$, $x > 5$, $x < -5$

Exercise 2.4

1. 8   **3.** $\sqrt{85}$ **5.** $4\sqrt{2}$ **7.** $\frac{4\sqrt{226}}{5}$ **9.** $\frac{\sqrt{1145}}{3}$ **11.** (6, 4)   **13.** $\left(\frac{1}{2}, \frac{7}{2}\right)$ **15.** $\left(\frac{-7}{2}, \frac{1}{2}\right)$

17. $(x + 4)^2 + (y + 5)^2 = \frac{9}{16}$ **19.** $(x - 5)^2 + y^2 = \frac{49}{16}$ **21.** (−3, 5); 6   **23.** (6, −2); $4\sqrt{5}$

25. $(7, -8); \sqrt{2}$

27. $(x + 1)^2 + (y - 3)^2 = 0$

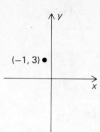

$(-1, 3) \bullet$

29. $(x - 6)^2 + (y - 1)^2 = 4$

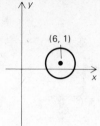

$(6, 1)$

31. $\left(x - \dfrac{1}{2}\right)^2 + (y + 3)^2 = \dfrac{-19}{4}$; No graph   **33.** $(x - 4)^2 + (y - 3)^2 = 0$ **35.** $x^2 + y^2 = 4$

37. $x'^2 + y'^2 = 9$

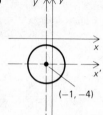

$(-1, -4)$

39. $x'^2 + y'^2 = 1$ **41.** $x'^2 + y'^2 = 49$ **43.** $x'^2 + y'^2 = 36$

45. $x'^2 + y'^2 = 64$ **47.** $y' = x'^2$

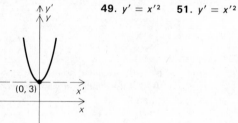

$(0, 3)$

49. $y' = x'^2$ **51.** $y' = x'^2$

53. $y' = |x'|$

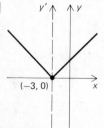

$(-3, 0)$

55. $y' = x'^3$

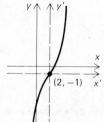

$(2, -1)$

57. $y' = 3x'$

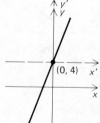

$(0, 4)$

59. $y' = -|x'|$

61. $y' = \dfrac{1}{x'}$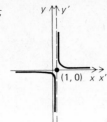

63. $y' = \dfrac{1}{x'}$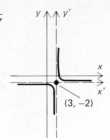

65. $y' = \dfrac{3x'^2}{x' - 1}$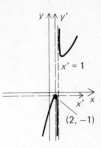

67. (4, 4)

Exercise 2.5

1. Vertex (0, 0); axis: $y = 0$; focus $(-2, 0)$; directrix: $x = 2$

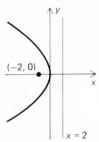

3. Vertex (0, 0); axis: $y = 0$; focus (2, 0); directrix: $x = -2$ **5.** Vertex (0, 0); axis: $y = 0$; focus $(-1, 0)$; directrix: $x = 1$ **7.** Vertex (0, 0); axis: $x = 0$; focus $(0, -1)$; directrix: $y = 1$ **9.** Vertex (0, 0); axis: $y = 0$; focus $\left(\dfrac{-1}{4}, 0\right)$; directrix: $x = \dfrac{1}{4}$

11. Vertex (5, 0); axis: $y = 0$; focus (2, 0); directrix: $x = 8$; $y'^2 = -12x'$

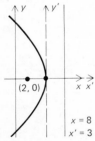

13. Vertex (0, 5); axis: $x = 0$; focus (0, 1); directrix: $y = 9$; $x'^2 = -16y'$ **15.** Vertex $(-1, 3)$; axis: $y = 3$; focus (1, 3); directrix: $x = -3$; $y'^2 = 8x'$ **17.** Vertex $(-3, 2)$; axis: $y = 2$; focus $\left(\dfrac{-5}{2}, 2\right)$; directrix: $x = \dfrac{-7}{2}$; $y'^2 = 2x'$ **19.** Vertex $(-4, -1)$; axis: $y = -1$; focus $\left(\dfrac{-15}{4}, -1\right)$; directrix: $x = \dfrac{-17}{4}$; $y'^2 = x'$

21. $(x - 1)^2 = 4(y + 2)$; vertex $(1, -2)$; axis: $x = 1$

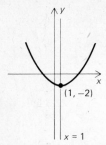

23. $(y + 1)^2 = \frac{3}{4}(x - 2)$; vertex $(2, -1)$; axis: $y = -1$ **25.** $(y - 5)^2 = \frac{7}{4}(x - 6)$; vertex $(6, 5)$; axis: $y = 5$

27. $(y - 5)^2 = \frac{5}{2}(x + 4)$; vertex $(-4, 5)$; axis: $y = 5$ **29.** $(x - 2)^2 = \frac{5}{2}(y + 1)$; vertex $(2, -1)$; axis: $x = 2$

31. Vertex $(5, 1)$; axis: $x = 5$

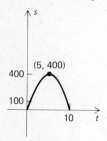

33. Vertex: $(-4, 3)$; axis: $x = -4$ **35.** Vertex $(-2, 1)$; axis: $x = -2$

37. Maximum height, $t = 5$; returns to ground, $t = 10$

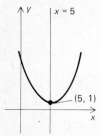

39. Maximum height, $t = \frac{3}{2}$; returns to ground, $t = 3$

41. $A(x) = 40x - 2x^2$; 10 m; 200 sq m   **43.** $P(x) = 12x - x^2$; 6

Exercise 2.6

1. Center $(0, 0)$

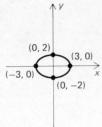

$(0, 2)$
$(3, 0)$
$(-3, 0)$
$(0, -2)$

3. Center $(0, 0)$ **5.** Center $(0, 0)$ **7.** Center $(0, 0)$

9. Center $(0, 0)$ **11.** Center $(-3, 4)$; $\dfrac{x'^2}{9} + \dfrac{y'^2}{16} = 1$

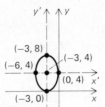

$(-3, 8)$
$(-6, 4)$ $(-3, 4)$
$(0, 4)$
$(-3, 0)$

13. Center $(-5, 2)$; $\dfrac{x'^2}{16} + \dfrac{y'^2}{4} = 1$ **15.** Center $(1, -2)$; $\dfrac{x'^2}{9} + \dfrac{y'^2}{1} = 1$ **17.** Center $(-3, 1)$; $\dfrac{x'^2}{\frac{25}{16}} + \dfrac{y'^2}{\frac{9}{4}} = 1$

19. Center $(-3, 5)$; $\dfrac{x'^2}{4} + \dfrac{y'^2}{16} = 1$ **21.** $\dfrac{(x-2)^2}{\frac{1}{4}} + \dfrac{(y+5)^2}{\frac{1}{9}} = 1$

23. $\dfrac{(x-3)^2}{9} + \dfrac{(y+1)^2}{4} = 1$ **25.** $\dfrac{(x-3)^2}{\frac{1}{4}} + \dfrac{(y+1)^2}{\frac{1}{9}} = 0$ **27.** $\dfrac{(x-5)^2}{81} + \dfrac{y^2}{9} = 1$

29. $\dfrac{x^2}{81} + \dfrac{(y-4)^2}{9} = 1$

31. Center $(0, 0)$; vertices $(3, 0)$, $(-3, 0)$; asymptotes $y = \dfrac{2x}{3}$, $y = \dfrac{-2x}{3}$

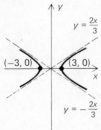

$y = \dfrac{2x}{3}$
$(-3, 0)$ $(3, 0)$
$y = -\dfrac{2x}{3}$

33. Center $(0, 0)$; vertices $\left(0, \frac{1}{3}\right)$, $\left(0, \frac{-1}{3}\right)$; asymptotes $y = \frac{2x}{3}$, $y = \frac{-2x}{3}$

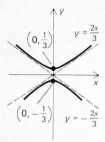

35. Center $(0, 0)$; vertices $\left(\frac{9}{4}, 0\right)$, $\left(\frac{-9}{4}, 0\right)$; asymptotes $y = \frac{2x}{3}$, $y = \frac{-2x}{3}$ **37.** Center $(0, 0)$; vertices $(0, 10)$, $(0, -10)$; asymptotes $y = 5x$, $y = -5x$ **39.** Center $(0, 0)$; vertices $(3, 0)$, $(-3, 0)$; asymptotes $y = \frac{x}{3}$, $y = \frac{-x}{3}$

41. Center $(-3, 4)$; vertices $(0, 4)$, $(-6, 4)$; $\frac{x'^2}{9} - \frac{y'^2}{16} = 1$; asymptotes $y' = \pm\frac{4x'}{3}$, $(y - 4) = \pm\frac{4(x + 3)}{3}$

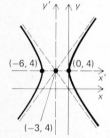

43. Center $(-5, -2)$; vertices $(-1, -2)$, $(-9, -2)$; $\frac{x'^2}{16} - \frac{y'^2}{4} = 1$; asymptotes $y' = \pm\frac{x'}{2}$, $y + 2 = \pm\frac{1}{2}(x + 5)$ **45.** Center $(6, -3)$; vertices $(7, -3)$, $(5, -3)$; $\frac{x'^2}{1} - \frac{y'^2}{9} = 1$; asymptotes $y' = \pm 3x'$, $y + 3 = \pm 3(x - 6)$ **47.** Center $(1, -4)$; vertices $(4, -4)$, $(-2, -4)$; $\frac{x'^2}{9} - \frac{y'^2}{4} = 1$; asymptotes $y' = \pm\frac{2x'}{3}$, $y + 4 = \pm\frac{2}{3}(x - 1)$ **49.** Center $(5, 4)$; vertices $(5, 14)$, $(5, -6)$; $\frac{y'^2}{100} - \frac{x'^2}{4} = 1$; asymptotes $y' = \pm 5x'$, $y - 4 = \pm 5(x - 5)$

51. $\frac{(x - 2)^2}{\frac{1}{4}} - \frac{(y + 5)^2}{\frac{1}{9}} = 1$

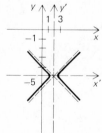

53. $\dfrac{(y + 1)^2}{\frac{1}{9}} - \dfrac{(x - 3)^2}{\frac{1}{4}} = 1$ **55.** $(x + 1)^2 - y^2 = 0$ **57.** $(x + 1)^2 - y^2 = 1$

59. $\dfrac{(y + 1)^2}{4} - \dfrac{(x - 3)^2}{9} = 1$ **61.** Ellipse   **63.** Circle   **65.** Hyperbola   **67.** Two intersecting lines

69. Point

Exercise 2.7

1. Not a function

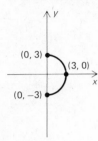

3. Not a function   **5.** Function   **7.** Function   **9.** Not a function   **11.** $\{(x, y) \mid 7y - 8x + 9 = 0\}$
13. $\{(x, y) \mid x^2 + y^2 = 9\}$ **15.** $\{(x, y) \mid x = y^2 - 6y + 9\}$ **17.** $\{(x, y) \mid x = 10^y\}$ **19.** $\{(x, y) \mid x = e^y\}$
21. $\{(x, y) \mid x = 3y - 7\}$; function   **23.** $\{(x, y) \mid 3y - 7x + 10 = 0\}$; function   **25.** $\{(x, y) \mid x = y^4\}$; not a
function   **27.** $\{(x, y) \mid x = \sqrt{y}\}$; function   **29.** $\{(x, y) \mid x = -y\}$; function   **31.** One-to-one
33. Not one-to-one   **35.** One-to-one   **37.** Not one-to-one   **39.** One-to-one   **41.** One-to-one
43. Not one-to-one   **45.** Not one-to-one   **47.** One-to-one   **49.** Not one-to-one

51. Domain $= \{x \mid x \geq 2\}$; range $= \{y \mid y \geq -1\}$; $f^{-1}(x) = \dfrac{x - 5}{3}$

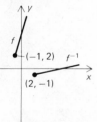

53. Domain $= \{x \mid x \geq 4\}$; range $= \{y \mid y \geq 2\}$; $f^{-1}(x) = \sqrt{x}$

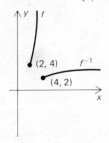

55. Domain $= \{x | x \geq -5\}$; range $= \{y | y \geq 0\}$; $f^{-1}(x) = \dfrac{x + 5}{2}$ **57.** Domain $= \{x | x \geq 0\}$;

range $= \{y | y \leq 0\}$; $f^{-1}(x) = -\sqrt{x}$ **59.** Domain $= \{x | x$ is a real number$\}$; range $= \{y | y$ is a real number$\}$;

$f^{-1}(x) = \dfrac{x + 4}{3}$

Exercise 2.8

1.
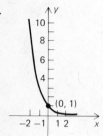

7.
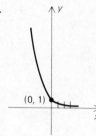

9.
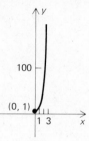

11. Range $= \{y | y$ is a real number$\}$; domain $= \{x | x > 0\}$; rule of f^{-1}: $y = \log_2 x$

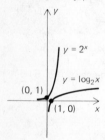

13. Range $= \{y | y$ is a real number$\}$; domain $= \{x | x > 0\}$; rule of f^{-1}: $y = \log_{1/3} x$

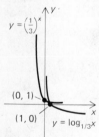

17. 0 **19.** 4 **21.** -2 **23.** -1 **25.** 1 **27.** 2 **29.** 4

31.
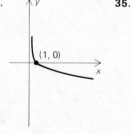
33.

35.
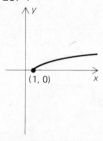

41. 180,000 **43.** $6.36

Exercise 2.9

1. $3^2 = 9$ 3. $5^2 = 25$ 5. $10^5 = x$ 7. $e^5 = x$ 9. $e^3 = x^2$ 11. $\log_2 y = 6$ 13. $\ln y = x^2$
15. $\log_2 8 = 3$ 17. $\ln y = x + h$ 19. $\ln y = m + n$ 21. u 23. 1 25. 0 27. x 29. x 31. x
33. x 35. $2x$ 37. -5 39. -5 41. $e^{x \ln 5}$ 43. $e^{x^2 \ln 2}$ 45. $e^{x \ln \sqrt{2}}$ 47. $e^{5 \ln x}$ 49. $e^{1/3 \ln x}$
51. Yes 53. Yes 55. No 57. Let $x = \ln M$. Then $M = e^x$. Let $y = \ln N$. Then $N = e^y$. So
$M/N = e^x/e^y = e^{x-y}$ or $x - y = \ln M/N$. That is, $\ln M - \ln N = \ln M/N$. 59. Let $x = \log M$. Then
$M = 10^x$. So $M^k = 10^{kx}$ and hence $M^k = 10^{k \log M}$ or $k \log M = \log M^k$. 61. 2.9652 63. $0.7619 - 4$
65. 4.7959 67. 5.8492 69. 3.2959 71. -2.0099 73. 3.3223 75. 2.7713
77. $(\log e)(\ln(3x + 1))$ 79. $(\log e)(\ln 38)$ 81. 8250 83. 35,700 85. 187,000 87. 256,000,000
89. 5.27 91. 185 93. 16.7

Exercise 2.10

1. $(f + g)(x) = e^{3x} + \sqrt{-x}$, domain $= \{x | x \leq 0\}$; $(f - g)(x) = e^{3x} - \sqrt{-x}$, domain $= \{x | x \leq 0\}$;
$(fg)(x) = e^{3x}\sqrt{-x}$, domain $= \{x | x \leq 0\}$; $\left(\dfrac{f}{g}\right)(x) = \dfrac{e^{3x}}{\sqrt{-x}}$, domain $= \{x | x < 0\}$; $(f \circ g)(x) = e^{3\sqrt{-x}}$,
domain $= \{x \leq 0\}$; $(g \circ f)(x) = \sqrt{-e^{3x}}$, domain is empty 3. $(f + g)(x) = \ln|x| + x^3$, domain $= \{x | x \neq 0\}$;
$(f - g)(x) = \ln|x| - x^3$, domain $= \{x | x \neq 0\}$; $(fg)(x) = x^3 \ln|x|$, domain $= \{x | x \neq 0\}$;
$\left(\dfrac{f}{g}\right)(x) = \dfrac{\ln|x|}{x^3}$, domain $= \{x | x \neq 0\}$; $(f \circ g)(x) = \ln|x^3|$, domain $= \{x | x \neq 0\}$; $(g \circ f)(x) = (\ln|x|)^3$,
domain $= \{x | x \neq 0\}$ 5. $(f + g)(x) = \ln x + 2x$, domain $= \{x | 0 < x \leq 1\}$; $(f - g)(x) = \ln x - 2x$,
domain $= \{x | 0 < x \leq 1\}$; $(fg)(x) = 2x \ln x$, domain $= \{x | 0 < x \leq 1\}$; $\left(\dfrac{f}{g}\right)(x) = \dfrac{\ln x}{2x}$,
domain $= \{x | 0 < x \leq 1\}$; $(f \circ g)(x) = \ln 2x$, domain $= \{x | 0 < x \leq 1\}$; $(g \circ f)(x) = 2 \ln x$,
domain $= \{x | e^{-1} \leq x \leq e\}$ 7. $(f + g)(x) = \sqrt{x + 3} + x^2$, domain $= \{x | x \geq -3\}$;
$(f - g)(x) = \sqrt{x + 3} - x^2$, domain $= \{x | x \geq -3\}$; $(fg)(x) = x^2 \sqrt{x + 3}$, domain $= \{x | x \geq -3\}$;
$\left(\dfrac{f}{g}\right)(x) = \dfrac{\sqrt{x + 3}}{x^2}$, domain $= \{x | x \geq -3 \text{ and } x \neq 0\}$; $(f \circ g)(x) = \sqrt{x^2 + 3}$, domain $= \{x | x \text{ is a real number}\}$;
$(g \circ f)(x) = x + 3$, domain $= \{x | x \text{ is a real number}\}$ 9. $(f + g)(x) = x^2 + x + 5$, domain $= \{x | x \leq 0\}$;
$(f - g)(x) = x^2 - x - 5$, domain $= \{x | x \leq 0\}$; $(fg)(x) = x^3 + 5x^2$, domain $= \{x | x \leq 0\}$;
$\left(\dfrac{f}{g}\right)(x) = \dfrac{x^2}{x + 5}$, domain $= \{x | x \leq 0 \text{ and } x \neq -5\}$; $(f \circ g)(x) = x^2 + 10x + 25$, domain $= \{x | x \leq -5\}$;
$(g \circ f)(x) = x^2 + 5$, domain $= \{x | x \leq 0\}$ 11. 11 13. 18 15. $\sqrt{10}$ 17. $\sqrt{65}$ 19. 1

Exercise 3.1

1. 7 3. 2 5. 4 7. Remainder 0, -8 is root; remainder $120 \neq 0$, -2 not root 9. Remainder 0,
$\frac{3}{2}$ is root; remainder $42 \neq 0$, 3 not root 11. $(x - 3)(x + 5)(x - 4) = 0$ 13. $(x^2 - 9)(x^2 - 16) = 0$
15. $(x^2 + 4)(x - 3) = 0$ 17. 0, 2, -4 19. 0, -3, 2 21. 0, multiplicity 2; -1, multiplicity 3;
4, multiplicity 5 23. 3, multiplicity 2; i, multiplicity 3; $-i$, multiplicity 3 25. 0, multiplicity 4;
2, multiplicity 3; -9, multiplicity 1 27. $(x + 5)^2(x + 4)^3(x + 1)^4 = 0$ 29. $(x - 4)^3(x - 3i)^5(x + 3i)^5 = 0$
31. 4, 4 33. 7, 7 35. 5, 5 37. 3 distinct roots: -1, 3, -4; 10 39. 4 distinct roots: -5, 2, i, $-i$; 7
41. $(-5i)^2 + 25 = 0$ 43. $(2 - 4i)^2 - 4(2 - 4i) + 20 = 0$ 45. $(1 + 2i)^2 - 2(1 + 2i) + 5 = 0$
47. $x^2 - 4x + 29$ 49. $x^2 - 4x + 40$ 51. 1, $\dfrac{-3 \pm \sqrt{11}i}{2}$ 53. -2, $\dfrac{-1 \pm \sqrt{7}i}{2}$ 55. 4, $-1 \pm 2i$
57. $\pm 1, \pm 2, \pm 4, \pm 8$ 59. $\pm 1, \pm 3, \pm 5, \pm 15, \pm \dfrac{1}{7}, \pm \dfrac{3}{7}, \pm \dfrac{5}{7}, \pm \dfrac{15}{7}$ 61. 1, $\dfrac{3}{2}$, -2 63. $-4, \dfrac{1}{5}, \dfrac{-5}{2}$
65. 2, 3 67. 3, -5, $\dfrac{1}{4}$ 69. 0, 4, 5 71. $\dfrac{2}{11}$, $\dfrac{-1 \pm \sqrt{11}i}{2}$ 73. -6, $\dfrac{-5 \pm \sqrt{21}}{2}$

75. $\dfrac{3}{8}, \dfrac{1 \pm \sqrt{2}i}{3}$ **77.** $P(0) = 12 > 0, P(1) = -36 < 0$ **79.** $P(0) = 9 > 0, P(2) = -3 < 0$

81. $P(0) = 15 > 0, P(2) = -21 < 0$ **83.** $P(-3) = 10 > 0, P(-4) = -36 < 0$

85. $P(0) = 60 > 0, P(-2) = -80 < 0$ **87.** 1.7   **89.** -1.9 **91.** 3.3   **93.** 0.4   **95.** 6.2

97. Rational root: -3; Irrational roots: $+\sqrt{5}, -\sqrt{5}$; No others   **99.** No rational roots; Irrational roots between 1 and 2, between -1 and -2, and nowhere else; 1.9

1. $x = 3, y = -1$ **3.** $x = -2, y = 3$ **5.** $x = -1, y = -4$ **7.** $x = 1, y = -1$ **9.** $x = -5, y = -2$

11. $x = -22, y = 3, z = -1$ **13.** $x = -1, y = -2, z = -3$ **15.** $A = 1, B = -2, C = 3,$

$D = -1, E = 4$ **17.** $x = 7, y = -5, z = 0$ **19.** $A = 2, B = 3, C = 4$ **21.** Not exactly one solution

23. $x = 2, y = 1$ **25.** No solution   **27.** No solution   **29.** No solution   **31.** Holstein: 6943 kg,

Jersey: 5488 kg   **33.** 2.8, 5.6   **35.** pullet: \$.50, large: \$.90

Exercise 3.3

1. -1 **3.** 4   **5.** 8   **7.** Second row, third column, minus   **9.** Fourth row, first column, minus

11. $\begin{vmatrix} 3 & 0 \\ 6 & -7 \end{vmatrix}$ **13.** $\begin{vmatrix} -1 & 5 \\ 0 & 4 \end{vmatrix}$ **15.** $\begin{vmatrix} -1 & 5 \\ -7 & 8 \end{vmatrix}$ **17.** -42 **19.** 9   **21.** -25 **23.** -292 **25.** -27

27. 686   **29.** 454   **31.** $(-)$ **33.** 2   **35.** 0

Exercise 3.4

1. $\begin{pmatrix} 28 & -6 \\ 45 & 9 \end{pmatrix}$ **3.** $\begin{pmatrix} -6 & 7 \\ 18 & 31 \end{pmatrix}$ **5.** $\begin{pmatrix} 9 & -8 \\ 7 & 6 \end{pmatrix}$ **7.** $\begin{pmatrix} -4 & 18 \\ 16 & 7 \end{pmatrix}$ **9.** $\begin{pmatrix} -1 & 3 & 2 \\ 14 & -4 & 0 \\ 13 & -11 & 6 \end{pmatrix}$ **11.** $\begin{pmatrix} 3x - 5y \\ -2x + 4y \end{pmatrix}$

13. $\begin{pmatrix} 2x + 3y \\ 4x \end{pmatrix}$ **15.** $\begin{pmatrix} x \\ y \end{pmatrix}$ **17.** $\begin{pmatrix} 2x - 4y + 5z \\ 3x - 7y + z \\ -2y + 6z \end{pmatrix}$ **19.** $\begin{pmatrix} 3x + 5y + 7z \\ x - 3y + 4z \\ x - 2y + 8z \end{pmatrix}$ **21.** $\begin{pmatrix} 6 & 12 \\ -3 & 9 \end{pmatrix}$

23. $\begin{pmatrix} 1 & 0 & 0 \\ 0 & 1 & 0 \\ 0 & 0 & 1 \end{pmatrix}$ **25.** $\begin{pmatrix} -3 & -6 & -8 \\ 2 & -4 & 1 \\ 3 & 0 & -5 \end{pmatrix}$ **27.** $AB = \begin{pmatrix} 1 & 0 \\ 0 & 1 \end{pmatrix}, BA = \begin{pmatrix} 1 & 0 \\ 0 & 1 \end{pmatrix}$

29. $AB = \begin{pmatrix} 7 & 6 \\ 5 & 4 \end{pmatrix}, BA = \begin{pmatrix} 7 & 6 \\ 5 & 4 \end{pmatrix}$ **31.** $\begin{pmatrix} 22 & -16 \\ -2 & 4 \end{pmatrix}$ **33.** $\begin{pmatrix} 22 & -16 \\ -2 & 4 \end{pmatrix}$ **35.** $\begin{pmatrix} 55 & -40 \\ -5 & 10 \end{pmatrix}$

37. $\begin{pmatrix} 3 & 6 \\ 1 & -1 \end{pmatrix}\begin{pmatrix} x \\ y \end{pmatrix} = \begin{pmatrix} 5 \\ 1 \end{pmatrix}$ **39.** $\begin{pmatrix} 1 & 1 & -5 \\ 3 & -1 & 2 \\ 1 & 1 & 1 \end{pmatrix}\begin{pmatrix} x \\ y \\ z \end{pmatrix} = \begin{pmatrix} 6 \\ -3 \\ 2 \end{pmatrix}$ **41.** $\begin{aligned} 2x - 7y &= -1 \\ 3x + y &= 5 \end{aligned}$ **43.** $\begin{aligned} 3x - 2y + 6z &= 2 \\ x + y - 3z &= -1 \\ 5x - 4y + 7z &= 3 \end{aligned}$

45. $\begin{aligned} x &= 3 \\ y &= -4 \\ z &= 7 \end{aligned}$ **47.** $\begin{pmatrix} 3 & -7 & 4 \\ 2 & 1 & 0 \\ 6 & -3 & 1 \end{pmatrix}$ **49.** $\begin{pmatrix} 1 & 0 \\ 0 & 1 \end{pmatrix}$ **51.** $AB = I, BA = I$ **53.** $AB = I, BA = I$

55. $AB = I, BA = I$ **57.** Not nonsingular   **59.** Not nonsingular   **61.** $\dfrac{1}{10}\begin{pmatrix} 1 & -3 \\ 2 & 4 \end{pmatrix}$ **63.** No inverse

65. $\dfrac{1}{6}\begin{pmatrix} -3 & 2 & 2 \\ 6 & -4 & 2 \\ 3 & 0 & 0 \end{pmatrix}$ **67.** $\dfrac{1}{7}\begin{pmatrix} -1 & 1 & -3 \\ -3 & 3 & -2 \\ 5 & 2 & 1 \end{pmatrix}$ **69.** $\dfrac{1}{12}\begin{pmatrix} 12 & 0 & 0 \\ 0 & 4 & 0 \\ 3 & -2 & 3 \end{pmatrix}$

71. $\begin{pmatrix} 2 & -3 \\ 1 & 7 \end{pmatrix}\begin{pmatrix} x \\ y \end{pmatrix} = \begin{pmatrix} 9 \\ -4 \end{pmatrix}; x = 3, y = -1$ **73.** $\begin{pmatrix} 4 & -3 \\ 1 & 1 \end{pmatrix}\begin{pmatrix} x \\ y \end{pmatrix} = \begin{pmatrix} -17 \\ 1 \end{pmatrix}, x = -2, y = 3$

75. $\begin{pmatrix} 5 & -4 \\ 3 & 6 \end{pmatrix}\begin{pmatrix} x \\ y \end{pmatrix} = \begin{pmatrix} 11 \\ -27 \end{pmatrix}; x = -1, y = -4$ **77.** $\begin{pmatrix} 11 & -9 \\ 6 & 5 \end{pmatrix}\begin{pmatrix} x \\ y \end{pmatrix} = \begin{pmatrix} 20 \\ 1 \end{pmatrix}; x = 1, y = -1$

79. $\begin{pmatrix} -1 & 5 \\ 2 & 3 \end{pmatrix}\begin{pmatrix} x \\ y \end{pmatrix} = \begin{pmatrix} -5 \\ -16 \end{pmatrix}; x = -5, y = -2$ **81.** $\begin{pmatrix} 1 & -6 & -4 \\ 1 & -3 & -1 \\ -1 & 9 & 8 \end{pmatrix}\begin{pmatrix} x \\ y \\ z \end{pmatrix} = \begin{pmatrix} -36 \\ 30 \\ 41 \end{pmatrix}; x = -22, y = 3, z = -1$

83. $\begin{pmatrix} 4 & 5 & 23 \\ 1 & 1 & 5 \\ -3 & -2 & -11 \end{pmatrix} \begin{pmatrix} x \\ y \\ z \end{pmatrix} = \begin{pmatrix} -83 \\ -18 \\ 40 \end{pmatrix}$; $x = -1, y = -2, z = -3$

85. $\begin{pmatrix} 1 & 0 & 1 & 0 & 0 \\ 3 & 1 & 2 & 1 & 0 \\ 3 & 3 & 1 & 1 & 1 \\ 1 & 3 & 0 & 0 & 0 \\ 0 & 1 & 0 & 0 & 0 \end{pmatrix} \begin{pmatrix} A \\ B \\ C \\ D \\ E \end{pmatrix} = \begin{pmatrix} 4 \\ 6 \\ 3 \\ -5 \\ -2 \end{pmatrix}$; $A = 1, B = -2, C = 3, D = -1, E = 4$

87. $\begin{pmatrix} -1 & -2 & -5 \\ 1 & 3 & 10 \\ -2 & -4 & -5 \end{pmatrix} \begin{pmatrix} x \\ y \\ z \end{pmatrix} = \begin{pmatrix} 3 \\ -8 \\ 6 \end{pmatrix}$; $x = 7, y = -5, z = 0$

89. $\begin{pmatrix} 1 & -3 & 1 \\ 1 & 0 & 1 \\ 3 & 1 & 0 \end{pmatrix} \begin{pmatrix} A \\ B \\ C \end{pmatrix} = \begin{pmatrix} -3 \\ 6 \\ 9 \end{pmatrix}$; $A = 2, B = 3, C = 4$

Exercise 3.5

1. $\dfrac{A}{x} + \dfrac{B}{x-3} + \dfrac{C}{x+7}$　　**3.** $\dfrac{A}{x-3} + \dfrac{B}{(x-3)^2}$　　**5.** $\dfrac{A}{x-1} + \dfrac{B}{(x-1)^2} + \dfrac{Cx+D}{3x^2+x+1}$

7. $\dfrac{1}{x} + \dfrac{3}{x-1} - \dfrac{2}{x-3}$　　**9.** $\dfrac{-1}{x-2} + \dfrac{x+3}{x^2+x+1} + \dfrac{2x}{(x^2+x+1)^2}$　　**11.** $x + 5 + \dfrac{3}{x+4} + \dfrac{1}{x-2}$

13. $\dfrac{3}{x} + \dfrac{3}{x+5} - \dfrac{5}{x-2}$　　**15.** $\dfrac{-1}{x-2} + \dfrac{2x+1}{x^2+4} + \dfrac{3x-1}{(x^2+4)^2}$

Exercise 3.6

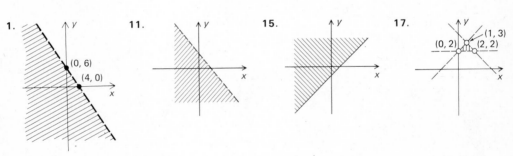

1.　　**11.**　　**15.**　　**17.** (0, 2)　(1, 3)　(2, 2)

21. Corn, 300 acres; rye, 200 acres　　**23.** A, 5000 boxes; B, 0 boxes　　**25.** Emergencies, 6; regular check-ups, 18; 9 hours

Exercise 3.7

1. $\{3n\}$; $f(n) = 3n$　　**3.** $\{n^2 + 1\}$; $f(n) = n^2 + 1$　　**5.** $\{2n^2 + 1\}$; $f(n) = 2n^2 + 1$　　**7.** 3, 17, 55, . . . , $2n^3 + 1$, . . .　　**9.** 2, 6, 12, . . . , $n(n + 1)$, . . .　　**11.** $\dfrac{3}{4}, \left(\dfrac{3}{4}\right)^2, \left(\dfrac{3}{4}\right)^3, \ldots, \left(\dfrac{3}{4}\right)^n, \ldots$　　**13.** 0.75, $(0.75)^2$, $(0.75)^3$, . . . , $(0.75)^n$, . . .　　**15.** 5, 8, 11, . . . , $5 + 3(n - 1)$, . . .

17. $3 + \dfrac{1}{5}, 3 + \dfrac{1}{25}, 3 + \dfrac{1}{125}, 3 + \dfrac{1}{625}, \ldots, 3 + \left(\dfrac{1}{5}\right)^n, \ldots$; $\lim\limits_{n \to \infty} 3 + \left(\dfrac{1}{5}\right)^n = 3$

19. $1 + \dfrac{1}{6}, 1 + \dfrac{1}{36}, 1 + \dfrac{1}{216}, 1 + \dfrac{1}{1296}, \ldots, 1 + \left(\dfrac{1}{6}\right)^n, \ldots$; $\lim\limits_{n \to \infty} 1 + \left(\dfrac{1}{6}\right)^n = 1$

21. $r = -2$; diverges   **23.** $r = -1$; diverges   **25.** $r = \frac{1}{2}$; converges to 0   **27.** $r = -1$; diverges

29. $r = 2$; diverges   **31.** $S_n = \frac{1 - 4^n}{-3}$; $S_5 = 341$ **33.** $S_n = \frac{1}{2} - \frac{1}{2}\left(\frac{1}{3}\right)^n$; $S_5 = \frac{121}{243}$

35. $S_n = \frac{-2 + 2(-2)^n}{3}$; $S_5 = -22$ **37.** $S_n = \frac{1 - (-1)^n}{2}$; $S_6 = 0$ **39.** $S_n = 8 - 8\left(\frac{1}{2}\right)^n$; $S_3 = 7$

41. $d = 2$; $S_n = n^2 + n$; $S_5 = 30$ **43.** $d = -2$; $S_n = -n^2 + 2n$; $S_5 = -15$

45. $d = -3$; $S_n = \frac{9n - 3n^2}{2}$; $S_5 = -15$ **47.** Geometric; $r = 2$ **49.** Geometric; $r = 4$

51. $\sum\limits_{n=1}^{\infty} 3 \times 2^{n-1}$ **53.** $\sum\limits_{n=1}^{\infty} n^2$ **55.** $\sum\limits_{n=1}^{\infty} n(n + 3)$ **57.** $3^4 + 3^5 + 3^6 + 3^7 + 3^8 + 3^9$

59. $\frac{1}{2} + \frac{1}{4} + \frac{1}{6} + \cdots + \frac{1}{2n} + \cdots$ **61.** $\frac{3}{10} + \frac{4}{17} + \frac{5}{26} + \frac{6}{37} + \frac{7}{50}$ **63.** $\frac{1}{3} + \frac{8}{9} + 1 + \cdots + \frac{n^3}{3^n} + \cdots$

65. $1 + \sqrt{2} + \sqrt{3} + 2 + \sqrt{5} + \sqrt{6} + \sqrt{7} + 2\sqrt{2} + 3$

67. $S_1 = 1$, $S_2 = 1 - \frac{1}{3}$, $S_3 = 1 - \frac{1}{3} + \frac{1}{9}$; $S_n = 1 - \frac{1}{3} + \frac{1}{9} + \cdots + \left(\frac{-1}{3}\right)^{n-1}$, $S_n = \frac{3}{4}\left[1 - \left(\frac{-1}{3}\right)^n\right]$; $\frac{3}{4}$

69. $S_1 = 1$, $S_2 = 1 + \frac{1}{6}$, $S_3 = 1 + \frac{1}{6} + \frac{1}{36}$; $S_n = 1 + \frac{1}{6} + \frac{1}{36} + \cdots + \left(\frac{1}{6}\right)^{n-1}$, $S_n = \frac{6}{5}\left[1 - \left(\frac{1}{6}\right)^n\right]$; $\frac{6}{5}$

71. Converges; $\frac{4}{3}$ **73.** Converges; $\frac{3}{4}$ **75.** Diverges   **77.** 2046   **79.** $100(1.06)^{10}$ dollars

81. 8 days; 1 day

Exercise 3.8

1. *Part I* $3 = \frac{3}{2}(3^1 - 1)$ since $3 = \frac{3 \times 2}{2}$.

Part II If $3 + 3^2 + \cdots + 3^k = \frac{3}{2}(3^k - 1)$, then $3 + 3^2 + \cdots + 3^k + 3^{k+1} = \frac{3}{2}(3^k - 1) + 3^{k+1}$

$$= \frac{3(3^k - 1) + 2 \times 3^{k+1}}{2}$$

$$= \frac{3(3^k - 1 + 2 \times 3^k)}{2}$$

$$= \frac{3(3 \times 3^k - 1)}{2}$$

$$= \frac{3}{2}(3^{k+1} - 1)$$

Check $S_n = \frac{3 - 3 \times 3^n}{1 - 3} = \frac{3(1 - 3^n)}{-2} = \frac{3}{2}(3^n - 1)$

21. *Part I* $2^3 > 2^2$ since $8 > 4$.

Part II $(k + 1)^3 = k^3 + 3k^2 + 3k + 1$
$> k^2 + 3k^2 + 3k + 1$, if $k^3 > k^2$
$= 4k^2 + 3k + 1$
$> k^2 + 2k + 1$, for $k \geq 2$
$= (k + 1)^2$
$(k + 1)^3 > (k + 1)^2$, if $k^3 > k^2$, $k \geq 2$

27. *Part I* $x^2 \cdot x^1 = x^{2+1}$ since $x^2 \cdot x^1 = x^3$.

Part II $x^2 \cdot x^{k+1} = x^2 \cdot x^k \cdot x$
$= x^{2+k} \cdot x$, if $x^2 \cdot x^k = x^{2+k}$
$= x^{2+k+1}$

31. *Part I* $(x - y)$ is a factor of $x^{2(1)} - y^{2(1)} = x^2 - y^2$ since $x^2 - y^2 = (x - y)(x + y)$.
Part II $x^{2(k+1)} - y^{2(k+1)} = x^{2k+2} - y^{2k+2}$
$$= x^{2k}x^2 - y^{2k}y^2 \text{ since } a^{m+2} = a^m \cdot a^2$$
$$= x^{2k}x^2 - x^2y^{2k} + x^2y^{2k} - y^{2k}y^2$$
$$= x^2(x^{2k} - y^{2k}) + y^{2k}(x^2 - y^2)$$
Note that $(x - y)$ is a factor of $x^2 - y^2$, so if $(x - y)$ is a factor of $x^{2k} - y^{2k}$, then $(x - y)$ is a factor of $x^{2(k+1)} - y^{2(k+1)}$.
33. *Part I* 1 is not greater than $1 + 3$.   *Part II* If we assume that $k > k + 3$, then adding 1 to both sides we have $k + 1 > (k + 1) + 3$.

Exercise 3.9

1. $x^8 + 8x^7 + 28x^6 + 56x^5 + 70x^4 + 56x^3 + 28x^2 + 8x + 1$ **3.** $x^4 + 8x^3y + 24x^2y^2 + 32xy^3 + 16y^4$
5. $1 + 6x + 15x^2 + 20x^3 + 15x^4 + 6x^5 + x^6$ **7.** $x^5 - 5x^4y + 10x^3y^2 - 10x^2y^3 + 5xy^4 - y^5$
9. $16a^4 + 96a^3b + 216a^2b^2 + 216ab^3 + 81b^4$ **11.** $a^{20} + 20a^{19}b + 190a^{18}b^2 + 1140a^{17}b^3 + \cdots$
13. $1 + 12x + 66x^2 + 220x^3 + \cdots$ **15.** $729x^6 - 1458x^5y + 1215x^4y^2 - 540x^3y^3 + \cdots$
17. $x^{10} - 10x^9y + 45x^8y^2 - 120x^7y^3 + \cdots$ **19.** $x^{30} + 30x^{29}y + 435x^{28}y^2 + 4060x^{27}y^3 + \cdots$
21. $56; 56x^3y^5$ **23.** $210; 210x^4y^6$ **25.** $435; 1740x^{28}y^2$ **27.** $66; 66a^2b^{10}$ **29.** $1820; 1820x^{12}y^4$
31. 6048 **33.** 1375 **35.** 823680 **37.** $\binom{6}{0}\binom{6}{1}\binom{6}{2}\binom{6}{3}\binom{6}{4}\binom{6}{5}\binom{6}{6}$ **39.** $\binom{8}{0}\binom{8}{1}\binom{8}{2}\binom{8}{3}\binom{8}{4}\binom{8}{5}\binom{8}{6}\binom{8}{7}\binom{8}{8}$
41. $\binom{4}{1} = 4, \binom{3}{0} = 1, \binom{3}{1} = 3; 4 = 1 + 3$ **43.** $\binom{6}{2} = 15, \binom{5}{1} = 5, \binom{5}{2} = 10; 15 = 5 + 10$
45. $\binom{7}{1} = 7, \binom{6}{0} = 1, \binom{6}{1} = 6; 7 = 1 + 6$ **47.** $\binom{4}{1} = 4, \binom{4}{3} = 4$ **49.** $\binom{6}{1} = 6, \binom{6}{5} = 6$
51. $9 \times 8 \times 7 \times 6! = 9 \times 8 \times 7 \times 6 \times 5 \times 4 \times 3 \times 2 \times 1 = 9!$
53. $10 \times 9 \times 8 \times 7 \times 6! = 10 \times 9 \times 8 \times 7 \times 6 \times 5 \times 4 \times 3 \times 2 \times 1 = 10!$
55. $11 \times 10 \times 9 \times 8 \times 7! = 11 \times 10 \times 9 \times 8 \times 7 \times 6 \times 5 \times 4 \times 3 \times 2 \times 1 = 11!$
57. $1 + \frac{1}{3}x - \frac{1}{9}x^2 + \frac{5}{81}x^3 + \cdots$ **59.** $1 + \frac{3}{2}x + \frac{3}{8}x^2 - \frac{1}{16}x^3 + \cdots$

Exercise 3.10

1. *abc, acb, cba, cab, bac, bca*
3. *pqrs, pqsr, prqs, prsq, psqr, psrq, qprs, qpsr, qrps, qrsp, qsrp, qspr, rpqs, rpsq, rsqp, rspq, rqps, rqsp, spqr, sprq, srqp, srpq, sqrp, sqpr*
5. 1234, 2134, 3124, 4123, 1243, 2143, 3142, 4132, 1324, 2341, 3241, 4231, 1342, 2314, 3214, 4213, 1423, 2413, 3412, 4321, 1432, 2431, 3421, 4312
7. $_6P_6 = 720, _6P_5 = 720$ **9.** $_5P_5 = 120, _5P_4 = 120$ **11.** 120   **13.** 60   **15.** 2520
17. *ab, ba, ca, da, ea, ac, bc, cb, db, eb, ad, bd, cd, dc, ec, ae, be, ce, de, ed*   **19.** 12, 21, 31, 13, 23, 32
21. 3; *bib, bbi, ibb*   **23.** 180   **25.** 34650   **27.** *abc, acd, abd, bcd*
29. *abc, abd, abe, acd, ace, ade, bcd, bce, bde, cde*   **31.** 10   **33.** 20   **35.** 56   **37.** 35   **39.** 84
41. $_4C_1 = 4, _3C_0 = 1, _3C_1 = 3; 4 = 1 + 3$ **43.** $_7C_6 = 7, _6C_5 = 6, _6C_6 = 1; 7 = 6 + 1$
45. $_8C_3 = 56, _8C_5 = 56; 56 = 56$ **47.** $_6C_0 \ _6C_1 \ _6C_2 \ _6C_3 \ _6C_4 \ _6C_5 \ _6C_6$
49. $_8C_0 \ _8C_1 \ _8C_2 \ _8C_3 \ _8C_4 \ _8C_5 \ _8C_6 \ _8C_7 \ _8C_8$ **51.** 22,100   **53.** (a) 1, (b) 495   **55.** 72
57. (2, 6), (3, 5), (4, 4), (5, 3), (6, 2); (3, 6), (4, 5), (5, 4), (6, 3); (4, 6), (5, 5), (6, 4); (5, 6), (6, 5); (6, 6)
59. 455   **61.** HH, HT, TH, TT   **63.** 15   **65.** 20   **67.** No, because the order makes a difference; 64,000
69. 12

71.

$$_nC_r = \frac{n!}{r!(n-r)!},$$

$$_{n-1}C_{r-1} = \frac{(n-1)!}{(r-1)!(n-r)!}$$

$$_{n-1}C_r = \frac{(n-1)!}{r!(n-1-r)!}$$

$$_{n-1}C_{r-1} + {}_{n-1}C_r = \frac{(n-1)!}{(r-1)!(n-r)!} + \frac{(n-1)!}{r!(n-1-r)!}$$

$$= \frac{r(n-1)! + (n-r)(n-1)!}{r!(n-r)!}$$

$$= \frac{(n-1)!(r+n-r)}{r!(n-r)!}$$

$$= \frac{n!}{r!(n-r)!}$$

$$= {}_nC_r$$

It states that any entry is the sum of the two entries in the preceding row that are located just to the left and right of it.

Exercise 3.11

1. $\frac{1}{2}$ **3.** (a) $\frac{5}{36}$ (b) $\frac{6}{36}$ (c) $\frac{5}{36}$ (d) $\frac{4}{36}$ **5.** 1   **7.** $\frac{68}{77}$ **9.** $\frac{72}{552}$ **11.** (a) $\frac{5376}{215940}$ (b) $\frac{496}{3599}$ **13.** $\frac{1}{36}$
15. $\frac{1}{12}$ **17.** $\frac{12}{2652}$ **19.** $\frac{64}{140608}$ **21.** $\frac{2}{6}$ **23.** $\frac{6}{36}$ **25.** $\frac{33}{36}$ **27.** $\frac{40183}{433160}$
29. (a) one to five   (b) five to one
31. (a) one to $_{52}C_{13} - 1 \approx$ one to 6.35×10^{11} (b) $_{52}C_{13} - 1$ to one $\approx 6.35 \times 10^{11}$ to one
33. $\frac{3}{28}$ **35.** (a) $\frac{351}{946}$ (b) 1 (c) $\frac{2723}{3311}$
37. $\dfrac{1}{(44)^5}$ **39.** $\frac{16}{75}$ **41.** $\frac{297}{325}$ **43.** (a) $\frac{1}{12}$ (b) $\frac{1}{6}$ **45.** $\frac{3}{4}$

Index

Abscissa, 165
Absolute value equations,
    109–120
  definition of, 113
Absolute value inequalities,
    127–133
  definition of, 127
Addition
  of complex numbers, 45
  of fractions, 73–74
  of like terms, 38
  of polynomials, 38
Antilogarithms, 308
Arithmetic progression, 424
Arithmetic sequence
  common difference of, 424
  definition of, 424
Arithmetic series, 428–430
Asymptotes, 211–216
  horizontal, 214–216
  of a hyperbola, 268
  vertical, 213–216

Base
  of an exponent, 16
  of a logarithm, 297
Binomial
  definition of, 35
  theorem, 454–468
    proof of, 458–460
    statement of, 457–458
Binomial coefficient
  definition of, 460
  rule, 456–457

Cartesian coordinate system
    164–165
Center
  of a circle, 229
  of an ellipse, 258
Characteristic, 308
Circles, equations of, 229–232
Coefficient, 35
Combinations
  definition of, 473
  of n things taken r at a time,
    474–475

Common binomial factors, 63
Common logarithms, 303
Common monomial factors, 58
Completing the square, 60, 98
Complex fractions, 78
Complex number(s), 8–10
  addition of, 45
  division of, 80
  multiplication of, 45–46
  subtraction of, 45
Composition of functions,
    320–321
Conformable matrices, 378
Conic sections, 274–276
  equations of, 274–276
Conjugate, 80
  of an imaginary number, 335
Conjugate root theorem, 335–336
Convergence of sequences,
    420–421
Coordinates, Cartesian 165
Counting numbers, 2
Cramer's rule, 368–371
Cube root, 16

Decimal
  repeating, 4
  terminating, 4
Degree
  of polynomial, 36
  of term, 36
Denominator, 3
  lowest common, 75
Dependent events, 489–490
Determinant
  definition of, 363
  expansion by minors, 364–365
  minors of, 364
  order of, 363
  properties of, 366–367
  value of $n \times n$, 363–365
  value of 2×2, 363
Diameter of a circle, 229
Difference of two cubes, 59
Difference of two squares, 58
Directrix of a parabola, 240

Distance between two points,
    226–228
Divergence
  of sequences, 421
  of series, 427
Dividend, 41
Division
  of complex numbers, 80
  of fractions, 72–73
  of imaginary numbers, 80
  of polynomials, 41
  synthetic, 43
Divisor, 41
Domain convention, 163
Domain of a function, 160

e, approximate value of, 5
Ellipse(s)
  definition of, 258
  equations of, 258–265
  foci of, 258
  major axis of, 261
  minor axis of, 261
Equation of a line
  intercept form of, 196
  point-slope form of, 193
  slope–y-intercept form of,
    193–194
Equations
  absolute value, 109–120
  of conic sections, 274–276
  of ellipses, 258–265,
    274–276
  equivalent, 90
  first-degree, 92
  fractional, 109–120
  graphs of, 168
  of hyperbolas, 265–273,
    275–276
  linear, 90–101
  linear in n variables, 135
  linear in two variables, 135
  nth-degree polynomial, 92
  polynomial, 332–352
  properties of, 90
  quadratic, 90–108
  radical, 109–120

solution of, 135
in two variables, 135
graphs of, 183
Linear functions, 196—206
definition of, 196
Linear inequalities, 122—133
in one variable, 122
in two variables, 406
Linear interpolation, 340—343
Linear programming, 408
Lines, 183—206
intersection of two, 184
oblique, 183
parallel to x-axis, 186—187
parallel to y-axis, 185—186
Location-of-roots theorem,
339—340
Logarithm(s)
characteristic of, 308
common, 303
computations with, 310—312
definition of, 297, 303
mantissa of, 308
natural, 304
properties of, 305
Logarithmic function, 296—302
graph of, 297—300
Logarithm tables, use of,
306—310
Lowest common denominator, 75
Lowest terms, reduction to, 70

Major axis of an ellipse, 261
Mantissa, 308
Mathematical induction, 441—453
principle of, 441—442
proof by, 441—443
Matrices
conformable, 378
equality of, 380—381
inversion of, 383—387
multiplication of, 376—379
row equivalent, 354
Matrix
columns of, 353
definition of, 353
elements of, 353
main diagonal of, 354
nonsingular, 384
row operations, 354
rows of, 353

square, 353
transpose of, 384
Midpoint of a line segment,
228—229
Minor axis of an ellipse, 261
Monomial, 35
Multiplication
of complex numbers, 45—46
of fractions, 72—73
of matrices, 376—379
of polynomials, 40
Mutually exclusive events, 491

Natural logarithms, 304
values of, 308—310
Natural numbers, 2
Nonsingular matrix, 384
Number line, 2
Number-of-roots theorem, 335
Numbers
common sets of, 2—15
complex, 8—10
counting, 2
imaginary, 8—10
irrational, 4
natural, 2
rational, 3
real, 5
whole, 2
Numerator, 3

Oblique line, 183
Odds
against E, 497
in favor of E, 497
One-to-one functions
definition of, 285
inverses of, 285
Operations on functions,
318—323
Ordered pairs, 162
Ordinate, 165
Origin of Cartesian
coordinates, 164

Parabola(s)
definition of, 240
directrix of, 240
equations of, 240—252,
275—276
focus of, 240
vertex of, 240

Parentheses, removal of, 39
Partial fractions, 399—404
Partial fraction theorem, 400—401
Pascal's triangle, 455—456
Path of a projectile, 251
Perfect-square trinomials, 59
Permutation(s)
definition of, 469
of n things not all different,
471—472
of n things taken r at a time,
470—471
Pi (π), 5
Point-slope form, 193
Polynomial equations, 332—352
imaginary roots of, 335—336
location of roots of, 339—340
number of roots of, 335
in one variable, 332
rational roots of, 337—339
roots of, 332—333
solutions of, 332—333
Polynomials, 35—56
addition of, 38
definition of, 35
degree of, 36
division of, 41
over the integers, 35
multiplication of, 40
in one variable, 35
over the reals, 35, 37
subtraction of, 39
in two variables, 35
Power, 16
Prime, 7
Prime polynomials, 57
Probability
definition of, 486
of dependent events, 490—491
of event certain to occur,
486—487
of independent events, 489
of mutually exclusive
events, 492
Projectile, path of, 251
Properties
of equations, 90
of graphs, 207—225
of logarithms, 305
derivation of, 306
of radicals, 29
Pythagorean theorem, 227

ABCDEFGHIJ–H–8210/79